AN INTRODUCTION TO ABSTRACT ALGEBRA
with Notes to the Future Teacher

Olympia E. Nicodemi
State University of New York, Geneseo

Melissa A. Sutherland
State University of New York, Geneseo

Gary W. Towsley
State University of New York, Geneseo

Upper Saddle River, New Jersey 07458

Library of Congress Cataloging-in-Publications Data

Nicodemi, Olympia
    An introduction to abstract algebra with notes to the future teacher/
    Olympia E. Nicodemi /Melissa A. Sutherland/ / Gary W. Towsley
    p. cm.
    Includes index
    ISBN 0-13-101963-5
      1. Algebra
CIP data available

Senior Editor: *Holly Stark*
Editorial Director: *Marcia Horton*
Production Editor: *Debbie Ryan*
Assistant Managing Editor: *Bayani Mendoza de Leon*
Senior Managing Editor: *Linda Mihatov Behrens*
Executive Managing Editor: *Kathleen Schiaparelli*
Manufacturing Buyer: *Maura Zaldivar*
Manufacturing Manager: *Alexis Heydt-Long*
Director of Marketing: *Patrice Jones*
Marketing Manager: *Halee Dinsey*
Art Director: *Jayne Conte*
Interior Designer: *Bayani Mendoza de Leon*
Cover Designer: *Bruce Kenselaar*
Art Editor: *Thomas Benfatti*
Creative Director: *Juan R. López*
Director of Creative Services: *Paul Belfanti*
Cover Photo: *Getty Images, Inc.*

© 2007 Pearson Education, Inc.
Pearson Prentice Hall
Pearson Education, Inc.
Upper Saddle River, NJ 07458

All rights reserved. No part of this book may be reproduced, in any form or by any means, without permission in writing from the publisher.

Pearson Prentice Hall™ is a trademark of Pearson Education, Inc.

Printed in the United States of America

10 9 8 7 6 5 4 3 2 1

ISBN 0-13-101963-5

Pearson Education, Ltd., *London*
Pearson Education Australia PTY. Limited, *Sydney*
Pearson Education Singapore, Pte., Ltd
Pearson Education North Asia Ltd, *Hong Kong*
Pearson Education Canada, Ltd., *Toronto*
Pearson Education de Mexico, S.A. de C.V.
Pearson Education – Japan, *Tokyo*
Pearson Education Malaysia, Pte. Ltd

To Mom, Aunt Don and Uncle Fred, titans all.
O.E.N.
To my family, especially my husband Don and my children Hunter and Emily.
M.A.S.
To my parents, Willard and Marguerite.
G.W.T.

Contents

Preface   ix

1 Topics in Number Theory   1

1.1 Preliminaries   1
1.2 Arithmetic and Divisibility   9
1.3 Greatest Common Divisors and Euclid's Algorithm   15
1.4 Primes and Unique Factorization   27
1.5 Worksheet 1: The Fibonacci Numbers   33
1.6 The Versatility of Euclid's Algorithm   36
1.7 Theorems of Euler and Fermat   42
1.8 In the Classroom: Problem Solving   48
1.9 From the Past: Fermat and Perfect Numbers   51
     Chapter 1 Highlights   56

2 Modular Arithmetic and Systems of Numbers   59

2.1 Congruences   59
2.2 Worksheet 2: Divisibility Tests   71
2.3 Worksheet 3: Public Key Cryptography   73
2.4 The Ring Z_m 75
2.5 Rings and Fields   82
2.6 The Field of Complex Numbers   92
2.7 In the Classroom: Why Is the Product of Two Negatives Numbers Positive?   100
2.8 From the Past: The Birth of the Complex Numbers   103
     Chapter 2 Highlights   107

3 Polynomials — 111

- 3.1 Polynomial Arithmetic 111
- 3.2 The Euclidean Algorithm for Polynomials 120
- 3.3 Worksheet 4: Derivatives 128
- 3.4 Factoring in $Z[x]$ and $Q[x]$ 130
- 3.5 The Fundamental Theorem of Algebra 136
- 3.6 Solving Cubic and Quartic Equations 146
- 3.7 Polynomial Congruences 150
- 3.8 Worksheet 5: Lagrange Interpolation 163
- 3.9 In the Classroom: Roots and Factoring 165
- 3.10 From the Past: Vieta and Symbolic Algebra 178
- Chapter 3 Highlights 183

4 A First Look at Group Theory — 187

- 4.1 Introduction to Groups 187
- 4.2 Groups and Geometry 196
- 4.3 Worksheet 6: Frieze Groups 201
- 4.4 Subgroups 207
- 4.5 Isomorphism and Homomorphism 216
- 4.6 Cyclic Groups 224
- 4.7 Permutation Groups 230
- 4.8 Worksheet 7: Groups and Counting 240
- 4.9 In the Classroom: Groups of Symmetries of Regular Polygons 246
- 4.10 From the Past: Évariste Galois and the Theory of Groups—Part 1 258
- Chapter 4 Highlights 266

5 New Structures from Old — 271

- 5.1 Cosets 271
- 5.2 Normal Subgroups and Quotient Groups 278

5.3 Conjugacy in S_n 285

5.4 Subrings, Ideals, and Quotient Rings   294

5.5 Worksheet 8: On the Ring of 2×2 Matrices over the Integers   304

5.6 Primes, Irreducibles, and the Gaussian Integers   306

5.7 Worksheet 9: Polynomial Ideals   313

5.8 In the Classroom: Why Rationalize?   315

5.9 From the Past: Évariste Galois and the Theory of Groups—Part 2   319

Chapter 5 Highlights   323

6 Looking Forward and Back   327

6.1 Extension Fields   327

6.2 The Degree of an Extension   336

6.3 Ruler and Compass Constructions   342

6.4 Worksheet 10: On the Construction of Regular Polygons   348

6.5 The Solvability of Polynomial Equations   351

6.6 In the Classroom: Constructions   359

6.7 From the Past: Quadratic Equations   367

Chapter 6 Highlights   372

Appendix   375

Bibliography   391

Selected Answers   395

Index   427

Preface

Students majoring in mathematics and students intending to teach mathematics at the secondary level learn mathematics in the same classroom. This book is for that classroom. It is for students of mathematics, many but not all of whom aspire to teaching.

For the mathematics major, this text offers a traditional course in abstract algebra. It begins with enough number theory to provide a solid background for group theory facts. Students with a good background in number theory can move through these sections quickly. Students who do not have the opportunity to take a full number theory course will get a good taste of that subject. Rings and in particular polynomial rings are introduced before groups because of the natural parallel of facts about integers to facts about polynomials. An introduction to groups follows that draws many examples from geometry. Then quotient structures for both groups and rings are explored. Finally, all topics of the text are engaged to confront the problem of the insolvability of quintic polynomials.

All students of mathematics appreciate links that connect the abstract mathematics encountered in upper-level courses to the familiar, more concrete mathematics of their earlier experiences. Such links are crucial to the future teacher as a bridge between the mathematics they are learning and the mathematics they will teach. To that purpose, most sections of our text finish with a short note called *To the Teacher*. It draws connections from the number theory or abstract algebra under consideration to secondary mathematics. Since we do not assume that the future teacher has had much, if any, pedagogical experience, most of our notes simply appeal to common elementary and high school experiences. So all students, future teachers or not, can find touchstones in these notes. Additionally, each chapter has a section dedicated to classroom concerns. Entitled *In the Classroom*, these sections are optional for students not aspiring to teach.

Historical context provides an extra dimension for all students. Instead of a sequence of historical snapshots, we have opted to provide each chapter with a section called *From the Past* in which the student experiences historical thinking and technique. Additionally, most chapters contain at least one section in the format of a *Worksheet*. The worksheets outline the skeleton of a topic, and the student provides the details and exposition. The worksheets are suitable for group work and for development as student presentations typical of a capstone experience.

Several documents that shape or will be shaping the undergraduate mathematics curriculum and the education of future school educators influence this text. Primary among them are the CUPM[1] recommendations, Principles and Standards for School Mathematics, published in 2000 by the NCTM[2] Standards, and especially

---

[1] CUPM (Committee on the Undergraduate Program in Mathematics) Curriculum Guide 2004, MAA Online
[2] National Council of Teachers of Mathematics

The Mathematical Education of Teachers (MET) report published by the Conference Board for the Mathematical Sciences (CBMS) in the fall of 2001. Elaborating on the specifics of curriculum for algebra and number theory, the MET report suggests the following agenda, which this text carries out.

> For all undergraduates, but especially for future high school teachers, [an] abstract algebra course can effectively build on familiar algebraic structures encountered in high school and other college mathematics courses. Examples of rings, integral domains, and fields familiar from high school are the most useful for future high school teachers. ... [G]roup theory would be closely connected to concrete examples such as isometry groups.
>
> Number theory has always been a popular elective in the mathematics major, especially among prospective teachers. Numbers are the most familiar of mathematical objects. The subject is a concrete setting for strengthening algebraic and proof-building skills. It is useful here again to explicitly examine mathematics underlying number theory concepts used in school mathematics.

HOW TO USE THIS BOOK

We assume that the student using this book has had some experience with proof, either from an elementary linear algebra course, discrete mathematics course, or a course specifically designed to teach proof. We have provided a brief appendix that reviews certain basics of logic, proof, set theory, and functions. Examples in the text are drawn from linear algebra. However, they can be skipped for students without that background. We also assume that the student will have access to a Computer Algebra System (CAS) such as Maple, or at the very least, a calculator such as the TI-89 that has CAS capabilities.

There are several different courses that can be taught from this book. First, please take a look at the Table of Contents. The following are the core sections of the text. Any section not listed can be skipped without loss of continuity.

Chapter 1, Sections 1, 2, 3, 4, 7
Chapter 2, Sections 1, 4, 5, 6
Chapter 3, Sections 1, 2, 4, 7 (The results of Sections 5 and 6 are necessary for Chapter 6.)
Chapter 4, Sections 1, 2, 4, 5, 6, 7
Chapter 5, Sections 1, 2, 3, 4
Chapter 6, Sections 1, 2, 5

We suggest that any course should contain at least some topics that are not core topics, perhaps for independent study or group work and presentation. Section 2.6 is a review of the arithmetic of complex numbers. Instructors can refer to it as needed. Here are some examples of different courses. We would like to note that Chapter 4 (A First Look at Group Theory) can be taught before Chapter 3 (Polynomials).

Course I. Number Theory, Rings and Groups. This course would be for students with no background in number theory. It would cover the core topics from Chapters 1, 2, 3, and 4, and also include Sections 5.1, 5.2, and 5.4.

Course II. Number Theory, Rings and Groups, with an emphasis on the concerns of future teachers. Like Course I, Course II would be for students with no background in number theory. It would cover the core topics from Chapters 1, 2, 3 and 4 and include time for student presentations and for discussion of the *In the Classroom* sections from these chapters.

Course III. Ring Theory and Group Theory for students with some familiarity with number theory. Instructors could review the material up to Section 2.4 very briefly or give it as a reading assignment. Beginning with Section 2.5, the course could cover the core topics from Chapters 3 through 5, include Section 6.1 and Section 6.2, and end with a look at Section 6.5.

Course IV. For experienced students (for instance, students with an undergraduate degree in mathematics, seeking certification in Secondary Mathematics). Such students should make sure to cover all the core sections of the book, including a serious study of Section 6.5. In that section, all their work in group theory and ring theory comes together and they will see a proof of the insolvability of the quintic. Such a class could allow time for extensive discussion of issues of pedagogy or allow for student presentation of optional sections.

There are many variations, depending on the degree of preparedness of the students and the number of excursions the course allows. Chapters 5 and 6 assume a slightly more sophisticated audience than Chapters 1 through 4.

ACKNOWLEDGMENTS

We offer our sincere thanks to administration and faculty of SUNY Geneseo for their support and encouragement. Our special thanks go to Stephen West, who chaired the mathematics department as we developed this project, and to Terry Holbrook, our indispensable secretary. We would like thank our students who used our material in rough form, especially Ryan Grover and Matthias Youngs, bright and avid readers and correctors. Most especially, we would like to thank Murray Wright, a superb mathematics teacher at Haverling High School in Bath, New York. While we have no doubt reintroduced errors, Murray's careful reading and thoughtful comments were crucial to the integrity of our endeavor.

We wish to thank the following reviewers for their many suggestions that helped shape this book:

Laurie Riggs, *California Polytechnic University*
Michael van Opstall, *University of Utah*
Jim Coykendall, *North Dakota State University*
Ferdinand Rivera, *San Jose State University*
Cheryl Roddick, *San Jose State University*

Robert Jamison, *Clemson University*
Peter Shiue, *University of Nevada at Las Vegas*
Anthony Vazzana, *Truman State University*
Charlotte Simmons, *University of Central Oklahoma*
Manfred Kolster, *McMaster University*
Mark L. Teply, *University of Wisconsin at Milwaukee*
Gregory P. Wene, *University of Texas at San Antonio*

At Prentice Hall, we would like to thank our editor, Sally Yagan, our marketing editor, Halee Dinsey, our production editor, Debbie Ryan, and our editorial assistant, Jennifer Urban.

Finally, we thank our families: You are wonderful!

Olympia E. Nicodemi
Melissa A. Sutherland
Gary W. Towsley

1

TOPICS IN NUMBER THEORY

Number theory is about the set of natural numbers $N = \{1, 2, 3, \ldots\}$ and the set of all integers $Z = \{\ldots, -2, -1, 0, 1, 2, 3, \ldots\}$. What could be easier? The subject is so familiar to us from school arithmetic. Elementary number facts, especially facts about addition, are easy to discover and understand. Since multiplication of natural numbers is simply repeated addition, e.g., $3 \times 2 = 2 + 2 + 2$, it too should be simple. But to determine basic facts about multiplication is more difficult. The questions "Which numbers are prime?" and "What is a number's prime factorization?" drive number theory. The answers are sometimes elusive, especially for large numbers. For instance, we will soon prove that the set of prime numbers is infinite, but mathematicians have no algorithm that produces all the primes, one after another. Factorization of small numbers is easy, just a matter of guessing and checking. But factoring a hundred-digit number can take a very long time even on a very fast computer. The topics of detecting primes and factoring numbers are thousands of years old and they are core topics in the training of very young students. Yet these same topics remain areas of active research. In the pages that follow we shall discover that the concepts of elementary number theory underlie almost every other topic in algebra, from group theory to factoring polynomials. It's really all about primes.

1.1 PRELIMINARIES

Our first section covers some frequently used tools for investigating the properties of numbers and algebraic systems. You will get a taste of how facts about numbers are proved with these tools. The appendix at the end of the text contains a more basic review of logic, sets, the language of proof and functions. It will help you hone your proving skills. So use it when you need it.

We begin with the **Well-Ordering Principle** for N:

Every nonempty set of natural numbers has a smallest member.

Examples are easy to think of. For instance, the set of prime numbers is a nonempty subset of the natural numbers. Its smallest member is 2. The set of natural numbers greater than the number e^{100} is not empty. The Well-Ordering Principle assures us that there is a smallest natural number greater than e^{100}. But to find out exactly what that natural number is can be a bit more challenging. The power of the Well-Ordering Principle is that it is a guarantee. It guarantees that there exists a smallest number in any nonempty set of natural numbers, even when we cannot explicitly name it.

The Well-Ordering Principle does not apply directly to the set $\mathbf{Z}$ of all integers. For example, its subset of even numbers, $\{\ldots, -4, -2, 0, 2, 4, \ldots\}$, does not have a smallest member. We shall need a slight generalization for the set $\mathbf{Z}$.

The General Well-Ordering Principle. Suppose that n_0 is an integer. Suppose that S is a nonempty subset of the integers and that every element of S is greater than or equal to n_0. Then S has a smallest element.

EXAMPLE 1

Suppose that a and b are integers with $a \geq b > 0$. Let $S = \{a - bx : x \in \mathbf{Z}$ and $a - bx \geq 0\}$. Notice that S is not empty because for $x = 0$, we see that a itself is in S. All members of S are greater than or equal to $n_0 = 0$. The Well-Ordering Principle guarantees that S must have a smallest member. Let's check the case where $a = 311$ and $b = 13$. Then $S = \{311 - 13x : x \in \mathbf{Z}$ and $311 - 13x \geq 0\}$. Letting $x = -2, -1, 0$, and 1, we find that $337, 324, 311$ and 298 are all members of S. So S is not empty. By letting x assume negative values, we see that S has no largest number. However, since all of its members are greater than or equal to $n_0 = 0$, the Well-Ordering Principle guarantees that S has a smallest element. In fact, $S = \{\ldots, 337, 324, 311, 298, 285, \ldots, 12\}$. Its smallest member is $12 = 311 - 23 \cdot 13$. Rearranged, we have $311 = 23 \cdot 13 + 12$. ∎

Next we will show how the familiar **Principle of Induction** can be deduced from the Well-Ordering Principle.

The Principle of Induction. Let S be a set of integers. Suppose that

i. $1 \in S$;
ii. if $x \in S$, then $x + 1 \in S$.

Then S contains all the natural numbers.

To deduce the Principle of Induction from the Well-Ordering Principle, we must prove the following:

If a set S satisfies conditions i and ii, then S contains the set of natural numbers $\mathbf{N}$.

To that end, let T be the subset of $\mathbf{N}$ of numbers **not** included in S. Assume that T is **not** the empty set. (We will get a contradiction.) The Well-Ordering Principle tells us that if T is not empty, then T has a smallest member, say x. Note that $x \neq 1$ from condition i. If x is the smallest natural number in T, then $x - 1$ is in S. But if $(x - 1) \in S$, then condition ii ensures that $(x - 1) + 1 = x$ is a member of S, contradicting our assumption that $x \notin S$. Thus T must be empty and $\mathbf{N}$ is therefore contained in S.

Induction proofs are of course based directly on the principle of induction. The bold type in the following simple example indicates where it is evoked.

EXAMPLE 2

Prove by induction that for all $n \in N$, $\sum_{i=1}^{n} i = \dfrac{n(n+1)}{2}$.

Proof. Let S be the set of integers for which the equality holds. **The integer $1 \in S$** since $\sum_{i=1}^{1} i = 1 = \dfrac{1(1+1)}{2}$. Now we must show that if $x \in S$ then $x+1$ is in S. **Suppose that $x \in S$**, which is to say that we assume that $\sum_{i=1}^{x} i = \dfrac{x(x+1)}{2}$. **We show that $x+1 \in S$.** To do so, we must prove that $\sum_{i=1}^{x+1} i = \dfrac{(x+1)((x+1)+1)}{2}$. Now $\sum_{i=1}^{x+1} i = \left(\sum_{i=1}^{x} i \right) + (x+1)$. By our assumption that $x \in S$, we can substitute $\dfrac{x(x+1)}{2}$ for $\sum_{i=1}^{x} i$. Thus we are left to show that $\dfrac{x(x+1)}{2} + (x+1) = \dfrac{(x+1)((x+1)+1)}{2}$. Simple arithmetic manipulation will show that both sides are equal to $\dfrac{(x+1)(x+2)}{2}$. **Since $1 \in S$ and since $x \in S$ implies that $x+1 \in S$, we can conclude that S contains all the natural numbers.** Rephrased: For all natural numbers n, it is true that $\sum_{i=1}^{n} i = \dfrac{n(n+1)}{2}$. ∎

In the course of proving various statements by induction, direct reference to the set S is usually suppressed. What follows is a typical, more streamlined, use of induction in a divisibility problem typical of elementary number theory.

EXAMPLE 3

We prove by induction that for each $n \in N$, $n^5 - n$ is a multiple of 5.

Proof. Let $n = 1$. Then $n^5 - n = 1 - 1 = 0 = 0 \cdot 5$. So the assertion is true for $n = 1$. Assume that $n^5 - n$ is a multiple of 5. We will prove that $(n+1)^5 - (n+1)$ is a multiple of 5. Expanding and regrouping, we obtain

$$(n+1)^5 - (n+1) = (n^5 + 5n^4 + 10n^3 + 10n^2 + 5n + 1) - (n+1)$$
$$= (5n^4 + 10n^3 + 10n^2 + 5n) + (n^5 - n).$$

By assumption, $(n^5 - n)$ is a multiple of 5. Since $(5n^4 + 10n^3 + 10n^2 + 5n) = 5(n^4 + 2n^3 + 2n^2 + n)$ is a multiple of 5, the sum $(5n^4 + 10n^3 + 10n^2 + 5n) + (n^5 - n)$ is a multiple of 5. Thus, for each $n \in N$, $n^5 - n$ is a multiple of 5. ∎

Some assertions require the following variation on induction often called "**The Strong Principle of Induction.**" Suppose $P(n)$ is the assertion that we wish to prove for $n \geq n_0$.

i. Suppose that $P(n_0)$ is true.
ii. Suppose that if $P(x)$ is true for all integers x such that $n_0 \leq x < n$, then $P(n)$ is also true.

Then $P(n)$ is true for all $n \geq n_0$.

EXAMPLE 4

We prove that every natural number $n \geq 12$ can be written as $n = 5s + 4t$, where s and t are nonnegative integers.

Proof. The assertion is true for $n_0 = 12$ because we can let $s = 0$ and $t = 3$. The assertion is also true for $13 = 5 \cdot 1 + 4 \cdot 2$, for $14 = 5 \cdot 2 + 4 \cdot 1$, and for $15 = 5 \cdot 3 + 4 \cdot 0$. Suppose that n is an integer that is greater than 15 and that the assertion is true for $12 \leq x < n$. If n is a multiple of either 5 or 4, there is nothing more to prove. If not, $n - 4 \geq 12$. By our assumption $n - 4 = 5s + 4t$ for some nonnegative values of s and t. Thus $n = 5s + 4(t + 1)$.

Note that the mechanism of the proof does not work for $n = 13$, 14, or 15 because in each case $n - 4$ is less than 12. That is why we singled out those cases for direct proof. ∎

Our investigation of number systems is often motivated by the need to find a set in which a given equation has a solution. For instance, the equation $x + 3 = 5$ has a solution in **N**, but to solve $x + 5 = 3$ we need to reach into the expanded number system **Z**. Similarly, $3x = -6$ has a solution in **Z**, but to solve $6x = -3$ we need to reach into the set of rational numbers **Q**. One of the principal tools for constructing new venues to do arithmetic is the **equivalence relation**. Since we shall make use of equivalence relations often in what follows, we now briefly review some of the key facts about relations and particularly about equivalence relations.

> **DEFINITION 1.** A **relation** on a set A is a subset R of $A \times A$. If $(x, y) \in R$, we write xRy.

EXAMPLE 5

Let $A = \{1, 2, 3, 4, 5\}$ and $R = \{(1, 1), (1, 2), (2, 3), (4, 5), (5, 4), (4, 4), (5, 5)\}$. In this relation, $1R2$ but $2\not{R}1$ since $(1, 2) \in R$ but $(2, 1) \notin R$. ∎

Let A be any set and R be a relation on A. We call R an **equivalence relation** if it has the following three properties:

1. R is **reflexive**. This means that for every $x \in A$, xRx.
2. R is **symmetric**. This means that for every pair of elements x and y, if xRy, then yRx.
3. R is **transitive**. This means that for every triple of elements x, y and z, if xRy and yRz, then xRz.

The relation given in Example 5 has none of the above properties. It is not reflexive because 2$\not\!R$2. It is not symmetric because 1R2 but 2$\not\!R$1. It is not transitive because 1R2 and 2R3, but 1$\not\!R$3.

EXAMPLE 6

Let A be the set of rational numbers $\mathbf{Q}$ and let $R = \{(x, y): x - y \in \mathbf{Z}\}$. So xRy if and only if $x - y$ is an integer. Thus $4.3R0.3$ because $4.3 - 0.3 = 4$, and $-0.3R1.7$ because $-0.3 - (1.7) = -2$. We now show that R is an equivalence relation.

1. Let $x \in \mathbf{Z}$. Since $x - x = 0$ and $0 \in \mathbf{Z}$, xRx. Thus R is reflexive.
2. Let x and y be in $\mathbf{Z}$ and suppose that xRy. Since $x - y \in \mathbf{Z}$, $y - x \in \mathbf{Z}$ and so yRx. Thus R is symmetric.
3. Let $x, y,$ and z be in $\mathbf{Z}$, and suppose that xRy and yRz. Since $(x - y)$ and $(y - z)$ are in $\mathbf{Z}$, $(x - y) + (y - z) = (x - z)$ is in $\mathbf{Z}$. Thus xRz and R is transitive.

Thus R is an equivalence relation. ∎

EXAMPLE 7

Let $A = \mathbf{Z} \times \mathbf{N}$. The elements of A are pairs (x, y) where x can be any integer and y can be any natural number. The following relation is an equivalence relation on A:

$$(x, y)R(s, t) \text{ if and only if } xt = ys.$$

Before we prove that it is an equivalence relation, let's see what is related to what. We have $(2, 4)R(3, 6)$ since $2 \cdot 6 = 4 \cdot 3$. Similarly, we have $(3, 6)R(4, 8)$ and $(-3, 1)R(-9, 3)$. However, $(1, 2)\not\!R(2, 1)$ because $2 \cdot 2 \neq 1 \cdot 1$.

1. The relation is reflexive. For each element (x, y) in A, we have $(x, y)R(x, y)$ since $xy = xy$.
2. The relation is symmetric. Suppose that $(x, y)R(s, t)$. Then $xt = ys$ and so $sy = tx$. Thus $(s, t)R(x, y)$.
3. The relation is transitive. Suppose that $(x, y)R(s, t)$ and that $(s, t)R(u, v)$. Then $xt = ys$ and $sv = tu$ so that $xtsv = ysut$. If $s \neq 0$, then we can cancel st from both sides to obtain $xv = yu$ and $(x, y)R(u, v)$. If $s = 0$, then both x and u must be 0 and we again have $xv = yu$. ∎

Suppose A is a set and R is an equivalence relation. The **equivalence class of $x \in A$** is the set $[x] = \{y \in A: yRx\}$. Thus $[x]$ is the set of all elements of A that are related to x. For example, let $x = 0.3$ and let R be the equivalence relation defined in Example 6. Then $[x] = \{\ldots, -1.7, -0.7, 0.3, 1.3, 2.3, \ldots\}$. If $x = (1, 2)$ and $y = (3, 2)$ and R is the relation given in Example 7, then $[x] = \{(1, 2), (2, 4), (3, 6), \ldots\}$ and $[y] = \{(3, 2), (6, 4), (9, 6), \ldots\}$. Notice that $[x]$ and $[y]$ have no members in common. Also, note that each element $x \in A$ is contained in its own equivalence class $[x]$ because equivalence relations are reflexive. The next proposition asserts that, in general, equivalence classes $[x]$ and $[y]$ are either disjoint and have no members in common, or they are identical and have exactly the same members. So every element of A is in *exactly* one equivalence class.

PROPOSITION 1 Suppose that R is an equivalence relation on the set A and that x and y are elements of A. If $[x] \cap [y]$ is not empty, then $[x] = [y]$.

Proof. Suppose that $z \in [x] \cap [y]$. Let $t \in [x]$. Since tRx and xRz, we have tRz by transitivity. Since tRz and zRy, we have tRy and $t \in [y]$. Thus $[x] \subseteq [y]$. Similarly, $[y] \subseteq [x]$ and therefore $[x] = [y]$. ▲

Suppose A is a set and that C is a collection of subsets of A. We call C a **partition** of A if every member of A is in exactly one member of the collection. For example, if $A = \{1, 2, 3, 4, 5, 6, 7, 8, 9, 10\}$, the collection $C = \{\{1, 5, 9\}, \{2, 6, 8\}, \{3, 4, 7, 10\}\}$ is a partition of A. What we proved in Proposition 1 is that the collection of equivalence classes of an equivalence relation on A forms a partition of A. We can work backward. If C is a partition of A, then we can find an equivalence relation R on A such that the members of C are its equivalence classes. We define R as follows: xRy whenever x and y are in the same member of C. For example, if C is the partition $\{\{1, 5, 9\}, \{2, 6, 8\}, \{3, 4, 7, 10\}\}$, then $1R5$ and $2R6$ but $2\cancel{R}1$. The proof that R is an equivalence relation is an exercise.

EXAMPLE 7 *(continued)*

The equivalence classes of the relation given in the first part of Example 7 can be thought of as a set of new mathematical objects that can be added and multiplied. Let's give the set a name. Let $Q = \{[(x, y)] : x \in Z \text{ and } y \in N\}$. Here's how addition and multiplication are defined.

$$[(x, y)] + [(s, t)] = [(xt + ys, yt)]$$

$$[(x, y)] \cdot [(s, t)] = [(xs, yt)]$$

There are many choices to be made when carrying out these operations. For instance, since $[(1, 2)] = [(2, 4)]$ and $[(3, 5)] = [(9, 15)]$, the product $[(1, 2)] \cdot [(3, 5)] = [(3, 10)]$ could also be computed by $[(2, 4)] \cdot [(9, 15)] = [(18, 60)]$. But these seemingly different answers are not actually different because $(3, 10)R(18, 60)$ and hence $[(3, 10)] = [(18, 60)]$. More generally, both operations are **well-defined**, meaning that the answers are independent of the choices we make to compute them. Suppose that $(x, y)R(s, t)$ and that $(u, v)R(w, z)$. To show that multiplication is well-defined, we must show that $[(x, y)] \cdot [(u, v)] = [(s, t)] \cdot [(w, z)]$, which is the same as showing $[(xu, yv)] = [(sw, tz)]$ or $xutz = yvsw$. We know that $xt = ys$ and $uz = vw$ so that $xtuz = ysvw$ as needed. The proof that addition is well-defined is left as an exercise. ■

We can change the notation in Example 7 from $[(x, y)]$ to $\frac{x}{y}$ by replacing the comma by the fraction bar and omitting various levels of parentheses. It then becomes obvious that, starting with the integers, we have constructed the set we usually call the set of rational numbers. We were able to define addition and multiplication on this "new" set of rational numbers Q so that it acquired an algebraic structure. You might think that this was cumbersome, but what is a rational number? It's not just one expression

$\frac{x}{y}$, but an object represented by many expressions that are somehow equivalent. That equivalence is made precise via the defining equivalence relation. We will use similar tools throughout the text to create new sets on which to do arithmetic and algebra. This example also serves to draw attention to the underlying complexity of the concepts and skills that we ask very young children to negotiate—mastering fractions.

1.1 Exercises

1. For each of the following sets, indicate if the Well-Ordering Principle guarantees a smallest member. Find the smallest member of each of the given sets if you can.
   i. All positive integers that can be written in the form $397x + 541y$, where x and y can be any integers.
   ii. All rational numbers $\frac{m}{n}$, where m and n can be any integers such that $0 < m < n$.

For Exercises 2 through 5, prove the given statement by induction.

2. For each natural number n, $1 + 3 + \ldots + (2n - 1) = n^2$.
3. For each natural number n and real number $r \neq 1$, $1 + r + r^2 + \ldots + r^n = \frac{1 - r^{n+1}}{1 - r}$.
4. For each natural number n, $n^3 + 2n$ is a multiple of 3.
5. If each of n people in a room shook hands with every other person, a total of $\frac{n(n-1)}{2}$ handshakes would have occurred.
6. A binary tree consists of a root node and a finite number of other nodes. (See Figure 1.) Every node either has two or no immediate offspring. Every node together with its entire offspring is itself a binary tree. Use strong induction to prove that the total number of nodes in a binary tree is odd.

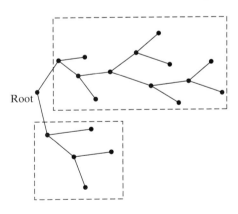

FIGURE 1

7. Define the relation R on the set of all ordered pairs of real numbers as follows: $(x, y)R(s, t)$ if and only if $2(x - s) = (y - t)$. Prove that R is an equivalence relation. Find the equivalence class of the point $(1, 1)$.

8. Prove that the operation of addition defined in Example 7 is well-defined.
9. Let C be a partition of a set A. Suppose that R is a relation defined by xRy whenever x and y are in the same member of C. Prove that R is an equivalence relation.
10. Show that in the statement of the Principle of Induction, we can replace the phrase "$1 \in S$" by the phrase "$n_0 \in S$", where n_0 can be any integer. The conclusion would then read, "Then S contains all integers $x \geq n_0$."
11. Prove that if we assume the Principle of Induction, then we can prove the Well-Ordering Principle as its corollary. (This exercise shows that the Principle of Induction and the Well-Ordering Principle are equivalent statements because each implies the other.)
12. The set of natural numbers is a subset of R, the set of all real numbers. Assume the (true) statement that given any real number x, there is a natural number n such that $n > x$. Using the standard arithmetic of inequalities, prove the following assertions.
    i. If $\varepsilon > 0$ is any real number, we can find a natural number n such that $0 < \frac{1}{n} < \varepsilon$.
    ii. If x and y are real numbers such that $0 < x < y$, we can find a rational number $\frac{m}{n}$ such that $x < \frac{m}{n} < y$. **Hint.** For part ii, let $\varepsilon = y - x$ and use part i to find n. Then use well-ordering to find m.
    iii. Extend part ii to any real numbers $x < y$.

To the Teacher

We, the authors, would like to welcome you into our text. We have set out a challenging mathematical agenda for you as students of mathematics. We have provided a historical context for your studies not just with names and dates, but with a taste of how the mathematical concepts that we study were implemented in their early stages. We have also tried to tie the mathematics you will study to the mathematics you will teach. Sometimes, you will be studying topics appropriate for the high school classroom. We will point those out and point to how they might be implemented. Other times, you will study topics that are more advanced. These will give you a deeper knowledge that will allow you see the high school curriculum in larger context. We will try to point out the connections.

The elementary mathematics studied in grades K through 12 considers some of the richest areas of our discipline: geometry, number theory and algebra. "Elementary" does not mean trivial or shallow. But having carried out its routine elements for many years, we often need to look again to see the depth and complexity that underlies elementary mathematics. For instance, in this section of the text, we reminded you of two of the principal tools that we will be using throughout the text: induction and the equivalence relation. But even these topics both tie in directly to school mathematics. Let's see how.

In its *Principles and Standards for School Mathematics*,[1] the National Council of Teachers of Mathematics (NCTM) emphasizes pattern recognition throughout the

---
[1] See Section 1.8, *In the Classroom*, for a fuller description of the mission of the NCTM.

pre K–12 curriculum. Students should be able to

- describe, extend, and make generalizations about geometric and numeric patterns;
- represent and analyze patterns and functions, using words, tables, and graphs;
- relate and compare different forms of representation for a relationship.

We know that recognizing a mathematical pattern and proving that a pattern must persist are two different aspects of the mathematical process, both indispensable. As a teacher of a high school class, you would be delighted to find that a student can see that the pattern of triangular numbers $\{1, 3, 6, 10, \ldots\}$ is obtained by adding $n + 1$ to the number of stars in the nth triangle. (See Figure 2.)

✩	✩ ✩ ✩	✩ ✩ ✩ ✩ ✩ ✩	✩ ✩ ✩ ✩ ✩ ✩ ✩ ✩ ✩ ✩
1	2	3	4

FIGURE 2

Another student might suggest the formula for the nth triangular number is $T(n) = n(n + 1)/2$. The teacher who appreciates the rigor of mathematical reasoning might not be ready to ask for a proof by induction but could prompt students to reconcile their answers: $T(n + 1) = T(n) + (n + 1)$ and $n(n + 1)/2 + (n + 1) = (n + 1)(n + 2)/2$. A teacher can impart inductive reasoning to the younger student without the formality of proof.

Also in this section, we constructed the set of rational numbers—fractions—by establishing an equivalence relation on the set $\mathbf{Z} \times \mathbf{N}$. This equivalence relation captures the fact that any given rational number x can be represented as a quotient in an infinite number of different ways. The number denoted by 1/2 is the same number denoted by 2/4 and 3/6, etc. Children come to internalize this through the social concerns that surround serving pizza. But a good teacher realizes that young students are asked to internalize some very difficult concepts such as how to know when two quotients represent the same number and to trust that arithmetic operations are well-defined: $\frac{2}{3} + \frac{3}{5} = \frac{10}{15} + \frac{9}{15}$. Your appreciation that there is a nontrivial, rigorous underpinning to the mathematics taught in grades K through 12 will help you to put the mathematics that you teach in a larger framework and to appreciate the challenges that mathematics presents at all levels.

Tasks

1. Explain why we "invert and multiply" when we divide fractions.
2. Show how you can use square arrays of dots to prove that $1 + 3 + 5 + \ldots + (2n - 1) = n^2$.

1.2 ARITHMETIC AND DIVISIBILITY

In all that follows, we shall denote the set of all integers $\{\ldots, -2, -1, 0, 1, 2, \ldots\}$ by $\mathbf{Z}$. This common usage comes from the fact that the German word for numbers is *Zahlen*. First we will summarize arithmetic properties of the integers. Though the arithmetic is basic, our summary gives us the opportunity to review key terms for future reference. In later chapters, we will investigate similar arithmetic properties as they arise in contexts less familiar than the set of integers. The focal point of this section is the Division Algorithm. If we divide an integer by a natural number b, the Division Algorithm guarantees that we will have a remainder r that lies between 0 and b. No surprise. But as you follow this text along, you might be tempted to rename it, "All Corollaries Great and Small of the Division Algorithm." The Division Algorithm will pop up everywhere. So explore it thoroughly.

Arithmetic Properties of the Integers

1. Addition and multiplication are **associative**. For any integers $a, b,$ and c, $a + (b + c) = (a + b) + c$ and $a(bc) = (ab)c$.
2. Addition and multiplication are **commutative**. For any integers a and b, $a + b = b + a$ and $ab = ba$.
3. There exists a unique **additive identity** 0 such that $0 + x = x$ for any integer x.
4. There exists a unique **multiplicative identity** 1 such that $1 \cdot x = x$ for any integer x.
5. Every integer x has an **additive inverse** $-x$ such that $x + (-x) = 0$.
6. Addition and multiplication satisfy the **distributive law**: For any integers $x, y,$ and z, $x(y + z) = xy + xz$.

The distributive law is the glue that relates multiplication and addition. For positive integers, we can interpret multiplication as repeated addition: 3 times 8 means "8 plus 8 plus 8." In this context, the distributive law is fairly easy to explain. (Try it!) But as soon as we introduce even a minor complication such as allowing negative integers, the validity of the distributive law becomes less obvious. (We will pick up this train of thought again in Section 2.5.) The proof of the next proposition demonstrates how to establish a familiar arithmetic fact assuming only properties 1 through 6 listed above. The fact we establish is simple: $0 \cdot a = 0$ for any integer a. If a is positive, we might explain the fact by saying if you have zero sets of a cookies, you have zero cookies! But it's a bit subtler with a negative. Notice that we are commenting on the multiplicative properties of the *additive* identity 0. So it is natural that the distributive law will be the key to the proof. Before we start, notice that because of property 2 (commutativity), $(y + z)x = x(y + z) = xy + xz = yx + zx$. So property 6 can also be expressed as follows:

For any integers $x, y,$ and z, $(y + z)x = yx + zx$.

PROPOSITION 1 For **any** integer a, it is true that $0 \cdot a = 0$.

Proof. Since 0 is the additive identity, $0 + 0 = 0$ (property 3). Thus,

(1) $$(0 + 0) \cdot a = 0 \cdot a.$$

By property 6 we can distribute (this is the key step) to obtain

(2) $$0 \cdot a + 0 \cdot a = 0 \cdot a.$$

Now, by property 5, every element has an additive inverse. The additive inverse of $0 \cdot a$ is denoted by $-(0 \cdot a)$. So we add $-(0 \cdot a)$ to both sides to obtain

(3) $$-(0 \cdot a) + (0 \cdot a + 0 \cdot a) = -(0 \cdot a) + 0 \cdot a.$$

Using associativity (property 1), we can rewrite (3) as

(4) $$(-(0 \cdot a) + 0 \cdot a) + 0 \cdot a = -(0 \cdot a) + 0 \cdot a.$$

Replacing $-(0 \cdot a) + 0 \cdot a$ with its sum 0 on each side, we are left with

(5) $$0 + (0 \cdot a) = 0.$$

Since $0 + (0 \cdot a) = 0 \cdot a$, we have $0 \cdot a = 0$, which is what we wanted to prove. ▲

Notice that we have referred only to the listed properties. So when we encounter other number systems like the set of complex numbers that also has properties 1 through 6, our proof will transfer directly.

Number theory concentrates on the study of primes and the factorization of composite numbers into primes. The essential information about any given integer x is the list of primes that divide it. So we now look at the basic properties of division.

> **DEFINITION 1.** Let a and b be integers with $b \neq 0$. We say that b **divides** a if there is an *integer* q such that $a = bq$.

When b divides a, we also say that a is **divisible** by b, or that a is a **multiple** of b, or that b is a **factor** of a. We sometimes write $b|a$ to mean that b divides a. Thus $3|6$ since $6 = 2 \cdot 3$ and also $3|0$ since $0 = 0 \cdot 3$. But it is not the case that $0|3$ since we cannot find a multiple of 0 that equals 3.

Caution: The expression $b|a$ is *very* different from b/a. The expression $b|a$ means that b is a factor of a. So $b|a$ is not a number itself, but just a way of describing the relation of a and b. But b/a is a rational number. It is the number obtained when b is divided by a. So $6/2 = 3$ but $6|2$ is a false statement because 2 is not an *integer* multiple of 6. Be careful!

The following theorem summarizes two familiar properties of division.

THEOREM 2 Let a, b and c be integers.

  i. If $a|b$ and $b|c$, then $a|c$.
  ii. If $a|b$ and $a|c$, then $a|(mb + nc)$ for any integers m and n.

Proof.

  i. Since $a|b$ and $b|c$, there are integers t and s such that $b = ta$ and $c = sb$. Substituting ta for b in the latter, we have $c = sta$. Let $m = st$. Then $c = ma$. Since m is an integer, a divides c.
  ii. The proof of Part ii is similar, and we leave it to you as an exercise. ▲

Geometrically, we can think of two **positive** integers a and b as two lengths, both multiples of a common unit length. Then b divides a when the length of a can be spanned exactly by q "rulers" of length b where q is an integer.

If b does **not** divide a, we can find a number q such that $qb < a$ (too short) but $(q+1)b > a$ (too long). If the units are notched on the ruler of length a, there will be an integer number r of unit lengths left over after we span a with q copies of b. Notice $0 < r < b$. This fact is traditionally called the **Division Algorithm** because of the underlying algorithm that finds the remainder r by repeatedly subtracting b from a. In Figure 1, $a = 8$ and $b = 3$. Since 2 copies of b fail to span a but 3 copies of b exceed a, we have $q = 2$ and $r = 2$.

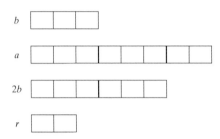

FIGURE 1

THEOREM 3 (***The Division Algorithm***) Let a and b be any integers with $b > 0$. There are unique integers q and r such that $a = qb + r$, where $0 \leq r < b$.

The proof is based on the Well-Ordering Principle. Before we present it formally, let's look at an example. Let $a = 43$ and $b = 7$. To look for the positive remainder of a after division by b, we consider how many times we can subtract 7 from 43, always keeping a nonnegative remainder.

$$43 - 0 \cdot 7 = 43$$
$$43 - 1 \cdot 7 = 36$$
$$43 - 2 \cdot 7 = 29$$
$$43 - 3 \cdot 7 = 22$$
$$43 - 4 \cdot 7 = 15$$
$$43 - 5 \cdot 7 = 8$$
$$43 - 6 \cdot 7 = 1$$
$$43 - 7 \cdot 7 = -6$$

Stop! We can subtract 7 six times from 43. Our remainder is 1. It is a unique positive remainder that is less than 7 and it has the form $43 - q \cdot 7$. Actually, there is an infinite set of nonnegative numbers expressible in the form $43 - q \cdot 7$. For instance, we have $43 - (-1) \cdot 7 = 50$, $43 - (-2) \cdot 7 = 57$, etc. The thing to notice is that $r = 1$ is the smallest **nonnegative** number in the set $S = \{\ldots, 50, 43, 36, 29, 22, 15, 8, 1\}$.

Proof of Theorem 3. Let a and b be any integers with $b > 0$ and let $S = \{a - qb : q \in \mathbf{Z} \text{ and } a - qb \geq 0\}$. S is the set of nonnegative remainders after the subtraction of integer multiples of b from a. To use the Well-Ordering Principle on S, we must first show that set S is not empty. There are two cases. If $a \geq 0$, let q be 0 so that $a = a - 0 \cdot b$ is in S. If $a < 0$, let $q = a$. Since $b > 0$, $a - ab \geq 0$. Thus $a - ab$ is in S and S is not empty. By the Well-Ordering Principle, S has a smallest member r. Let q be the integer such that $r = a - qb$. Rearranging, $a = qb + r$.

To finish the proof, we must show that r is less than b and that r is unique. Suppose $r \geq b$. (We will get a contradiction.) Then

$$r - b = a - qb - b = a - (q+1)b$$

is also nonnegative and hence $r - b$ is in S. But $r - b < r$, contradicting our choice of r as the *smallest* member of S. Thus $0 \leq r < b$ and $a = bq + r$. To show uniqueness, we show that if $a = qb + r$ and $a = q_1 b + r_1$ with both $0 \leq r < b$ and $0 \leq r_1 < b$, then $r = r_1$. Suppose that $b > r \geq r_1 \geq 0$. Subtracting, we have

$$a - a = 0 = (q - q_1)b + (r - r_1)$$

or

$$(q_1 - q)b = (r - r_1).$$

Since $0 \leq r - r_1 < b$, $q_1 - q$ cannot be negative and cannot be greater or equal to 1. Thus $q_1 - q = 0$ and $r = r_1$. ▲

EXAMPLES

1. Let $a = 31$ and $b = 5$. Then $q = 6$ and $r = 1$ since $31 = 6 \cdot 5 + 1$ and $0 \leq 1 < 5$.
2. Let $a = -31$ and $b = 5$. Then $q = -7$ and $r = 4$ since $-31 = -7 \cdot 5 + 4$. (Remember, we must have $0 \leq r < b$ so that r is always nonnegative.) ■

SUMMARY

After reviewing some elementary properties of the set of integers, this section focused on divisibility. The key definition is that of the phrase "***b* divides *a*.**" Remember that we say that b divides a if a is an *integer* multiple of b. No fractions allowed! Be careful not to confuse the notation $b|a$ with b/a. The key theorem is Theorem 3, the **Division Algorithm**. It characterizes the quotient and remainder in an integer division problem. Elementary number theory is based on this simple theorem.

1.2 Exercises

For Exercises 1 through 4, model your proofs on the proof of Proposition 1.
1. Let a, b and c be integers. Prove that if $a + b = a + c$, then $b = c$.
2. For any integer a, prove that $(-1) \cdot a = -a$.
3. For any integer a, prove that $-(-a) = a$.
4. Let a and b be integers. Prove that $(-a) \cdot b = -(ab)$.
5. For any integers a, b and c, prove that if $ab = ac$ and $a \neq 0$, then $b = c$. What additional arithmetic fact beyond properties 1 through 6 did you need?

6. Prove part ii of Theorem 2.
7. For each pair a and b given below, find q and r as in the **Division Algorithm** and express a as $a = bq + r$.
   i. $a = 335$ and $b = 17$
   ii. $a = -335$ and $b = 17$
   iii. $a = 21$ and $b = 13$
   iv. $a = 13$ and $b = 8$
8. Prove that if a divides b and c divides d, then ac divides bd.
9. Suppose that $b > 0$ and that $a = qb + r$. Find the quotient and remainder when $-a$ is divided by b.
10. Prove that 3 divides $n(n + 1)(n + 2)$ for each integer n.
11. Prove that for each integer $n \geq 0$, the integer $2^{n+1} + 3^{3n+1}$ is divisible by 5.

To the Teacher

The Division Algorithm doesn't seem like an algorithm at all. But it really is an algorithm if we think of division as repeated subtraction. For simplicity, suppose (as Euclid would have) that a and b are both natural numbers. As an algorithm, the instructions to find r and q would be as follows.

1. Set x equal to a and set $q = 0$. (Initialize.)
2. If $x < b$, set r equal to x and stop. If $x \geq b$, reset x to $x - b$ and q to $q + 1$. Return to the beginning of step 2.

The key is iteration. The number of times we return to the beginning of step 2 is q and $r = a - qb$. But when we think of an algorithm for division, we usually think about the **long division** algorithm. Many of us still groan at having had to carry out long division on very big numbers by hand all too frequently. Such tedium—certainly not a good invitation to mathematics—should be put to rest by the calculator. But let's look at what we can still learn (and teach) from the process of long division. The example we will pull apart is 35780 divided by 57. The answer is 627 with a remainder of 41.

$$\begin{array}{r} 627 \\ 57{\overline{\smash{\big)}\,35780}} \\ \underline{-342} \\ 158 \\ \underline{-114} \\ 440 \\ \underline{-399} \\ 41 \end{array}$$

- Estimation and base-10 decomposition.
  The first step is about order of magnitude. Are there 1000 copies of 57 in 35780? No, since $57000 > 35780$. Are there at least 100 copies of 57? Yes, since $5700 < 35780$. How many hundreds of copies? Now we invoke the Goldilocks Principle: 500 is too small because $35780 - 57 \cdot 500 = 7280$. But 700 is too big since $35780 - 57 \cdot 700 = -4120$. So 600 is just right. Such considerations

give the young student fundamental practice in developing "number sense." Estimation skills are honed and the base-10 place system reinforced. Students must engage in the ever-crucial activity of applying well-reasoned feedback to a mathematical problem in progress.

$$\text{Step 1:} \quad 57 \overline{\smash{)}35780} \\ \phantom{\text{Step 1:} \quad 57)}\underline{-34200} \\ \phantom{\text{Step 1:} \quad 57)\,\,}1580$$

(quotient 600)

- Iteration is fundamental to any algorithm. Practice with long division can help to develop a student's algorithmic sense. It is not just a matter of rote. It is the recognition that at the second and subsequent steps of an algorithm, we return to an analogous but simpler problem. In our example, the next step is to divide 1580 by 57. In iterating the procedure, we ask the same kinds of questions. Are there 100 copies of 57 in 1580? No. There are in fact 20 copies of 57 in 1580 with a remainder of 440. Iterate again: there are 7 copies of 57 in 440 with a remainder of 41. End because there are no copies of 57 in 41.

$$\text{Step 2:} \quad 57 \overline{\smash{)}1580} \\ \phantom{\text{Step 2:} \quad 57)}\underline{-1140} \\ \phantom{\text{Step 2:} \quad 57)\,\,}440$$

(quotient 20)

$$\text{Step 3:} \quad 57 \overline{\smash{)}440} \\ \phantom{\text{Step 3:} \quad 57)}\underline{-399} \\ \phantom{\text{Step 3:} \quad 57)\,}41$$

(quotient 7)

So our answer is $600 + 20 + 7$ with a remainder of 41.

- The process of long division uses all the other arithmetic operations and it demands a certain facility with them.

Technology is a bonus in time and accuracy. Because it is there, we as teachers must reflect carefully about what we want a given lesson to convey mathematically. The option of simply teaching mathematics as a series of routines is not there. On the other hand, what seems like just routine can sometimes carry valuable lessons.

Tasks

1. Carry out long division of $a = 42321$ divided by $b = 341$ where both numbers are given in base 5. For a more challenging task, carry out the division in base 12.
2. Show how to model long division with base-10 blocks.

1.3 GREATEST COMMON DIVISORS AND EUCLID'S ALGORITHM

In the previous section, we explored what it means for one integer to divide another. In this section, we explore how to determine if two integers share common divisors. Our principal tool is Euclid's Algorithm, an ancient and simple method that has proven to be extremely useful in a variety of mathematical settings. We will continue to call upon it throughout this text. The power of Euclid's Algorithm comes from the fact that it finds the common factors or common divisors of two numbers without the knowledge of the divisors of either of the two numbers. With Euclid's Algorithm, there's no factoring involved in finding common factors!

> **DEFINITION 1.** Let a and b be integers, not both zero. The **greatest common divisor** of a and b, denoted by **gcd(a, b)**, is the largest integer that divides both a and b.

EXAMPLE 1

i. $\gcd(12, 18) = 6$. The set of divisors of 12 is $S_{12} = \{-12, -6, -4, -3, -2, -1, 1, 2, 3, 4, 6, 12\}$. The set of divisors of 18 is $S_{18} = \{-18, -9, -6, -3, -2, -1, 1, 2, 3, 6, 9, 18\}$. The set of *common divisors* is $S_{12} \cap S_{18} = \{-6, -3, -2, -1, 1, 2, 3, 6\}$. The *largest* integer in $S_{12} \cap S_{18}$ is 6. Thus $\gcd(12, 18) = 6$.

ii. $\gcd(3, -3) = 3$ because the set of common factors is $\{-3, -1, 1, 3\}$.

iii. $\gcd(0, 3) = 3$. Since 3 divides 3 and 3 divides 0, the common factors are $\{-3, 3\}$.

iv. $\gcd(87141987598750913580985, 93149583591309510580) = ????$. ∎

The greatest common divisor (gcd) of two integers a and b is always positive because if a negative number x divides a and b, then $-x$, a positive number greater than x, also divides a and b. The problem of finding the greatest common divisor of two small integers is easy, simply a matter of listing and comparing their small sets of divisors. But if we rely only on our ability to factor, the task of finding the gcd of two large numbers could be formidable because factoring very large numbers is very time consuming, even for the fastest computers. In what follows, we develop facts about greatest common divisors that do not rely on factorization. Our next theorem may seem somewhat surprising. It shows that the gcd of two numbers a and b can be expressed as a weighted sum of a and b. (A weighted sum of a and b is a sum of the form $ma + nb$ where m and n are integers.)

THEOREM 1 Let a and b be integers, not both equal to zero. Then $\gcd(a, b)$ is the smallest positive integer that can be expressed as a sum $ma + nb$, where m and n are also integers.

Before we start the proof, let's observe the relation of $\gcd(a, b)$ to sums of the form $sa + tb$ where s and t can be any integers. Let d denote $\gcd(a, b)$. Then $a = qd$ and $b = pd$ for some integers q and p. Any sum $sa + tb$ is a multiple of d because $sa + tb = sqd + tpd = d(sq + tp)$. Now let's list some strictly positive sums of the form $sa + tb$ for a specific pair of integers, $a = 6$ and $b = 15$ with $d = \gcd(6, 15) = 3$.

$$
\begin{aligned}
s = 1, \quad & t = 1, \quad & sa + tb = 21 \\
s = 3, \quad & t = 0, \quad & sa + tb = 18 \\
s = -1, \quad & t = 1, \quad & sa + tb = 9 \\
s = 3, \quad & t = -1, \quad & sa + tb = 3
\end{aligned}
$$

All the sums are multiples of 3, but what is more important, 3 itself is on the list. So d is of the form $sa + tb$. In other words, the set $S = \{sa + tb : sa + tb > 0, s$ and t in $\mathbf{Z}\}$ contains the $\gcd(a, b)$. Note that d is the smallest number in the set because every other number in the set is a multiple of d. The proof of Theorem 1 shows that what we have discovered for 6 and 15 holds in general. We will use the Well-Ordering Principle to pick d out of the set S.

Proof of Theorem 1. Suppose $a \neq 0$. Let $S = \{sa + tb : sa + tb > 0, s$ and t in $\mathbf{Z}\}$. First we show that the set S is not empty. If $a > 0$, we let $s = 1$ and $t = 0$ to show that a itself is in S. If $a < 0$, let $s = -1$ and $t = 0$. This shows that $-a$, a positive integer, is in S. Since S is a nonempty set of positive integers, the Well-Ordering Principle guarantees that S has a smallest member. Let d denote the smallest member of S.

We shall now show that $d = \gcd(a, b)$. Suppose that m and n are the integers such that $d = ma + nb$. If $a = ct$ and $b = cs$, then $d = c(mt + ns)$. So any common divisor c of a and b divides d. Thus d is at least as large as any common divisor of both a and b. By showing that d itself divides both a and b, we will show that d is the largest common divisor of a and b, i.e. that $d = \gcd(a, b)$. Suppose that d fails to divide a. (We will obtain a contradiction.) By the Division Algorithm, we would have $a = qd + r$ with $0 < r < d$. Solving for r and substituting $ma + nb$ for d, we would have

$$r = a - qd = a - q(ma + nb) = (1 - qm)a + (-qn)b.$$

Let $s = 1 - qm$ and $t = -qn$. Thus $r = sa + tb$, a positive number **less than** d, would also be in S, contradicting our choice of d. So d must divide both a and b. Thus $d = \gcd(a, b)$. ▲

Because any common divisor of both a and b is a divisor of $\gcd(a, b)$, the set of factors of $\gcd(a, b)$ is a complete list of the common divisors of a and b. Let $d = \gcd(a, b)$. Since $d = ma + nb$ for some integers m and n, $td = tma + tnb$. So any multiple of d can be expressed as a weighted sum of a and b. The next corollary summarizes this observation.

COROLLARY 2 An integer x is of the form $ma + nb$, with $m, n \in \mathbf{Z}$, if and only if x is a multiple of $\gcd(a, b)$.

EXAMPLE 2

Let $a = 114$ and $b = 66$. Since $114 = 2 \cdot 3 \cdot 19$ and $66 = 2 \cdot 3 \cdot 11$, we see that $\gcd(114, 66) = 6$. By Corollary 2, any multiple of 6 can be expressed as $m \cdot 114 + n \cdot 66$. For instance, a bit of experimenting will show that $6 = -4 \cdot 114 + 7 \cdot 66$, and so 30 can be written as

$$30 = 5 \cdot 6 = 5(-4 \cdot 114 + 7 \cdot 66) = -20 \cdot 114 + 35 \cdot 66.$$

Here $m = -20$ and $n = 35$. ■

DEFINITION 2. Let a and b be two integers, not both zero. We say that a and b are **relatively prime** if $\gcd(a, b) = 1$.

If $\gcd(a, b) = 1$, then the only common positive factor is 1. Thus 9 and 10 are relatively prime numbers although neither number is itself prime. The following proposition shows that if you divide two numbers by their gcd, the resulting numbers are integers that are relatively prime (i.e., they have no common factors). For instance, $\gcd(20, 16) = 4$ and $\gcd(5, 4) = 1$.

PROPOSITION 3 Suppose that $\gcd(a, b) = d$. Then $\gcd\left(\frac{a}{d}, \frac{b}{d}\right) = 1$.

Proof. From Theorem 1, we can express d as $d = ma + nb$. Since $a = dp$ and $b = dq$ for some integers p and q, we can express d as

$$d \cdot 1 = mdp + ndq = d(mp + nq).$$

Since $d \neq 0$, we can cancel d to obtain

$$1 = mp + nq.$$

Again from Theorem 1, we have that $\gcd(p, q) = 1$. Since $p = \frac{a}{d}$ and $q = \frac{b}{d}$, our theorem is proved. ▲

The following result, known as **Euclid's Lemma**, is an important tool for investigating divisibility and factorization.

THEOREM 4 (*Euclid's Lemma*) Suppose that $\gcd(a, b) = 1$ and that a divides bc. Then a divides c.

Proof. Since $\gcd(a, b) = 1$, we can find m and n such that $1 = ma + nb$. Multiplying through by c, we obtain $c = mac + nbc$. Since a divides a, a divides mac. Since, by hypothesis, a divides bc, a divides nbc. Thus a divides the sum $mac + nbc$. Since $c = mac + nbc$, a divides c. ▲

The focus of $\gcd(a, b)$ is on what numbers divide both a and b. It picks out the largest of these. Notice that Euclid's Lemma changes the focus to what happens when a divides a multiple of b. Euclid's Lemma addresses what is divisible by a instead of what divides a.

Euclid's Algorithm

Euclid's Algorithm is an ancient tool. It allows us to find the gcd of two numbers without factoring them. It is a very efficient and versatile algorithm. (We will see one of its many applications at the end of this section.) The following two propositions form the theoretical foundations for the algorithm. They will help us find the gcd of two numbers using repeated division rather than factorization.

PROPOSITION 5 If $a > 0$ and a divides b, then $\gcd(a, b) = a$.

Proof. Exercise 6.

PROPOSITION 6 Let a and b be integers and suppose that $a = qb + r$. Then $\gcd(a, b) = \gcd(b, r)$.

Proof. Let S be the set of common divisors of a and b and let T be the set of common divisors of b and r. Any common divisor of a and b must divide $a - bq$ and hence divide r. So $S \subseteq T$. Similarly, any common divisor of b and r must divide $qb + r$ and hence a. So it follows that $T \subseteq S$ and that $S = T$. Since the set of common divisors of a and b is exactly the set of common divisors of r and b, the largest element of S, which is $\gcd(a, b)$, must equal the largest element of T, which is $\gcd(b, r)$. ▲

Let's use Propositions 5 and 6 to help us find the greatest common divisor of 134 and 123 through repeated division.

Since $134 = 1 \cdot 123 + 11$, Proposition 6 says $\gcd(134, 123) = \gcd(123, 11)$.
Since $123 = 11 \cdot 11 + 2$, Proposition 6 says $\gcd(123, 11) = \gcd(11, 2)$.
Since $11 = 5 \cdot 2 + 1$, Proposition 6 says $\gcd(11, 2) = \gcd(2, 1)$.
Since 1 divides 2, Proposition 5 says that the $\gcd(2, 1) = 1$. Thus the $\gcd(134, 123) = 1$.

The procedure just illustrated is formalized in **Euclid's Algorithm**.

Euclid's Algorithm:

Let $a \geq b > 0$ be two positive integers. Set $x = a$ and $y = b$.

1. Find r such that $x = yq + r$ and $0 \leq r < y$.
2. If $r = 0$, then $\gcd(a, b) = y$. End the search.
3. If not, set x equal to y and y equal to r (i.e., $x := y$ and $y := r$).[1] Return to Step 1.

EXAMPLE 3

i. We find the greatest common divisor of 203 and 91 using Euclid's Algorithm. In each case, y is assigned the underlined number.

$$203 = 2 \cdot \underline{91} + 21$$
$$91 = 4 \cdot \underline{21} + 7$$
$$21 = 3 \cdot \underline{7} + 0$$

Thus, the $\gcd(91, 203) = 7$. Note that 7 is the last value assigned to y before the algorithm halts when 0 is assigned to r.

ii. We show that 21 and 34 are relatively prime using Euclid's Algorithm.

$$34 = 1 \cdot \underline{21} + 13$$
$$21 = 1 \cdot \underline{13} + 8$$
$$13 = 1 \cdot \underline{8} + 5$$

---

[1] The symbols "$x := y$" means, "Assign to x the value currently held by y."

$$8 = 1 \cdot \underline{5} + 3$$
$$5 = 1 \cdot \underline{3} + 2$$
$$3 = 1 \cdot \underline{2} + 1$$
$$2 = 2 \cdot \underline{1} + 0$$

Thus $\gcd(21, 34) = 1$. You might have noticed the pattern of Fibonacci numbers in the sequence of remainders. You can explore them further in Section 1.5. ∎

EXAMPLE 4

In this example, we show how we can use the Euclidean Algorithm (in reverse) to express the greatest common divisor of two integers x and y as a weighted sum $mx + ny$. Theorem 1 of this section guarantees that such an expression exists but does not give us a procedure to find it. We shall demonstrate a procedure through an example. Our problem is to use Euclid's Algorithm to find $\gcd(309, 21)$ and then to express $\gcd(309, 21)$ in the form $m309 + n21$ where m and n are integers.

First we apply Euclid's Algorithm to find $\gcd(309, 21)$.

$$309 = 14 \cdot \mathbf{21} + 15$$
$$21 = 1 \cdot \mathbf{15} + 6$$
$$15 = 2 \cdot \mathbf{6} + 3$$
$$6 = 2 \cdot 3 + 0$$

So $\gcd(309, 21) = 3$. Now we solve for the remainders in each step in which the remainder is not equal to 0. **Don't simplify or multiply out!**

$$309 - 14 \cdot \mathbf{21} = 15$$
$$21 - 1 \cdot \mathbf{15} = 6$$
$$15 - 2 \cdot \mathbf{6} = 3$$

Notice that each remainder appears in the line below it. Starting with the bottom equation (here's where we reverse), substitute in for the remainder in bold from the line above it. Repeat until the top line is reached. **Remember, don't simplify or multiply out!**

$$15 - 2 \cdot \mathbf{6} = 3$$
$$\mathbf{15} - 2 \cdot (21 - 1 \cdot \mathbf{15}) = 3$$
$$(309 - 14 \cdot \mathbf{21}) - 2 \cdot (21 - 1 \cdot (309 - 14 \cdot \mathbf{21})) = 3$$

Now group the copies of 309 and the copies of 21 to get

$$3 \cdot 309 - 44 \cdot 21 = 3.$$

Letting $m = 3$ and $n = -44$, we have $309m + 21n = 3$. ∎

EXAMPLE 5

Here's another example, with fewer intervening instructions. Our goal is to express $\gcd(21, 8)$ as $21m + 8n$.

Euclid's Algorithm applied to 21 and 8:

$$21 = 2 \cdot 8 + 5$$
$$8 = 1 \cdot 5 + 3$$
$$5 = 1 \cdot 3 + 2$$
$$3 = 1 \cdot 2 + 1.$$

So $\gcd(21, 8) = 1$. Solving for the remainders, we have

$$21 - 2 \cdot 8 = 5$$
$$8 - 1 \cdot 5 = 3$$
$$5 - 1 \cdot 3 = 2$$
$$3 - 1 \cdot 2 = 1.$$

Now, starting from the last line, substitute for the remainders. This time we will gather terms as we proceed, still making sure not to multiply out copies of the remainders that we will substitute for.

$$3 - 1 \cdot 2 = 1$$

Substitute for 2:
$$3 - 1 \cdot (5 - 1 \cdot 3) = 1 \text{ or } 2 \cdot 3 - 5 = 1$$

Substitute for 3:
$$2 \cdot (8 - 1 \cdot 5) - 5 = 1 \text{ or } 2 \cdot 8 - 3 \cdot 5 = 1$$

Substitute for 5:
$$2 \cdot 8 - 3(21 - 2 \cdot 8) = 1 \text{ or } 8 \cdot 8 - 3 \cdot 21 = 1$$

Thus $1 = 21m + 8n$, where $m = -3$ and $n = 8$. ∎

Next we investigate the relationship between the greatest common divisor of two integers and the least common multiple of two integers.

> **DEFINITION 3.** Let a and b be two nonzero integers. The **least common multiple** of a and b, denoted **lcm**(a, b), is the smallest positive integer m that is a multiple of both a and b.

For example, the positive multiples of 12 are $\{12, 24, 36, 48, \ldots\}$ and the positive multiples of 18 are $\{18, 36, 54, \ldots\}$. The smallest common positive multiple is 36 and thus $\text{lcm}(12, 18) = 36$. Similarly, $\text{lcm}(7, 9) = 63$. As with the gcd, it could be quite difficult to find the lcm of two large numbers if we approached the task through searching lists of multiples or through factorizations. As the next proposition shows, Euclid's Algorithm allows us to avoid this difficulty.

PROPOSITION 7 Let a and b be two positive integers. Then $\text{lcm}(a,b) = \dfrac{ab}{\gcd(a,b)}$.

Proof. Let $d = \gcd(a,b)$ and $m = \frac{ab}{d}$. Then $a = q_1 d$ and $b = q_2 d$. Thus $m = d q_1 q_2$ is a multiple of both a and b. To show that m is the smallest positive integer multiple of both a and b we will show that *any* positive multiple of both a and b is also a multiple of m. To that end, let y be a positive multiple of both a and b so that $y = sa = tb$ for some integers s and t. Then $\frac{y}{d} = \frac{sa}{d} = \frac{tb}{d}$. By Proposition 3 of this section, $\gcd\left(\frac{a}{d}, \frac{b}{d}\right) = 1$. Since $\frac{tb}{d}$ is a multiple of $\frac{a}{d}$, Euclid's Lemma tells us that the integer $\frac{a}{d}$ divides t. So $t = \frac{qa}{d}$ for some integer q. Substituting $\frac{qa}{d}$ for t, we have $y = \frac{qab}{d}$. Thus y is a multiple of $m = \frac{ab}{d}$. ▲

EXAMPLE 6

We can apply Euclid's Algorithm to $x = 51811$ and $y = 142283$ to find that the $\gcd(x,y) = 263$. Thus, $\text{lcm}(x,y) = \dfrac{xy}{\gcd(x,y)} = \dfrac{7371824513}{263} = 28029751$. ■

Diophantine Equations

The most basic **Diophantine equation** has the form $ax + by = c$, where a, b and c are integers. We look for all **integer values** of x and y that satisfy the equation. Geometrically, we can think of the expression $ax + by = c$ as defining a line in the xy-plane. The solution to the Diophantine equation is the set of all points with integer coordinates through which the line passes. Figure 1 illustrates the solution set to the Diophantine equation $2x - y = -3$. The large dots are points that have integer coordinates and satisfy the equation.

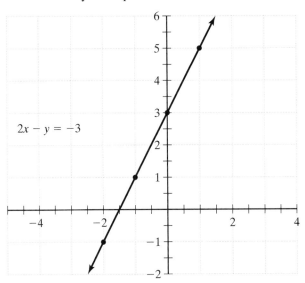

FIGURE 1

EXAMPLE 7

One solution to the Diophantine equation $y - 2x = 3$ is $x = 1$ and $y = 5$. Other solutions are of the form $x = n$ and $y = 2n + 3$ where n can be any integer. For instance, if $n = 7$, then $x = 7$ and $y = 17$ satisfy $y - 2x = 3$. There are an infinite number of integer pairs that satisfy the equation $y - 2x = 3$. Example 8 shows how to find one of them systematically. Proposition 8 below tells us how to find the rest of the solutions. ∎

Since all numbers of the form $ax + by$ are multiples of $\gcd(a, b)$, a Diophantine equation $ax + by = c$ has a solution if and only if c is a multiple of $\gcd(a, b)$. Euclid's Algorithm, used in reverse, provides a way of finding a solution whenever c is a multiple of $\gcd(a, b)$.

EXAMPLE 8

In this example, we solve the Diophantine equation $309x + 21y = 12$. The $\gcd(309, 21) = 3$ and $12 = 4 \cdot 3$. So the equation has a solution. To find it, we first solve $309m + 21n = 3$ for m and n. In Example 4, we used Euclid's Algorithm to show that $3 \cdot 309 - 44 \cdot 21 = 3$ so that $m = 3$ and $n = -44$. We then multiply m and n by 4 to get a solution to get $x = 12$ and $y = -176$. Thus $309x + 21y = 12$. ∎

When there is one solution to an equation of the form $ax + by = c$, there are infinitely many solutions. Next we show how to obtain them.

PROPOSITION 8 Let $d = \gcd(a, b)$ and let the pair x_0 and y_0 be a solution to $ax + by = c$. Then all other solutions are of the form $x = x_0 + \frac{bt}{d}$ and $y = y_0 - \frac{at}{d}$, where t is any integer in $\mathbf{Z}$.

Proof. Substituting x and y into the given equation shows that the pair $x = x_0 + \frac{bt}{d}$ and $y = y_0 - \frac{at}{d}$ is indeed a solution if the pair x_0 and y_0 is a solution. We need to show that all solutions are of the indicated form. Suppose that the pair x_1 and y_1 is also solution to $ax + by = c$. So we have $ax_0 + by_0 = c$ and $ax_1 + by_1 = c$. Subtracting, we obtain $a(x_0 - x_1) + b(y_0 - y_1) = 0$. Let $d = \gcd(a, b)$ and divide through by d to obtain

$$(1) \qquad \frac{a}{d}(x_0 - x_1) = \frac{-b}{d}(y_0 - y_1).$$

The $\gcd\left(\frac{a}{d}, \frac{b}{d}\right) = 1$ and $\left(\frac{-b}{d}\right)(y_0 - y_1)$ is a multiple of $\frac{a}{d}$. By Euclid's Lemma, we have $(y_0 - y_1) = \frac{at}{d}$ for some t in $\mathbf{Z}$. Substituting $\frac{at}{d}$ for $(y_0 - y_1)$ in equation (1) and canceling $\frac{a}{d}$ from both sides, we find that $x_0 - x_1 = -\frac{bt}{d}$ or $x_1 = x_0 + \frac{bt}{d}$. Similarly, $y_1 = y_0 - \frac{at}{d}$. ▲

EXAMPLE 9

In Example 8 we found that $x_0 = 12$ and $y_0 = -176$ is a solution to the Diophantine equation $309x_0 + 21y_0 = 12$. Proposition 8 tells us that all other solutions are of the form $x = 12 + \frac{21}{3}t$ and $y = -176 - \frac{309}{3}t$ where t can be any integer. Let's check it for $t = 2$. We get $x = 26$ and $y = -382$. Indeed, $26 \cdot 309 - 382 \cdot 21 = 12$. ∎

EXAMPLE 10 (optional)

(This is an example for readers who appreciate formulas.) Now that we have run Euclid's Algorithm in reverse to solve the Diophantine equation $sa + tb = \gcd(a, b)$, let's summarize the process with a recursive procedure to find s and t. First we display Euclid's Algorithm when it is applied to finding the $\gcd(r_0, r_1)$ where $0 < r_1 < r_0$ so that we may establish notation for the quotients and remainders that ensue.

$$r_0 = q_2 r_1 + r_2$$
$$r_1 = q_3 r_2 + r_3$$
$$\vdots$$
$$r_{n-2} = r_{n-1} q_n + r_n$$
$$r_{n-1} = r_n q_{n+1} + 0$$

With this notation, $\gcd(r_0, r_1) = r_n$. The goal is to express r_n as a sum $r_n = s_n r_0 + t_n r_1$. Notice that if $r_2 = 0$, then $r_1 = \gcd(r_0, r_1)$ and $r_1 = 0 \cdot r_0 + 1 \cdot r_1$. So $s_1 = 0$ and $t_1 = 1$. Now suppose that $r_2 = \gcd(r_0, r_1)$. Then $r_2 = r_0 - q_2 r_1$. In this case, we have $s_2 = 1$ and $t_2 = -q_2$. These are the values we need to begin our recursive procedures. Here they are:

$$s_1 = 0, \; s_2 = 1, \; t_1 = 1, \; t_2 = -q_2;$$
$$s_n = -q_n \cdot s_{n-1} + s_{n-2};$$
$$t_n = -q_n \cdot t_{n-1} + t_{n-2} \text{ for } n > 2.$$

The first few values generated for s_i are as follows:

$$s_3 = -q_3$$
$$s_4 = 1 + q_4 q_3$$
$$s_5 = -q_3 - q_5(1 + q_4 q_3) = -q_3 - q_5 - q_5 q_4 q_3.$$

The first few values generated for t_i are as follows:

$$t_3 = 1 + q_3 q_2$$
$$t_4 = -q_2 - q_4(1 + q_3 q_2) = -q_2 - q_4 - q_4 q_3 q_2$$
$$t_5 = q_3 q_2 + 1 - q_5(-q_2 - q_4(1 + q_3 q_2))$$
$$= q_3 q_2 + 1 + q_5 q_2 + q_5 q_4 + q_5 q_4 q_3 q_2.$$

Let's apply our formulas to the task of expressing $\gcd(45, 33)$ as the weighted sum of 45 and 33. Here $r_0 = 45$ and $r_1 = 33$.

$$45 = 1 \cdot 33 + 12$$
$$33 = 2 \cdot 12 + 9$$
$$12 = 1 \cdot 9 + 3$$

Section 1.3 Greatest Common Divisors and Euclid's Algorithm 25

We have $q_2 = 1, q_3 = 2, q_4 = 1$. Thus $s_4 = 3$ and $t_4 = -4$. Checking, $3 = 3 \cdot 45 - 4 \cdot 33$. ∎

SUMMARY

The most important concept of this section is that of the **greatest common divisor** (gcd) of two integers a and b. The tool to compute it, **Euclid's Algorithm**, does so via the Division Algorithm rather than factoring. **Euclid's Lemma** (Theorem 4) will be key to proving the fact that all natural numbers can be factored uniquely as the product of primes, a fact that we all take for granted, but have not actually used or proved yet. In this section, we also addressed the relation of the gcd of two integers a and b to the **least common multiple** (lcm) of a and b. Finally, we used the gcd of a and b to characterize all integer solutions to **Diophantine equations** of the form $ax + by = c$.

1.3 Exercises

1. Find the greatest common divisors of the following pairs of numbers without using Euclid's Algorithm.
    i. 38 and 108
    ii. 51 and 187
    iii. -113 and 173
    iv. 462 and 1105
2. Show the following.
    i. $\gcd(m, n) = \gcd(-m, -n)$
    ii. $\gcd(n, n) = |n|$, where $|n|$ denotes the absolute value of n
    iii. $\gcd(1, n) = 1$
3. Prove that every common divisor of a and b divides $\gcd(a, b)$.
4. Suppose that a divides b and that c divides b. Suppose also that $\gcd(a, c) = 1$. Prove that ac divides b. (Give some examples before proving the result.)
5. Let x, y and m be integers. Suppose that $\gcd(x, m) = 1$ and $\gcd(y, m) = 1$. Prove that $\gcd(xy, m) = 1$.
6. Prove Proposition 5.
7. Use Euclid's Algorithm to find the gcd of the following pairs of numbers:
    i. 23 and 13
    ii. 1234 and 123
    iii. 442 and 289
8. Find the least common multiple of each of the following pairs:
    i. 462 and 442
    ii. 705 and 1175
    iii. 173 and 229
9. Prove that if n is odd, then $\gcd(3n, 3n + 2) = 1$.
10. Find all solutions to the following Diophantine equations.
    i. $35x + 25y = 8$
    ii. $35x + 25y = 15$
    iii. $47x + 19y = 4$
11. Determine the positive integer solutions to $18x - 5y = 15$.

12. Here's a classic: If a cock is worth 5 coins and a hen 3 coins, and 3 chicks together worth 1 coin, how many cocks, hens and chicks totaling 100 can be bought for 100 coins? Remember: no negative or fractional coins or fowl.

13. Use the formulas obtained in Example 10 to express the gcds found in Exercise 7 as weighted sums of the given integers.

To the Teacher

The NTCM's *Principles and Standards for School Mathematics* urge that students "understand how mathematical ideas interconnect and build on one another to produce a coherent whole." Part of understanding interconnection is to understand how concepts and results can be generalized. Let's take a look at how a teacher might extend Euclid's Algorithm by applying it to two positive rational numbers a and b, instead of two natural numbers. Euclid's Algorithm depends on the Division Algorithm to generate remainders. But when we divide a by b, we find that $q = a/b$, where q is another rational number, and $a = qb$. There is no remainder. No remainder, no Euclid's Algorithm! So let's first retool the Division Algorithm.

A Division Algorithm for Rational Numbers.

Let a and b be positive rational numbers. There is an *integer* q such that $a = qb + r$, where $0 \leq r < b$.

The integer q is the maximal number of times we can add b to itself and have $0 \leq b + b + b + \ldots + b \leq a$ and $r = a - qb$. With this extension of the Division Algorithm, we can apply Euclid's Algorithm to any two positive rational numbers directly.

EXAMPLE 11

Let $a = 4/5$ and $b = 2/3$.

$$4/5 = 1 \cdot 2/3 + \mathbf{2/15}$$
$$2/3 = 5 \cdot 2/15 + 0$$

Thus, $\gcd(4/5, 2/3) = 2/15$.

This should raise both computer science questions and mathematical questions.

- Will the procedure always halt?
- If so, what will it compute?
- How can we interpret the outcome?

The first two questions can be answered by expressing a and b as $a = m/c$ and $b = n/c$, where c is a common denominator. Euclid's Algorithm applied to a and b yields the $\gcd(m, n)/c$. As for the last question, $\gcd(m, n)/c$ is the largest rational number of which both a and b are *integer* multiples. In this sense, the $\gcd(a, b)$ can still be interpreted as the largest quantity of which both a and b are **integer** multiples. Euclid's Algorithm will appear many times throughout this text. For a discussion that extends the questions raised in this note, see Section 1.6.

Tasks

1. Apply Euclid's Algorithm to the following pairs of rational numbers. What does the final nonzero remainder represent?

   (a) $\frac{1}{2}$ and $\frac{1}{3}$

   (b) $\frac{3}{8}$ and $\frac{5}{6}$

2. Prove that Euclid's Algorithm applied to a and b must halt and that it yields $\gcd(m, n)/c$, where $a = m/c$ and $b = n/c$.

3. Let $a = m/c$ and $b = n/c$. Prove that $\gcd(m, n)/c$ is the largest rational number of which both a and b are *integer* multiples.

1.4 PRIMES AND UNIQUE FACTORIZATION

In the previous sections, we investigated two questions: When does one number divide another and when do two numbers share a common divisor? These were questions about the relation between two numbers. Now we focus on the prime numbers, the building blocks of the integers.

> **DEFINITION 1.** An integer $p > 1$ is called **prime** if its only positive divisors are itself and 1.

Any number greater than 1 that is not prime is called a **composite number**. Notice that 1 is neither prime nor composite. The prime numbers less than 20 are 2, 3, 5, 7, 11, 13, 17, and 19. It is an easy concept and not difficult to check for small numbers. But questions abound. How many primes are there? Is there a pattern to the distribution of primes? How can we determine quickly if a given number is a prime or not? The proof of Theorem 1 will show that there are an infinite number of primes, a familiar but not an altogether obvious fact. We shall prove it in much the same way that Euclid did more than 2000 years ago.

THEOREM 1 The number of prime numbers is infinite.

Proof. Suppose $S = \{p_1, p_2, \ldots, p_n\}$ is the set of the first n primes. Let $q = (p_1 \cdot p_2 \cdots p_n) + 1$. Then q is not a multiple of any prime p_i in S. For if p_i divides $p_1 \cdots p_n$ and p_i divides q, then p_i divides $q - p_1 \cdot p_2 \cdots p_n$, which means that p_i divides 1. The latter is impossible and so the finite set S cannot include all primes. Since any finite set of prime numbers cannot be the complete set of prime numbers, the set of prime numbers must be an infinite set. ▲

At first glance, it would seem that the proof of Theorem 1 is giving us a recipe for a new prime. It looks as if given the first n primes $p_1, p_2, \ldots, p_n$, then $q = (p_1 \cdot p_2 \cdots p_n) + 1$ is prime. After all, $2 \cdot 3 + 1 = 7$ is prime and $2 \cdot 3 \cdot 5 + 1 = 31$ is prime. However, $2 \cdot 3 \cdot 5 \cdot 7 \cdot 11 \cdot 13 + 1 = 30031$, but $30031 = 59 \cdot 509$. (It's not that easy to generate new primes!) The proof of the theorem says simply that q is not divisible by any of the primes $p_1, p_2, \ldots, p_n$. So there must be some prime *between* the largest prime of the set $\{p_1, p_2, \ldots, p_n\}$ and q. In the case of $q = 30031$, both 59 and 509 are primes larger than 13 but less than 30031.

Our next goal is to prove that any natural number greater than 1 can be factored into primes in only one way. Experience tells us that this is true for small numbers. To prove it true for all numbers, we will need the following variation of Euclid's Lemma (Theorem 4 of Section 1.3).

PROPOSITION 2 Let p be a positive integer. Then p is prime if and only if for all integers a and b, if p divides ab then either p divides a or p divides b.

Proof. Suppose p is prime and that a and b are integers such that p divides ab. If p divides a then we are done. If not, then $\gcd(a, p) = 1$ because the only common divisor of both p and a is 1. Thus, by Euclid's Lemma, p divides b. Conversely, if p is not prime, then we can find integers a and b such that $p = ab$ where neither a nor b is divisible by p. ▲

The following result follows by induction.

PROPOSITION 3 Suppose that p is prime and that p divides $a_1 a_2 a_3 \cdots a_n$. Then p divides a_k for some k such that $1 \leq k \leq n$. Also, if a_k is prime, then $p = a_k$.

Our definition of a prime number p is in terms of how p can (or cannot) be divided. But Euclid's Lemma gives us a characterization of prime numbers in terms of how they divide other numbers. We will exploit this dual characterization in the proof of the Fundamental Theorem of Arithmetic, which says that all positive integers can be factored uniquely into the product of primes. The fact that all positive integers can be factored into primes in some way is intuitively obvious: We can factor a composite number and then keep factoring its factors. (We shall use induction to mimic this reasoning.) But uniqueness—that there is only one way to factor a number into primes—is less obvious, especially for large numbers.

THEOREM 4 (*The Fundamental Theorem of Arithmetic*) Let n be any integer greater than 1. There are distinct primes $p_1, p_2, \ldots, p_m$ and positive exponents $k_1, k_2, \ldots, k_m$ such that $n = p_1^{k_1} p_2^{k_2} \cdots p_m^{k_m}$. Furthermore, such a factorization is unique except for the order of the factors.

The theorem says is that there is only one way to factor a positive integer into primes. For instance, if $n = 3375$, we can factor n as $5^3 3^3$ or as $3^3 5^3$ or as $3 \cdot 5 \cdot 3 \cdot 5 \cdot 3 \cdot 5$. But we will always end up with exactly 3 copies of 3 and 3 copies of 5 in its factorization. So if we expressed 3375 as $2^{n_1} 3^{n_2} 5^{n_3} 7^{n_4}$, we know that $n_1 = 0, n_2 = 3, n_3 = 3$ and $n_4 = 0$. There are no options!

Proof. We proceed by induction on n. The statement is true for $n = 2$ because 2 is prime and we can let $p_1 = 2$ and $k_1 = 1$. (In fact, the statement is true if n is any prime.) Let $n > 2$ and assume that the theorem holds for any integer less than n. If n is prime we are done. If not, we can factor n as $n = xy$, where $1 < x < n$ and $1 < y < n$. By the induction hypothesis we can find two sets of primes, $\{q_1, q_2, \ldots, q_i\}$ and $\{s_1, s_2, \ldots, s_j\}$, and two sets of exponents, $\{e_1, e_2, \ldots, e_i\}$ and

$\{h_1, h_2, \ldots, h_j\}$, such that $x = q_1^{e_1} q_2^{e_2} \cdots q_i^{e_i}$ and $y = s_1^{h_1} s_2^{h_2} \cdots s_j^{h_j}$. Thus $n = x \cdot y = q_1^{e_1} q_2^{e_2} \cdots q_i^{e_i} s_1^{h_1} s_2^{h_2} \cdots s_j^{h_j} = p_1^{k_1} p_2^{k_2} \cdots p_m^{k_m}$, where we have grouped like terms by adding their exponents.

Uniqueness will follow from Proposition 2 and induction. The assertion is true if $n = 2$ (or any other prime) because clearly, there is no other way to factor 2. Assume that unique factorization holds for any number less than n. Now suppose that n factors into primes as follows: $n = p_1^{k_1} p_2^{k_2} \cdots p_m^{k_m}$ and $n = q_1^{j_1} q_2^{j_2} \cdots q_s^{j_s}$. Then, by Proposition 2, we know that p_1 must divide some q_j. Assume that we have numbered the factors so that that p_1 divides q_1 and hence $p_1 = q_1$. Let $t = \frac{n}{p_1}$. Then $t = p_1^{k_1-1} p_2^{k_2} \cdots p_m^{k_m} = q_1^{j_1-1} \cdots q_s^{j_s}$. Since $0 < t < n$, unique factorization holds for t by our induction hypothesis. Thus $p_i = q_i$ and $k_i = j_i$ for $1 \le i \le t$. Thus we have shown that n can be uniquely factored as the product of primes. ▲

The Fundamental Theorem of Arithmetic might seem to be a fact only about integers. However, it has consequences for rational and irrational numbers as well. In the Exercises, you will be asked to prove the following corollary.

COROLLARY 5   Every positive rational number can be expressed uniquely as $\frac{m}{n}$, where m and n are relatively prime natural numbers.

The corollary expresses the familiar fact that every fraction can be expressed in "lowest terms." As a consequence of Corollary 5, we now prove that if x is any rational number, then $x^2 \ne 2$. So $\sqrt{2}$ cannot be expressed as the ratio of two integers. It must be ir-ratio-nal.

THEOREM 6   There is no rational number whose square is 2.

Proof.   (By contradiction.) Suppose to the contrary that $x \in \boldsymbol{Q}$ and that $x^2 = 2$. Suppose also that $x = \frac{m}{n}$ and that m and n are relatively prime. So $\frac{m^2}{n^2} = 2$ and $m^2 = 2n^2$. Since 2 is prime and since 2 divides m^2, 2 must divide m. So $m^2 = 2^2 p$ for some integer p. Thus $2^2 p = 2n^2$, or $2p = n^2$, and we see that 2 divides n as well. But this contradicts our assumption that m and n are relatively prime. Therefore there is no rational number x such that $x^2 = 2$. ▲

Suppose that you had a ruler, some fixed length that you used as a reference for a unit length, like a meter or a kilometer. Suppose also that you had some other length like the length of a football field. You would not expect to lay out m copies of your ruler end to end, starting at one end of the football field and ending exactly at the other end. But what if you could first divide the ruler into n equal, perhaps very small pieces? Could you lay out m copies of such small pieces and span the field exactly? Your first instinct was probably to say, "Yes." But the answer is, "No!" If your football field had the length of the diagonal of a square that had the length of its side equal to the length of your ruler, then the length of the field would be $\sqrt{1^2 + 1^2} = \sqrt{2}$. You could not lay out m copies of pieces of length $\frac{1}{n}$ and have $\sqrt{2} = \frac{m}{n}$.

Just what are the numbers in the real number line, the "continuum"? How do we characterize all possible lengths? What is the relation between number and geometry?

These questions were raised by the ancient Greeks and addressed by mathematicians such as Euclid, but they were not resolved until the late nineteenth century through the work of Peano and Dedekind.

Let's get back to questions about primes. Suppose we want to list the prime numbers that lie between 2 and a fixed number n. An ancient method to do this that works well if n is not too large is called the **Sieve of Eratosthenes** (c. 200 B.C.E.). It is an algorithm. To use the sieve, first list the numbers from 2 to n in order.

Instructions.

1. Put a star on the first number on the list that is neither starred nor crossed out. Call this number p.
2. Cross out all numbers listed after p that are multiples of p and not yet crossed out.
3. If all of the numbers on the list are either starred or crossed out, stop. The starred numbers are a complete list of prime numbers less than or equal to n. If not, return to step 1.

Suppose that you wanted to use the sieve method to find all the primes less than or equal to 50. List all the numbers from 2 to 50:

$$2\ 3\ 4\ 5\ 6\ 7\ 8\ 9\ 10\ 11\ 12\ 13\ 14\ 15\ldots.$$

To start, no numbers are starred or crossed out so that $p = 2$. Star 2. Now cross out all multiples of 2.

$$2^*\ 3\ \cancel{4}\ 5\ \cancel{6}\ 7\ \cancel{8}\ 9\ \cancel{10}\ 11\ \cancel{12}\ 13\ \cancel{14}\ 15\ldots$$

Since not all numbers are starred or crossed out, we return to the beginning of the list. Now 3 is the first number not starred or crossed out. Put a star on 3. Let $p = 3$ and cross out all multiples of p. After the second iteration of step 2, the list should look like this:

$$2^*\ 3^*\ \cancel{4}\ 5\ \cancel{6}\ 7\ \cancel{8}\ \cancel{9}\ \cancel{10}\ 11\ \cancel{12}\ 13\ \cancel{14}\ \cancel{15}\ \cancel{16}\ 17\ldots.$$

Repeat steps 1 and 2 until all numbers are starred or crossed out. When done, the primes are the starred numbers. If we **erase** all the crossed out numbers, the list will look like Table 1. Only primes between 2 and 50 remain. They don't fall through the sieve! The primes less than 50 are 2, 3, 5, 7, 11, 13, 17, 19, 23, 29, 31, 37, 41, 43, and 47.

The list is interesting and raises questions. For instance, sometimes there are big spaces between primes. But sometimes we find pairs of "twin primes" like 11 and

TABLE 1

	2*	3*		5*		7*			
11*		13*				17*		19*	
		23*						29*	
31*						37*			
41*		43*				47*			

13 and 41 and 43, pairs of primes with difference 2. Are there an infinite number of pairs of twin primes? Do the primes become rarer as we go further out on the number line? Number theory is full of questions that are easy to state but difficult to answer. The second question has been answered affirmatively in a statistical fashion in what has come to be known as the Prime Number Theorem. It states that, on average, there are about $\frac{n}{\ln(n)}$ primes between 1 and n. For instance, the theorem predicts that between 10000 and 11000, a span of 1000 integers, there should be about $\frac{11000}{\ln(11000)} - \frac{10000}{\ln(10000)} = 96.34$ primes. There are actually 106 such primes. But between 100000 and 101000, it predicts 79.28 primes and there are 81 such primes. Primes definitely get rarer. But statistical scarcity does not mean uniform scarcity. The diminishing average frequency with which primes appear does not rule out the appearance of very large pairs of twin primes. The question of whether there are an infinite number of pairs of twin primes remains unanswered, open to present and future researchers.

SUMMARY

This section addressed the most basic facts about **prime numbers**. The **Fundamental Theorem of Arithmetic** tells us that every integer greater than 1 can be factored into primes in exactly one way and Theorem 1 tells us that there are infinitely many prime numbers. The **Sieve of Eratosthenes** gives us a simple way of listing all primes less than a given value of n. It is practical only for relatively small n. But the Sieve Method prompts questions about the prime numbers, some of which remain open to research today. If there is a fixed pattern underlying the distribution of the primes, it has eluded mathematical research for thousands of years.

1.4 Exercises

1. Try to find the prime factorization of 12347983 without using a calculator. (Not easy!)
2. Let a and b be positive integers. Suppose that the complete list of primes that divide either a or b is $\{p_1, p_2, \ldots, p_n\}$ and that $a = p_1^{m_1} p_2^{m_2} \cdots p_n^{m_n}$ and $b = p_1^{k_1} p_2^{k_2} \cdots p_n^{k_n}$, respectively. (Any of the exponents can be zero.)
   i. Characterize the gcd of a and b in terms of the factorization of a and b.
   ii. Find $\gcd(a, b)$ if $a = 2^5 3^2 5^0 7^1 11^2 13^3$ and $b = 2^2 3^5 5^1 7^2 11^0 13^0$.
3. Repeat Exercise 2 replacing gcd with lcm.
4. Prove that any prime of the form $3m + 1$ for some $m \in \mathbf{N}$ is also of the form $6n + 1$ for some $n \in \mathbf{N}$.
5. Prove that if n is composite, then some prime p that divides n satisfies $p \leq \sqrt{n}$. Determine if 541 is prime by testing its divisibility by all primes less than $\sqrt{541}$.
6. Show that, when finding the primes from 2 to n using the Sieve of Eratosthenes, we can stop crossing out once $p \geq n/2$.
7. Suppose that $\{p_1, p_2, \ldots, p_n\}$ is a list of the first n primes. Can you find a number of the form $p_1 p_2 \cdots p_n + 1$ that is greater than $2 \cdot 3 \cdot 5 \cdot 7 \cdot 11 \cdot 13 + 1 = 30031$ and that *is* prime? (It is not known whether

the set of primes of the form $p_1p_2 \cdots p_n + 1$ is an infinite set or whether the set of composite numbers of the form $(p_1p_2 \cdots p_n) + 1$ is an infinite set!)

8. Prove Corollary 5.
9. Express each even number n from 2 to 24 as the sum of two numbers x and y that are either primes or 1. For example, $8 = 1 + 7$ or $8 = 3 + 5$ and $6 = 3 + 3$. The **Goldbach Conjecture** asserts that every positive even integer can be expressed this way. But it remains a conjecture, unproven.
10. Prove (by induction) that if p_n is the nth prime, then $p_n \leq 2^{2^{n-1}}$.
11. Use unique factorization to prove that the square root of any prime is irrational. (Model this on the proof of Theorem 6.)

To the Teacher

To develop a number sense and to develop skill in detecting primes, students need a reason to practice and a goal. The same questions that intrigue and motivate the erudite number theorist can and will intrigue the curious student. There are two conjectures about prime numbers, still unproven and avidly researched, that can be easily understood by high school students. The first is the **Twin Prime Conjecture**. It asserts that there are an infinite number of pairs of twin primes, pairs like 5 and 7 and like 17 and 19, that are just two units apart on the number line. The second is **Goldbach's Conjecture**. It asserts that every even number greater than or equal to 4 can be expressed as the sum of two primes. For instance, $8 = 5 + 3$ and $16 = 5 + 11$. A teacher might challenge a class (or competing teams within a class) with the following tasks.

> Challenge 1. Find as many twin prime pairs as you can that lie between 400 and 500. (One such pair is 419 and 421.)
>
> Challenge 2. Express 200 as the sum of primes in as many ways as possible. (For example, $200 = 197 + 3$ and $200 = 193 + 7$, etc.)

In each case, the search for primes is the principal hurdle. Students' skills are honed in the search.

Problem Solving: To find primes between 400 and 500, students will recognize how to reduce to the problem significantly–eliminate all even numbers, multiples of 5, and multiples of 10. If students know divisibility tests (see Section 2.2), other numbers might be quickly eliminated.

Number Sense: A composite number has a factor and hence a prime factor less than its square root. So to eliminate a number x as nonprime efficiently, students will realize that they need only check its divisibility by relatively small primes. (For 400, they would need to check only primes less than 20.)

Beyond simply acquiring or reinforcing skills, solving such problems gives students an appreciation and ownership of the material. The numbers 419 and 421 graduate from boring to interesting. The relation between adding and multiplying—why should an even number have a Goldbach expression at all?—opens up new avenues of inquiry. Even the difficulties are instructive. Factoring large numbers is hard, even for the fastest computers. And the fact that simple mathematical questions can remain unresolved undoes the impression that all answers lie in some great teacher's manual in the sky.

Task

Complete Challenges 1 and 2.

1.5 *WORKSHEET 1: THE FIBONACCI NUMBERS*

"Fibonacci," really Leonardo da Pisa, was born in Pisa around 1170, a time of intense activity and growth in Europe, intellectually, commercially, and socially. (The famous tower in Pisa was built in 1173 and began to lean almost immediately!) Because of his father's business connections, Leonardo knew Arabic and studied the highly developed mathematics of Arab scholars. He was very influential in the fields of geometry and trigonometry. It was Leonardo who initiated the wide spread use of the Arabic numerals and the current decimal system. (The arithmetic of bookkeeping became much easier and the computations became much more transparent. There was some resistance to the decimal system.)

In 1228, Fibonacci wrote his *Liber Abacci*, a textbook for his students. In it he poses the following "real world" problem to students:

> A man put one pair of rabbits in a certain place entirely surrounded by a wall. How many pairs of rabbits can be produced in a year, if the nature of these rabbits is such that every month each pair bears a new pair, a male and a female, which from their second month on, becomes productive?

As with any applied problem, we must first add a few assumptions before we can model the problem mathematically. The assumptions are the following: No rabbits die in Pisa; the initial pair was newly born; a pair becomes productive in its second month of life and produces a new pair each month. We shall extend the scope of the problem slightly, and ask how many rabbits are present after n months, for any integer $n \geq 0$. Table 1 summarizes the first five months and gives the number of rabbits at the beginning of each month. (Month 0 is included for computational convenience later on.)

TABLE 1

Month Number	0	1	2	3	4	5	6
Mature Pairs	0	0	1	1	2	3	5
Immature Pairs	0	1	0	1	1	2	3
Total Pairs	0	1	1	2	3	5	8

Table 2 extends the computations for the first year. We let i denote the month number, and F_i denote the total number of pairs of rabbits in month i.

TABLE 2

i	F_i
0	0
1	1
2	1
3	2
4	3
5	5
6	8
7	13
8	21
9	34
10	55
11	89
12	144

To extend the chart beyond the first year, note that for each $i > 1$, we have the recurrence relation, $F_i = F_{i-1} + F_{i-2}$. This is justified biologically because in any month i, all F_{i-1} pairs of rabbits that were alive in the previous month continue to live and, in addition, we have F_{i-2} pairs of newborn rabbits. Remember that only the rabbits alive in month $i-2$ reproduce in month $i-1$ to contribute to the total number of rabbits in month i. (The rabbits born in month $i-1$ are not yet productive in month i.) We can extend the chart as far as we wish. The sequence of numbers we obtain is called the **Fibonacci sequence**. The sequence provides the answer to each of the following problems, which are drawn from very different contexts.

a. How many ways can we express n as the sum of 1s and 2s? For example, the number 3 can be expressed as
$$3 = 1 + 1 + 1 = 2 + 1 = 1 + 2.$$
(We count the two latter expressions as different.)

b. In how many ways can an elf climb a set of n stairs if the elf can jump either one step or two steps at a time?

c. A drone bee has only one parent, a mother, whereas a female bee has both a mother and a father. Find the number of (great)^n-grandmothers that a drone has.

Task 1

Justify why the Fibonacci sequence provides the answers to each of the above problems.

Task 2

In the following sequence of propositions, aspects of the number theory of the Fibonacci sequence are developed. The proof of Proposition 2 is provided to serve as a model. Your job is to prove the remaining propositions. They build up to an interesting theorem that you shall prove as a grand finale.

Before starting, prove the following lemma based on the material of Section 1.3.

LEMMA 1 If c divides b, then $\gcd(a+b, c) = \gcd(a, c)$.

PROPOSITION 2 For every $n > 0$, $\gcd(F_n, F_{n+1}) = 1$.

Proof. The proof is by induction on n. If $n = 1$, $F_1 = F_2 = 1$ and $\gcd(1, 1) = 1$. Assume that the statement is true for all k such that $0 < k < n$. We show that the statement must hold for $k = n$. From the definition of the Fibonacci sequence, we know that $F_{n+1} = F_{n-1} + F_n$. Thus $\gcd(F_{n+1}, F_n) = \gcd(F_{n-1} + F_n, F_n)$. Since $F_n | F_n$, we can apply Lemma 1 to obtain that $\gcd(F_{n-1} + F_n, F_n) = \gcd(F_{n-1}, F_n)$. By the induction hypothesis, $\gcd(F_{n-1}, F_n) = 1$. ▲

Alternatively, you can prove Proposition 2 using Euclid's Algorithm. In applying the algorithm you will find that at each step division looks like

$$F_i = 1 \cdot F_{i-1} + F_{i-2} \text{ until } F_3 = 2F_2 + 0.$$

The last nonzero remainder is $F_2 = 1$.

PROPOSITION 3 For all $m > 0$, $F_{m+n} = F_{m-1}F_n + F_m F_{n+1}$.
  Hint. Fix m and do induction on n.

PROPOSITION 4 For $m > 0$ and $n > 0$, F_{mn} is divisible by F_m.
  Hint. Do induction on n. Assume true for mk and show true for $m(k+1) = mk + m$ by applying Proposition 3.

PROPOSITION 5 If $m = qn + r$, then $\gcd(F_m, F_n) = \gcd(F_n, F_r)$.
  Hint. Apply Propositions 2, 3 and 4 to $\gcd(F_m, F_n) = \gcd(F_{qn+r}, F_n)$.

The goal of the above set of propositions was to pave the way for you to prove the following theorem. It reminds us yet again that the Fibonacci sequence is a source of endlessly surprising relations.

THEOREM 6 The $\gcd(F_m, F_n) = F_{\gcd(m,n)}$.

Here are some additional Fibonacci identities to prove.

1. Prove that for $n > 1$, $F_n^2 - F_{n+1}F_{n-1} = (-1)^{n-1}$.
2. In Task 1 we saw that the number of distinct (counting order of addends) ways to express n as the sum of 1s and 2s is F_{n+1}. Show that the number of ways to express n with exactly k 2s is $\binom{n-k}{k}$ and thus $F_{n+1} = \sum_{k=0}^{n} \binom{n-k}{k}$ where we take $\binom{j}{i} = 0$ if $i > j$. Find this fact as it is expressed in Pascal's triangle.

The list of propositions that you proved in this worksheet provides a skeleton around which you can build an oral or written presentation. To flesh it out, you will need to do a few things. First, you must of course supplement the propositions with examples. Second, you need to provide clear, instructive proofs that are neither too brief nor unduly long. To find this balance, you need to keep the skill level of your audience in mind. Are they novices who need all the details? Are they as adept as you are at the underlying number theory skills so that you can write or speak in a more abbreviated manner? Most important, you must provide a narrative for this material, a story line that makes it interesting to the reader or listener. (This worksheet did NOT do that!) Note that the propositions build up to a very interesting theorem. As a writer or speaker, you need to highlight that goal, perhaps within a set of introductory examples surrounded by prose. If you are simulating a classroom situation in which you are a motivating teacher, you might want to lay out a sequence of examples that leads your audience to discover Theorem 6. The spare and efficient theorem–proof style of presenting mathematics often hides its beauty and depth, a flaw easily remedied by thoughtful, not necessarily abundant, prose.

1.6 THE VERSATILITY OF EUCLID'S ALGORITHM

In this section we investigate what Euclid's Algorithm can do in other contexts. First let's see how Euclid formulated his algorithm. Here's what Euclid tells us to do in Book VII, Proposition 2, of his *Elements*:[1]

> [I]t is required to find the greatest common measure of *AB*, *CD*. If now *CD* measures *AB*—and it also measures itself—*CD* is a common measure of *CD*, *AB*. And it is manifest that it is also the greatest; for no greater number than *CD* will measure *CD*. But, if *CD* does not measure *AB*, then the lesser of the numbers *AB*, *CD* being continually subtracted from the greater, some number will be left which will measure the one before it.

Suppose that x and y represent lengths that are integral multiples of a fixed unit length. We can think of a ruler of length x (meaning x times the given unit length). When we can measure length y with exactly an integral number of this ruler, then "x measures y." So, for natural numbers x and y—the only numbers Euclid would have considered—the phrase "x measures y" means that y is an integer multiple of x or equivalently, x is a factor of y. Euclid asserts that when x is a factor of y, then $\gcd(x, y) = x$ because x is certainly the largest factor of itself and hence the largest common factor of the pair x and y. Then he tells us what to do if x does not measure y. In that case, we measure off and subtract copies of x from y until we have a remainder r that is less than x. If we subtract x off q times, we have $y = qx + r$. We continue the process with x and r, "subtracting the lesser from the greater," to obtain $x = q_1 r + r_1$. We repeat the process until a remainder measures (is a factor of) the number before it. That occurs when one more application of repeated subtraction (division) results in a remainder of 0. So the common measure of x and y is the last nonzero remainder.

Now let's look again at Euclid's statement, this time interpreting the word "magnitude" as "arbitrary length." If we start with two strips of paper of arbitrary lengths

---
[1]Heath, T., Translator. *The Thirteen Books of the Element's*, Euclid, Vol. 2, Books III–X, Dover, New York, 1956.

$x < y$, we can mark off the length of x on the y-strip, and use scissors to "subtract the smaller from the greater." Let's try it on strips that have the lengths of the sides of a Golden Rectangle, as depicted in Figure 1.

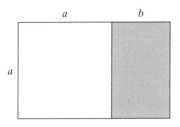

FIGURE 1

A rectangle is *golden* if the small rectangle that remains after removing the square constructed on the shorter side is similar to the original rectangle. In terms of the labeling in Figure 1, a rectangle is golden if $\frac{a+b}{a} = \frac{a}{b}$. The smaller, shaded rectangle is therefore also golden. The *Golden Ratio*, the ratio of the larger side to the smaller side, is traditionally denoted by φ. Since $(a+b)b = ab + b^2 = a^2$, we know that $b < a$ and $1 < \varphi < 2$. (We are deliberately suppressing the numerical value of φ to work with its geometrical properties.)

Now cut two strips whose lengths are the same as the sides of our Golden Rectangle. Use hypothetical scissors rather than division to implement the Euclidean Algorithm on the strips of lengths $a+b$ and a. On the second iteration, apply the scissors to lengths a and b, then to a and $(a-b)$ on the third iteration, and so on. (See Figure 2.)

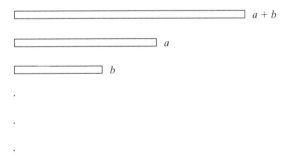

FIGURE 2

The thickness of your scissors and the delicacy of your paper no doubt determined the stage at which you stopped executing the algorithm. If the scissors were infinitely fine and the paper infinitely divisible, you would still be cutting! The algorithm cannot terminate. Let's see why. Since $b < a$, we can only subtract one copy of a from $a+b$ in the first iteration of the algorithm. Our first remainder is thus b. The second iteration of the algorithm is applied to strips of length a and b, the longer and shorter sides of the shaded rectangle in Figure 1. But that rectangle is golden too! Look at Figure 3.

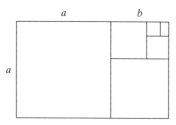

FIGURE 3

We see we have an infinite regression of similar rectangles, each with sides in the Golden Ratio. Such self-similar images are said to be "fractal." At each step in the Euclidean Algorithm, the remainder is the smaller side of a golden rectangle and we are left to apply the algorithm to two new strips that are in **exactly the same proportion** as the two previous strips. The number of steps it takes for the algorithm to terminate depends not on the absolute lengths of the strips, but on the proportion of the lengths to each other. At each stage of the algorithm, we replace two lengths with a new pair in the same proportion. If the algorithm terminated in n steps when applied to $a+b$ and a, it would still require n steps to terminate when applied to the next pair a and b. That is to say, if the algorithm terminated in n steps, we would have $n = n - 1$. Impossible!

If the Golden Ratio φ were a rational number, say $\frac{m}{n}$, the rectangle with sides m and n would be golden. The Euclidean Algorithm would of course terminate when applied to m and n as it would for any pair of integers. So the fact that it **cannot** terminate when applied to the sides of the Golden Rectangle tells us that φ is **not** a rational number.

If you have already set $b = 1$ and solved for a to find that $\varphi = \frac{1+\sqrt{5}}{2}$, you might be saying, "Of course φ is irrational!" But think of all you have to know to prove that there is no rational number x such that $x^2 = 5$. With the Euclidean Algorithm and the fractal picture in Figure 3, we can deduce the irrationality of φ without even having a numerical expression for φ. In ancient Greek mathematics, proof was diagram. Our argument that φ is irrational is very close in spirit to pre-Euclidean mathematics!

Continued Fractions

Euclid's Algorithm is the key tool to a classical problem, that of expressing a number as a continued fraction. A number expressed as a **simple finite continued fraction** has the form

(1) $$a_0 + \cfrac{1}{a_1 + \cfrac{1}{a_2 + \cfrac{\ddots}{ + \cfrac{1}{a_{n-1} + \cfrac{1}{a_n}}}}},$$

where a_0 is any integer and, for $i > 0$, each a_i is a positive integer.

EXAMPLE 1

$$\frac{38}{11} = 3 + \cfrac{1}{2 + \cfrac{1}{5}} \quad \text{and} \quad \frac{36}{115} = \cfrac{1}{3 + \cfrac{1}{5 + \cfrac{1}{7}}}$$

∎

A convenient notation for the continued fraction in expression (1) is $\{a_0; a_1, a_2, \ldots, a_n\}$, where a_0 can be any integer and, for each $i > 0$, each a_i is a strictly positive integer. With this notation, $38/11 = \{3; 2, 5\}$ and $36/115 = \{0; 3, 5, 7\}$.

Let a and b be integers with $b > 0$. Euclid's Algorithm gives us a simple way to express the rational number a/b as a continued fraction. Applying the algorithm to a and b we obtain the following.

$$a = q_0 b + r_1$$
$$b = q_1 r_1 + r_2$$
$$r_1 = q_2 r_2 + r_3$$
$$\vdots$$
$$r_{n-1} = q_n r_n + 0$$

In each of the above equations, divide through by the divisor (the first by b, the next by r_1, etc.) to obtain

$$a/b = q_0 + r_1/b = q_0 + 1/(b/r_1)$$
$$b/r_1 = q_1 + r_2/r_1 = q_1 + 1/(r_1/r_2)$$
$$r_1/r_2 = q_2 + r_3/r_2 = q_2 + 1/(r_2/r_3)$$
$$\vdots$$
$$r_{n-1}/r_n = q_n + 0.$$

Now substitute to find that

$$\frac{a}{b} = q_0 + \cfrac{1}{\cfrac{b}{r_1}} = q_0 + \cfrac{1}{q_1 + \cfrac{1}{\left(\cfrac{r_1}{r_2}\right)}} = \cdots = q_0 + \cfrac{1}{q_1 + \cfrac{1}{q_2 + \cfrac{\cdot}{\cdot \cdot + \cfrac{1}{q_{n-1} + \cfrac{1}{q_n}}}}}$$

$$= \{q_0; q_1, q_2, \ldots, q_n\}.$$

EXAMPLE 2

To express $\frac{17}{7}$ as a continued fraction, we apply Euclid's Algorithm to 17 and 7 to obtain

$$17 = 2 \cdot 7 + 3 \; (q_0 = 2)$$
$$7 = 2 \cdot 3 + 1 \; (q_1 = 2)$$
$$3 = 3 \cdot 1 + 0 \; (q_3 = 3).$$

Thus $\frac{17}{7} = \{2; 2, 3\} = 2 + \frac{1}{2+\frac{1}{3}}$. ∎

Now let's take an **irrational** number x and use Euclid's Algorithm with x and 1. The algorithm will not terminate and so we will simply stop after a few iterations.

EXAMPLE 3

Let's apply Euclid's Algorithm to $\sqrt{2} = \frac{\sqrt{2}}{1}$. This time, we will let our calculators help.

$$\sqrt{2} = 1 \cdot 1 + (\sqrt{2} - 1)$$
$$1.41421\ldots = 1 \cdot 1 + 0.41421\ldots$$
$$1 = 2 \cdot 0.41421\ldots + 0.171573\ldots$$
$$0.41421 = 2 \cdot 0.171573\ldots + 0.07068\ldots \text{ etc.}$$

A **finite continued fraction approximation** to $\sqrt{2}$ is $\{1; 2, 2\} = 1 + \frac{1}{2+\frac{1}{2}} = 1.4$. We would obtain a better approximation if we had carried out a few more iterations of the algorithm. In the language of calculus, if $c_n = \{a_0; a_1, a_2, \ldots, a_n\}$ is the continued fraction approximation to x, then $\lim_{n \to \infty} c_n = x$. ∎

Let's try $x = \pi$.

$$\pi = \{3; 7, 15, 1, \ldots\}$$

$$\{3; 7\} = 3 + \frac{1}{7} = \frac{22}{7}$$

$$\{3; 7, 15\} = 3 + \frac{1}{7 + \frac{1}{15}} = \frac{333}{106}$$

$$\{3; 7, 15, 1\} = 3 + \frac{1}{7 + \frac{1}{15 + \frac{1}{1}}} = \frac{355}{113}$$

The approximations to π that we obtain are quite good:

$$|\pi - 22/7| \approx 0.00126$$
$$|\pi - 333/106| \approx 0.0000832$$
$$|\pi - 355/113| \approx 0.00000027$$

In fact, given the size of the denominator of each approximating fraction, we cannot do better. If $\frac{p}{q}$ is a finite continued fraction approximation to an irrational number x, then there is no rational number $\frac{m}{n}$ closer to x for $n \leq q$. So among the fractions with denominator less than or equal to 113, the fraction nearest to π is $\frac{355}{113}$, which is quite close indeed.

SUMMARY

This section saw Euclid's Algorithm take some unusual turns. We used it to prove that the **Golden Ratio** is an irrational number, and we did it without even computing the value of the Golden Ratio! Although the subject is very old, the reasoning is very modern: when applied to the Golden Ratio and 1, Euclid's Algorithm does not terminate. Then we used Euclid's Algorithm to express rational numbers or approximate irrational numbers by **continued fractions**. We indicated that a continued fraction approximation to an irrational number was in some sense an optimal rational approximation. Again, an old technique produced results very useful to current mathematical concerns.

1.6 Exercises

1. Apply Euclid's Algorithm to the following pairs of rational numbers. What does the final nonzero remainder represent?
   i. $\frac{1}{2}$ and $\frac{1}{3}$
   ii. $\frac{3}{8}$ and $\frac{5}{6}$
2. Prove that Euclid's Algorithm must terminate when applied to any two nonzero rational numbers.
3. Express the following as a fraction:
   i. $\{0; 1, 2, 3\}$
   ii. $\{3; 1, 2, 1, 2, 1, 2\}$
4. What familiar ratios are $\{0; 1, 1\}$, $\{0; 1, 1, 1\}$ and $\{0; 1, 1, 1, 1\}$?
5. Express the following as continued fractions:
   i. $13/37$
   ii. $23/11$
   iii. $45/37$
6. Express 3.14 and 3.1416 as continued fractions.
7. Find a continued fraction approximation to the Golden Ratio $\varphi = \frac{1+\sqrt{5}}{2}$ of the form $\{a_0; a_1, a_2, \ldots, a_n\}$ for $n = 3, 4, 5$, etc.
8. Find a continued fraction approximation to the natural number e of the form $\{a_0; a_1, a_2, \ldots, a_n\}$ for $n = 3$ and $n = 6$. How well do these approximate your calculator's approximation to e?
9. Suppose that the continued fraction expression for a rational number x is $\{a_0; a_1, a_2, \ldots, a_n\}$.
   i. Prove that if $a_n \neq 1$, then x can also be expressed as $\{a_0; a_1, a_2, \ldots, a_n - 1, 1\}$.
   ii. Prove that if $a_n = 1$, then x can also be expressed as $\{a_0; a_1, a_2, \ldots, a_{n-1} + 1\}$.

To the Teacher

Euclid's Algorithm is an amazingly rich resource. The aspects of its uses discussed in this section are only the tip of the iceberg. In fact, we shall revisit the Algorithm in Chapter 3, where we will apply it to polynomials. Its uses in that context are equally rich. Each of the two themes of this section is very appropriate for the high school student, perhaps in an exploratory class or as an enrichment activity. Why do it? The topics are eye openers. That Euclid's Algorithm should say something about anything but integers is a surprise! So much of what we do when we teach or learn from a textbook (even this one) is tailored to and focused on the job at hand. Learn it. Do it. We use Euclid's Algorithm for gcd's and we don't look right or left. But lots of new and exciting mathematics comes from turning familiar mathematics inside-out and finding what else it can do when we drop the harnesses. We as teachers need to keep our sense of play and exploration and share that with our students. That is as important as it's flip side: the careful agenda that we all need to successfully carry a student through a year's curriculum.

Tasks

1. List the different themes from high school mathematics that are pulled together for this section.
2. Develop an activity through which high school students can explore continued fractions.
3. Find out about the origins of continued fractions—try the Web!

1.7 THEOREMS OF EULER AND FERMAT

The Division Algorithm tells us about the remainder of an integer a when divided by a positive integer b. In this section, we again investigate these remainders, but this time, we look at what happens when a *power* of a is divided by b. Let's preview what happens when $a = 5$ and $b = 7$. Suppose that $5^i = 7q_i + r_i$. Table 1 lists the remainders r_i for $1 \leq i \leq 14$.

TABLE 1

i	r_i
1	5
2	4
3	6
4	2
5	3
6	1
7	5
8	4
9	6
10	2
11	3
12	1
13	5
14	4

Since 5 and 7 are relatively prime, no power of 5 is divisible by 7. So it is not surprising that the list of reminders does not include 0. Notice that the remainders cycle periodically so that, for instance, $r_2 = r_{6+2} = r_{6+6+2}$ and, more generally, $r_i = r_{6q+i}$. Assuming (correctly) that the pattern started in the table continues, we can use the pattern to find the remainder of 5^{538} after division by 7. By the Division Algorithm, $538 = 6 \cdot 89 + 4$. So $r_{538} = r_{6 \cdot 89+4} = r_4 = 2$. That's a nice way to find the desired remainder and yet avoid computing the very large number 5^{538}.

In this section, we shall prove **Euler's Theorem** and its corollary, known as **Fermat's Little Theorem**. The essence of these theorems is that what holds in our example is true in general. These beautiful facts about arithmetic are extremely useful. For instance, Euler's Theorem is at the heart of the widely used RSA encryption scheme, which you can explore in Section 2.3. We begin by investigating the Euler phi-function, a well-used tool of number theory that we will need to formulate Euler's Theorem. Given a positive integer n, the **Euler phi-function** counts the number of integers relatively prime to n that are smaller than n.

> **DEFINITION 1.** For $n \geq 1$, $\varphi(n)$ denotes the number of positive integers less than or equal to n that are relatively prime to n. We call $\varphi(n)$ the **Euler phi-function**.

For example, to compute $\varphi(8)$, we note 1, 3, 5, 7 are the numbers that are positive, less than or equal to 8, and relatively prime to 8. Since there are exactly four such numbers, $\varphi(8) = 4$. Similarly, $\varphi(17) = 16$. Notice that 17 is prime and so all 16 natural numbers less than 17 are relatively prime to 17. Proposition 1 and Theorem 2 will help us to compute $\varphi(17)$ more efficiently.

PROPOSITION 1 If p is a prime and $i \geq 1$, then $\varphi(p^i) = p^i - p^{i-1} = p^i\left(1 - \frac{1}{p}\right)$.

Proof. All the numbers between 1 and $p-1$ are relatively prime to p. So $\varphi(p) = p-1$. Similarly, when $i > 1$, all positive integers less than p^i are relatively prime to p^i except for the multiples of p. We can list the multiples of p as follows: $1p, 2p, \ldots, p^{i-1}p$, which shows that there are exactly p^{i-1} multiples of p that are less than or equal to p^i. So there are a total of $p^i - p^{i-1}$ numbers between 1 and p^i that are relatively prime to p. ▲

EXAMPLES

1. $\varphi(5^3) = 5^3(1 - \frac{1}{5}) = 5^3 \cdot \frac{4}{5} = 100$
2. $\varphi(7) = 7 - 1 = 6$

By listing the numbers less than 15, we can verify that $\varphi(15) = 8 = \varphi(3) \cdot \varphi(5)$. However, $\varphi(12) = 4$ but $\varphi(2) \cdot \varphi(6) = 2$. The critical difference is that, in the case of 15, the factors 3 and 5 are relatively prime whereas, in the case of 12, the factors

2 and 6 are not relatively prime. (Notice that $\varphi(12) = 4 = \varphi(4) \cdot \varphi(3)$ and 3 and 4 are relatively prime.)

> **DEFINITION 2.** We say that $f: Z \to Z$ is **multiplicative** if $f(xy) = f(x)f(y)$ whenever x and y are relatively prime.

In the next theorem we will show that the Euler phi-function is multiplicative. But first we put the property to use.

EXAMPLE 3

To compute $\varphi(72)$ we first factor 72 into primes as $72 = 2^3 \cdot 3^2$. Thus $\varphi(72) = \varphi(2^3)\varphi(3^2) = 2^3(1 - 1/2) \cdot 3^2(1 - 1/3) = 4 \cdot 6 = 24$. Similarly, $966219013 = 29^3 \cdot 173 \cdot 229$ and so $\varphi(966219013) = \varphi(29^3)\varphi(173)\varphi(229) = 29^3(1 - 1/29) \cdot 172 \cdot 228 = 923458368$. ■

THEOREM 2 Let m and n be positive integers that are relatively prime. Then

$$\varphi(mn) = \varphi(m)\varphi(n).$$

Proof. Arrange the integers from 1 to mn into n rows and m columns as follows:

1	2	...	r	...	m
$m+1$	$m+2$	...	$m+r$	...	$2m$
$2m+1$	$2m+2$	...	$2m+r$	...	$3m$
...	...	...	...	...	...
$(n-1)m+1$	$(n-1)m+2$	...	$(n-1)m+r$	...	mn

There are $\varphi(m)$ integers relatively prime to m in the first row. In any other row, we have an entry x less than or equal to mn expressed as $qm + r$. Since $\gcd(x, m) = \gcd(r, m)$, there are exactly $\varphi(m)$ integers relatively prime to m in each row. We will show that each of the columns labeled with an r relatively prime to m contains exactly $\varphi(n)$ entries relatively prime to n for a total of $\varphi(m) \cdot \varphi(n)$ entries that are relatively prime to both m and n. Since m and n are relatively prime to each other, these are exactly the integers that are relatively prime to the product mn.

Pick a column r such that $\gcd(r, m) = 1$ and let $y = pm + r$ and $z = qm + r$ be entries in column r. Suppose that $p > q$. By the Division Algorithm, we can express y and z as $y = sn + r_1$ and $z = tn + r_2$ respectively. The numbers y and z cannot have the same remainders after division by n. For if $r_1 = r_2$, we could subtract to find that n divides $(y - z)$. In turn, since y and z both have remainder r when divided by m, we would have that n divides $(p - q)m$. But that is impossible since $0 < p - q < n$ and $\gcd(m, n) = 1$. So, after division by n, there are $n - 1$ distinct remainders between 0 and $n - 1$ in the rth column. Thus the remainders assume all the values between 0 and $n - 1$. There are exactly $\varphi(n)$ integers relatively prime to n in the rth column. Counting the $\varphi(n)$ numbers in each of the $\varphi(m)$ columns, we conclude that there are exactly $\varphi(m)\varphi(n)$ numbers relatively prime to both m and n. ▲

Now we are ready for Euler's Theorem.

THEOREM 3 (*Euler's Theorem*) If $m > 1$ and a are integers that are relatively prime to each other, then m divides $(a^{\varphi(m)} - 1)$.

Before we prove Euler's Theorem, let's verify the theorem for $a = 15$ and $m = 8$. Then $\varphi(8) = 4$ and $15^4 = 50625$. Certainly, 8 divides $50625 - 1 = 50624$. Now we will reach the same conclusion indirectly, without direct computation. In so doing, we will anticipate the steps in the proof that follows. The set of numbers that are relatively prime to 8 is $S = \{1, 3, 5, 7\}$. For each x in S, we shall apply the Division Algorithm to ax.

(1)
$$15 \cdot 1 = 1 \cdot 8 + 7$$
$$15 \cdot 3 = 5 \cdot 8 + 5$$
$$15 \cdot 5 = 9 \cdot 8 + 3$$
$$15 \cdot 7 = 13 \cdot 8 + 1$$

Notice that *all* the numbers in S appear as remainders. The product of the left sides of the equations in (1) is

(2) $\qquad\qquad 15^4(1 \cdot 3 \cdot 5 \cdot 7)$ or $15^{\varphi(8)}(1 \cdot 3 \cdot 5 \cdot 7)$.

The product of the right sides is $(1 \cdot 8 + 7)(5 \cdot 8 + 5)(9 \cdot 8 + 3)(13 \cdot 8 + 1)$. If we gather the coefficients of 8, we can express the product in expression (2) as $8t + (1 \cdot 3 \cdot 5 \cdot 7)$ for some integer t. So $15^{\varphi(8)}(1 \cdot 3 \cdot 5 \cdot 7) = 8t + (1 \cdot 3 \cdot 5 \cdot 7)$. Equivalently,

(3) $\qquad 15^{\varphi(8)}(1 \cdot 3 \cdot 5 \cdot 7) - (1 \cdot 3 \cdot 5 \cdot 7) = (15^{\varphi(8)} - 1)(1 \cdot 3 \cdot 5 \cdot 7) = 8t.$

Clearly, 8 divides $(15^{\varphi(8)} - 1)(1 \cdot 3 \cdot 5 \cdot 7)$ but since 8 and $1 \cdot 3 \cdot 5 \cdot 7$ are relatively prime, Euclid's Lemma guarantees that 8 divides $15^{\varphi(8)} - 1$, as Euler's Theorem asserts. The key is that *all* the remainders in S appear on the right side of the equations in (1). The proof of Euler's Theorem that comes next will prove that this must happen whenever a and m are relatively prime.

Proof of Euler's Theorem. Let a and $m > 1$ be integers such that $\gcd(a, m) = 1$. Let $S = \{n_1, \ldots, n_{\varphi(m)}\}$ be the set of integers between 1 and m that are relatively prime to m. For each $x \in S$, we use the Division Algorithm to express the product ax as $ax = q_x m + r_x$, where $0 < r_x < m$. Because both x and a are relatively prime to m, their product ax is also relatively prime to m. It follows that $\gcd(m, r_x) = 1$ because $1 = \gcd(ax, m) = \gcd(q_x m + r_x, m) = \gcd(m, r_x)$. Thus the remainder $r_x \in S$. Now suppose that $y \in S$ and that $y > x$. The remainders r_x and r_y are not equal. If they were, then m would divide the difference $ya - xa = (y - x)a$. Since $\gcd(a, m) = 1$, we can apply Euclid's Lemma to conclude that m divides $(y - x)$. But that is impossible since $0 < y - x < m$. So the $\varphi(m)$ remainders are all distinct elements of S. Thus $\{r_x : x \in S\} = S$. Let $s = n_1 n_2 \cdots n_{\varphi(m)}$ which is to say that s is the product of the positive integers less than m and relatively prime to m. What we have just shown is that s is also equal to the product of the remainders of the numbers ax, $x \in S$, after division by m. Now let's consider the product of the numbers ax, $x \in S$.

$$\prod_{x \in S} ax = a \cdot n_1 \cdot a \cdot n_2 \cdots a \cdot n_{\phi(m)} = s a^{\phi(m)}$$

$$= \prod_{x \in S} (q_x m + r_x) = tm + s$$

for some integer t. To see this, note that m is a divisor of every term in the product except for the last term, which is the product of the remainders. Thus m divides $(sa^{\varphi(m)} - s)$ or, equivalently, m divides $s(a^{\varphi(m)} - 1)$. Since $\gcd(m, s) = 1$, we can apply Euclid's Lemma again to see that m divides $(a^{\varphi(m)} - 1)$, which is what we set out to prove. ▲

EXAMPLES

4. Let $m = 7$ and $a = 4$. Then $\gcd(7, 4) = 1$ and $\varphi(7) = 6$. We see that $4^6 - 1 = 4095 = 7 \cdot 585$, which is indeed divisible by 7.
5. Let $a = 15$ and $m = 14$. Then $\gcd(15, 14) = 1$ and $\varphi(14) = 6$. We have $15^6 - 1 = 11390624 = 813616 \cdot 14$. The theorem tells us that the remainder of 15^6 after division by 14 is 1. ∎

The special case of Euler's Theorem when m is prime is known as **Fermat's Little Theorem.**

COROLLARY 4 (*Fermat's Little Theorem*) Let p be a prime number and suppose that a is any integer. Then p divides $a^p - a$.

Proof. If p divides a, then p divides $(a^p - a)$. If p does not divide a, then p and a are relatively prime. By Euler's Theorem, p divides $(a^{p-1} - 1)$ and thus p divides $(a^p - a)$. ▲

EXAMPLE 6

It is not difficult to show that if x has remainder 1 after division by m and y has remainder r after division by m, then xy has remainder r after division by m. So suppose we want to find the remainder of 25^{26} after division by 7. First we divide 26 by $\varphi(7) = 6$, a much easier task, to obtain $26 = 4 \cdot 6 + 2$, so that $25^{26} = 25^2 (25^6)^4$. By Euler's Theorem (or Fermat's), 25^6 has a remainder of 1 after division by 7 and thus so does $(25^6)^4$. Now the remainder of 25^2 after division by 7 is 2 and so the remainder of 25^{26} after division by 7 is 2. ∎

SUMMARY

This section focuses on **Euler's Theorem** and its corollary, **Fermat's Little Theorem.** Both address what happens when a power of an integer a is divided by another integer m. Both are very useful in the arithmetic of large numbers. On the way to Euler's Theorem, we investigated the **Euler phi-function.** We discovered that $\varphi(n)$ is an example of a **multiplicative** function, a fact that allowed us to compute $\varphi(n)$ easily when the factorization of n is known.

1.7 Exercises

1. Let $p = 5$. Verify that p divides $14^4 - 1$ but that p does not divide $15^4 - 1$. Does the latter contradict Fermat's Little Theorem?
2. Let $p = 11$ and $a = 4$. For $x = 0, 1, 2, \ldots, 10$, let r_x be the remainder after ax is divided by p. Verify that the set $\{r_0, r_1, \ldots, r_{10}\}$ is a reordering of the set $\{0, 1, 2, \ldots, 10\}$.
3. Suppose $\gcd(a, 35) = 1$. Show that $a^{12} - 1$ is divisible by 35. (***Hint.*** Factor 12 as $2 \cdot 6$ and apply Fermat's Theorem for $p = 7$. Then factor 12 as $3 \cdot 4$ and apply the theorem for $p = 5$.)
4. Use Fermat's Little Theorem to find the unit digit of 3^{100}. (***Hint.*** Use $p = 5$ and consider parity.)
5. **Primality Testing** is a process by which one determines if an odd number q is prime or not. Is Fermat's Little Theorem a primality tester? In other words, is it always true that if $a^{q-1} - 1$ is divisible by q for some integer $a > 1$, then q is prime? ***Hint.*** Investigate the results by letting $q = 91$ and $a = 3$. How can we use the theorem to test for primes?
6. Let p be an odd prime and a be an integer that is not a multiple of p. Let $q = \frac{p-1}{2}$. Show that p divides $(a^q - 1)$ or p divides $(a^q + 1)$, but not both.
7. Use Proposition 1 and Theorem 2 to compute $\varphi(138915) = \varphi(7^3 \cdot 5 \cdot 3^4)$.
8. Give examples to show that for each positive integer n, $n = \sum_{d|n} \phi(d)$, where d ranges over all the positive divisors of n. The proof is a bit challenging so here are some hints. First show that $\sum_{d|n} \phi\left(\frac{n}{d}\right) = \sum_{d|n} \phi(d)$. So it is equivalent to prove that $n = \sum_{d|n} \phi\left(\frac{n}{d}\right)$. To prove the latter, let $c(d) =$ the number of integers x, $1 \leq x \leq n$, such that $d = \gcd(x, n)$. Prove that $\sum_d c(d) = n$ and that $c(d) = \phi\left(\frac{n}{d}\right)$.
9. Let n be a positive integer. Define $\tau(n)$ to be the number of positive divisors of n. Define $\sigma(n)$ to be the sum of the positive divisors of n. Suppose that $n = p_1^{n_1} p_2^{n_2} \cdots p_q^{n_q}$.
    i. Compute $\tau(n)$ and $\sigma(n)$ for $n = 2, 10, 28$.
    ii. Prove that $\tau(n) = \prod_{i=1}^{q} (n_i + 1)$ where the n_i are the exponents in the prime factorization of n for $i = 1, \ldots, q$.
    iii. Prove that $\sigma(n) = \prod_{i=1}^{q} \frac{1 - p_i^{n_i+1}}{1 - p_i}$.
    iv. Determine $\sigma(n)$ and $\tau(n)$ for $n = 7^3 13^2 19^5$.

48 Chapter 1 Topics in Number Theory

10. Based on Exercise 9, prove that τ and σ are multiplicative. That is, prove that if m and n are relatively prime, then $\tau(mn) = \tau(m)\tau(n)$ and $\sigma(mn) = \sigma(m)\sigma(n)$.

11. Prove that if $2^n - 1$ is prime, then $\sigma((2^n - 1)(2^{n-1})) = (2^n - 1)2^n$. A number that is equal to the sum of its positive divisors less than itself is called a **perfect number**. Find several perfect numbers with this formula and verify.

12. Show that $a^{37} - a$ is divisible by 1729 for any integer a.

To the Teacher

The material in this section is challenging but basic to the study of number theory. It probably will not enter the high school curriculum directly. However, Euler's Theorem serves as the basis of an important modern application of number theory, RSA encryption, which is a method of secretly encoding data for transmission. Section 2.3 is a worksheet that offers a simple experience in this important topic. It is readily adaptable to the high school classroom and can be carried out with a calculator. As a teacher you will want to be sure that you know both the computational aspects of how to encode and decode, but you will want to be a few steps ahead by becoming comfortable with the theory behind the method. Students who investigate RSA encoding will see that mathematics that they can use and understand is alive, well and very relevant to their lives. Mathematically, students will develop a sense of inverse processes through the relation of encoding to decoding. An excursion into encryption exercises can also serve as an excuse to discuss computer ethics. Discussions can go from downloading copyrighted music to whether or not a purely mathematical result—like a theorem about prime numbers—can or should be kept secret because it might be useful to a national adversary.

1.8 IN THE CLASSROOM: PROBLEM SOLVING

In April 2000, the National Council of Teachers of Mathematics (NCTM) released its *Principles and Standards for School Mathematics*. This book amplifies and continues the NCTM publications *Curriculum and Evaluation Standards for School Mathematics* (1989) and the *Professional Standards for Teaching Mathematics* (1991) and a decade of feedback and research about how students learn mathematics and learn about the role of mathematics in our current world.

The *Principles and Standards for School Mathematics* establishes six principles and ten standards for mathematics education from prekindergarten to grade 12. Five of the standards are **content standards**, that is, they deal with what we teach rather than how we teach. The content areas are number and operations, algebra, geometry, measurement, and data analysis and probability. Five are **process standards**: problem solving, reasoning and proof, communication, connections, and representation. Our *In the Classroom* sections will connect the mathematics from our chapters to the secondary curriculum in the light of the NCTM content and process standards. We chose problems that align with NCTM's vision of problem solving and worthwhile mathematical tasks as set out in the following two standards.

Problem Solving Standard (NCTM, 2000, p. 52)

Instructional programs from prekindergarten through grade 12 should enable all students to

- *build new mathematical knowledge through problem solving;*
- *solve problems that arise in mathematics and in other contexts;*
- *apply and adapt a variety of appropriate strategies to solve problems;*
- *monitor and reflect on the process of mathematical problem solving.*

Standard 1: Worthwhile Mathematical Tasks (NCTM, 1991, p. 25)

The teacher of mathematics should pose tasks that are based on

- *sound and significant mathematics;*
- *knowledge of students' understandings, interests, and experiences;*
- *knowledge of the range of ways that diverse students learn mathematics;*

and that

- *engage students' intellect;*
- *develop students' mathematical understandings and skills;*
- *stimulate students to make connections and develop a coherent framework for mathematical reasoning;*
- *promote mathematics as an ongoing human activity;*
- *display sensitivity to, and draw on, students' diverse background experiences and dispositions;*
- *promote the development of all students' dispositions to do mathematics.*

What is a Problem?

In most school textbooks, solving problems has meant little more than the following: Read the word problem, translate it into an arithmetic sentence or two, do a computation, and get the answer. Such "problems" would be better referred to as "exercises." When NCTM discusses student learning through problem solving, they are not talking about such exercises. To be called a "problem," a mathematical task should have several characteristics. It should be interesting to the students. It should involve some perplexing situation that students are drawn to solve. Techniques or procedures to solve a problem may not be obvious. There may be a period of time that is uncomfortable for students who like to memorize formulas and apply them to find answers as if by reflex. What constitutes a problem for one student may be an exercise to another student with more experience or skills. As in real life, assumptions might have to be made to focus the problem and there may be multiple solution paths. Students may choose one approach over another depending on their ability level and previous experience. Some students may choose a more algebraic approach. Others may draw pictures or choose to make guesses, check their guesses, and revise them based on logical thought along the way. Rather than one correct answer, there may be multiple solutions depending on the assumptions brought to the problem.

Students should be encouraged to make use of tools, such as manipulatives, calculators, computer algebra systems, and dynamic geometry software to aid them in their problem solving. Calculators and computers these days can do much of the work to get the answer for a mere arithmetic exercise. But they are just tools for a true problem. Computation should not be a barrier to problem solving.

Students should be encouraged to work with one another to solve problems. But it is often appropriate to give each student a period of time alone with the problem. This allows the individual time to come up with some initial thoughts for solving the problem that they can share with others. There are a variety of excellent resources on establishing effective cooperative learning experiences in your classroom. (See Hagelgans et al., 1995, which contains an extensive bibliography.)

With these ideas in mind, we pose the following interesting problem based on one that appeared in Martin Gardner's "Mathematical Games" section of *Scientific American* magazine (Gardner, 1978).

The Social Security Number Problem
I have an interesting social security number; its nine digits include every digit from 1 through 9, and they form a number in which the first two digits (reading from left to right) make a number that is divisible by 2, the first three digits make a number divisible by 3, the first four digits make a number divisible by 4, and so on until the entire number is divisible by 9. What is my social security number?

This problem addresses NCTM's Number and Operations Standard which calls for students to "understand numbers, ways of representing numbers, relationships among numbers, and number systems..." (NCTM 2000, p. 32), as well as the Problem Solving Standard (NCTM 2000, p. 52). It is challenging enough to do with students in high school and above while still accessible to middle school students, where students might first encounter the various divisibility rules. This problem can be used as a culminating activity following coverage of divisibility material or as a motivator to learn the divisibility rules. Such rules will help students limit the possibilities for the various positions in the number rather than randomly guessing and checking values.

You are encouraged to start working on this problem before reading further. Keep a log for yourself detailing your progress with the problem. As you work on it, think about your approaches as the problem solver and how, as a future teacher, you might guide your students' exploration of the problem.

Prior Knowledge Needed to Do the Problem

There is not much needed to begin work on this problem. Students will need to have a good understanding of Definition 1 (Section 1.2):

> Let a and b be integers with $b \neq 0$. We say that b **divides** a (a is **divisible** by b) if there is an integer q such that $a = bq$.

Facility with various divisibility rules will be very helpful. (By doing the worksheet on Divisibility Tests in Section 2.2, you can review various divisibility rules and learn some new ones.)

Getting Students Started on the Problem

As future teachers, you need to look carefully at this problem and think of ways that you might help students get started on the problem while at the same time letting them work (and work hard) on solving it themselves. Think about how you began your work when you tried the problem yourself. What was helpful to you? Where did you begin? Here are some ideas to help students (or yourself) get started. Have students make a frame for the number with nine blanks:

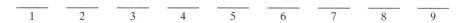

Give them some time to grapple with the problem. You can ask them some of the following questions:

Is there any divisibility rule that can easily be applied to fix one of these nine positions?

Can you tell where the even or where the odd numbers from 1 to 9 must be positioned?

Have we narrowed individual positions enough so that it makes sense to look at clusters of the positions and try to eliminate some of the possibilities for these clusters?

A discussion of this problem and its solution including an interesting sample dialogue that developed when this problem was used with actual grade 9–12 students is in the NCTM 1983 Yearbook, (Schlesinger, 1983). (But don't peek until you have solved the problem yourself.) We can extend the problem with questions such as the following: Is this number unique? Can we formulate some similar problems in base five or base twelve?

Bibliography

Gardner, Martin. "Mathematical Games." *Scientific American*, December 1978, p. 23.

Hagelgans, Nancy L., Reynolds, Barbara E., Schwingendorf, Keith E., Vidakovic, Draga, Dubinsky, Ed, Shahin, Mazen, and Wimbish, Jr., G. Joseph. *A Practical Guide to Cooperative Learning in Collegiate Mathematics*. Mathematical Association of America Notes Series, 37, Mathematical Association of America, 1995.

National Council of Teachers of Mathematics. *Curriculum and Evaluation Standards for School Mathematics*. Reston, Va.: National Council of Teachers of Mathematics, 1989.

_____. *Principles and Standards for School Mathematics*. Reston, Va.: National Council of Teachers of Mathematics, 2000.

_____. *Professional Standards for Teaching Mathematics*. Reston, Va.: National Council of Teachers of Mathematics, 1991.

Pólya, George. *How to Solve It: A New Aspect of Mathematical Method*, 2nd edition. Princeton, N.J.: Princeton University Press, 1957.

Schlesinger, Beth M. "A Senior High School Problem-solving Lesson." In *The Agenda in Action*, 1983 Yearbook of the National Council of Teachers of Mathematics, edited by Gwen Shufelt and James R. Smart, pp. 70–78. Reston, Va.: National Council of Teachers of Mathematics, 1983.

1.9 FROM THE PAST: FERMAT AND PERFECT NUMBERS

When Pierre de Fermat (1601–1656) described himself, it was as a jurist—a position that was part judge and part lawyer. However, to a small number of scientists and intellectuals throughout Europe, he was a mathematician of the first rank. His fundamental contributions to number theory are well recognized today, with his name attached to several important theorems and definitions. His mathematics was communicated to the world through letters to other mathematicians, and in challenges to offer or solve problems. We will consider one of these problems here.

Fermat's interest in number theory began with perfect numbers. The notion of a perfect number arose in Greece and was continuously studied through the Middle Ages, up to the time of Fermat. A **perfect number** is a positive integer that equals the sum of its proper, positive factors. (A factor of a positive integer n is proper if it is smaller than n.) Thus 6 is a perfect number since $6 = 1 + 2 + 3$ and $\{1, 2, 3\}$ is the set of proper factors of 6. Four perfect numbers had been known since ancient times. They are 6, 28, 496, and 8128. Many open questions surrounded these numbers. Is the set of perfect numbers infinite? Do their last digits continue to alternate between 6 and 8? Is there always a perfect number between adjacent powers of ten? (6 is between 1 and 10, 28 is between 10 and 100, etc.) How does one find perfect numbers? Are there any odd perfect numbers? The problem of perfect numbers led Fermat to a theorem that bears his name today. It is a very important result in modern number theory.

In the *Elements* (Book IX, Proposition 36), Euclid proved a theorem concerning perfect numbers. What follows is a modern wording of Euclid's Theorem.

THEOREM 1 (*Euclid*) If n is a positive integer such that $2^n - 1$ is a prime number, then $N = 2^{n-1}(2^n - 1)$ is a perfect number.

Proof. The proof is a computation. First, what are the factors of N? Consider 2^{n-1}. Its factors are $\{1, 2, 4, \ldots, 2^{n-1}\}$. Since $2^n - 1$ is assumed to be a prime, the complete list of the proper factors of N is

$$\{1, 2, 4, \ldots, 2^{n-1}\} \cup \{2^n - 1, 2(2^n - 1), 4(2^n - 1), \ldots, 2^{n-2}(2^n - 1)\}.$$

The sum of the first list is

$$1 + 2 + 4 + \ldots + 2^{n-1} = \frac{2^n - 1}{2 - 1} = 2^n - 1.$$

The sum of the second list is

$$(2^n - 1)(1 + 2 + 4 + \ldots + 2^{n-2}) = (2^n - 1)\left(\frac{2^{n-1} - 1}{2 - 1}\right)$$

$$= (2^n - 1)(2^{n-1} - 1).$$

The sum of all the proper factors of N is

$$(2^n - 1) + (2^n - 1)(2^{n-1} - 1) = (2^n - 1)(1 + 2^{n-1} - 1)$$

$$= (2^n - 1)2^{n-1} = N. \quad \blacktriangle$$

Fermat knew Euclid's result but had little else in his search for perfect numbers. He needed better procedures for determining when $2^n - 1$ was prime, procedures that he proceeded to invent. (Few other mathematicians in Europe were working on such problems.) In a letter to Frenicle, another French mathematician, he stated three results that helped in his search for primes of the form $2^n - 1$.

1. If n is a composite integer (always assumed positive), then $2^n - 1$ is composite.
2. If $n > 2$ is a prime, then all prime factors of $2^n - 1$ are of the form $2kn + 1$ for some integer k.
3. If n is a prime, then n divides $2^n - 2$.

Fermat gave no proofs for these results. We simply do not know how he arrived at them. But based on other writings of Fermat, the following demonstrations of these facts are in line with Fermat's thought.

Result 1.

The first result follows from the factorization: $x^n - 1 = (x - 1)(x^{n-1} + x^{n-2} + \ldots + x + 1)$, which holds for all $n > 0$. Suppose that n is not prime and let $n = pq$, where $p > 1$ and $q > 1$. Then $2^n - 1 = 2^{pq} - 1 = (2^p)^q - 1 = (2^p - 1)((2^p)^{q-1} + (2^p)^{q-2} + \ldots + 1)$, which is a nontrivial factorization of $2^n - 1$.

EXAMPLE 1

For example, $2^{15} - 1$ has the following factorizations:
$$2^{15} - 1 = 32767 = (2^3)^5 - 1 = (2^3 - 1)((2^3)^4 + (2^3)^3 + \ldots + 1)$$
$$= 7 \cdot (4096 + 512 + 64 + 8 + 1) = 7 \cdot 4681.$$
Also,
$$2^{15} - 1 = 32767 = (2^5)^3 - 1 = (2^5 - 1)(2^{10} + 2^5 + 1) = 31 \cdot 1057.$$ ■

For $2^n - 1$ to be prime, it is necessary that n be prime, but it is not sufficient. For example, $2^{11} - 1 = 2047 = 23 \cdot 89$. So $n = 11$ is prime, but $2^{11} - 1$ is not. This left the following problem: For which primes n is $2^n - 1$ prime? Fermat needed new ways to factor numbers and to cut down on the work involved in determining whether a number was prime. This led to his second result. But to prove the second result we need the third result. (Our numbering here preserves Fermat's numbering.)

Result 3.

This is a special case ($a = 2$) of what has become known as Fermat's Little Theorem. It is presented and proved in Section 1.7. There it is stated as follows:

> Let p be a prime number and suppose that a is any integer. Then $p \mid (a^p - a)$.

This very important tool in number theory was precipitated and necessitated by Fermat's work on perfect numbers. Fermat might have reasoned as follows:

Let a be thought of as a sum of a copies of 1, $a = 1 + 1 + \ldots + 1$. Then $a^n = (1 + 1 + \ldots + 1)^n$. Expanding, $a^n = 1^n + 1^n + \ldots + 1^n +$ terms divisible by n. (Fermat knew the basic facts about what we call multinomial coefficients, in particular that $\frac{n!}{r_1! r_2! \cdots r_n!}$ is divisible by the prime n for $1 \leq r_i \leq n-1$.) Thus $a^n - a$ is divisible by n. If a is not a multiple of the prime n, then n divides $a^{n-1} - 1$.

In a later letter to Frenicle, Fermat elaborated on this theorem. His assertion was more nuanced.

If n is prime and a is an integer relatively prime to n, then n divides some member of the set $\{a^r - 1 : r = 1, 2, \ldots\}$. Let s be the smallest positive integer for which n divides $a^s - 1$. Then n divides $a^t - 1$ if and only if t is a positive multiple of s. Since $a | (a^{n-1} - 1)$, it follows that $s | (n-1)$.

Again, we do not know how Fermat arrived at this conclusion but he may have used the following line of reasoning:

Let a be an integer relatively prime to prime n. Consider $\{a, a^2, a^3, \ldots\}$, the set of powers of a. Upon division of each of these by n, there are at most $n-1$ different remainders. Suppose that a^k and a^m are the first two in the list whose remainders are equal and assume that $k < m$. Then the remainder of $a^m - a^k$ when divided by n is 0. Since $a^m - a^k = a^k(a^{m-k} - 1)$ and since n does not divide a^k, it must be that n divides $(a^{m-k} - 1)$. Let $s = m - k$. Since n divides $a^s - 1$, n also divides $a(a^s - 1)$ which equals $a^{s+1} - a^1$. Thus the k and m are 1 and $s + 1$, respectively. The sequence of remainders is periodic with period s. Let r be the remainder of a divided by n. Then it is also the remainder when a^{s+1} is divided by n. Multiplying both a and a^{s+1} by a gives a^2 and a^{s+2}, which have the same remainder when divided by n. By repeatedly multiplying by a, we find that the remainders repeat with period s. Since n divides $a^{n-1} - 1$, the periodicity implies that s divides $n-1$.

EXAMPLE 2

Let $n = 7$ and $a = 2$. Then 7 must divide a number in the sequence

$$\{2^1 - 1 = 1, \ 2^2 - 1 = 3, \ 2^3 - 1 = 7, \ 2^4 - 1 = 15, \ldots\}.$$

We see that 7 divides the third term since $2^3 - 1 = 7$. Let $3c$ be a positive multiple of 3. Then $2^{3c} - 1 = (2^3)^c - 1 = (2^3 - 1)(2^{3(c-1)} + 2^{3(c-2)} + \ldots + 1)$ which is indeed a multiple of 7. (Notice that $3 | (7-1)$ as Fermat claimed it must.) If we look at the sequence of numbers 2^c and find the remainders after division by 7, we have the following: $\{2, 4, 1, 2, 4, 1, \ldots\}$. It is a periodic sequence of period 3.

Now let $n = 11$ and $a = 2$. Then 11 must divide some number in the sequence

$$\{2^1 - 1, 2^2 - 1, \ldots\} = \{1, 3, 7, 15, 31, 63, 127, 255, 511, 1023, \ldots\}.$$

This time 11 divides the 10th term in the sequence, 1023. In this case s is actually equal to $n - 1$.

Result 2.

It isn't too difficult to derive the second result from the third. Suppose that $n > 2$ is prime and that p is a prime factor of $2^n - 1$. We know that $p|(2^n - 1)$ and from Fermat's Little Theorem, we know that $p|(2^{p-1} - 1)$. From Fermat's elaboration on the Little Theorem, we know that both exponents n and $p - 1$ must be multiples of the smallest s for which $2^s - 1$ is divisible by p. Since n is prime, $s = n$ and so $n|(p - 1)$. Since $p - 1$ is even, $p - 1 = 2kn$ for some integer k. Thus $p = 2kn + 1$.

EXAMPLE 3

Let $n = 11$. The prime factors of $2047 = 2^{11} - 1$ are 23 and 89. Now $23 = 2 \cdot 1 \cdot 11 + 1$ and $89 = 2 \cdot 4 \cdot 11 + 1$. For $n = 5$ we have $2^5 - 1 = 31$. Its only prime factor is $2 \cdot 3 \cdot 5 + 1$. ∎

EXAMPLE 4

Consider $N = 2^{13} - 1 = 8191$. According to Fermat's result, to check if 8191 is prime it suffices to check for divisibility by all primes of the form $2 \cdot 13 \cdot k + 1$ up to $\sqrt{8191} < 91$. These primes are 53 and 79. Since neither of these divides 8191, Fermat knew that $2^{13} - 1$ was a prime and hence that $2^{12}(2^{13} - 1) = 33550336$ is a perfect number, a perfect number not known to the ancients. ∎

1.9 Exercises

1. Find a prime factor of $2^{35} - 1$.
2. Find the four odd primes that Fermat would divide $2^{17} - 1$ by in order to show that $2^{17} - 1$ was a prime. Since $2^{17} - 1$ is a prime, we know that $2^{16}(2^{17} - 1) = 8589869056$ is a perfect number.
3. Show that $2^{23} - 1$ is not prime by finding a small prime factor using Fermat's method.
4. For each of the following primes n, find the smallest value of r such that $2^r - 1$ is a multiple of n. In each case show that the r you found is a factor of $n - 1$:

$$n = 13$$
$$n = 17$$
$$n = 19.$$

5. Fermat stated his third result for any base rather than just 2. That is, if n is a prime and a is an integer relatively prime to n, then n divides some member of the set $\{a^r - 1 : r = 1, 2, 3, \ldots\}$. Repeat Exercise 4 for $n = 5, 7,$ and 11 using $3^r - 1$ instead of $2^r - 1$.
6. Fermat also considered numbers of the form $2^n + 1$. He stated that in order for $2^n + 1$ to be prime, n must be a power of 2. Using the following factorization, $x^n + 1 = (x + 1)(x^{n-1} - x^{n-2} + x^{n-3} - \ldots - x + 1)$, which holds for

all positive odd integers n, show that $2^n + 1$ is composite if $n = pq$, where p is an odd prime. Fermat found the first few primes of the form $2^n + 1$ when $n = 2^0, 2^1, 2^2, 2^3,$ and 2^4.

$$2^1 + 1 = 3$$
$$2^2 + 1 = 5$$
$$2^4 + 1 = 17$$
$$2^8 + 1 = 257$$
$$2^{16} + 1 = 65537$$

These are called Fermat primes today. Fermat conjectured that every number of this form was prime. Euler showed that the next case, $2^{32} + 1 = 4294967297$ factored as $(641)(6700417)$. No other prime of this form has ever been found.

Bibliography

A wonderful mathematical biography of Fermat, with detailed treatment of all aspects of his work, is

M. Mahoney, *The Mathematical Career of Pierre de Fermat, 1601–1665*, 2nd rev. ed. Princeton, N.J.: Princeton University Press, 1994.

Chapter 1 Highlights

This chapter is about division. Its most important theorem is the **Division Algorithm** which tells us that given integers a and b with $b > 0$, we can write a as $qb + r$ in one way such that $0 \leq r < b$. When $r = 0$, we say that b **divides** a or that b is a **factor** of a or that b is a **divisor** of a. **Euclid's Algorithm** uses the Division Algorithm to determine the **greatest common divisor** of two integers a and b. Euclid's Algorithm is important because it determines the common factors of a and b *without* the knowledge of the individual factors of a and b. The steps of the algorithm also enable us to express $\gcd(a, b)$ as $ma + nb$, a weighted sum of a and b which, in turn, enables us to find all solutions to **Diophantine equations** of the form $ax + by = c$. When a and b are **relatively prime**, and a divides bc, **Euclid's Lemma** guarantees that a must divide c. This important result is used to prove the **Fundamental Theorem of Arithmetic**, which asserts that every integer greater than 1 can be factored into primes in exactly one way. **Euclid's Theorem** and **Fermat's Little Theorem** extend our study of divisibility to what happens when powers of a number a are divided by a number m that is relatively prime to a. These theorems are formulated in terms of the **Euler phi-function** $\varphi(n)$.

Chapter Questions

1. Give the definitions of the following terms or phrases.

    i. a divides b
    ii. relatively prime

iii. prime
iv. unique factorization
v. gcd(a, b)
vi. lcm(a, b)

2. Give the statements of the following propositions or theorems.

    i. The Division Algorithm
    ii. Euclid's Lemma
    iii. The Fundamental Theorem of Arithmetic
    iv. Euclid's Theorem
    v. Fermat's Little Theorem

3. Express a as $qb + r$ as in the Division Algorithm.

    i. $a = 2$ and $b = 17$
    ii. $a = 1503$ and $b = 57$
    iii. $a = -1503$ and $b = 57$

4. Use Euclid's Algorithm to find the gcd of the following pairs of numbers.

    i. 234 and 456
    ii. 589403 and 93840
    iii. Any Fibonacci number and its successor

5. Find all solutions to the following Diophantine equations.

    i. $30x + 12y = 27$
    ii. $31x + 17y = 12$

6. Use Fermat's Little Theorem to find the remainder of 12^{7651} divided by 7.
7. Let $x = 5^4 7^5 13^2$. Find $\varphi(x)$.
8. For each of the following, find the missing number.

    i. $xy = 35712$ $\gcd(x, y) = 6$ $\text{lcm}(x, y) = $ _____
    ii. $xy = $ _____ $\gcd(x, y) = 17$ $\text{lcm}(x, y) = 7684$
    iii. $xy = 8087040$ $\gcd(x, y) = $ _____ $\text{lcm}(x, y) = 56160$

9. Outline the strategy of the proof of each theorem given in Question 2.
10. Suppose that $xs = yt$ and that x and y are relatively prime. Prove that x divides t and that y divides s.

MODULAR ARITHMETIC AND SYSTEMS OF NUMBERS

> In this chapter we begin the journey toward algebra that is more abstract. First we will study the arithmetic of the remainders obtained when we divide by a fixed number m. Following the language that Gauss used, we will call m "the modulus" and we will study "modular" arithmetic. Then we make a leap into abstraction—the remainders modulo m become a new system of numbers. We will add and multiply them, but the new system has its own rules. Nonetheless, these rules share a lot in common with the ordinary arithmetic of the integers. That similarity is captured in the definition of the abstract algebra structure known as a ring. The set of integers and its rules of arithmetic serve as the prototype of a ring. What other rings do we know of? The set of rational numbers, the set of real numbers, and the set of complex numbers are the most familiar. But there are many others, as we shall see.
>
> The applications of modular arithmetic abound. Young children internalize it when they learn about telling time. "Clock arithmetic" is arithmetic modulo 12. The worksheet that follows (Section 2.2) will show that modular arithmetic is behind some of our familiar and not so familiar divisibility tests. The *To the Teacher* note at the end of this section shows how modular arithmetic enters the grocery store through the familiar UPC symbols.

2.1 CONGRUENCES

In his *Disquisitiones Arithmeticae* of 1801, Gauss wrote, "If a number n measures the difference between two numbers a and b, then a and b are said to be congruent with respect to n; if not, incongruent." In modern idiom, we use the word "divide" instead of "measure." From Gauss's words, we formulate the following definition.

60 Chapter 2 Modular Arithmetic and Systems of Numbers

> **DEFINITION 1.** Let m be a positive integer. Two integers a and b are **congruent modulo** m if $a - b = qm$ for some integer q. We write $a \equiv b$ mod m.

The number m is called the **modulus**. Since $a - b = qm$ if and only if $a = b + qm$, we see that $a \equiv b$ mod m whenever a is obtained from b by the addition of some integer multiple of m.

EXAMPLE 1

i. In Definition 1, let $m = 5$, $a = 32$ and $b = 17$. Then $32 \equiv 17$ mod 5 since $32 - 17 = 15 = 3 \cdot 5$. Note $32 = 17 + 3 \cdot 5$, which is 17 plus a multiple of 5.

ii. In Definition 1, let $m = 5, a = 42$ and $b = 17$. Then $42 \equiv 17$ mod 5 since $42 - 17 = 25 = 5 \cdot 5$. So both 42 and 32 are congruent to 17 modulo 5 and they are congruent to each other: $42 \equiv 32$ mod 5.

iii. Let $m = 7$. Then $38 \equiv 17$ mod 7 since $38 - 17 = 3 \cdot 7$. Notice that both 38 and 17 have a remainder 3 when divided by 7.

The next proposition shows that part iii of Example 1 holds more generally. Congruence modulo m identifies numbers that have the same remainders when divided by m.

PROPOSITION 1 Two integers a and b are congruent mod m if and only if they have the same remainder after division by m.

Proof. Suppose that $a = qm + r_1$ and $b = pm + r_2$, where $0 \leq r_i < m$ for $i = 1$ and 2. Suppose also that $r_1 \geq r_2$ so that $0 \leq r_1 - r_2 < m$. If a and b have the same remainder after division by m, then $a - b = (q - p)m + (r_1 - r_2) = (q - p)m + 0$. So $a - b$ is divisible by m and a is congruent to b mod m. Conversely, suppose that a is congruent to b mod m so that $a - b$ is divisible by m. Since $a - b$ is divisible by m, $r_1 - r_2$ is also divisible by m. Since $0 \leq r_1 - r_2 < m$, $r_1 - r_2 = 0$. ▲

Now let's turn to the arithmetic of congruences. The symbol $\equiv$ looks like an equal sign and, in many ways, it behaves like one. For instance, consider $2 \equiv 13$ mod 11. We can multiply each side by a constant and the congruence still holds: $3 \cdot 2 \equiv 3 \cdot 13$ mod 11 since $6 - 39$ is divisible by 11. Now consider a second congruence, $23 \equiv 34$ mod 11. We can add its left side and right side to the left and right of $2 \equiv 13$ mod 11 to obtain the congruence $25 \equiv 47$ mod 11. Similarly, we can multiply both sides to obtain the congruence $46 \equiv 442$ mod 11. The next theorem formalizes these basic arithmetic properties.

THEOREM 2 (*Fundamental Properties*) Let a, b, c, d and k be any integers and let $m > 0$ be an integer.

i. If $a \equiv b$ mod m, then $ka \equiv kb$ mod m for every integer k.

ii. If $a \equiv b$ mod m and $c \equiv d$ mod m, then $a + c \equiv b + d$ mod m.

iii. If $a \equiv b \bmod m$ and $c \equiv d \bmod m$, then $ac \equiv bd \bmod m$.
iv. If $a \equiv b \bmod m$, then $a^n \equiv b^n \bmod m$ for any integer $n \geq 0$.

Proof. We shall prove part iii, and indicate the proof of part iv. The other parts are left as exercises.

iii. Let $a \equiv b \bmod m$ and $c \equiv d \bmod m$. Let p and q be the integers such that $a = b + pm$ and $c = d + qm$. Multiplying the left sides of the equations together, and then the right sides, we obtain $ac = bd + m(dp + bq + pqm)$. Thus $ac \equiv bd \bmod m$.

iv. In part iii, we can certainly let $a = c$ and $b = d$. Then $ac \equiv bd \bmod m$ becomes $a^2 \equiv b^2 \bmod m$. By induction, $a^n \equiv b^n \bmod m$ for any integer $n \geq 0$. ▲

EXAMPLE 2

Let's find the remainder of 93^{25} after division by 7 without evaluating 93^{25}. First, note that $93 \equiv 2 \bmod 7$. By part iv of Theorem 2, we have $93^{25} \equiv 2^{25} \bmod 7$. Now notice that $2^3 \equiv 1 \bmod 7$ and that $2^{25} = 2 \cdot 2^{24}$. Since $(2^3)^8 \equiv 1 \bmod 7$, we can use Theorem 2 again to obtain $2 \cdot (2^3)^8 \equiv 2 \bmod 7$. Thus $2^{25} \equiv 2 \bmod 7$, which shows that $93^{25} \equiv 2 \bmod 7$. ■

A comment on notation. On calculators and in computer algebra systems (CAS), we often find a function "$x \bmod m$" that returns the remainder after division by m. That is to say, the command "$x \bmod m$" returns r_x where $x = qm + r_x$ as in the Division Algorithm. (Syntax may vary.) Thus 17 mod 5 returns 2. Using this notation, we can rephrase Proposition 1 as follows:

(1)   "$a \equiv b \bmod m$ if and only if $a \bmod m = b \bmod m$."

The expression "$a \equiv b \bmod m$" is a statement about the divisibility of the difference $a - b$ by m, whereas the expression "$a \bmod m = b \bmod m$" is a statement about the remainders of a and b after division by m. The distinction is subtle and the notation is sometimes confusing. As a consequence of Proposition 1, we can restate Theorem 2 using the "$x \bmod m$" notation as follows. (Keep in mind that "$x \bmod m$" is a number; it is the remainder after division by m.)

If $a \bmod m = b \bmod m$ and $c \bmod m = d \bmod m$, and if k is any integer, then

i. $(ka) \bmod m = (kb) \bmod m$;
ii. $(a + c) \bmod m = (b + d) \bmod m$;
iii. $(ac) \bmod m = (bd) \bmod m$.

Since $x \equiv (x \bmod m) \bmod m$—again, remember $(x \bmod m)$ is a number—we have

iv. $kx \bmod m = (k(x \bmod m)) \bmod m$;
v. $(x + y) \bmod m = (x \bmod m + y \bmod m) \bmod m$;
vi. $xy \bmod m = (x \bmod m \cdot y \bmod m) \bmod m$.

It might seem that there are just too many uses of the phrase "mod m." But read carefully. To perform $(x + y)$ mod m, we take two integers x and y, add them and take the remainder of the sum after division by m. But on the right, in $(x \bmod m + y \bmod m) \bmod m$, we first add the remainders of x and y after division by m: $x \bmod m + y \bmod m$. That sum might be greater than m. So we must reduce again modulo m. Thus $(x + y) \bmod m = (x \bmod m + y \bmod m) \bmod m$. The fundamental arithmetic properties, expressed in terms of the mod m notation, are illustrated in Example 3.

EXAMPLE 3

i. Let $k = 3$ and note that 8 mod 5 = 13 mod 5. To illustrate property i, note that $(3 \cdot 8) \bmod 5 = 4$ and $(3 \cdot 13) \bmod 5 = 4$. Thus $(3 \cdot 8) \bmod 5 = (3 \cdot 13) \bmod 5$.

ii. To illustrate property ii, note that 8 mod 5 = 3 and 19 mod 5 = 4. Thus $(8 + 19) \bmod 5 = (3 + 4) \bmod 5 = 2$.

iii. To illustrate property iii, note that 13 mod 5 = 8 mod 5 and that 4 mod 5 = 19 mod 5. So $(13 \cdot 4) \bmod 5 = (8 \cdot 19) \bmod 5 = 2$. ∎

More of the general properties of congruences emerge when we take a look at the sets of numbers congruent to x modulo 5 for $x = 0, 1, 2, 3, 4$, respectively. These sets are

$\{\ldots, -15, -10, -5, 0, 5, 10, 15, \ldots\}$ for $x = 0$,
$\{\ldots, -14, -9, -4, 1, 6, 11, 16, \ldots\}$ for $x = 1$,
$\{\ldots, -13, -8, -3, 2, 7, 12, 17, \ldots\}$ for $x = 2$,
$\{\ldots, -12, -7, -2, 3, 8, 13, 18, \ldots\}$ for $x = 3$,
$\{\ldots, -11, -6, -1, 4, 9, 14, 19, \ldots\}$ for $x = 4$.

The sets sort the integers out according to their remainders after division by 5. Every integer appears in one and exactly one of the above sets so that the collection of sets forms a partition of **Z**. This tells us that congruence is an equivalence relation. (You may wish to review the material in Section 1.1 that deals with equivalence relations and partitions.) In general, an equivalence relation on a set A relates two elements that are the same with respect to some aspect or property of the members of A. For congruence, the property that remains the same when two numbers are congruent is that they have the same remainder modulo m. Theorem 3 formalizes these ideas.

THEOREM 3   Let m be a positive integer. Let R be the relation on **Z** that is defined as follows: Given two integers a and b, aRb when $a \equiv b$ mod m. The relation R is an equivalence relation.

Proof.   To prove that R is an equivalence relation, we will show that it is reflexive, symmetric, and transitive.

i. Since $x - x = 0 \cdot m$, we have $x \equiv x$ mod m for all integers x. Since xRx for all x in **Z**, the relation is reflexive.

ii. Since $x - y = qm$ if and only if $y - x = -qm$, we see that $x \equiv y$ mod m if and only if $y \equiv x$ mod m. Thus, for all x and y in **Z**, xRy if and only if yRx and so the relation is symmetric.

iii. To see that the relation is transitive, let $a \equiv b$ mod m and $b \equiv c$ mod m. Then we can find integers p and q such that $a - b = qm$ and $b - c = pm$. Adding the left and right sides of the equations, we have $a - c = (q + p)m$, which proves that $a \equiv c$ mod m. Thus, for all a, b, and c in $\mathbf{Z}$, if $a \equiv b$ mod m and $b \equiv c$ mod m, then $a \equiv c$ mod m and so the relation is transitive. ▲

For the mod m relation, the equivalence class of an integer x is called its **congruence class**. We denote the congruence class of x by $[x]_m$. Thus $[x]_m = \{y \in \mathbf{Z} : y \equiv x \mod m\} = \{y \in \mathbf{Z} : y = x + qm \text{ for some } q \in \mathbf{Z}\} = \{\ldots, x - 2m, x - m, x, x + m, x + 2m, \ldots\}$.

EXAMPLE 4

i. For $m = 2$, $[0]_2 = \{\ldots, -2, 0, 2, 4, 6, \ldots\}$, and $[1]_2 = \{\ldots, -3, -1, 1, 3, 5, 7, \ldots\}$. Notice that $[0]_2 = [2]_2 = [2346]_2$ and that $[1]_2 = [17]_2 = [-33]_2$. Also notice that $[0]_2 \cup [1]_2 = \mathbf{Z}$.

ii. For $m = 3$, $[0]_3 = \{\ldots, -6, -3, 0, 3, 6, 9, \ldots\}$, $[1]_3 = \{\ldots, -5, -2, 1, 4, 7, 10, \ldots\}$ and $[2]_3 = \{\ldots, -4, -1, 2, 5, 8, 11, \ldots\}$. These three congruence classes form a partition of $\mathbf{Z}$. ■

For each $m > 0$, there are exactly m congruences classes, $[0]_m, [1]_m, \ldots, [m - 1]_m$. Thus $\mathbf{Z} = [0]_m \cup [1]_m \cup \cdots \cup [m - 1]_m$. Each congruence class can appear under an infinite number of different names. For instance, $[1]_5 = [6]_5 = [12111]_5 = [-199]_5$. However, $[x]_m$ has only one member r such that $0 \leq r < m$. We call this special member of $[x]_m$ the **principal representative of $[x]_m$**. It is the remainder of x after division by m. For example, the principal representative of $[380]_7$ is 2 because $380 = 54 \cdot 7 + 2$ and so $[380]_7 = [2]_7$.

So far we have investigated congruence as relation. Now let's do algebra. A **linear congruence** is an expression of the form $ax \equiv b$ mod m. The solution to $ax \equiv b$ mod m is the set of all values of x for which the expression is true.

EXAMPLE 5

i. The solution set of the linear congruence $2x \equiv 3$ mod 7 is the set $\{\ldots, -2, 5, 12, \ldots\}$ or $[5]_7$. To check, let $x = 12$, a member of $[5]_7$. Since $2 \cdot 12 - 3 = 21$, 12 is indeed a solution to $2x \equiv 3$ mod 7.

ii. The solution set of $3x \equiv 3$ mod 9 is the set $\{\ldots, -2, 1, 4, 7, \ldots\} = [1]_9 \cup [4]_9 \cup [7]_9$.

iii. However, $4x \equiv 1$ mod 6 has no solution since $4x - 1$ is odd for all integer values of x, and hence not a multiple of 6. The solution set is empty. ■

Example 5 shows that, in each case, the solution to $ax \equiv b$ mod m is an equivalence class, the union of equivalence classes, or empty. In what follows, we investigate exactly how to determine which situation holds. We will also look for similarities and differences between linear congruences and linear (degree one) equations.

64 Chapter 2 Modular Arithmetic and Systems of Numbers

PROPOSITION 4 Let z be any solution to $ax \equiv b$ mod m. Then all the integers in the set $[z]_m$ are also solutions to $ax \equiv b$ mod m.

Proof. If z is a solution to $ax \equiv b$ mod m, we can find an integer p such that $az - b = pm$. Suppose $y \in [z]_m$ so that $y = z + qm$ for some integer q. Substituting $y - qm$ for z, we have $a(y - qm) - b = pm$. Regrouping, $ay - b = (p + aq)m$. So y is also a solution to $ax \equiv b$ mod m. ▲

Proposition 4 shows that the solution set to a linear congruence will be a union of equivalence classes mod m because once x_0 is a solution to $ax \equiv b$ mod m, every member of $[x_0]_m$ is a solution. It also gives us a preliminary strategy for finding all solutions to a congruence because every congruence class has a principal representative.

Strategy: To solve $ax \equiv b$ mod m, look among the numbers $0, 1, \ldots, m-1$ for a solution. Any solution will be congruent to one of these numbers. Once one solution x_0 from this restricted set is found, all members of its congruence class, namely $[x_0]_m$, are solutions.

EXAMPLE 6

To solve $3x \equiv 6$ mod 9, we check among the numbers $\{0, 1, 2, \ldots, 8\}$ and find that $2, 5$ and 8 are solutions. The solution set is $[2]_9 \cup [5]_9 \cup [8]_9$. ■

In part iii of Example 5, we saw that it is possible that a linear congruence has no solution at all. The next proposition tells us exactly when a congruence does or does not have a solution.

PROPOSITION 5 The congruence $ax \equiv b$ mod m has a solution if and only if $\gcd(a, m)$ divides b.

Proof. Suppose that $\gcd(a, m) = d$ and that d divides b so that $b = qd$ for some integer q. From Theorem 1 of Section 1.3, we can find integers s and t such that $d = as + tm$. Thus $b = aqs + tqm$. Setting $z = qs$, we have $az \equiv b$ mod m. Conversely, suppose that z is a solution to $ax \equiv b$ mod m. Then there is an integer q such that $b = az + qm$. Since d divides a and d divides m, we see that d divides b. ▲

We have seen the argument of the proof of Proposition 5 before. It was in our discussion of Diophantine equations in Section 1.3. The congruence $ax \equiv b$ mod m has a solution if and only if we can find values of x and q such that the Diophantine equation $ax - qm = b$ has a solution. (It is the value of x that we want.) We already know that $ax - qm = b$ has a solution if and only if $\gcd(a, m)$ divides b. Knowing one solution to $ax \equiv b$ mod m, we can find all solutions. The next proposition shows us how.

PROPOSITION 6 Suppose that $d = \gcd(a, m)$ and that d divides b. Suppose that x_0 is a solution to $ax \equiv b$ mod m. Then all solutions are of the form $x_i = x_0 + \frac{im}{d}$, where $i \in \mathbf{Z}$.

Proof. Exercise. (***Hint.*** Since we know that $ax \equiv b$ mod m has a solution if and only if we can solve the Diophantine equation $ax - qm = b$ for q and x, we can use Proposition 8 of Section 1.3.) ▲

COROLLARY 1 There are d equivalence classes in the solution set of $ax \equiv b$ mod m.

Proof. Let x_0 be a solution to $ax \equiv b$ mod m. Let i and j be integers such that $0 \leq i < j < d$. From Proposition 6, we know that both $x_0 + \frac{im}{d}$ and $x_0 + \frac{jm}{d}$ are solutions to $ax \equiv b$ mod m. But they are not congruent to each other because $\left(x_0 + \frac{jm}{d}\right) - \left(x_0 + \frac{im}{d}\right) = \frac{(j-i)m}{d} < m$. The d solutions of the form $x_0 + \frac{im}{d}$ for $0 \leq i < d$ are distinct. If $i \geq d$, then $i = kd + i_0$ for $0 \leq i_0 < d$ and $x = x_0 + \frac{im}{d}$ is congruent $x_0 + \frac{i_0 m}{d}$. Thus there are exactly d distinct equivalence classes in the solution set to $ax \equiv b$ mod m. ▲

EXAMPLE 7

To solve $4x \equiv 8$ mod 12, first note that $d = \gcd(4, 12) = 4$ and that 4 divides 8. So solutions exist. There are four solutions between 0 and 11. One solution is $x = 2$. Since $\frac{m}{d} = \frac{12}{4} = 3$, the other solutions are $2 + 1 \cdot 3, 2 + 2 \cdot 3$, and $2 + 3 \cdot 3$, or 5, 8 and 11. So the solution set is $[2]_{12} \cup [5]_{12} \cup [8]_{12} \cup [11]_{12}$. Notice that we cannot cancel the 4 from both sides of the congruence. If we had canceled and tried to solve $x \equiv 2$ mod 12, we would have obtained only $[2]_{12}$, not the entire solution. ■

As mentioned in Example 7, we would run into trouble if we tried to simplify the job of solving a congruence such as $2x \equiv 2$ mod 4 by canceling the 2 and solving $x \equiv 1$ mod 4 instead. When can we cancel? The corollary to the following proposition addresses the cancellation question.

PROPOSITION 7 Suppose $r \neq 0$ is an integer and suppose that $rx \equiv rb$ mod m. Then
$$x \equiv b \mod \left(\frac{m}{\gcd(m, r)}\right).$$

Proof. Exercise 12. (***Hint.*** You can use Euclid's Lemma to prove this.) ▲

COROLLARY 2 If $\gcd(r, m) = 1$ and $rx \equiv rb$ mod m, then $x \equiv b$ mod m.

EXAMPLE 8

i. $4 \cdot 2 \equiv 4 \cdot 5$ mod 6 but it is not the case that $2 \equiv 5$ mod 6. However, $\gcd(4, 6) = 2$, and it is true that $2 \equiv 5 \mod\left(\frac{6}{\gcd(4,6)}\right)$, that is, $2 \equiv 5$ mod 3.

ii. $6x \equiv 27$ mod 11, we could cancel 3 from both sides safely. The congruence $2x \equiv 9$ mod 11 has the same solution set because $\gcd(3, 11) = 1$. ■

When we solve a linear equation $ax = b, a \neq 0$, we multiply both sides by a^{-1} to obtain $x = a^{-1}b$. In what follows, we look to see how we can generalize this basic algebra step to linear congruences.

> **DEFINITION 2.** We call x a **multiplicative inverse of** a **modulo** m if $ax \equiv 1 \bmod m$.

Proposition 5 tells us that we can solve $ax \equiv 1 \bmod m$ if and only if $\gcd(a, m)$ divides 1, which means that a and m must be relatively prime. Thus we can find a multiplicative inverse to a modulo m if and only if $\gcd(a, m) = 1$.

EXAMPLE 9

Let's look at $3x \equiv 1 \bmod 7$. Notice that 3 and 7 are relatively prime and so 3 has a multiplicative inverse modulo 7. Since $x = 5$ is a solution to $3x \equiv 1 \bmod 7$, we see that 5 is a multiplicative inverse to 3 modulo 7. However, 4 has no inverse modulo 6 since 4 and 6 are not relatively prime. The congruence $4x \equiv 1 \bmod 6$ has no solution. ∎

If a has a multiplicative inverse modulo m, we can solve $ax \equiv b \bmod m$ by multiplying both sides by that inverse just as we would with a linear equation. For example, in Example 9 we noted that the inverse of 3 mod 7 is 5. Since 5 is relatively prime to 7, $3x \equiv 4 \bmod 7$ has the same set of solutions as $15x \equiv 20 \bmod 7$. But $15 \equiv 1 \bmod 7$ and $20 \equiv 6 \bmod 7$. So solving $15x \equiv 20 \bmod 7$ is equivalent to solving $x \equiv 6 \bmod 7$. Thus the solution set is $[6]_7$.

EXAMPLE 10

In this example, we will find the multiplicative inverse of 53 modulo 60 and use it to solve $53x \equiv 27 \bmod 60$. Since 53 and 60 are relatively prime, the inverse exists. To find it, we must solve $53x \equiv 1$ modulo 60. This is equivalent to solving the Diophantine equation $53x + 60q = 1$ for x and q. We need the value of x. By Euclid's Algorithm,

$$60 = 1 \cdot 53 + 7$$
$$53 = 7 \cdot 7 + 4$$
$$7 = 1 \cdot 4 + 3$$
$$4 = 1 \cdot 3 + 1.$$

We can solve for 1 as a weighted sum of 60 and 53 directly or call on the formulas of Section 1.3, Example 7. Either way, $x = 17$ and $q = -15$. So the inverse of 53 mod 60 is 17. To solve $53x \equiv 27 \bmod 60$, we multiply both sides by 17 to get $x \equiv 459 \bmod 60$. The solution to the congruence is $[459]_{60}$ or, using the principal representative of the equivalence class, i.e., the remainder of 459 after division by 60, the answer is $[39]_{60}$. ∎

Just as we can solve sets of linear equations simultaneously, we can solve sets of congruences simultaneously. For instance, one solution to the pair of congruences

$x \equiv 2$ mod 5 and $x \equiv 3$ mod 7 is $x = 17$. All the members of equivalence class $[17]_{35}$ are also solutions to the pair of congruences. For example, you can check that -18 is a member of $[17]_{35}$ and it is a solution to both congruences. There are no other solutions other than the members of $[17]_{35}$. The full story of solving simultaneous congruences is found in the **Chinese Remainder Theorem** that follows.

THEOREM 8 (*The Chinese Remainder Theorem*) Let $m_1, m_2, \ldots, m_n$ be a set of pairwise relatively prime integers that are greater than 1 and let $a_1, a_2, \ldots, a_n$ be any integers. The set of congruences

$$x \equiv a_1 \mod m_1$$
$$x \equiv a_2 \mod m_2$$
$$\vdots$$
$$x \equiv a_n \mod m_n$$

has a solution x_0. Furthermore, z is a solution if and only if $z \equiv x_0 \mod M$, where $M = m_1 m_2 \cdots m_n$.

Proof. For each i, $1 \leq i \leq n$, let $N_i = M/m_i$. Since N_i and m_i are relatively prime, we can find a solution y_i to the congruence $N_i y \equiv 1 \mod m_i$. Let

$$x_0 = a_1 N_1 y_1 + a_2 N_2 y_2 + \ldots + a_n N_n y_n.$$

Note that if $i \neq j$, then $N_j \equiv 0 \mod m_i$ so that for each i, $x_0 \equiv a_i \mod m_i$. An integer z is also a solution if and only if $z \equiv x_0 \mod m_i$ for each i, which is to say that $z - x_0$ is divisible by each m_i. Since the m_i are mutually relatively prime, $z \equiv x_0 \mod m_i$ if and only if $z \equiv x_0 \mod M$. ▲

EXAMPLE 11

We solve the system:
$$x \equiv 2 \mod 3$$
$$x \equiv 4 \mod 5$$
$$x \equiv 2 \mod 7.$$

Multiplying moduli, we have $M = 105$, and $N_1 = 35, N_2 = 21$, and $N_3 = 15$. First we solve the following congruences.

$$35y \equiv 1 \mod 3$$
$$21y \equiv 1 \mod 5$$
$$15y \equiv 1 \mod 7$$

Solutions are $y_1 = 2, y_2 = 1$, and $y_3 = 1$, respectively. Thus $x_0 = 2 \cdot 35y_1 + 4 \cdot 21y_2 + 2 \cdot 15y_3 = 254$. The solution set is $[254]_{105} = [44]_{105}$. ■

SUMMARY

In Chapter 1, we began our study of divisibility and remainders through the Division Algorithm. In this section, we continued that theme with the notion of **congruence**. We proved that two integers a and b are **congruent modulo** m or $a \equiv b$ **mod** m if and only if they have the same remainder after division by m (i.e., $a \bmod m = b \bmod m$). Congruence is an equivalence relation that sorts the integers into **congruence classes** of integers with the same remainders modulo m. After investigating the arithmetic of the congruences, we turned our attention to solving **linear congruences** of the form $ax \equiv b$ **mod** m. If x is a solution to $ax \equiv b$ mod m, then so are all the elements in $[x]_m$. So we can think of the object $[x]_m$ as one solution to $ax \equiv b$ mod m. We will pick up that line of thought in the next section when the congruences classes themselves are elements we add and subtract. Lastly, we learned how to solve systems of linear congruences through the **Chinese Remainder Theorem.**

2.1 Exercises

1. List seven members of each of the following congruence classes: $[13]_9$, $[3]_{10}$, $[4]_{11}$.
2. Prove part i of Theorem 2.
3. Prove part ii of Theorem 2.
4. Prove part iv of Theorem 2 by induction.
5. Compute 103^{45} mod 5 as in Example 2.
6. Compute 58^{29} mod 11 as in Example 2.
7. Suppose that $\gcd(a, 5) = 1$. Prove that $a^{4i} \equiv 1$ mod 5 for $i \in \mathbf{N}$.
   (*Hint.* Consider terminal digits.)
8. Reformulate Euler's Theorem and Fermat's Little Theorem in terms of congruences. (These are found in Section 1.7, Theorem 3, and Corollary 4.)
9. Find the solution set of the following congruences. Express your answers as unions of congruence classes.
   i. $2x \equiv 1$ mod 4
   ii. $4x \equiv 4$ mod 8
   iii. $6x \equiv 3$ mod 12
   iv. $8x \equiv 4$ mod 5
   v. $3x \equiv 7$ mod 11
10. Prove Proposition 6.
11. For each of the following congruences, find the number of distinct congruence classes in the solution set. Find the smallest nonnegative solution and use Proposition 6 to find all solutions x such that $0 \leq x < m$.
    i. $6x \equiv 9$ mod 15
    ii. $15x \equiv 20$ mod 35
    iii. $15x \equiv 16$ mod 19

12. Prove Proposition 7.
13. In each equation of the form $ax \equiv b \bmod m$, determine if a has a multiplicative inverse. If so, use it to solve the congruence.
    i. $7x \equiv 5 \bmod 11$
    ii. $8x \equiv 2 \bmod 6$
    iii. $5x \equiv 3 \bmod 12$
14. Use Euclid's Algorithm (in reverse) to help solve the following congruences.
    i. Solve $31x \equiv 1 \bmod 53$. Then solve $31x \equiv 23 \bmod 53$.
    ii. Solve $23x \equiv 1 \bmod 31$. Then solve $23x \equiv 15 \bmod 31$.
15. Compute the following.
    i. $2^{20} \bmod 7$
    ii. $5^{100} \bmod 13$
    iii. Find the multiplicative inverse of $2^{20} \bmod 7$.
16. Prove the following lemma and Wilson's Theorem.
    i. **Lemma**. If p is prime and $a^2 \equiv 1 \bmod p$, then $a \equiv 1 \bmod p$ or $a \equiv -1 \bmod p$.
    From the lemma, a is its own multiplicative inverse mod p if and only if $a \equiv 1 \bmod p$ or $a \equiv p - 1 \bmod p$.
    ii. **Wilson's Theorem**. Let p be a prime. Then $(p - 1)! \equiv -1 \bmod p$.
    Hint. Verify the Theorem for $p = 2$ and $p = 3$ directly. Let $p > 3$ and note that $p - 3$ is even. Each of the $p - 3$ numbers in the set $\{2, 3, \ldots, p - 2\}$ has an inverse distinct from itself in the set. Take the product of the members in the set.
17. Solve the following system of congruences.
    $x \equiv 1 \bmod 5$
    $x \equiv 2 \bmod 6$
    $x \equiv 3 \bmod 7$
18. Solve the following system of congruences.
    $x \equiv 3 \bmod 7$
    $x \equiv 7 \bmod 11$
    $x \equiv 2 \bmod 13$
19. *Ancient Chinese Problem.* A band of 17 pirates stole a sack of coins. When they tried to divide them equally among them, 3 coins remained. In the brawl that followed, one pirate was killed. Again, they tried to divide the coins equally among themselves only to find 10 coins remaining. Another pirate was killed! Now, upon dividing the coins, none was left. What is the minimum number of coins that the pirates could have been distributing?
20. Prove that the congruences $x \equiv b \bmod m$ and $x \equiv a \bmod n$ have a common solution if and only if $\gcd(m, n)$ divides $a - b$. Confirm that any two solutions are congruent mod $\text{lcm}(n, m)$.

To the Teacher

Modular arithmetic often enters the school curriculum as "clock arithmetic." The clocks can be a bit weird, having 5 or 7 or n hours (depending on the modulus in use) with high noon occurring at the 0th hour. And the clocks have just one hand! (See Figure 1.)

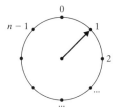

FIGURE 1

An experience with clock arithmetic should

- Reinforce students' number sense and number skills. The Division Algorithm is up front in all calculations. Students will begin to group integers into remainder classes. With a 7-hour clock for instance, where only nonnegative integers in the set $\{0, 1, \ldots, 6\}$ are available, the fact that 16 and -16 have remainders 2 and 5, respectively, when divided by 7 (rather that 2 and -2) is inescapable.
- Put familiar operations in a new light. For instance, to negate 5 mod 7 is now not simply placing a minus sign in front of it. A student must think about $5 + ? =$ noon. To subtract is to go counterclockwise: $3 - 6 = 4$ mod 7. Or a student can add 1 to 3 because 1 is the additive inverse of 6.
- Refocus the activity of solving equations. Finding the solution set to $ax = b$ in algebra-as-usual is a rote process for most students. The solution set always has just one element. The discovery that in clock arithmetic, there can be one, many, or no solutions focuses the student on the thought process behind solving equations rather than on rote calculation.

Since real clocks only come mod 12 or 24, why do clock arithmetic for other moduli? Students and teachers encounter (literally!) modular arithmetic every time a UPC symbol (bar code) is scanned.

FIGURE 2

Codes come in many varieties but the one given in Figure 2 is typical. The vertical bands and spaces represent the 12 digits that are written below. The fields (groups of digits) can represent the product, its manufacturer and its price. But the last digit is special. It's called a "check digit." It is there to help insure that the previous digits have been scanned correctly. The value of the check digit depends on the previous 11 digits as follows: Three times the sum of the digits in the previous odd positions added to the sum of the digits in the even positions must be a multiple of 10. In other words, choose the check digit d_{12} so that

$$3(d_1 + d_3 + d_5 + d_7 + d_9 + d_{11}) \\ + (d_2 + d_4 + d_6 + d_8 + d_{10} + d_{12}) \equiv 0 \mod 10. \tag{1}$$

Let's check that the check digit in Figure 2 satisfies formula (1):

$$3(0 + 2 + 4 + 6 + 8 + 0) + (1 + 3 + 5 + 7 + 9 + 5) = 90 \equiv 0 \mod 10.$$

If a digit is scanned incorrectly, it is unlikely that the sum computed on the left side of formula 1 would remain 0 mod 10. The error would be flagged and the product rescanned. There are other schemes, but most are variations of the one given here. For more information, check out the following article:

Gallian, J. A., "The Mathematics of Identification Numbers," *The College Mathematics Journal* 22 (1991) 194–202.

Tasks

1. Compute the check digit for the code 2-23431-25630-d_{12}.
2. There are scanning mistakes that the check digit will not pick up. Find two valid codes that differ only in two positions.
3. Inspect the bar code on your favorite cereal box. How does it compare to our examples?

2.2 WORKSHEET 2: DIVISIBILITY TESTS

Tricks of the Trade Before There Were Calculators

You can tell if a number is divisible by nine or three by looking at the sum of its digits. In this worksheet, you will prove why this familiar arithmetic trick actually works. You might also add a new trick or two to your bag of skills. Before you start working, prove the following helpful proposition and its corollary. They are based on the material found in Section 2.1.

PROPOSITION A For any $n \geq 0$, $10^n \equiv 1 \mod 9$.

COROLLARY 1 For any integer a and for $n \geq 0$, $a \cdot 10^n \equiv a \mod 9$.

Now for our bag of tricks. The first four tricks test for divisibility by 9, 3, 11 and 7 respectively. The last trick is really an error detection test. **Your task is to prove how and why each trick works.** Before you start working on the proofs of the four divisibility tests, use each test on the following numbers: $a = 682632027$ and $b = 184581410$. The answer is yes for a and no for b in all four cases.

Trick 1. A number x is divisible by 9 if and only if the sum of its digits is divisible by 9.

    Hint. To prove that Trick 1 works, start by writing x as $x_n 10^n + x_{n-1} 10^{n-1} + \ldots + x_0$ where x_0 is the digit in the one's place of x, x_1 is the digit in the tens place of x, etc. Then apply Proposition A and Corollary B.

Trick 2. A number x is divisible by 3 if and only if the sum of its digits is divisible by 3.

The next two tricks may not be familiar.

Trick 3. A number x is divisible by 11 if and only if 11 divides the alternating sum of the digits of x. For example, $715 = 65 \cdot 11$ and $7 - 1 + 5 = 11$, and 11 is divisible by 11.

    Hint. Note that $10 \equiv -1 \mod 11$ and so $10^n \equiv (-1)^n \mod 11$. Check it out!

Trick 4. To determine if a number x is divisible by seven, remove the last digit, double it and subtract it from the truncated original number. Iterate this procedure until you are left with a single or double digit number y. The original number x is divisible by 7 if and only if this last number y is divisible by 7.

As an example, we'll test 7783 for divisibility by 7. Twice the last digit is 6, which will be subtracted from 778, the original number with the last digit removed. Then we will iterate.

$$778 - 6 = 772$$
$$77 - 4 = 73$$
$$7 - 6 = 1$$

Since 1 is not divisible by 7, then neither is 7783.

    Hint. To show that this trick works, first uncover the magic of the trick. When you subtract 6 from 778, you are really subtracting 63 from 7783, and then deleting the terminal zero. In general, you are really subtracting a multiple of 7 from the original number and ignoring a terminal zero.

Trick 5. (Casting out nines) This test checks for errors in the addition or multiplication of two integers x and y. Suppose you think that $x + y = z$ or that $xy = w$, but you want to check. If you are correct, the sum of the digits of x plus the sum of the digits of y must equal to the sum of the digits of z. Similarly, to check w, multiply the sum of the digits of x by the sum of the digits of y and call the product q. The sum of the digits of q must equal the sum of the digits of w.

As an example, let $x = 234$ and $y = 54$ so that $x + y = 288$ and $xy = 12636$. For addition, we check that $(2 + 3 + 4) + (5 + 4) = (2 + 8 + 8)$. For multiplication, we check $(2 + 3 + 4)(5 + 4) = 9 \cdot 9 = 81$. The sum of the digits of 81 is equal to 9. The sum of the digits of $12636 = (1 + 2 + 6 + 3 + 6) = 18$ and $1 + 8 = 9$. So this test detects no errors in our computation of $x + y$ and xy.

Reminder: Your task is to prove that Trick 5 works.

To conclude, we pose a few additional tasks that extend and complement your work.

1. Invent a test for divisibility by four in base five arithmetic. Similarly, invent a test for divisibility by eleven in base twelve arithmetic. Can you generalize?
2. Develop a test for divisibility by six in base five arithmetic. Remember, in base 5, six is represented by "11."
3. Does "casting out nines" detect all errors in addition and multiplication? Explain exactly how the test can mislead you.
4. Where do the nines go when "casting out nines"?
   Hint. $324 = 3 \cdot 99 + 3 + 2 \cdot 9 + 2 + 4$.

The material in this worksheet covers mathematics you might have learned in an early grade, but it covers it from an advanced point of view. You have discovered not just how these tests work but why. If you are presenting this worksheet as an oral presentation, think about who your audience is. In a presentation to your peers, you would review familiar tests briefly and concentrate on bringing your audience to an understanding of the underlying reasons for their validity. If the audience is a tenth grade class, you would emphasize proof less and skill and discovery more. But don't neglect the "why's" behind the "how's." What did you enjoy discovering? Enable your audience to enjoy the same. If this worksheet is an expository writing exercise, pretend you are the author of the ideal textbook, and write it up as a chapter in a textbook that you want to read! Be clear in exposition, generous with examples, and efficient with proofs. Keep your audience in mind at all times!

2.3 WORKSHEET 3: PUBLIC KEY CRYPTOGRAPHY

A current, widely used scheme for encrypting text and data is known as RSA Public Key Cryptography. It was developed in the 1970s by mathematicians R. L. Rivest, A. Shamir, and L. Adelman.[1] It is based on very simple number theory. Its security is based upon the fact that the task of factoring very large numbers is very time-consuming even for the fastest computers. The central idea is that the number schemes (keys) needed to encode messages are publicly known: Anyone can encrypt and send a message. But only the receiver who generated the keys knows how to decode because, in practice, decoding requires the factorization of very large numbers. Here are the basics.

Part 1: Inventing the public key

1. Find two large primes p and q. (It's a good idea to have them bigger than the digits to be encoded.)
2. Let $n = pq$ and $m = \text{lcm}(p - 1, q - 1)$.
3. Find k such that $1 < k < m$ and $\gcd(m, k) = 1$.

---
[1] Rivest, R. L., Shamir, A. and Adelman, L. *On Digital Signatures and Public Key Cryptosystems*, MIT Laboratory for Computer Science Technical Memorandum 82, April 1977.

The pair n and k is the "public key" because they are not secret. Next we show how to encode messages with n and k. Anyone can encode a message. We will assume that the messages to be encoded have been transformed into numerical data. For instance, a primitive scheme would be to assign a numerical value to each symbol in the alphabet, perhaps expanded to include numbers, punctuation, etc.

Part 2: Encoding a number M (M is the numerical value of a message unit.)

1. Suppose that M is the numerical value of the message to be encoded. Raise M to the k power and set $r = M^k \mod n$.
2. Your encoded message is r.

Part 3: Decoding

1. Find j such that $jk \equiv 1 \mod m$.
2. Raise r to the jth power. Your message unit: $M = r^j \mod n$.

Sample: (In practice, the primes should be much, much larger.) Let's try an RSA code with $p = 17$, $q = 13$ so that $n = 221$. Then $m = \text{lcm}(12, 16) = 48$. We pick $k = 19$. Suppose our message unit is $M = 25$. Then our encoded message is

$$r = 25^{19} \mod 221 = 155.$$

To decode (i.e., get 25 back), first solve $19j \equiv 1 \mod 48$. We get $j = 43$. Then $155^{43} \mod 221 = 25$. We decoded correctly!

Here are some handy facts for Maple users.
 First load the number theory package with the command "with(numtheory)."
 To find $x \mod m$, use the command "x mod m" in Maple.
 To solve $ax \equiv b \mod m$, use "msolve(a*x = b, m)".
 To find primes, use ithprime(n). This command returns the nth prime.

To encode a message, we must convert text to numbers. As mentioned, the easiest way is to assign a number to each letter. We'll omit all punctuation. You can extend the code for punctuation and spacing.

A = 1 B = 2 C = 3 D = 4 E = 5 F = 6 G = 7 H = 8 I = 9 J = 10
K = 11 L = 12 M = 13 N = 14 O = 15 P = 16 Q = 17 R = 18 S = 19 T = 20
U = 21 V = 22 W = 23 X = 24 Y = 25 Z = 26 blank = 27

The command THINK becomes 20 8 9 14 11. With n, m, k and j as given in the sample code, THINK encoded becomes

$$214 \quad 70 \quad 100 \quad 92 \quad 158$$

because $20^{19} \mod 221 = 214$ and $8^{19} \mod 221 = 70$, etc.

Task 1

Check that 214 70 100 92 158 **decodes** properly as THINK.

Task 2

Still using the sample code, decode 59 125 92 30 14 70 125 181 16

Task 3

Use the sample code to encode a short secret message like "MATH IS FUN" and have someone in the class decode it.

Task 4

Use the public key: $n = 7957, k = 83$. Encode the message, "SO WHAT." Decode: 7352, 4585, 6397, 2984, 4585, 2852, 6674.

In practice, groups of letters—say sequences of 2, 3 or more letters—are assigned numerical values so as to avoid statistical deduction as to what the code numbers stand for. The RSA coding looks pretty easy to crack! All we need to do is factor n into p and q, compute m and solve for j. That's correct! The problem is factoring n. In practice it is not hard to generate primes with one hundred or more digits, but it takes time on the order of decades to factor composite numbers that have such large primes as their factors. By that time, the message is out of date!

Task 5

Public Key $n = 168149075693$ and $k = 569$. Decode:

   83568369728, 63187316702, 66500245442, 167425612598, 84479136007, 1

Devise what you think is an unbreakable code. Challenge the class to break it.

Task 6

(Why RSA coding works)
Prove that the decoding procedure works, namely that $(M^k)^j = M$ mod n.
   Hint. Let $d = \gcd((p-1)(q-1))$. Write $kj = 1 + s(p-1)(q-1)/d$. Use Fermat's Little Theorem.

MORE CRYPTIC: Eliminate spaces in your messages. (Spies don't do grammar!) Chunk the letters in groups of two for starters. For instance, EAT AT JO's becomes EA TA TJ OS = 0501 2001 2010 1519. Now use the algorithm on these larger numbers. In practice, much larger chunks are used. Note that p and q must be large enough to accommodate these large numbers with no redundancy. For these bigger numbers, Maple needs the "big number" exponential operation, &^ instead of ^.

Task 7

Develop a code that accommodates two-letter chunks.

2.4 THE RING Z_m

In this section, we take another big step in the direction of abstraction. We distill our notions of modular arithmetic and invent a new context in which to do arithmetic: We will add and subtract congruence classes. The congruence class, which we usually think of as an infinite set of different numbers, becomes a single object that can be added to and multiplied by similar objects. Most of the properties of arithmetic in this new context will be familiar. For instance, addition and multiplication are each commutative and associative, and the distributive law holds. But some facts will be surprising. For instance, if $x + x = y + y$, it is not always the case that $x = y$. This section anticipates and links us to the next section in which we begin our investigation of the abstract algebraic structure called a "ring."

We know that for a fixed integer $m > 0$, each integer $x \in \mathbf{Z}$ is congruent mod m to exactly one of the numbers in the set $0, 1, \ldots, m - 1$ because this set forms the complete set of possible remainders (or **residues** as we sometimes say) after division by m. Thus the entire set of congruence classes modulo m can be listed as $\{[0]_m, [1]_m, [2]_m, [3]_m, \ldots, [m-1]_m\}$.

DEFINITION 1. Let $m \in \mathbf{Z}$ be greater than 0. The set of all congruences classes modulo m, $\mathbf{Z_m}$ is the set $\{[0]_m, [1]_m, [2]_m, [3]_m, \ldots, [m-1]_m\}$.

EXAMPLE 1

i. $\mathbf{Z_5} = \{[0]_5, [1]_5, [2]_5, [3]_5, [4]_5\}$
ii. $\mathbf{Z_3} = \{[0]_3, [1]_3, [2]_3\}$
iii. $\mathbf{Z_7} = \{[0]_7, [1]_7, [2]_7, [3]_7, [4]_7, [5]_7, [6]_7\}$ ∎

We could also list the congruence classes of $\mathbf{Z_5}$ as $\{[5]_5, [11]_5, [-103]_5, [33]_5, [114]_5\}$ because $[0]_5 = [5]_5$ and $[1]_5 = [11]_5$, etc. But most often a congruence class will be indicated by its unique principal representative that lies between 0 and $m - 1$. For example, $\mathbf{Z_5} = \{0, 1, 2, 3, 4\}$.

To give $\mathbf{Z_m}$ an algebraic structure, we need to say how to add and multiply its elements. The operations $+$ and $\cdot$ are defined on $\mathbf{Z_m}$ as follows.

Addition: $[a]_m + [b]_m = [a+b]_m$

Multiplication: $[a]_m \cdot [b]_m = [a \cdot b]_m$

This means that to add $[a]_m + [b]_m$, we pick a number from each congruence class, say a and b. We add these numbers. The sum is the congruence class of $a + b$. Multiplication is similar.

EXAMPLE 2

i. Let $m = 7$. Then $[5]_7 + [4]_7 = [5+4]_7 = [9]_7 = [2]_7$. Also, $[5]_7 \cdot [4]_7 = [5 \cdot 4]_7 = [20]_7 = [6]_7$.
ii. Let $m = 8$. Then $[5]_8 + [4]_8 = [5+4]_8 = [9]_8 = [1]_8$. Also, $[5]_8 \cdot [4]_8 = [5 \cdot 4]_8 = [20]_8 = [4]_8$. ∎

To add $[5]_7$ and $[4]_7$, we could have picked the pair $12 \in [5]_7$ and $25 \in [4]_7$ and computed $[5]_7 + [4]_7$ as $[12 + 25]_7$ instead. Notice that the outcome is the same: $[5 + 4]_7 = [9]_7 = [2]_7$ and similarly $[12 + 25]_7 = [37]_7 = [2]_7$. The operations of addition and multiplication are said to be **well-defined**, which means that no matter what two representatives $x \in [a]_m$ and $y \in [b]_m$ we choose to add (or multiply), the resulting sum (or product) is in the same equivalence class as $a + b$ (or ab). This is formally proved in the next proposition.

PROPOSITION 1 The operations $+$ and $\cdot$ defined on Z_m are well-defined.

Proof. Suppose that $[a]_m = [x]_m$ and that $[b]_m = [y]_m$. To show that addition is well-defined, we must show that $[a + b]_m = [x + y]_m$. Since $[a]_m = [x]_m$ and $[b]_m = [y]_m$, we know that $a \equiv x \bmod m$ and $b \equiv y \bmod m$. By Theorem 2, part ii, of Section 2.1, we have $(a + b) \equiv (x + y) \bmod m$. Thus $[a + b]_m = [x + y]_m$. The proof that multiplication is well-defined is similarly based on Theorem 2 of Section 2.1. ▲

The operations $+$ and $\cdot$ on Z_m inherit their basic properties such as commutativity from addition and multiplication on Z. For instance, the computation $[5]_7 \cdot [4]_7 = [5 \cdot 4]_7 = [4 \cdot 5]_7 = [4]_7 \cdot [5]_7$ verifies that multiplication is commutative.

PROPOSITION 2 Let $m > 0$. For any integers $a, b, c,$ and d the following properties hold.

   i. (Commutativity of addition): $[a]_m + [b]_m = [b]_m + [a]_m$
  ii. (Commutativity of multiplication): $[a]_m \cdot [b]_m = [b]_m \cdot [a]_m$
 iii. (Associativity of addition): $([a]_m + [b]_m) + [c]_m = [a]_m + ([b]_m + [c]_m)$
  iv. (Associativity of multiplication): $([a]_m \cdot [b]_m) \cdot [c]_m = [a]_m \cdot ([b]_m \cdot [c]_m)$
   v. (The Distributive Law): $[a]_m([b]_m + [c]_m) = [a]_m \cdot [b]_m + [a]_m \cdot [c]_m$

Proof. The fact that $[a]_m \cdot [b]_m = [ab]_m = [ba]_m = [b]_m \cdot [a]_m$ shows that multiplication is commutative (part ii). The proofs of the other parts are left as exercises. ▲

EXAMPLE 3

Here are examples that illustrate parts iii and v of Proposition 2.

   iii. $([6]_{11} + [10]_{11}) + [7]_{11} = [16]_{11} + [7]_{11} = [23]_{11} = [1]_{11}$ and
        $[6]_{11} + ([10]_{11} + [7]_{11}) = [6]_{11} + [17]_{11} = [23]_{11} = [1]_{11}$
    v. $[3]_5([4]_5 + [2]_5) = [3]_5 \cdot [6]_5 = [18]_5 = [3]_5$ and
        $[3]_5 \cdot [4]_5 + [3]_5 \cdot [2]_5 = [12]_5 + [6]_5 = [18]_5 = [3]_5$ ■

Suppose that m is fixed and that we agree that we will always denote a congruence class by its principal representative. Then our notation becomes much simpler if we drop the subscripts and omit the square brackets. The set Z_m is the set $\{0, 1, \ldots, m - 1\}$. Addition and multiplication can also be defined more simply as follows. For a

and b in $\mathbf{Z}_m$,
$$a + b = (a+b) \bmod m,$$
$$a \cdot b = (a \cdot b) \bmod m.$$

EXAMPLE 4

i. In $\mathbf{Z}_7$, we have $5 \cdot 6 = 2$ because $30 \bmod 7 = 2$. Also, $4^3 = 1$ because $64 \bmod 7 = 1$.

ii. In $\mathbf{Z}_{13}$, we have $5 \cdot 6 = 4$ because $30 \bmod 13 = 4$. Also $4^3 = 12$ because $64 \bmod 13 = 12$. ∎

Let's look at the addition and multiplication tables for $\mathbf{Z}_m$ when $m = 5$ given in Figure 1. We will use the abbreviated notation with subscripts and brackets omitted.

+	0	1	2	3	4		·	0	1	2	3	4
0	0	1	2	3	4		0	0	0	0	0	0
1	1	2	3	4	0		1	0	1	2	3	4
2	2	3	4	0	1		2	0	2	4	1	3
3	3	4	0	1	2		3	0	3	1	4	2
4	4	0	1	2	3		4	0	4	3	2	1

FIGURE 1 Addition and multiplication in $\mathbf{Z}_5$

Some observations about $\mathbf{Z}_5$:

1. 0 is the additive identity since $0 + x = x$ for all $x \in \mathbf{Z}_5$. (This holds in any $\mathbf{Z}_m$.)
2. 1 is the multiplicative identity since $1 \cdot x = x$ for all $x \in \mathbf{Z}_5$. (This holds in any $\mathbf{Z}_m$.)
3. For each x there is a y such that $x + y = 0$. We call y the **additive inverse of x** and denote it by the symbol $-x$. For instance, in $\mathbf{Z}_5$, 3 is the additive inverse of 2 because $2 + 3 = 0$. So in $\mathbf{Z}_5$, $3 = -2$. Similarly, $4 = -1$ in $\mathbf{Z}_5$ because $4 + 1 = 0$. More generally, the additive inverse of any x in $\mathbf{Z}_m$ is $(m - x) \bmod m$ since $(x + (m - x)) \bmod m = m \bmod m = 0$.
4. If $x \cdot y = 1$ in $\mathbf{Z}_m$, then the element y is called the **multiplicative inverse** of x. We denote y by the symbol x^{-1}. In $\mathbf{Z}_5$, every nonzero element has a multiplicative inverse. For instance, $2 \cdot 3 = 1$ and so $2^{-1} = 3$. (Also $3^{-1} = 2$). The multiplication table for $\mathbf{Z}_6$ is given in Figure 2. Again, brackets and subscripts are omitted. It shows that it is *not* always the case that a nonzero element has a multiplicative inverse.

·	0	1	2	3	4	5
0	0	0	0	0	0	0
1	0	1	2	3	4	5
2	0	2	4	0	2	4
3	0	3	0	3	0	3
4	0	4	2	0	4	2
5	0	5	4	3	2	1

FIGURE 2 The multiplication table for Z_6

Multiplying in Z_6 is quite different from multiplying in Z_5. The nonzero elements 2, 3, and 4 do not have multiplicative inverses. Further, $2 \cdot 3 = 0$ even though neither factor is 0. Similarly, $4 \cdot 3 = 0$. A nonzero element x in Z_m is called a **zero divisor** if there is a nonzero value y such that $x \cdot y = 0$. The elements 2, 3, and 4 are all zero divisors in Z_6. An element a in Z_m that has a multiplicative inverse is called a **unit**. For a to be a unit we must find x in Z_m such that $a \cdot x = 1$. In terms of congruences, we must be able to solve for x in $ax \equiv 1 \mod m$. We know from Proposition 5 of Section 2.1 that this can be done if and only if $\gcd(a, m) = 1$. Let's check in Z_6. The $\gcd(5, 6) = 1$ and 5 is indeed a unit since $5 \cdot 5 \mod 5 = 1$. But $\gcd(4, 6) = 2$ and 4 fails to have an inverse.

EXAMPLE 5

i. In Z_8, the units are 1, 3, 5, and 7. Each of these numbers is relatively prime to 8 and each has a multiplicative inverse: $1^{-1} = 1, 3^{-1} = 3, 5^{-1} = 5$, and $7^{-1} = 7$.
ii. The zero divisors in Z_8 are 2, 4 and 6 since $2 \cdot 4 = 0$ and $6 \cdot 4 = 0$.
iii. In Z_7, the units are 1, 2, 3, 4, 5, and 6 and there are no zero divisors. ∎

Additive and multiplicative inverses are the tools for solving for x in equations such as $a + x = b$ and $ax = b$. In other words, we can now do algebra in Z_m as demonstrated in the next example.

EXAMPLE 6

i. To solve the equation $4 + x = 2$ in Z_6, we add -4 to both sides. In Z_6, $-4 = 2$. So we add 2 to both sides.

$$2 + (4 + x) = 2 + 2$$
$$(2 + 4) + x = 4$$
$$0 + x = 4$$
$$x = 4$$

The solution is $x = 4$. In what follows, we will be a bit less detailed.

ii. To solve $5x = 2$ in Z_6, we multiply both sides by $5^{-1} = 5$ to obtain $x = 5 \cdot 2 \mod 6 = 4$.

iii. The equation $2x = 1$ has no solution in Z_6. You can check the table in Figure 2 to see that 1 does not appear in the row or column labeled 2. But better, notice that solving $2x = 1$ is the same as solving the congruence $2x \equiv 1 \bmod 6$. Since $2 = \gcd(2, 6)$, which does not divide 1, there is no solution. Note that 2 does not have a multiplicative inverse in Z_6. But that alone is not enough to preclude a solution to $2x = b$ in Z_6 as the next computation shows.

iv. The equation $2x = 2$ has more than one solution in Z_6 but we cannot obtain them through the use of a multiplicative inverse since 2 does not have a multiplicative inverse. So instead we must solve the congruence $2x \equiv 2 \bmod 6$. The solution set of the congruence is $[1]_6 \cup [4]_6$. So the solution set in Z_6 is $\{1, 4\}$. Notice that in Z_6 we cannot cancel the 2 on both sides of the equation $2x = 2$. If we did, we would miss the solution $x = 4$. ∎

SUMMARY

In this section, we saw how to make the set $Z_m = \{[0]_m, [1]_m, [2]_m, [3]_m, \ldots, [m-1]_m\}$ into an algebraic structure. The operations of **addition and multiplication on Z_m**, defined as $[a]_m + [b]_m = [a+b]_m$ and $[a]_m \cdot [b]_m = [ab]_m$, respectively, are both **well-defined**. The arithmetic that we can do with these operations has many of the familiar properties of ordinary arithmetic. For instance, the distributive law holds. The set Z_m joins the list of contexts where the operations of addition and multiplication are natural operations that have similar rules. Such structures all come under the umbrella term of "ring," which is the topic of the next section. However, there were some surprises in the ring Z_m. For instance, we found that some elements were **zero divisors** and some were **units** and that cancellation was not always possible.

2.4 Exercises

1. Write out the addition and multiplication tables for Z_3 and Z_4.
2. Prove that multiplication in Z_m is well-defined.
3. Perform the indicated arithmetic.
    i. $6 + 7$ in Z_{13}
    ii. $6 \cdot 7$ in Z_{13}
    iii. $4 + 8$ in Z_{10}
    iv. $4 \cdot 8$ in Z_{10}
4. Prove that the distributive law holds on Z_m.
5. Solve for x in the following expressions.
    i. $3 + x = 2$ in Z_5
    ii. $3 \cdot x = 4$ in Z_5
    iii. $12 + x = 3$ in Z_{13}
    iv. $12 \cdot x = 3$ in Z_{13}
    v. $8 + x = 4$ in Z_{12}
    vi. $4 \cdot x = 8$ in Z_{12}
6. Find the units in Z_{12} and find their multiplicative inverses. Do the same for Z_9 and Z_{10} and Z_{11}.

7. Let p be prime. Prove that every nonzero element in Z_p has a multiplicative inverse.
8. Prove that a is a unit in Z_m if and only if a and m are relatively prime.
9. Prove that a nonzero element a in Z_m is a zero divisor if and only if a and m are **not** relatively prime. (Together, Exercises 8 and 9 prove that every nonzero element in Z_n is either a unit or a zero divisor.)
10. Let U_m denote the set of units mod m. How many elements are there in U_m? Write out the multiplication tables for U_{12} and U_8.
11. Can you "cancel" in Z_6? That is, if $xy = xz$, is $y = z$?
12. Find a value of m and values of x and y in Z_m such that $x + x = y + y$ but $x \neq y$.
13. Find a value of m and values of x and y in Z_m such that $x + x + x = y + y + y$ but $x \neq y$.
14. Find an element x of Z_7 such that every nonzero element in Z_7 can be written as $x^n = x \cdot x \cdot x \cdots x$ for some value of n. Can you do the same for Z_8? Z_{10}? Z_{11}?

To the Teacher

The jump from clock arithmetic to the addition and multiplication tables of mod n arithmetic is not a big leap for the high school student. But producing such tables and inspecting them can provide very valuable lessons about what is the same as and what is different from ordinary integer arithmetic.

In Z and in Z_n,

- The operations are both commutative and associative.
- Addition and multiplication both have identity elements.
- All elements have additive inverses.
- We can always solve $a + x = b$ for x.

But in Z_n,

- $1 + 1 + \ldots + 1$ (n times) equals 0.
- Sometimes $ab = 0$ even though neither a nor b is zero.
- Some nonzero numbers do not have multiplicative inverses.
- Sometimes you can't cancel. For example, you can't cancel the 2 from the expression $2 \cdot 4 = 2 \cdot 1$ in Z_6.
- Sometimes you can't solve $ax = b$.

Not only do these observations prepare you and the student for more abstract mathematics such as what follows in the next chapters, they solidify the mathematical facts that underlie their everyday arithmetic. For instance, the symbol 3^{-1} in Z_n points to the symbol x in Z_n (if it exists) such that $x \cdot 3 = 1$. (It does not just mean the ratio 1/3.) In $Z_7, 3^{-1} = 5$ because $3 \cdot 5 = 1$. Similarly, $2 \div 3 = 2 \cdot 3^{-1} = 2 \cdot 5 = 3$ in Z_7. A symbol like x^{-1} asks for a number with a certain property that does not necessarily exist in a given number system. The questions, "What does $\frac{1}{\pi}$ mean?"

and "Does $\frac{1}{\pi}$ exist within the set of real numbers?" become two distinct questions. The answer to the latter is not easy to come by! It's not just a matter of notation.

Tasks

1. Develop a lesson in which students discover that things go wrong when you try to put the elements of Z_n in size order with the relation $<$. (You might want to invoke rocks, scissors, and paper.)
2. Explain why questions, "What does $\frac{1}{\pi}$ mean?" and "Does $\frac{1}{\pi}$ exist within the set of real numbers?" are two distinct questions.

2.5 RINGS AND FIELDS

The familiar settings for arithmetic such as Z, Q, R, and Z_m all come equipped with two operations, addition and multiplication, that are related through a distributive law. They are examples of the algebraic structure called a **ring**. In this section, we explore the abstract properties of rings. Along the way, we will find new ways to explain arithmetic facts that you might just take for granted, like why the product of two negative numbers must be positive. We start with the definition of a ring. The definition distills the properties that are common to doing arithmetic in a great variety of mathematical settings.

DEFINITION 1. A **ring** is a set R equipped with two binary operations,[1] here denoted by $+$ and $*$, that have the following properties.

1. $(a+b)+c = a+(b+c)$ for all a, b and c in R (addition is associative).
2. $a+b = b+a$ for all a and b in R (addition is commutative).
3. There is an element $0 \in R$ such that $0+x = x+0 = x$ for all $x \in R$.
4. For each element $x \in R$, there is a unique element $y \in R$ such that $x+y = y+x = 0$. (We denote y by $-x$.)
5. $(a*b)*c = a*(b*c)$ for all a, b, and c in R (multiplication is associative).
6. (The distributive law) $a*(b+c) = a*b + a*c$ and $(b+c)*a = b*a + c*a$ for all $a, b,$ and c in R.

A **ring with unity** satisfies the following additional property.

7. There is an element $1 \in R$, $1 \neq 0$, such that $1*x = x*1 = x$ for all $x \in R$.

We shall usually write ab rather that $a*b$. The operation of **subtraction** $x - y$ is defined by
$$x - y = x + (-y).$$

---
[1]The appendix at the end of this text reviews binary operations.

Multiplication is **not** necessarily commutative. When it is, we call R a **commutative ring**. The multiplicative identity element 1 required in Property 7 is called the **unity** of the ring R. Some rings do not have a unity as we shall see in the examples that follow.

EXAMPLES

1. Z, R, Q, and C are all commutative rings with unity under their usual operations of addition and multiplication.
2. The set Z_m is a commutative ring with unity under addition and multiplication modulo m for each $m > 1$.
3. The set of even integers is defined as $2Z = \{2x : x \in Z\}$. The sum of two even integers is even, and the product of two even integers is even. So addition and multiplication are both closed on the set $2Z$ and hence they are binary operations on $2Z$. Note that 0 is an even integer because $0 = 2 \cdot 0$. Since properties 1 through 6 listed in Definition 1 apply to *all* integers in Z, they hold when applied to the even integers. Thus $2Z$ is a ring. However, since 1 is not an even integer, property 7 of Definition 1 does not apply. The ring $2Z$ fails to have a multiplicative identity element. We say that $2Z$ is a **ring without unity**. More generally, for each $n \in Z$, the set $nZ = \{nx : x \in Z\}$ is a ring. If $n \neq \pm 1$, then nZ is a ring *without* unity.

   Note: Some authors require a ring to have a unity. For them, the set nZ would not be considered a ring.
4. The set of all polynomials with rational coefficients, denoted by $Q[x]$, is a ring when we add and multiply polynomials in the standard manner. The ring $Q[x]$ will be the focus of many of our later investigations. We will generalize to consider polynomials with coefficients in other rings. For instance, $C[x]$ denotes the ring of polynomials with complex coefficients. Similarly, $Z_5[x]$ denotes the ring of polynomials with coefficients in Z_5. In that ring, we add and multiply coefficients mod 5. For instance,

$$(2x + 2) + (3x + 2) = (5 \bmod 5)x + 4 \bmod 5 = 0x + 4 = 4$$

$$(2x + 2)(3x + 2) = (6 \bmod 5)x^2 + (10 \bmod 5)x + (4 \bmod 5)$$

$$= x^2 + 0x + 4 = x^2 + 4.$$

5. Let $M_{2,2}$ be the set of 2×2 matrices with real entries, i.e., entries from the ring R. It is a ring when addition and matrix multiplication are defined in the standard manner, which is as follows:

$$\begin{bmatrix} a & b \\ c & d \end{bmatrix} + \begin{bmatrix} e & f \\ g & h \end{bmatrix} = \begin{bmatrix} a+e & b+f \\ c+g & d+h \end{bmatrix}$$

$$\begin{bmatrix} a & b \\ c & d \end{bmatrix} \times \begin{bmatrix} e & f \\ g & h \end{bmatrix} = \begin{bmatrix} ae+bg & af+bh \\ ce+dg & cf+dh \end{bmatrix}.$$

The additive identity of $M_{2,2}$ is the matrix $\begin{bmatrix} 0 & 0 \\ 0 & 0 \end{bmatrix}$. Its unity is the identity matrix $I = \begin{bmatrix} 1 & 0 \\ 0 & 1 \end{bmatrix}$ because $I \times A = A \times I$ for every matrix A in $M_{2,2}$. Matrix addition and multiplication are both associative and the distributive law holds. However, matrix multiplication is not commutative. For example,

$$\begin{bmatrix} 1 & 2 \\ 5 & 2 \end{bmatrix} \times \begin{bmatrix} 1 & -4 \\ -1 & 2 \end{bmatrix} = \begin{bmatrix} -1 & 0 \\ 3 & -16 \end{bmatrix} \text{ but } \begin{bmatrix} 1 & -4 \\ -1 & 2 \end{bmatrix} \times \begin{bmatrix} 1 & 2 \\ 5 & 2 \end{bmatrix} = \begin{bmatrix} -19 & -6 \\ 9 & 2 \end{bmatrix}. \blacksquare$$

In the following proposition, we prove some common arithmetic facts about how addition and multiplication are related. It is interesting to see how to deduce these facts from the abstract properties of rings rather than from the elementary cookie-counting arguments that we usually use to explain the arithmetic of the natural numbers.

PROPOSITION 1 Let a and b be elements of a ring R.
  i. $a0 = 0$ and $0a = 0$
 ii. $a(-b) = -ab$ and $(-a)b = -ab$
iii. $(-a)(-b) = ab$
 iv. $(-1)a = -a$ (in rings with unity)

Proof.
  i. $a0 = a(0+0) = a0 + a0$ by the distributive law. Adding $-a0$ to both sides of the equation $a0 = a0 + a0$, we obtain $0 = a0$. A similar argument shows that $0a = 0$.
 ii. By part i, we have $0 = a(b + (-b)) = ab + a(-b)$. Adding $-ab$ to both sides, we obtain: $(-ab) = a(-b)$. Similarly, $(-a)b = -ab$.

We leave the proofs of parts iii and iv as exercises. ▲

A word about notation:

If x is any element of any ring R, the sum $(x + x + \ldots + x)$ is denoted by nx where n is the number of summands. We use this notation even though n itself may not be an element of the ring R. For instance, if $R = M_{2,2}$ and $x = \begin{bmatrix} 1 & -1 \\ 0 & 3.5 \end{bmatrix}$, then $2x = \begin{bmatrix} 1 & -1 \\ 0 & 3.5 \end{bmatrix} + \begin{bmatrix} 1 & -1 \\ 0 & 3.5 \end{bmatrix} = \begin{bmatrix} 2 & -2 \\ 0 & 7 \end{bmatrix}$. Thus $2x$ is an element of $M_{2,2}$ even though the number 2 is not. Similarly, we use the notation x^n to denote the product $x \cdot x \cdots x$, where n is the number of factors. For example, if $x = \begin{bmatrix} 1 & -1 \\ 0 & 3.5 \end{bmatrix}$ in $M_{2,2}$, then $x^2 = \begin{bmatrix} 1 & -1 \\ 0 & 3.5 \end{bmatrix} \times \begin{bmatrix} 1 & -1 \\ 0 & 3.5 \end{bmatrix} = \begin{bmatrix} 1 & -4.5 \\ 0 & 12.25 \end{bmatrix}$. The usual rules of exponents hold. For instance, $x^n x^m = x^{m+n}$.

In Proposition 1, we established some standard arithmetic facts for rings. However, some procedures and assumptions that are valid for $\mathbf{Z}$ and $\mathbf{Q}$ do not hold in general. One example is cancellation. In $\mathbf{Z}_6$, for instance, we have $2 \cdot 4 = 2 \cdot 1$ but $4 \neq 1$. We cannot "cancel" the 2 and remain with a true equation. Also, in $\mathbf{Z}$ we know that the product of two nonzero numbers is not zero. However, that assertion does not hold for all rings. In $\mathbf{Z}_6$, we have $2 \cdot 3 = 0$ but neither 2 nor 3 is equal to zero in $\mathbf{Z}_6$.

> **DEFINITION 2.** Let R be a ring. Suppose that a and b are nonzero elements in R such that either $ba = 0$ or $ab = 0$. Then a and b are called **zero divisors**.

EXAMPLES

6. The zero divisors in $\mathbf{Z}_{12}$ are 2, 3, 4, 6, 8, 9, and 10. For instance, since $9 \cdot 4 = 0$ and neither 9 nor 4 is zero, both 9 and 4 are zero divisors.
7. In $\mathbf{Z}_4[x]$, the nonzero polynomial $2x + 2$ is a zero divisor because $(2x + 2)^2 = 4x^2 + 8x + 4 = 0$.
8. In $\mathbf{M}_{2,2}$, let $a = \begin{bmatrix} 1 & 2 \\ 1 & 2 \end{bmatrix}$ and $b = \begin{bmatrix} 2 & -4 \\ -1 & 2 \end{bmatrix}$. Because $ab = \begin{bmatrix} 0 & 0 \\ 0 & 0 \end{bmatrix}$, both a and b are zero divisors. Note that $ba = \begin{bmatrix} -2 & -4 \\ 1 & 2 \end{bmatrix}$. In a noncommutative ring it may be the case that $ab = 0$, but $ba \neq 0$. Nonetheless, a and b are zero divisors. ∎

For a and b in $\mathbf{Q}$, $a \neq 0$, we solve the equation $ax = b$ for x by multiplying both sides a by a^{-1}. Thus the solution is $x = a^{-1}b$. The solution is unique because if $ay = b$, we can again multiply by a^{-1} to find that $y = a^{-1}b$. But if the ring is $\mathbf{Z}$ and $a = 9$ and $b = 3$, the equation $ax = b$ has no solution. Note that 9^{-1} does not exist in $\mathbf{Z}$. However, in the ring $\mathbf{Z}_{12}$, there is no solution to $9x = 1$ and so 9^{-1} does not exist in $\mathbf{Z}_{12}$. But the equation $9x = 3$ has three different solutions, namely 3, 7, and 11. Finding these solutions is not always as easy as multiplying through by a^{-1}.

Suppose that x and y and z are elements of a ring R with unity 1. Suppose also that $xy = yx = 1$ and that $xz = zx = 1$. Then $yxz = (yx)z = z$ and $yxz = y(xz) = y$. So $y = z$. So for any $x \in R$, there is at most one element y that satisfies the equation $xy = yx = 1$.

> **DEFINITION 3.** Let R be a ring with unity element 1. Let x and y be members of R. If $xy = yx = 1$, then y is called the **multiplicative inverse** of x. We denote y by the symbol x^{-1}.

EXAMPLES

9. Let p be a prime number. By Proposition 5 of Section 2.1, the congruence $ax \equiv 1 \bmod p$ has a solution for any value of a that is not divisible by p. Thus every nonzero member of $\mathbf{Z}_p$ has a multiplicative inverse.

10. No integers other than 1 and -1 have multiplicative inverses in **Z**.
11. In $\mathbf{M}_{2,2}$, the set of 2×2 matrices with entries in **Q**, some nonzero matrices do not have multiplicative inverse while others do. For instance, the matrix $\begin{bmatrix} 1 & 1 \\ 0 & 0 \end{bmatrix}$ has no multiplicative inverse because no choice of f and h can satisfy $\begin{bmatrix} 1 & 1 \\ 0 & 0 \end{bmatrix} \times \begin{bmatrix} e & f \\ g & h \end{bmatrix} = \begin{bmatrix} 1 & 0 \\ 0 & 1 \end{bmatrix}$. However, the matrix $A = \begin{bmatrix} 1 & 2 \\ 1 & 3 \end{bmatrix}$ has a multiplicative inverse because $\begin{bmatrix} 1 & 2 \\ 1 & 3 \end{bmatrix} \times \begin{bmatrix} 3 & -2 \\ -1 & 1 \end{bmatrix} = \begin{bmatrix} 3 & -2 \\ -1 & 1 \end{bmatrix} \times \begin{bmatrix} 1 & 2 \\ 1 & 3 \end{bmatrix} = \begin{bmatrix} 1 & 0 \\ 0 & 1 \end{bmatrix}$. ∎

In most rings, some elements have multiplicative inverses and others do not. When an element a does have a multiplicative inverse, the equation $ax = b$ has a unique solution, namely $x = a^{-1}b$. The following definition gives a name to elements with multiplicative inverses.

> **DEFINITION 4.** An element x in a ring R with unity is called a **unit** if it has a multiplicative inverse.

EXAMPLES

12. In **Z** the only units are 1 and -1.
13. In $\mathbf{Z}_6$, the elements 1 and 5 are units, but the elements 0, 2, 3, and 4 are not. An element $a \in \mathbf{Z}_m$ is a unit if there exists an element $x \in \mathbf{Z}_m$ such that $ax \bmod m = 1$. Thus there are $\varphi(m)$ units in $\mathbf{Z}_m$. (Recall that φ is the Euler phi-function.)
14. In the ring **Q**, every nonzero element is a unit. ∎

An element in a ring can be neither a unit nor a zero divisor (e.g., $5 \in \mathbf{Z}$ is neither). But no element can be *both* a unit and a zero divisor. For if x is a unit and $xy = 0$, then $y = (x^{-1}x)y = x^{-1}(xy) = 0$. So $y = 0$ and thus x is not a zero divisor. In **Z**, our prototypical ring with unity, multiplication is commutative and there are no zero divisors. The rings **Q**, **R**, and **C** are similar. The following definition groups such rings under one name.

> **DEFINITION 3.** An **integral domain** is commutative ring with unity that has no zero divisors.

EXAMPLES

15. Suppose that p is prime, and x and y are elements of $\mathbf{Z}_p$. Then $xy = 0$ if and only if xy is divisible by p. Since p is prime, x or y must be divisible by p. So $x = 0$ or $y = 0$ in $\mathbf{Z}_p$. Thus the ring $\mathbf{Z}_p$ is an integral domain when p is prime.

16. The ring $M_{2,2}$ is not an integral domain. It fails because multiplication is not commutative and because it has zero divisors.
17. The ring $Q[x]$ of polynomials with rational coefficients is an integral domain, as we shall show in Section 3.1. ∎

Integral domains share another familiar arithmetic property with the integers: We can cancel in integral domains. The following proposition shows why.

PROPOSITION 2 Let R be an integral domain and let a, b be elements in R and let c be an element of R that is not equal to 0. If $ac = bc$, then $a = b$.

Proof. If $ac = bc$, then $ac - bc = (a - b)c = 0$. Since $c \neq 0$ and R is an integral domain, it must be case that $a - b = 0$ and $a = b$. ▲

Notice that elements in an integral domain do not necessarily have multiplicative inverses. So in the proof of Proposition 2, we could not assume that we could simply multiply both sides of the equation $ac = bc$ by c^{-1} to show that $a = b$. But in rings such as Q and Z_5, that simpler reasoning is available because every nonzero element is a unit.

DEFINITION 5. An integral domain F in which every nonzero element is a unit is called a **field**.

Since a field is an integral domain, it has unity, multiplication is commutative, and there are no zero divisors. Additionally, every nonzero element is a unit.

EXAMPLES

18. The rational numbers Q, the real numbers R and the complex numbers C are all fields. The set of integers Z is an integral domain but not a field.
19. The ring Z_m is a field if and only if m is a prime number. If m is composite, the ring Z_m has zero divisors, which cannot be units. Thus Z_5 and Z_7 are fields while Z_6 and Z_8 are not. The fields Z_p are our most important examples of finite fields.
20. The set $Q(\sqrt{2}) = \{a + b\sqrt{2} : a, b \in Q\}$ is a subset of the field of real numbers R. It is also a sub-field, which is to say that it is closed[2] under addition and multiplication, and the additive and multiplicative inverses of each element of $Q(\sqrt{2})$ are also in $Q(\sqrt{2})$. To illustrate that it is closed under addition and multiplication, let $x = 2 + 3\sqrt{2}$ and $y = 1 - \sqrt{2}$. Then $x + y = 3 + 2\sqrt{2}$ and $xy = -4 + \sqrt{2}$, which are both elements of $Q(\sqrt{2})$. We will show that each nonzero element of $Q(\sqrt{2})$ has an inverse that is also of the form $a + b\sqrt{2}$. Suppose that x and y are rational numbers that are not both 0 and let $z = x + y\sqrt{2}$. Since $\sqrt{2}$ is not rational, $x^2 - 2y^2$ is a rational number that is not equal to 0. Let $w = \frac{x}{x^2 - 2y^2} + \left(\frac{-y}{x^2 - 2y^2}\right)\sqrt{2} = \frac{x - y\sqrt{2}}{x^2 - 2y^2}$. Multiplying, we see that

---
[2]See the appendix if you are unfamiliar with this term.

$w = z^{-1}$ because $zw = (x + \sqrt{2}y)\left(\frac{x-\sqrt{2}y}{x^2-2y^2}\right) = \frac{x^2-2y^2}{x^2-2y^2} = 1$. Fields like $Q(\sqrt{2})$ will be important in our study of polynomials. ∎

The ring Z_5 and the ring Z behave very differently when we sum a finite number of 1s. In Z_5, $1 + 1 + 1 + 1 + 1 = 0$, whereas in the rings Z and Q, no finite sum of 1s is equal to 0. The difference is captured in the following definition.

> **DEFINITION 6.** Let R be a ring. Suppose that for some integer $n > 0$, the sum $1 + 1 + \ldots + 1$ (n-summands) $= 0$ but, if $m < n$, the sum $1 + 1 + \ldots + 1$ (m summands) $\neq 0$. Then we call n the **characteristic** of the ring R. If no such n exists, we say that R has **characteristic 0**.

EXAMPLES

21. The ring Z_m has characteristic m. The rings Z and Q have characteristic 0.
22. The ring $Z_m[x]$ and the ring of 2×2 matrices with entries in Z_m are both rings of characteristic m. ∎

PROPOSITION 3 If R is an integral domain with characteristic $n \neq 0$, then n is prime.

Proof. Suppose that R has characteristic n and that $n = pq$ is composite. Let $x = 1 + 1 + \ldots + 1$ (p times) and let $y = 1 + 1 + \ldots + 1$ (q times). Neither x nor y is zero. Using the distributive law, we see that $xy = 1 + 1 + \ldots + 1$ (pq times) $= 0$, which is impossible in a integral domain. ▲

Since every field is an integral domain, we have the following corollary.

COROLLARY 1 If F is a finite field, then the characteristic of F is a prime number.

In Chapter 3, we will investigate rings of polynomials more closely. We will broaden the context of that study beyond that encountered in secondary school by considering polynomials that have coefficients in rings such as Z_m. The emphasis will be on divisibility and factoring, just as it was in Chapter 1. In fact, many of the techniques of Chapter 1 will carry over to these new rings.

SUMMARY

In this section we introduced the algebraic structure called a **ring**. The set of integers with its two familiar operations of addition and multiplication serves as a prototype. But there are nuances in all the examples. A ring can be **commutative** or not. It can have a **unity** or not. In rings with unity, some elements called **units** have multiplicative inverses. Other elements are **zero divisors**. A commutative ring with

Section 2.5   Rings and Fields   89

no zero divisors is most like the integers. We call such rings **integral domains**. An integral domain in which all the nonzero elements are units is called a **field**. The arithmetic of fields is similar to the arithmetic of the rational numbers. But unlike our most familiar examples of fields such as Q, R, and C, some fields such as Z_7 do not have **characteristic zero**, but rather have prime **characteristic**.

2.5 Exercises

1. Prove parts iii and iv of Proposition 1.
2. Prove that a ring has at most one unity.
3. Prove that if an element x has a multiplicative inverse in a ring R, then it has only one inverse. (Remember that a ring can be noncommutative.)
4. Prove that a nonzero element x in Z_m is a zero divisor if and only if $\gcd(x, m) \neq 1$.
5. In which of the rings Z_4, Z_5, Z_{20} or Z_{15} is it true that if $2x = 2y$, then $x = y$?
6. Give an example from a noncommutative ring that shows that $(x + y)^2$ is not necessarily equal to $x^2 + 2xy + y^2$. What is the correct expression for $(x + y)^2$?
7. Characterize the units in Z_m. Prove your conclusions.
8. What are the units in $Q[x]$, the ring of polynomials with rational coefficients?
9. Prove that the ring $R[x]$ of polynomials with coefficients in R is an integral domain if and only if R is an integral domain.
10. Explain why the ring $R[x]$ of polynomials is not a field even if R is a field.
11. Is there a solution to the equation $x^2 + x + 4 = 0$ in Z_5? Repeat the question for Z_7.
12. Let $M_{2,2}$ denote the ring of 2×2 matrices with real valued entries. Characterize the units of $M_{2,2}$.
13. Show that every nonzero element in the set $Q(\sqrt{5}) = \{a + b\sqrt{5} : a, b \in Q\}$ has a multiplicative inverse that is also in $Q(\sqrt{5})$.
14. Let $Z_3[i] = \{a + bi : a \text{ and } b \text{ are elements of } Z_3 \text{ and } i^2 = -1\}$ Add and multiply elements of $Z_3[i]$ as if they were binomials reducing coefficients mod 3 and replacing i^2 by -1. Show that $Z_3[i]$ is a field. What is the multiplicative inverse of the element $2 + i$? Solve the equation $(2 + i)x = 1 + 2i$ in $Z_3[i]$.
15. Is the ring $Z_5[i] = \{a + bi : a \text{ and } b \text{ are elements of } Z_5 \text{ and } i^2 = -1\}$ an integral domain?
16. Show that in any integral domain, the equation $x^2 = x$ has exactly two solutions.
17. Prove that if the ring R has characteristic p, then the ring $R[x]$ also has characteristic p.
18. Prove that any finite integral domain is a field. (**Hint.** For each $x \neq 0$, show that the map $T_x : R \rightarrow R$ defined by $T_x(y) = xy$ is injective or "one-to-one." Since R is finite, T_x must then be surjective or "onto.")

To the Teacher

Many mathematics majors find proving Proposition 1 difficult because its assertions are so familiar that they seem to need no proof. But to the young learner, it may not be obvious that a times the additive inverse of b must equal the additive inverse of the product ab (i.e., $a(-b) = -(ab)$). (In fact, it wasn't particularly obvious to us until we used symbols instead of words!) The *In the Classroom* essay at the end of this section addresses ways to communicate the meaning behind the symbols to young learners.

For older students, the NCTM Standards for grades 9 through 12 suggest that students should "develop an understanding of, and representations for, the addition and multiplication of vectors and matrices." In other words, they should become acquainted with the ring $\mathbf{M}_{n,n}$ of $n \times n$ matrices with real valued entries in advanced algebra or precalculus. Matrix addition is natural but motivating the definition of matrix multiplication in this new ring is a bit more challenging. It seems artificially complicated. What follows is an application that might make the definition seem more natural.

Suppose that we had a large cage for some unusually brilliant psychology department mice. Suppose the cage had several feeding stations, represented by the labeled dots in Figure 1. Suppose that to get from one station to another, a mouse must pass through a one-way door, represented by an arrow. So a mouse could go directly from station 1 to 2, but not from 2 to 1. The information as to which two stations are connected by a door is stored in the matrix A. Its rows and columns are labeled with the names of the feeding stations. If there is a door that lets a mouse go from station i to station j, then $a_{i,j}$, the entry in the row labeled i and column labeled j of A, is set to 1. Otherwise, $a_{i,j}$ is set to 0.

	1	2	3	4	5	6
1	0	1	0	1	0	0
2	0	0	1	0	0	0
3	0	0	0	1	0	0
4	0	1	0	0	1	0
5	0	0	1	0	0	0
6	1	0	0	0	1	0

$$A = \begin{bmatrix} 0 & 1 & 0 & 1 & 0 & 0 \\ 0 & 0 & 1 & 0 & 0 & 0 \\ 0 & 0 & 0 & 1 & 0 & 0 \\ 0 & 1 & 0 & 0 & 1 & 0 \\ 0 & 0 & 1 & 0 & 0 & 0 \\ 1 & 0 & 0 & 0 & 1 & 0 \end{bmatrix}$$

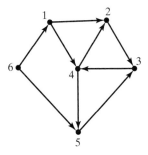

FIGURE 1

The matrix tells us when a mouse can get from station i to station j by going through exactly one door. Our first question is how many ways can a mouse get from station i to station j going through *exactly* two doors. To do this, there must be a third station k and doors from i to k and from k to j. In the matrix A, we must have $a_{i,k} = 1$ and $a_{k,j} = 1$. To find out how many ways there are to get from station i to station j in total, we sum the following:

(1) $$a_{i,1}a_{1,j} + a_{i,2}a_{2,j} + \ldots + a_{i,k}a_{k,j} + \ldots + a_{i,n}a_{n,j}.$$

The only nonzero contributions to the sum will be from terms of the form $a_{i,k}a_{k,j}$, where both $a_{i,k}$ and $a_{k,j}$ are not zero. The sum in expression (1) is exactly the i, jth entry in the matrix A^2. Thus the information as to how many ways each pair of stations is connected via exactly 2 doors is stored in matrix A^2. The exponent carries a double meaning. It can denote the number of doors or the product of A with itself.

$$A^2 = \begin{bmatrix} 0 & 1 & 1 & 0 & 1 & 0 \\ 0 & 0 & 0 & 1 & 0 & 0 \\ 0 & 1 & 0 & 0 & 1 & 0 \\ 0 & 0 & 2 & 0 & 0 & 0 \\ 0 & 0 & 0 & 1 & 0 & 0 \\ 0 & 1 & 1 & 1 & 0 & 0 \end{bmatrix}$$

Notice the 2 in the 4th row, 3rd column. A mouse can get from station 4 to station 3 via two doors by going through either station 2 or station 5. With our small set of doors, the entries in A^2 would have been fairly easy to determine by inspecting the diagram. But let's go on. How many ways are there to get from station i to station j using exactly 3 doors? The answer is stored in the i, jth entry of $A^3 = A^2 \times A$. Why? We have to get somewhere using two doors first! As the number of doors gets high, using only the diagram gets more difficult, but matrix multiplication does not: the number of ways to get from station i to station j using q doors is stored in A^q.

Finally, it may not be possible to get from station i to station j by any route. We would know this if the entry in the i, jth position of each of the matrices A^k was 0 for $k = 1$ through $n - 1$, where n is the total number of stations. Instead of inspecting each matrix, we can simply add $A^1 + A^2 + \ldots + A^{n-1}$. If position i, j is 0, then you can't get from i to j.

Tasks

1. The matrix B indicates the door set up of another cage. Draw the set up of the cage and determine the number of ways the mouse can get between each ordered pair of stations that use exactly 2 doors. Repeat for 3 and 4 doors.

$$B = \begin{bmatrix} 0 & 0 & 0 & 1 & 0 & 1 & 0 \\ 1 & 0 & 1 & 0 & 0 & 0 & 1 \\ 0 & 0 & 0 & 1 & 0 & 0 & 1 \\ 0 & 1 & 0 & 0 & 1 & 0 & 0 \\ 1 & 0 & 1 & 0 & 0 & 1 & 0 \\ 0 & 1 & 1 & 0 & 1 & 0 & 0 \\ 0 & 0 & 0 & 0 & 1 & 0 & 0 \end{bmatrix}$$

2. Explain why a mouse never needs to use more than $n - 1$ doors if it can get from station i to station j in a cage with n feeding stations. (The smarter mice will figure this out!)
3. Can a mouse always get back to where it started?
4. Can the number of ways to get between two stations ever be infinite? Explain.
5. Can the number of paths between two stations fail to be infinite? Explain.
6. Indicate other scenarios that might utilize a similar modeling scheme.
7. Suppose that addition and multiplication were carried out using the following tables. What information would the matrices A^2, A^3, etc. give?

+	0	1
0	0	1
1	1	1

*	0	1
0	0	0
1	0	1

2.6 THE FIELD OF COMPLEX NUMBERS

The simple polynomial $x^2 - 2$ is an element of $Q[x]$, the ring of all polynomials with rational numbers as coefficients. But it does not have roots in Q. Its roots $\pm\sqrt{2}$ lie in the bigger ring of real numbers R, a field that contains Q. The roots of $x^2 + 1$, a simple polynomial in $R[x]$, lie outside of R, in C, the set of complex numbers. The set $C[x]$ is different from both $Q[x]$ and $R[x]$ in the following way: Every polynomial in $C[x]$ has all its roots in C. So every polynomial with complex coefficients factors completely into linear factors. That is the essence of the Fundamental Theorem of Algebra, the focus of Section 3.5. In this section we provide a brief introduction to the arithmetic of the field of complex numbers so that we are better equipped to probe these issues.

To construct the complex number system from the real numbers R, we first define i to be $\sqrt{-1}$. Thus i is a symbol for a number such that $i^2 = -1$. The set of complex numbers is the set $C = \{z = a + bi : a \in R \text{ and } b \in R\}$. If $z = a + bi$, we call a the **real part** of z and b the **imaginary part** of z, denoted $Re(z)$ and $Im(z)$, respectively. Thus $z = 3 - \pi i$ is a complex number with $Re(z) = 3$ and $Im(z) = -\pi$. Note that both $Re(z)$ and $Im(z)$ are *real* numbers. The addition and multiplication of the complex numbers $a + bi$ and $c + di$ are defined as follows.

(1) $$(a + bi) + (c + di) = (a + c) + (b + d)i$$

(2) $$(a + bi)(c + di) = (ac - db) + (ad + cb)i$$

Since $(0 + i)^2 = i^2 = -1$, we can add and multiply complex numbers as if they were binomials in the variable i, and then replace i^2 by -1. For example, if we let $x = 3 - i$ and $y = 2 + 5i$, then $x + y = 5 + 4i$ and $xy = (3 - i)(2 + 5i) = 6 + 13i - 5i^2 = 6 + 13i + 5 = 11 + 13i$. Addition and multiplication are both associative and commutative. The numbers $0 + 0i$ and $1 + 0i$ are additive and multiplicative identities, respectively. Each number $x + iy$ has an additive inverse. The distributive law holds. Thus C is a commutative ring with unity.

To show that C is a field, we show that, for each nonzero complex number $z = a + bi$, we can find $x + yi$ such that $(x + yi)(a + bi) = (ax - by) + (bx + ay)i = 1 + 0i$. By equating real and imaginary parts, we obtain and solve the simultaneous equations $\begin{Bmatrix} ax - by = 1 \\ bx + ay = 0 \end{Bmatrix}$ to find that $x = \frac{a}{a^2+b^2}$, and $y = \frac{-b}{a^2+b^2}$. Since $a^2 + b^2 \neq 0$ and $(x + yi)(a + bi) = 1 + 0i$, the multiplicative inverse of z is $x + yi$. By identifying the real number x with the complex number $x + 0i$, we can regard C as a field that contains R.

EXAMPLE 1

Let $z = 2 + 3i$. Then $z^{-1} = \frac{2}{2^2+3^2} + \frac{(-3)}{2^2+3^2}i = \frac{2}{13} - \frac{3}{13}i$. Let $w = -1 + 5i$. To divide w by z, we multiply w by z^{-1}:

$$\frac{w}{z} = wz^{-1} = (-1 + 5i)\left(\frac{2}{13} - \frac{3}{13}i\right) = \frac{13}{13} + \frac{13}{13}i = 1 + i.$$

The **complex conjugate** of $z = a + bi$ is the complex number $\bar{z} = a - bi$. The **modulus** or **absolute value** of z, denoted by $|z|$, is the real number $\sqrt{a^2 + b^2}$. For example, if $z = \frac{1}{2} - 5i$, then $\bar{z} = \frac{1}{2} + 5i$ and $|z| = \sqrt{\frac{1}{4} + 25} = \sqrt{\frac{101}{4}} = \frac{\sqrt{101}}{2}$.

Plotting Complex Numbers

We plot complex numbers in the xy-plane by associating $z = a + bi$ with the ordered pair (a, b). The point (a, b) defines the vector $\mathbf{v} = [a, b]$. We have the following relationships between the complex numbers and the geometry of the xy-plane. (See Figure 1.)

i. The addition of complex numbers $z = a + bi$ and $w = c + di$ corresponds to vector addition.
ii. The modulus of z is the length of the associated vector.
iii. The conjugate of z is obtained by reflecting the point (a, b) across the x-axis.
iv. The negative of z is obtained by reflecting the point (a, b) across the origin on the line through the origin and (a, b).

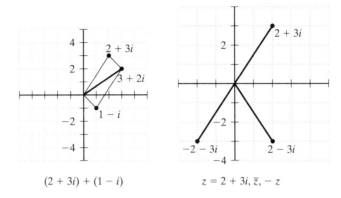

FIGURE 1

Complex Numbers and Polar Coordinates

A set of polar coordinates of a point (x, y) in the plane is the pair (r, θ) where $r = \sqrt{x^2 + y^2}$ and θ is any angle such that $x = r\cos(\theta)$ and $y = r\sin(\theta)$. If (x, y) is the graph of the complex number $z = x + yi$ in the xy-plane, then r is the modulus of z so that $r = |z|$. The angle θ is called the **argument** of z or **arg(z)**. Note that $\arg(z)$ is not unique. Values of $\arg(z)$ can differ by integer multiples of 2π. The product of two complex numbers $z = r\cos(\theta) + r\sin(\theta)i$ and $z_1 = r_1\cos(\theta_1) + r_1\sin(\theta_1)i$ has a particularly simple expression in terms of the moduli and the arguments of z and z_1. To obtain the product of z and z_1, we multiply moduli and add angles. Here's why.

$$zz_1 = rr_1[(\cos(\theta)\cos(\theta_1) - \sin(\theta)\sin(\theta_1)) + (\cos(\theta)\sin(\theta_1) + \cos(\theta_1)\sin(\theta))i]$$

From trigonometry, we recall that

$$\cos(\theta)\cos(\theta_1) - \sin(\theta)\sin(\theta_1) = \cos(\theta + \theta_1)$$

and

$$\cos(\theta)\sin(\theta_1) + \cos(\theta_1)\sin(\theta) = \sin(\theta + \theta_1).$$

Thus

(1) $$zz_1 = rr_1(\cos(\theta + \theta_1) + \sin(\theta + \theta_1)i).$$

The modulus of the product zz_1 is the product of the moduli of z and z_1. The argument of zz_1 is the sum of the arguments of z and z_1 so that $\arg(zz_1) = \arg(z) + \arg(z_1)$.

EXAMPLE 2

Let $z = 1 + i$ and $z_1 = \sqrt{3} + i$. Then $|z| = \sqrt{2}$ and $|z_1| = 2$. We can take $\arg(z) = \pi/4$ and $\arg(z_1) = \pi/6$. Thus $zz_1 = 2\sqrt{2}(\cos(\frac{5\pi}{12}) + \sin(\frac{5\pi}{12})i)$. (See Figure 2.) ∎

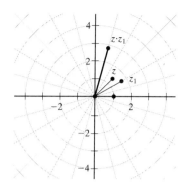

FIGURE 2

From (1) and induction we obtain **de Moivre's formula**:

$$z^n = r^n \cos(n\theta) + r^n \sin(n\theta)i \text{ for } n = 1, 2, \ldots.$$

EXAMPLE 3

Let $z = \sqrt{3} + i = 2\left(\cos\left(\frac{\pi}{6}\right) + \sin\left(\frac{\pi}{6}\right)i\right)$. Then $z^5 = 32\left(\cos\left(\frac{5\pi}{6}\right) + \sin\left(\frac{5\pi}{6}\right)i\right) = 16(-\sqrt{3} + i)$. ∎

Finding the nth roots of $z = r(\cos(\theta) + \sin(\theta)i)$ with de Moivre's formula is just as easy as finding powers. We want to find $w = b(\cos(\varphi) + \sin(\varphi)i)$ such that $w^n = z$. From de Moivre's formula, we know that $b^n = r$. So set b equal to $\sqrt[n]{r}$, the positive real nth real root of the positive real number r. The angle φ must satisfy $n\varphi = \theta + 2\pi k$ for some integer k. Solving for φ in $n\varphi = \theta + 2\pi k$, we have

$$\varphi = \frac{\theta}{n} + \frac{2\pi k}{n} \text{ for } k = 0, 1, 2, \ldots, (n-1).$$

(If we took $k > n - 1$, we would redundantly add multiples of 2π to angles already in the set.) Thus, any nonzero complex number z has exactly n distinct n^{th} roots.

EXAMPLE 4

To find the square roots of i, we first express i in polar coordinates as $i = \cos(\pi/2) + \sin(\pi/2)i$. The modulus of i is 1 and so the modulus of any root is also 1. The argument of i is any angle of the form $\theta = \pi/2 + 2\pi k$, for $k = 0, 1, \ldots$ and so the distinct arguments modulo 2π of the square roots of i are $\varphi = \pi/4 + \pi k$, for $k = 0, 1$. Thus the square roots of i are $a = \cos(\pi/4) + \sin(\pi/4)i$ and $b = \cos(5\pi/4) + \sin(5\pi/4)i$, or $\pm\left(\frac{\sqrt{2}}{2} + \frac{\sqrt{2}}{2}i\right)$. ∎

EXAMPLE 5

We can solve $z^4 + 16 = 0$ by finding the 4th roots of -16. In polar notation,

$$-16 = 16(\cos(\pi) + \sin(\pi)i).$$

So the 4th roots of -16 are $z_k = 2[(\cos(\pi/4 + \pi k/2) + \sin(\pi/4 + \pi k/2)i]$ for $k = 0, 1, 2, 3$.

$$z_0 = \sqrt{2} + \sqrt{2}i, \quad z_1 = -\sqrt{2} + \sqrt{2}i,$$
$$z_2 = -\sqrt{2} - \sqrt{2}i, \quad z_3 = \sqrt{2} - \sqrt{2}i.$$

■

EXAMPLE 6 *(The Roots of Unity)*

The n solutions to the equation $z^n = 1$ are called **the nth roots of unity**. They are the roots of the polynomial $z^n - 1 = (z - 1)(z^{n-1} + z^{n-2} + \ldots + 1)$. From de Moivre's formula, we see that distinct nth roots of unity are

$$\xi_k = \cos\left(\frac{2\pi k}{n}\right) + \sin\left(\frac{2\pi k}{n}\right)i \text{ for } k = 0, 1, \ldots, n-1.$$

The modulus of each of these roots is 1 and so each of them lies on the unit circle. (See Figure 3.) When we plot them on the circle of radius 1, we have the vertices of a regular polygon with n sides.

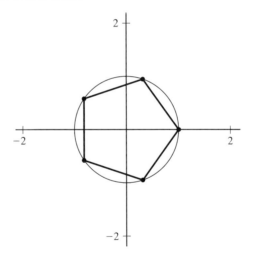

FIGURE 3 The 5th roots of unity

For each k, $\xi_k = (\xi_1)^k$. Set ξ_1 equal to ξ. The n roots of unity are the members of the set $\{1, \xi^1, \xi^2, \ldots, \xi^{n-1}\}$. Notice that the product $\xi^i \cdot \xi^j = \xi^{(i+j) \bmod n}$.   ■

2.6 Exercises

1. Let $z = (1 - 3i)$ and $w = (2 + i)$. Find $z + w$, zw, z/w, w^2, and z^3.
2. Solve for z in the equation $(2i + 1) + (3i)z = 5i$.
3. Let $x = a + bi$, $y = c + di$, and $z = e + fi$. Verify the distributive law: $x(y + z) = xy + xz$.

4. Suppose that $w = a + bi$ and $z = c + di$, $z \neq 0$. Show that $\frac{w}{z} = \frac{ac+bd}{c^2+d^2} + \frac{bc-ad}{c^2+d^2}i$.
5. Find all powers of i.
6. Let $z = 3 + 4i$. Find the complex conjugate and modulus of z, $1/z$, and z^2.
7. Prove that $|z|^2 = \bar{z}z$.
8. Let z and w, $w \neq 0$, be complex numbers. Prove the following statements.
   i. $\bar{z} + \bar{w} = \overline{(z+w)}$
   ii. $\bar{z}\,\bar{w} = \overline{(zw)}$
   iii. $\overline{\left(\frac{1}{w}\right)} = \frac{1}{\bar{w}}$
9. Find $(\sqrt{3} + i)^5$ and $(1+i)^n, n = 1, 2, \ldots$.
10. Find all the cube roots of 1 by solving the equation $x^3 - 1 = (x-1)(x^2 + x + 1) = 0$. Then find the cube roots of 1 with de Moivre's formula. Show that your results agree.
11. Find all the sixth roots of 1.
12. Find the square roots of $1 + \sqrt{3}i$.
13. Find the cube roots of $1 + i$.
14. Why must $x^2 + x + 1$ be a factor of $x^5 + x^4 + x^3 + x^2 + x + 1$?
15. If n is not prime, show that $x^{n-1} + x^{n-2} + \ldots + 1$ can be properly factored.
16. Note that $x^{12} + 1 = (x^4 + 1)(x^8 - x^4 + 1)$ and show that the 24th root of unity $\cos\left(\frac{\pi}{12}\right) + \sin\left(\frac{\pi}{12}\right)i$ is a root of $x^8 - x^4 + 1$.
17. Use the quadratic formula to solve the equation $z^2 + 3iz + (2i) = 0$.

To the Teacher

Complex numbers enter high school mathematics through the quadratic formula. We introduce them when we find that, in solving for the roots of $ax^2 + bx + c$, the discriminant $b^2 - 4ac$ is negative. We need complex numbers to make sense of the square root of a negative number. In the next paragraphs, we make sense of the logarithm of a negative number. The story takes us to the edge of the precalculus curriculum and it provides the secondary teacher with a broader prospective of familiar territory.

What function fits the following description: $f(a)f(b) = f(a+b)$? The most natural answer for a high school math student would be an exponential function, in particular, $f(x) = e^x$ since $e^a e^b = e^{a+b}$. But a student in precalculus could answer correctly with the function

$$g(\theta) = \cos(\theta) + i\sin(\theta).$$

De Moivre's formula, a standard precalculus topic, guarantee us that $g(a)g(b) = g(a+b)$. The two answers seem to differ greatly. The function $y = e^x$ is a one-to-one, unbounded function while $g(\theta)$ is a complex valued, periodic function and $|g(\theta)| = 1$ for all θ. However, our two answers are bound together in Euler's extension of the exponential function to the complex domain. It is defined as follows. For $z = a + bi$,

$$e^z = e^{a+bi} = e^a(\cos(b) + i\sin(b)).$$

EXAMPLE 1

a. Let $z = 2 + 3i$. Then $e^z = e^2(\cos(3) + i \sin(3))$.

b. Let $z = \frac{\pi i}{4}$. Then $e^z = e^0 \left(\cos\left(\frac{\pi}{4}\right) + i \sin\left(\frac{\pi}{4}\right) \right) = \frac{\sqrt{2}}{2} + i \frac{\sqrt{2}}{2}$.

c. If $z = r(\cos(\theta) + i \sin(\theta))$, and $r > 0$, we can express z as $z = e^{\ln(r) + i\theta}$. ∎

EXAMPLE 2

The nth roots of unity can be expressed very compactly using exponential notation. They are $z_k = e^{\left(\frac{2\pi k}{n}\right)i}$, $k = 0, 1, \ldots, n - 1$. ∎

When a function's domain is extended to a larger set, its values on the original domain must remain unchanged. Let's check. When $b = 0$, so that $a + ib$ is a real number, $e^{a+bi} = e^a(\cos(0) + i \sin(0)) = e^a$. Thus e^z is unchanged for real values of z. One salient feature of the exponential function evaluated at real numbers is that $e^x e^y = e^{x+y}$. That property holds for complex values as well, namely, $e^z e^w = e^{z+w}$.

To explore this new function, we can't just graph it because a graph would require four dimensions, two for the complex domain and two for the complex co-domain. So we investigate its behavior on various lines in the complex plane. First, we keep a constant and let b vary. The vertical line $x = a$ is the set of points (a, t), where t can be any real value. These points in the complex plane are mapped to the values $e^a(\sin(t) + i \cos(t))$, which is a circle of radius e^a. The line wraps around the circle once for segment of length of 2π. When $a = 0$, the line $x = 0$ is wrapped around the unit circle. For $a > 0$, the circles are bigger. For $a < 0$, the circles are smaller and have positive radius less than 1. All complex numbers *except* the origin lie on one of these circles. Figure 4 shows the graph of the lines $x = -1, x = 0, x = 1$ and $x = 2$. These lines are mapped onto the circles $r = e^{-1} = .367\ldots, r = e^0 = 1$, $r = e = 2.718\ldots$, and $r = e^2 = 7.389\ldots$, respectively. In each line, any two points with y-coordinates that differ by a multiple of 2π map to the same point on the circle.

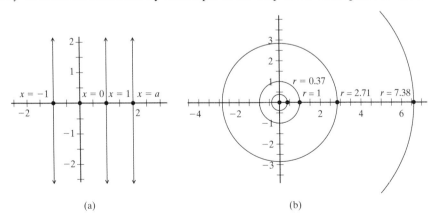

FIGURE 4

Now let's keep b constant and let a vary. All points on the horizontal line $y = b$ are mapped onto the ray $\theta = b$. Points (a, b) with $a < 0$ are mapped onto the segment of the ray that lies within the unit circle and those with $a > 0$ are mapped outside. (See Figure 5.) No point is mapped to $(0, 0)$. So the complex exponential function both wraps around like a trigonometric function and grows exponentially.

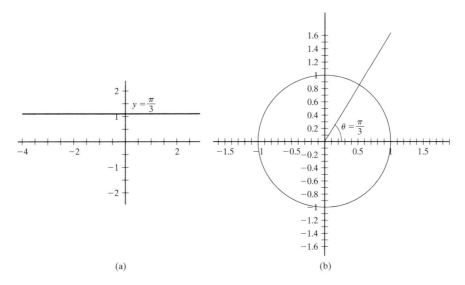

FIGURE 5

For any complex number $z \neq 0$, Arg(z) denotes the unique angle $-\pi < \theta \leq \pi$ such that $z = |z|(\cos(\theta) + i \sin(\theta))$. Now if $z = e^{a+ib}$ and $w = e^{s+it}$, then Arg(zw) = (Arg(z) + Arg(w)) mod 2π. This resembles a logarithm function and that's how we use it. Suppose that $w = e^{a+bi}$. We define Log(w) as follows:

$$\text{Log}(w) = \ln(|w|) + i\,\text{Arg}(w).$$

Then $a = \ln(|w|)$ and Arg(w) = b mod 2π. Just as arcsin (x) is the inverse of sin(x) restricted to a portion of its domain, Log(w) is the inverse of e^{a+ib} on a portion of its domain on which it is one-to-one, namely on the strip $-\pi < b \leq \pi$.

Note: "Log" does *not* denote $\log_{10}$. Log is the name we give to the extension of the natural log function to the complex domain.

Let's compute.

i. Log($3 + 0i$) = ln(3). No change!
ii. Log(-3) = ln($|3|$) + $i\pi$. To check, we see that $e^{\ln(|3|)+i\pi} = 3(\cos(\pi) + i \sin(\pi)) = -3$.
iii. Log(-1) = $i\pi$.

Any complex number except for $z = 0$ can be expressed in exponential form as $z = e^{a+bi}$ by letting $a = \ln|z|$ and $b = \text{Arg}(z)$.

Tasks

1. Let $z = \sqrt{3} + i$. Find a and b such that $z = e^{a+bi}$.
2. Express all the fifth roots of unity in the form e^{a+bi}.
3. Find the $\text{Log}(z)$ for each of the following values of z:

    i. $z = 2 + 3i$
    ii. $z = -e$
    iii. $z = i$

4. Discuss the similarities and differences of the function $w = \text{Log}(z)$ and the function $y = \arctan(x)$.

2.7 IN THE CLASSROOM: WHY IS THE PRODUCT OF TWO NEGATIVES NUMBERS POSITIVE?

Elementary school students are introduced to negative numbers by extending the number line and through familiar applications, such as temperatures below 0. In middle school, students further develop their understanding of the ring of integers and arithmetic operations on them. One thing that students often memorize by rote is the fact that the product of two negative numbers is positive. This result is Proposition 1, part iii, in Section 2.5 when a, b are positive whole numbers. We give suggestions for teachers to provide explanations of this fact at various levels for the middle school student. It is likely that if you teach multiplication of integers that you will see one or more of the methods that follow in your class textbook.

Explanation 1

Pedagogically, we try to build on prior knowledge that our students have. Our first explanation is based on repeated addition as a common model for multiplication of whole numbers. For example, $2 \times 3 = 3 + 3 = 2 + 2 + 2 = 6$, with 2 groups of 3 positives or 3 groups of 2 positives, for a total of 6 positives:

Consider extending this model with the inclusion of negative integers. As long as one of the values is positive, such as 2×-3, we can think about having 2 groups of 3 negatives,

which gives a total of 6 negatives, so $2 \times -3 = -3 + -3 = -6$.

Now, is there a way to continue to extend that model once both values are negative? How can we justify that $-2 \times -3 = 6$? We thought of 2×-3 as *having* 2 groups of -3. Likewise, we can think of -2×-3 as *taking away* 2 groups of -3 (the $-$ in front of the 2, meaning "taking away"). In both situations, you start with nothing, or chips worth 0, and either give yourself groups or take groups away. For -2×-3, we start with 0, represented by

(a positive charge and a negative charge is a 0 charge, $1 + (-1) = 0$).

Then we can take away 2 groups of 3 negatives, leaving 6 positives

and therefore, $-2 \times -3 = 6$.

Explanation 2

In middle school, many explanations are based on students observing and extending patterns. Our second explanation, which is seen in many current textbooks, makes use of patterns. For the following, students need to have an understanding, as described above, that the product of a positive and a negative number is negative. The teacher constructs a series of calculations beginning with one involving a product of a positive and a negative. For example, to find -2×-3, the series of problems can start with 4×-3. The calculations keep the negative value constant and decrease the positive value by one. Students look for the pattern that develops: The products are increasing by 3 each time. (Some students will incorrectly say "decreasing by 3" since they will ignore the negative signs and look only at the 12, 9, 6, and 3.) The pattern suggests that the next products, involving multiplying by 0 and multiplying by another negative number, should be $-3 + 3 = 0, 0 + 3 = 3, 3 + 3 = 6$, giving us $-2 \times -3 = 6$.

$$4 \times -3 = -12$$
$$3 \times -3 = -9$$
$$2 \times -3 = -6$$
$$1 \times -3 = -3$$
$$0 \times -3 = \underline{0}$$
$$-1 \times -3 = \underline{3}$$
$$-2 \times -3 = \underline{6}$$

Explanation 3

We can also extend the number line model for adding and subtracting integers. Begin at 0. The sign of the first integer determines whether you move left (when negative) or right (when positive) and the sign of the second integer determines whether you want to know where you will be in the future (positive) or where you were in the past (negative).

For example,

2×3 means that if you are now at 0 and are going right 2 units per move, where you will be three moves from now? You will be at 2 after one move, 4 after two moves, and 6 after three moves.

2×-3 means that if you are now at 0 and are going right 2 units per move, where were you three moves ago? You were at -2 one move ago, -4 two moves ago, and -6 three moves ago.

-2×3 means that if you are now at 0 and are going left 2 units per move, where will you be three moves from now? You will be at -2 after one move, -4 after two moves, and -6 after three moves.

-2×-3 means that if you are now at 0 and are going left 2 units per move, where were you three moves ago? You were at 2 one move ago, 4 two moves ago, and 6 three moves ago.

Explanation 4

There are also examples of justifying the sign of the product of two integers based on videotaping people walking. If you videotape people walking forward (first integer is positive) and you play it backward (second integer is negative), then the people on the tape are moving backward (negative). If you videotape people walking backward (first integer is negative) and you play it backward (second integer is negative), then the people on the tape will be moving forward (positive).

Explanation 5

None of the explanations above would be considered a proof in your typical college abstract algebra class. But each is likely to convince middle school students that the rule is true. Below is a proof that is based on the properties, structures, and relationships found in rings.

> Let $a, b \in \mathbf{Z}$. Then $a + -a = 0$, and $-b(a + -a) = -b(0)$, by multiplying both sides by $-b$. By the distributive law and part i of Proposition 1 of Section 2.5, $(-b)a + (-b)(-a) = 0$. By part ii of the same proposition and the commutativity of multiplication of integers, $-ab + (-a)(-b) = 0$. Since additive inverses are unique (check this!) $(-a)(-b) = ab$, the additive inverse of $-ab$. Thus, when $a, b > 0$, we have that the product of two negatives is positive.

2.7 Exercises

1. Look at various middle school textbooks and see what types of explanation are given for the product of two negative numbers. Make a note of any new ones you find.
2. Definition 1 in Section 2.5 states that in a ring R, there is an element $0 \in R$ such that $0 + x = x + 0 = x$ for all x in R. Prove that this 0 is unique.

3. Definition 1 in Section 2.5 states that each element x in a ring R has an element y (also in R), such that $x + y = y + x = 0$. Prove that each element x has a unique such element y.

Bibliography

Bassarear, Tom. *Mathematics for Elementary School Teachers*. Boston: Houghton Mifflin Company, 2005.

Bennett, Jennie M., David J. Chard, Audrey Jackson, Jim Milgram, Janet K. Scheer, and Bert K. Waits. *Middle School Math Course 1*. Austin: Holt, Rinehart and Wilson, 2004.

Chapin, Suzanne H., Theodore J. Gardella, Mark Illingworth, Marsha S. Landau, Joanna O. Masingila, and Leah McCracken. *Middle Grades Math: Tools for Success Course 1*. Upper Saddle River, N.J.: Prentice Hall, 1999.

Lappan, Glenda, James Fey, William M. Fitzgerald, Susan N. Friel, and Elizabeth Difanis Phillips, *Accentuate the Negative: Integers, Connected Mathematics Grade 7*. Upper Saddle River, N.J.: Prentice Hall, 2002.

2.8 FROM THE PAST: THE BIRTH OF THE COMPLEX NUMBERS

In the first 70 years of the sixteenth century, complex numbers went from an essentially unknown concept to an idea forced upon Girolamo Cardano by his method of solution of a particular form of the cubic equation. It was the solution of cubic equations that gave rise to the use of complex numbers in algebra. The stage was set for this development when Luca Pacioli published his *Summa de Arithmetica, Geometrica, Proportioni et Proportionalita* in 1509. This book presented most of the mathematics known in Europe at the time. It contained both practical and theoretical mathematics; it moved algebra forward as a branch of mathematics. Pacioli developed several abbreviations for words used in algebraic expressions that were to become standard over the next century. For example, he used p for addition and m for subtraction, these coming from the Italian words *piu* (more or plus) and *meno* (less or minus). After showing how to solve quadratic equations, he commented that it seemed impossible to do the same for cubic equations. But within 20 years most types of cubic equations had been solved, proving Pacioli wrong. Scipione del Ferro (1465–1526) was the first to discover a method to solve a certain class of cubic equations. In modern notation, these were equations of the form $x^3 + px = q$ and $x^3 + q = px$. We would consider these as the same form of equation because we are comfortable with negative numbers. However, del Ferro was dealing with relations between geometric quantities rather than algebraic equations. For example, his expression for $x^3 + px = q$ would have been something like, "Find a cube such that its volume added to the volume of the solid whose base had area p and height equal to the side of the cube equals a given volume q."

Del Ferro's solution was not made public, but instead he passed it on to his students. Being in control of such a technique allowed one to compete in mathematical tournaments at a considerable advantage. Such tournaments or mathematical problem solving contests were public displays of mathematical talent. To good practitioners they were also a source of income. Another mathematician, Niccolo Tartaglia, was also working on cubic equations and independently found a solution for the form $x^3 + px^2 = q$. Tartaglia was easily able to defeat Antonio Maria Fiore, a student of del Ferro's, in a contest of equation solving using his new method. Tartaglia also kept his methods secret. However, his secret was presented to the public by Girolamo Cardano in his work *Artis Magnae Sive de Regulis Algebraicis*. (For various reasons Cardano felt that he could publish what had been offered him in private. He found that del Ferro had discovered the solution of one form of the cubic before Tartaglia and since he (Cardano) could essentially turn all cubics to the form that del Ferro solved, he no longer felt that he needed to keep Tartaglia's method secret.) The book, which in English is called *The Great Art*, contains much more than just the solution of cubic equations. Cardano, a very capable mathematician, had fleshed out the whole field of cubic equations. His work showed how to remove the x^2 term from a cubic and then gave recipes to solve the various remaining cases. Cardano was interested in the structure of the solutions. Which equations, for example, had only one root and which had two or three? He also considered negative roots, or what he called false roots. He gave examples of equations with positive roots which could be transformed into equations with negative roots. For example, in Chapter I he says,

> For instance, although $x^3 + 21 = 2x$ lacks a true (positive) solution, yet it has a fictitious (negative) one, -3, and this (3) is a the true solution for $x^3 = 2x + 21$.

Cardano does not mention complex roots for cubic equations yet he clearly understands that the square root of a negative number can arise in the solution to a particular equation. In Chapter XII, *On the Cube Equal to the First Power and Number*, he is dealing with equations of the form $x^3 = px + q$. He gives the following rule for solution of this form:

> The rule therefore is: When the cube of one-third the coefficient of x is not greater than the square of one-half the constant of the equation, subtract the former from the latter and add the square root of the remainder to one-half the constant of the equation and, again, subtract it from the same half, and you will have, as was said, a *binomium* and its *apotome*, the sum of the cube roots of which constitutes the value of x.

The rule is rather dense reading but it gives a formula for solving cubics. First one must check that $\left(\frac{p}{3}\right)^3 \leq \left(\frac{q}{2}\right)^2$. This is because the square root of the remainder is $\sqrt{\frac{q^2}{4} - \frac{p^3}{27}}$ and so the quantity $\frac{q^2}{4} - \frac{p^3}{27}$ must be nonnegative. The *binomium* and the *apotome* refer to terms of the form $a + \sqrt{b}$ and $a - \sqrt{b}$. The whole formula is thus

$$x = \sqrt[3]{\frac{q}{2} + \sqrt{\frac{q^2}{4} - \frac{p^3}{27}}} + \sqrt[3]{\frac{q}{2} - \sqrt{\frac{q^2}{4} - \frac{p^3}{27}}}.$$

Cardano then says that if the coefficients do not satisfy the necessary inequality, the solution can be found in another of his books on geometric problems by using something called the aliza problem. Unfortunately, this book on geometric problems has been lost. Cardano doesn't mention the problem of the square root of a negative at this point but it clearly was something he thought about. In Chapter XXXVII, *On the Rule for Postulating a Negative*, he writes

> The second species of negative assumption involves the square root of a negative. I will give an example: If it should be said, divide 10 into two parts the product of which is either 30 or 40, it is clear that this case is impossible. Nevertheless, we will work thus: We divide 10 into two equal parts, making each 5. These we square, making 25. Subtract 40, if you will, from the 25 thus produced, ..., leaving a remainder of -15, the square root of which added to or subtracted from 5 gives parts the product of which is 40. These will be $5 + \sqrt{-15}$ and $5 - \sqrt{-15}$. ... Putting aside the mental tortures involved, multiply $5 + \sqrt{-15}$ by $5 - \sqrt{-15}$, making $25 - (-15)$ which is 40. Hence this product is 40.

Cardano leaves the square root of negative numbers at this point commenting, "So progresses arithmetic subtlety the end of which, as is said, is as refined as it useless." Rafael Bombelli (1526–1572) of the generation after Cardano took up the cause of complex numbers and showed how to produce a solution from Cardano's method even though the square root of a negative number was involved. In his book, *Algebra*, he presents an example of the equation that Cardano could not solve. Consider $x^3 = 15x + 4$. Since $\left(\frac{15}{3}\right)^3 > \left(\frac{4}{2}\right)^2$ Cardano would have used some other method to solve it. Using the method Cardano presented in Chapter XII, Bombelli found that $x = \sqrt[3]{2 + \sqrt{-121}} + \sqrt[3]{2 - \sqrt{-121}}$. Bombelli presents rules for computation with numbers involving the square root of a negative. Using these rules he assumes that $\sqrt[3]{2 + \sqrt{-121}} = A + \sqrt{-B}$ and that $\sqrt[3]{2 - \sqrt{-121}} = A - \sqrt{-B}$. Multiplying the left sides and the right sides together yields $\left(\sqrt[3]{4 + 121}\right) = 5 = A^2 + B$. Cubing both sides of the first equation gives $\left(\sqrt[3]{2 + \sqrt{-121}}\right)^3 = 2 + \sqrt{-121} = (A + \sqrt{-B})^3 = A^3 + 3A^2\sqrt{-B} + 3A(-B) - B\sqrt{-B}$. Collecting terms not involving the square root of the negative yields $2 = A^3 - 3AB$. So $A = 2$ and $B = 1$ is a solution to the pair of equations. Thus $\sqrt[3]{2 + \sqrt{-121}} + \sqrt[3]{2 - \sqrt{-121}} = 2 + \sqrt{-1} + 2 - \sqrt{-1} = 4$. Note that 4 is a root of the original cubic equation. Notice that Bombelli had found that the square root of a negative could be useful if one simply used it like a number.

Cardano evidently explored how such numbers could enter calculations but did not apply them to the cubic equations themselves. Bombelli pushed forward and found that complex numbers could be useful in solving equations, leaving aside any notion of what they might mean. They both treated them as "numbers" that could enter into calculations but didn't speculate on their actual meaning. It would be several hundred years before new models of the complex numbers allowed them to become a part of algebra.

In the 1700s, Leonhard Euler and others would find new uses for the complex numbers by giving them geometric meaning as points in a plane. Euler gave the complex numbers a kind of legitimacy in his *Introduction to Algebra*, saying,

> Since all numbers which it is possible to conceive are either greater or less than 0, or are 0, it is evident that we cannot rank the square root of a negative number amongst possible numbers, and we must therefore say that it is an impossible quantity. In this manner we are led to the idea of numbers, which from their nature are impossible; and therefore they are usually called *imaginary quantities,* because they exist merely in the imagination. All such expressions as $\sqrt{-1}, \sqrt{-2}, \ldots$ are consequently impossible, or imaginary numbers, since they represent the roots of negative quantities; ... But notwithstanding this these numbers present themselves to the mind; they exist in our imagination, and we still have a sufficient idea of them; since we know that by $\sqrt{-4}$ is meant a number which, multiplied by itself, produces -4; for this reason also, nothing prevents us from making use of imaginary numbers, and employing them in calculation.

Gauss found the ultimate importance of the complex numbers to algebra when he proved the Fundamental Theorem of Algebra. It might have seemed that moving from whole numbers, to the rationals and the integers, and to the complexes would be just steps on a never-ending trail of numbers. But the fundamental theorem shows that in one real sense the complex numbers are an endpoint. Every polynomial with complex coefficients can be fully factored over the complexes and hence a polynomial of degree n has exactly n roots, counting multiplicities. The complex numbers are one of the most natural arenas for algebra.

2.8 Exercises

1. Cardano solved the problem of breaking 10 into two parts whose product is 40. Solve the problem if the product is to be 30.
2. Cardano claims that the equation $x^3 + px = q$ always has a true (positive) solution. It is assumed here that p and q are positive numbers. Justify this claim.
3. Cardano also claims that if the equation $x^3 = px + q$ has a true (positive) solution a, then the equation $x^3 + px = q$ has the false (negative) solution $-a$. Justify this claim.
4. Use Bombelli's method to show that $\sqrt[3]{-4 + \sqrt{-200}} + \sqrt[3]{-4 - \sqrt{-200}} = 4$.
5. The use of complex numbers necessitated some new caution in algebraic computation. We certainly view the two rational numbers $\frac{-4}{1}$ and $\frac{4}{-1}$ as equal. We also use the fact that $\sqrt{\frac{p}{q}} = \frac{\sqrt{p}}{\sqrt{q}}$ quite freely. Show that the use of this identity leads to two different answers when applied to $\frac{-4}{1}$ and $\frac{4}{-1}$.

respectively. (Note that if $a > 0$, then $\sqrt{a}$ denotes the positive square root of a. Use the fact that $\frac{1}{i} = -i$.)

Bibliography

Cardano's work on algebra is an early original source that can be read without too much difficulty. Consulting a general history like that of Katz [9] can smooth the reading.

Cardano, Girolamo. *The Great Art, or the Rules of Algebra*, translated and edited by T. Richard Witmer. Cambridge MASS.: MIT Press, 1968.

Euler's work is similarly accessible.

Euler, Leonhard, *Elements of Algebra*, translated by John Hewlett. New York: Springer-Verlag, 1984.

Chapter 2 Highlights

In this chapter, we climbed a ladder of abstraction. The number theory we did in Chapter 1 was based on the Division Algorithm, a statement about quotients and remainders. **Modular arithmetic** distills these results into an arithmetic of remainders. To say that a **congruence** $ax \equiv b$ mod m has a solution means that we can find an integer x such that ax and b have the same remainders after division by m. The role of equality of familiar arithmetic equations is replaced by the equivalence relation of "having the same remainder." The facts of number theory translate into important facts of modular arithmetic.

- If $a \equiv b$ mod m and $c \equiv d$ mod m, then $a + c \equiv b + d$ mod m, $ac \equiv bd$ mod m, and $ra \equiv rb$ mod m.
- The congruence $ax \equiv b$ mod m has a solution if and only if the $\gcd(a, m)$ divides b.
- The solution set of $ax \equiv b$ mod m is the union of d **congruence classes** when d divides b, where d is the $\gcd(a, m)$.

Abstracting further, we introduced the arithmetic operations of addition and multiplication mod m on the set $Z_m = \{0, 1, 2, \ldots, m - 1\}$, the set of remainders after division by m as prescribed by the Division Algorithm. In this way, the remainders themselves become the objects we add and multiply. We noticed similarities and differences with ordinary integer arithmetic.

The progression of abstraction culminated with the introduction of the term **ring**. A ring is a set with an addition operation and a multiplication operation that are related via a distributive law. The definition of a ring distills the arithmetic properties common to familiar sets such as the integers and the real numbers and some less familiar sets like Z_m and $M_{2,2}$. Some rings are **commutative rings**, others not. Some rings have a **unity**, others not. In rings with unity, some elements can be **units** and others not. Some rings have **zero divisors**, others not. A commutative ring with unity and without zero divisors is much like the set of integers. We call such a ring an

integral domain. An integral domain in which every nonzero element is a unit is called a **field**. The rings Q, R, and C are our most familiar examples of fields. If p is prime, Z_p is a field. Unlike Q, R, and C, which have **characteristic 0**, the field Z_p has characteristic p because the sum of 1 with itself p times is 0.

We ended the chapter with a brief review of the arithmetic of the field of complex numbers. Its importance in analysis and algebra comes in part because every polynomial with complex coefficients has a complex root. We begin our study of polynomials in the next chapter.

Chapter Questions

1. Write the definitions of the following terms.
    i. $a \equiv b \mod m$
    ii. congruence class
    iii. principal representative
    iv. linear congruence
    v. ring
    vi. ring with unity
    vii. unit
    viii. zero divisor
    ix. integral domain
    x. field
    xi. characteristic of a ring

2. Prove that congruence is an equivalence relation using Proposition 1 of Section 2.1. (You will then have an alternate proof to Theorem 3 of Section 2.1.)

3. If x is the principal representative of $[y]_m$, what is the principal representative of $[-y]_m$?

4. Use Theorem 2 of Section 2.1 to find $42^{3335} \mod 13$ without evaluating 42^{3335}.

5. Express the solutions to the following linear congruences in terms of unions of congruence classes.
    i. $6x \equiv 9 \mod 15$
    ii. $7x \equiv 2 \mod 12$
    iii. $3x \equiv 7 \mod 12$
    iv. $23x \equiv 1 \mod 211$

6. Give an example that shows that the following statement is *not* true: $(a + b) \mod m = a \mod m + b \mod m$.

7. Solve the following system of simultaneous congruences.
$$x \equiv 5 \mod 7$$
$$x \equiv 4 \mod 8$$
$$x \equiv 1 \mod 11$$

8. Let $a \equiv b \mod m$ and $a \equiv b \mod n$. Prove that $a \equiv b \mod(\text{lcm}(m, n))$.

9. Perform the indicated arithmetic.

   i. $16 + 17$ in Z_{33}
   ii. $16 \cdot 17$ in Z_{33}
   iii. $14 + 15$ in Z_{20}
   iv. $14 \cdot 15$ in Z_{20}

10. Find the multiplicative inverse of 13 in Z_{17}.
11. Solve for x in the following expressions.

    i. $13 \cdot x = 4$ in Z_{17}
    ii. $4 \cdot x = 2$ in Z_{13}
    iii. $6 \cdot x = 3$ in Z_9

12. Find the units in Z_{15} and find their multiplicative inverses.
13. Find the multiplicative inverse of $m - 1$ in Z_m for several values of m. Find a formula for $(m - 1)^{-1}$ and prove that your result holds in general.
14. Characterize the zero divisors of the ring Z_m and find all the zero divisors of the rings Z_{17} and Z_{20}.
15. Characterize the units of the ring Z_m and find all the units of Z_{17} and Z_{20}.
16. Find the multiplicative inverse of $1 + 3\sqrt{2}$ in the ring $Q(\sqrt{2})$ and use it to solve the equation $(1 + 3\sqrt{2})x = 1 - 5\sqrt{2}$.
17. Let $x = \begin{bmatrix} 1 & -2 \\ 3 & 1 \end{bmatrix}$ and $y = \begin{bmatrix} 4 & 0 \\ 7 & 1 \end{bmatrix}$. Evaluate the expression $(x + y)^2$.
18. Let $Z_p[i]$ denote the ring $\{a + bi : a, b \in Z_p \text{ and } i^2 = -1\}$.
   i. Show that if p is not prime, then $Z_p[i]$ is not an integral domain.
   ii. Assume p is prime. Show that every nonzero element in $Z_p[i]$ is a unit if and only if $x^2 + y^2 \neq 0 \mod p$ for any pair of elements x and y in Z_p.
19. Let $z = (2 - 5i)$ and $w = (1 - 2i)$. Find $\bar{z}, z + w, zw, z/w, w^2$, and z^3.
20. Solve for z in the equation $z^2 + (2i + 1)z + i$.
21. Express all the eighth roots of unity in terms of radicals. Note, $x^4 + 1 = (x^2)^2 - (i^2)$.
22. Use de Moivre's formula to find cube roots of i.

3
POLYNOMIALS

> Solving polynomial equations was the principal focus of algebraic activity before the time of Abel and Galois (c. 1830). The main goal was to find an expression for the roots of a given polynomial in terms of sums, products and roots of its coefficients. The quadratic formula accomplishes this goal for quadratic polynomials. The question was whether similar formulas could be found for polynomials of higher degrees, and if so, to determine in what number system would the roots lie. (The coefficients of $x^2 + 2$ are in the ring of integers but its roots are complex numbers, not in the same ring as its coefficients.) In this chapter, we pursue these questions and related issues. Along the way, we will repeat much of the agenda of Chapter 1 as we develop a theory of division, remainders, and factorization for polynomials. Our principal tools turn out to be a Division Algorithm for Polynomials and Euclid's Algorithm for Polynomials.

3.1 POLYNOMIAL ARITHMETIC

In high school algebra, the polynomials studied usually had coefficients that were either integers or rational numbers. We will extend the scope of that investigation to consider polynomials with coefficients in other commutative rings.

> **DEFINITION 1.** A **polynomial** $p(x)$ **with coefficients in a commutative ring** R is an expression of the form $a_n x^n + a_{n-1} x^{n-1} + \ldots + a_0$, where $a_i \in R$ for $i = 0, 1, \ldots, n$ and x is a variable. The set of all polynomials with coefficients in R is denoted by $R[x]$.

The variable x is simply a placeholder; it is not assumed to stand for an element of the ring R. Two polynomials $p(x)$ and $g(x)$ in $R[x]$ are **equal** if and only if all their coefficients are identical. Often, we will work with polynomials simply as algebraic symbols. But we will also think of a polynomial $f(x) \in R[x]$ as defining a function $f : R \to R$ in the following way: If $f(x) = b_n x^n + \ldots + b_0$ and $a \in R$, then $f(a) = b_n a^n + \ldots + b_0$, which is an element of R. If $f(a) = 0$, then a is called a **root** of $f(x)$ or a **zero** of $f(x)$.

EXAMPLES

1. Let $p(x) = x^2 + \sqrt{2}x + (3+i)$ be a polynomial in $C[x]$, the set of polynomials with complex coefficients. Here, $a_2 = 1, a_1 = \sqrt{2}$, and $a_0 = 3+i$. By applying the quadratic formula, we find that its roots are $\dfrac{-\sqrt{2} \pm \sqrt{-10 - 4i}}{2}$. Since the square root of any complex number is another complex number, the roots of any quadratic polynomial with coefficients in C are also in C. The Fundamental Theorem of Algebra, which is proved in Section 3.5, guarantees that all polynomials in $C[x]$ have roots in C.

2. The expression $g(x) = 2x^3 + 3x + 3$ can be regarded as a polynomial in $Z[x], Q[x], R[x]$, or $C[x]$. However, this polynomial has no roots in Z or in Q. It has one root in R and two additional roots in C. Later in this chapter, we will develop the algebraic means to determine facts like these.

3. Regarded as members of $Z_3[x]$, the polynomials $x^3 + 2x + 1$ and $4x^3 - x - 2$ are identical polynomials because their coefficients are equal in Z_3.

4. The polynomials $g(x) = x^5 + x + 1$ and $h(x) = x^3 + x^2 + 1$ are distinct polynomials in $Z_2[x]$ because they have different coefficients. However, as functions on Z_2, they are identical since $g(0) = h(0) = 1$ and $g(1) = h(1) = 1$. There is a subtle but critical difference between polynomials as functions and polynomials as algebraic objects. ∎

Let R be any commutative ring. Addition and multiplication of polynomials in $R[x]$ is carried out in the manner familiar from high school algebra. For example, let $p(x) = 3x^3 + 2x + 3$ and $g(x) = x^2 + 3x + 3$ be polynomials in $Z_5[x]$. Then

$$p(x) + g(x) = 3x^3 + x^2 + 0x + 1. \text{ (Add coefficients mod 5.)}$$

The product is

$$p(x)g(x) = 3x^5 + 4x^4 + x^3 + 4x^2 + 0x + 4. \text{ (Multiply and add coefficients mod 5.)}$$

More formally, let $p(x) = a_n x^n + a_{n-1} x^{n-1} + \ldots + a_0$ and $q(x) = b_m x^m + b_{m-1} x^{m-1} + \ldots + b_0$ be polynomials in $R[x]$. Let $k = \max(m, n)$ and let $s = m + n$. Assume that $a_i = 0$ if $i > n$ and $b_i = 0$ if $i > m$. The **sum** of $p(x)$ and $q(x)$ is defined by

$$p(x) + q(x) = (a_k + b_k)x^k + (a_{k-1} + b_{k-1})x^{k-1} + \ldots + (a_0 + b_0).$$

The **product** of $p(x)$ and $q(x)$ is defined by

$$p(x)q(x) = c_s x^s + c_{s-1} x^{s-1} + \ldots + c_0,$$

where

$$c_t = a_t b_0 + a_{t-1} b_1 + \ldots + a_1 b_{t-1} + a_0 b_t = \sum_{i+j=t} a_i b_j \text{ for } t = 0, 1, \ldots, s.$$

It is straightforward to check that $R[x]$ is itself a commutative ring when its ring of coefficients R is a commutative ring. If R is a ring with unity, then so is $R[x]$. Its unity element is the polynomial $p(x) = 1$, where 1 is the unity of R.

> **DEFINITION 2.** The **degree** of a polynomial $p(x) = a_n x^n + a_{n-1} x^{n-1} + \ldots + a_0$ is n if $a_n \neq 0$ and $a_i = 0$ for all $i > n$.

We call a_n the **leading coefficient** of $p(x)$ and we write $\deg(p(x)) = n$. If $a_n = 1$ (the unity of R), we say that $p(x)$ is **monic**. The polynomial $p(x) = 0$ does not have a degree.

EXAMPLES

5. In $Q[x]$, the polynomial $p(x) = 3$ has degree 0. The degree of $4x^3 + 3$ is 3. The polynomial $q(x) = x^2 + 0.5x - 3$ is monic of degree 2. The polynomial $r(x) = 0$ does not have a degree.
6. The polynomial $p(x) = 3x^3 + 3x + 3$, regarded as a polynomial in $Z[x]$, has degree 3 and it is not monic. Regarded as a polynomial in $Z_3[x]$, $p(x)$ does not have a degree since its coefficients are all equal to 0 modulo 3. Regarded as a polynomial in $Z_2[x]$, $p(x)$ is monic of degree 3 since $3 \equiv 1 \bmod 2$.
7. Let $p(x) = 2x^2 + 2x + 2$ and $q(x) = 2x^2 + x + 1$. If we regard $p(x)$ and $q(x)$ as polynomials in Z, then $p(x) + q(x) = 4x^2 + 3x + 3$ and $p(x)q(x) = 4x^4 + 6x^3 + 8x^2 + 4x + 2$. However, if we carry out the addition and multiplication in $Z_4[x]$, $p(x) + q(x) = 3x + 3$ and $p(x)q(x) = 2x^3 + 2$. In the latter case, the degrees of the sum and product are less than expected. ∎

In general, if p and q are nonzero polynomials such that $p + q = 0$, then $p + q$ has no degree and similarly for $p \cdot q = 0$. Otherwise,

$$\deg(p(x) + q(x)) \leq \max(\deg(p(x)), \deg(q(x))),$$

and

$$\deg(p(x)q(x)) \leq \deg(p(x)) + \deg(q(x)).$$

Recall that an element $z \neq 0$ in a ring R is a **zero divisor** if there is an element $y \neq 0$ in R such that $zy = 0$ or $yz = 0$. For instance, in Z_6 the element 3 is a zero divisor since $2 \cdot 3 = 0$. If R has no zero divisors, then $\deg(p(x)q(x)) = \deg(p(x)) + \deg(q(x))$ because the product of the leading coefficients cannot be zero. Thus, if R is an integral domain, so is $R[x]$.

Recall also that an element $u \in R$ is a **unit** if there is a value of $y \in R$ such that $uy = yu = 1$. So 2 is a unit in Z_5 since $2 \cdot 3 = 1$ in Z_5. Let R be an integral domain. Then the only units in $R[x]$ are the degree zero polynomials $p(x) = c$ for which c is a unit in R. For these polynomials, we let $q(x) = d$ where $cd = 1$. Then $p(x)q(x) = 1$. Any other nonzero polynomial would be of the form $f(x) = a_n x^n + \ldots + a_0$ where $n \geq 1$ and $a_n \neq 0$. If $g(x) = b_k x^k + \ldots + b_0$ with $k \geq 0$ and $b_k \neq 0$, then $f(x)g(x) = a_n b_k x^{n+k} + \ldots + a_0 b_0$. Since $a_n b_k \neq 0$, $\deg(f(x)g(x)) \geq n \geq 1$, and we see that $f(x)g(x) \neq 1$. Hence $R[x]$ is not a field even if R is a field.

The Division Algorithm for Polynomials

The Division Algorithm and Euclid's Algorithm are two of the principal tools of elementary number theory. We now develop an analogous division algorithm for polynomials. (Euclid's Algorithm for Polynomials is the topic of the next section.) Its proof recaptures the process of polynomial division familiar from high school mathematics. Let's look at the first step that we take when we divide $2x^5 + x - 1$ by $3x^2 + 2$ in $Q[x]$:

$$\begin{array}{r} \frac{2}{3}x^3 \ldots \\ 3x^2 + 2 \overline{)\, 2x^5 + x - 1.} \end{array}$$

We must use the reciprocal of 3. In general, when dividing $g(x) = ax^n + \ldots + a_0$ by $p(x) = b_m x^m + \ldots + b_0$, the first term in the quotient is $\frac{a_n}{b_m} x^{n-m}$, which requires the ring R to contain b_m^{-1}. Thus, to carry out the division of $g(x)$ by $p(x)$ in $R[x]$, the leading coefficient of $p(x)$ must be a unit in R. To assure this, we shall assume that the coefficients of our polynomials come from a field for the remainder of this section.

THEOREM 1 *(The Division Algorithm for Polynomials)* Let F be a field and let $p(x)$ and $g(x)$ be two polynomials in $F[x]$ with $p(x) \neq 0$. There are polynomials $q(x)$ and $r(x)$ such that $g(x) = p(x)q(x) + r(x)$ and either $r(x) = 0$ or $\deg(r(x)) < \deg(p(x))$. Furthermore, $q(x)$ and $r(x)$ are unique. (We call $q(x)$ the **quotient** and call $r(x)$ the **remainder**.)

Proof. First we show how to find the quotient and remainder, $q(x)$ and $r(x)$. If $g(x) = 0$ or if $\deg(g(x)) < \deg(p(x))$, we let $q(x) = 0$ and $r(x) = g(x)$. Then

$$g(x) = 0 \cdot p(x) + g(x)$$

as required by the theorem. Now suppose that $g(x) \neq 0$ and that $\deg(g(x)) \geq \deg(p(x))$. We proceed by induction on $\deg(g(x))$. Suppose that $\deg(g(x)) = 0$ so that $g(x) = a_0$, a nonzero constant polynomial. Since $\deg(p(x)) \leq \deg(g(x))$, the degree of $p(x)$ is also 0 so that $p(x) = b_0$. Neither a_0 nor b_0 is 0. Since the coefficients are from a field, we can set $q(x) = \frac{a_0}{b_0} = q_0$ and $r(x) = 0$. Thus $g(x) = q_0 b_0 + 0$ as required by the theorem. Now let $n > 0$ be an integer and suppose that the theorem is true for polynomials of degree less than n. Let $g(x) = a_n x^n + a_{n-1} x^{n-1} + \ldots + a_0$ and let $p(x) = b_m x^m + b_{m-1} x^{m-1} + \ldots + b_0$, where $m \leq n$ and $b_m \neq 0$. Let

$$g_1(x) = g(x) - \frac{a_n}{b_m} x^{n-m} p(x).$$

(This mimics the first step of long division.) The polynomial $g_1(x)$ has degree strictly less than n. By our induction hypothesis, the theorem holds for $g_1(x)$. So we can find polynomials $q_1(x)$ and $r(x)$ such that $g_1(x) = q_1(x) p(x) + r(x)$. Substituting $g(x) - \frac{a_n}{b_m} x^{n-m} p(x)$ for $g_1(x)$ and solving for $g(x)$, we have

$$g(x) = \left(q_1(x) + \frac{a_n}{b_m} x^{n-m} \right) p(x) + r(x).$$

Set $q(x) = q_1(x) + \frac{a_n}{b_m}x^{n-m}$. Thus $g(x) = q(x)p(x) + r(x)$ with either $r(x) = 0$ or $\deg(r(x)) < \deg(p(x))$ as required by the theorem.

To show uniqueness, assume that $g(x) = q(x)p(x) + r(x) = q_1(x)p(x) + r_1(x)$. Regrouping, we obtain

$$(q(x) - q_1(x))p(x) = r_1(x) - r(x).$$

If $q(x) - q_1(x) \neq 0$, the degree of $(q(x) - q_1(x))p(x)$ would be greater than or equal to $\deg(p(x))$. However, on the other side of the equation, $\deg(r_1(x) - r(x)) < \deg(p(x))$. Thus $q(x) = q_1(x)$ and $r(x) = r_1(x)$ and the theorem is proved. ▲

EXAMPLE 8

To finish the problem that introduced this section, we divide $2x^5 + x - 1$ by $3x^2 + 2$ in $Q[x]$:

$$2x^5 + x - 1 = (3x^2 + 2)\left(\frac{2}{3}x^3 - \frac{4}{9}x\right) + \left(-1 + \frac{17}{9}x\right).$$

In this example, $q(x) = \left(\frac{2}{3}x^3 - \frac{4}{9}x\right)$ and $r(x) = \left(-1 + \frac{17}{9}x\right)$. ■

EXAMPLE 9

We carry out the division of $x^3 + x + 3$ by $2x + 1$ in $Z_7[x]$. Recall that in Z_7, $2^{-1} = 4$ since $2 \cdot 4 \equiv 1 \mod 7$.

$$\begin{array}{r}
4x^2 + 5x + 5 \\
2x+1{\overline{\smash{\big)}\,x^3 + x + 3}} \\
\underline{x^3 + 4x^2} \\
3x^2 + x + 3 \\
\underline{3x^2 + 5x} \\
3x + 3 \\
\underline{3x + 5} \\
5
\end{array}$$

In this example, $q(x) = 4x^2 + 5x + 5$ and $r(x) = 5$. ■

If the remainder is $r(x) = 0$ when $f(x)$ is divided by $p(x)$, then $f(x) = q(x)p(x)$. In this case we say that **$p(x)$ divides $f(x)$** or that **$p(x)$ is a factor of $f(x)$**. When $p(x)$ **divides $f(x)$** we sometimes write $p(x)|f(x)$.

EXAMPLE 10

If we divide $x^4 - x^3 - x + 1$ by $x^2 + x + 1$ in $Q[x]$, the results are

$$x^4 - x^3 - x + 1 = (x^2 - 2x + 1)(x^2 + x + 1) + 0.$$

Thus $(x^2 + x + 1)$ divides $(x^4 - x^3 - x + 1)$ in $Q[x]$. ■

The Division Algorithm for Polynomials has several immediate and important corollaries for polynomials, thought of as functions.

COROLLARY 1 (*The Remainder Theorem*) Let $f(x)$ be any polynomial in $F[x]$ and let $p(x) = x - a$. Suppose $f(x) = q(x)(x - a) + r_0$ as in the Division Algorithm. Then $r_0 = f(a)$.

Proof. Because $\deg(x - a) = 1$, the remainder must equal 0 or have degree 0 and hence be a constant r_0. Evaluating $f(x)$ at $x = a$, we find that $f(a) = q(a)(a - a) + r_0 = r_0$. ▲

COROLLARY 2 (*The Root Theorem*) For any polynomial $f(x)$ in $F[x]$ and any $a \in F$, $f(a) = 0$ if and only if $x - a$ is a factor of $f(x)$.

The proof is an exercise. (Use Corollary 1.)

EXAMPLE 11

Let $g(x) \in Q[x]$ be the polynomial $x^7 + 3x^2 + 2x$. The remainder of $g(x)$ after division by $x - 1$ is 6 because $g(1) = 6$. Thus $x = 1$ is not a root of $g(x)$. However, since $g(-1) = 0$, -1 is a root of $g(x)$ and $x - (-1) = x + 1$ is a factor of $g(x)$. In fact, $g(x) = x(x + 1)(x^5 - x^4 + x^3 - x^2 + x + 2)$. ■

EXAMPLE 12

Let $f(x) = x^6 + 16x^5 - 76x^4 + 79x^3 - 16x^2 + 76x - 80$ and consider $f(x)$ to be a polynomial in $Q[x]$. A lucky guess shows that $f(1) = 0$ and that $(x - 1)$ is a factor of $f(x)$. We can use the Division Algorithm to divide $f(x)$ by $(x - 1)$ to obtain $f(x) = (x - 1)g(x)$ where $g(x) = x^5 + 17x^4 - 59x^3 + 20x^2 + 4x + 80$. If we can find a root of $g(x)$, say $x = a$, we can iterate this procedure to find that $g(x) = (x - a)h(x)$, where $h(x)$ has degree 4. Indeed, $g(2) = 0$. So $g(x) = (x - 2) \cdot h(x)$ where $h(x) = (x^4 + 19x^3 - 21x^2 - 22x - 40)$ and $f(x) = (x - 1)(x - 2) \cdot (x^4 + 19x^3 - 21x^2 - 22x - 40)$. We have found two roots of $f(x)$. Clearly, we can expect at most 4 more roots this way, for a maximum of six roots for $f(x)$, a polynomial of degree six. Actually, as a polynomial in $Q[x]$, $f(x) = (x - 1)(x - 2)^2 \cdot (x + 20)(x^2 + x + 1)$. The quadratic $x^2 + x + 1$ has no roots in Q. Thus $f(x)$ has four roots in Q, counting **multiplicity**, i.e., counting the root $x = 2$ twice because $(x - 2)^2$ is a factor of $f(x)$. ■

As Example 12 suggests, a polynomial in $F[x]$ of degree n has at most n roots, counting multiplicities. The following corollary formalizes this fact. The procedure found in Example 12 (divide and conquer!) points to a proof by induction.

COROLLARY 3 Let F be a field. A polynomial $g(x) \in F[x]$ of degree $n \geq 1$ has at most n roots in F, counting multiplicities.

The proof is an exercise. (Use induction on n.)

SUMMARY

In this section, we expanded our acquaintance with polynomials to include **polynomials with coefficients in an arbitrary commutative ring R**. We defined the **degree** of a polynomial and the **addition** and **multiplication** of polynomials in a manner familiar from high school. We noted that the ring of polynomials $R[x]$ is an integral domain if R is an integral domain, but it is not a field, even if R is a field. With the notion of degree giving us a sense of size for polynomials, we formulated the **Division Algorithm for Polynomials** with coefficients in a field. Its corollaries, the **Root and Remainder Theorems**, are extremely useful for factoring polynomials. We concluded with the observation that a polynomial of degree n with coefficients in a field can have at most n roots. But finding the roots of a polynomial and its factorization can present quite a challenge. We explore these issues further in the next sections.

3.1 Exercises

1. Write down two polynomials in $Z[x]$ that are not equal in $Z_5[x]$ but are equal in $Z_3[x]$.

2. How many polynomials are there of degree 2 in $Z_2[x]$? $Z_5[x]$? $Z_n[x]$? Of these, how many are monic?

3. Add and multiply the given polynomials:
   i. $p(x) = 2x^2 + 3x + 1$ and $q(x) = 3x + 2$ in $Z[x]$
   ii. $p(x) = 2x^2 + 3x + 1$ and $q(x) = 3x + 2$ in $Z_5[x]$
   iii. $p(x) = 2x^2 + 2x + 1$ and $q(x) = 2x + 2$ in $Z_3[x]$
   iv. $p(x) = (2 + 3i)x^2 + 3x + i$ and $q(x) = 3x + 2i$ in $C[x]$

4. Let $p(x) = a_m x^m + \ldots + a_0$ and $q(x) = b_k x^k + \ldots + b_0$ be polynomials in $Z[x]$ and let $p_n(x)$ and $q_n(x)$ be the polynomials $p(x)$ and $q(x)$ with coefficients reduced mod n. For example, if $p(x) = 7x^2 + 13x - 7$, then $p_5(x) = 2x^2 + 3x + 3$. Prove that $(p(x) + q(x))_n = (p_n(x) + q_n(x))$ mod n and $(p(x)q(x))_n = (p_n(x)q_n(x))$ mod n. That is, show that the resulting polynomials are the same whether we reduce the coefficients before adding and multiplying or after.

5. For each of the following, find $q(x)$ and $r(x)$ as in the Division Algorithm for Polynomials for the given $g(x)$ and $p(x)$.
   i. $g(x) = x^3 - 1$, $p(x) = x - 1$ in $Z[x]$
   ii. $g(x) = x^n - 1$, $p(x) = x - 1$, $n = 1, 2, \ldots$, in $Z[x]$
   iii. $g(x) = 2x^4 + x^3 + 3x + 1$, $p(x) = x^2 + x + 1$ in $Z_5[x]$
   iv. $g(x) = x^5 + x + 1$, $p(x) = x^2 + x + 1$ in $Z_2[x]$
   v. $g(x) = x^3 + (1 + 2i)x + 1$, $p(x) = x - 1$ in $C[x]$
   vi. Learn how to carry out the above exercises in your favorite CAS (computer algebra system).

6. Let R be a ring. Prove that the ring $R[x]$ has no zero divisors if and only if R has no zero divisors.

7. Determine if $p(x)$ divides $f(x)$ in $Q[x]$.
   i. $f(x) = x^5 - x^3 + x^2 - 2x + 1$ and $p(x) = x^2 + 1$
   ii. $f(x) = x^6 - x^4 + 2x^3 - x + 2$ and $p(x) = x^3 + 1$
   iii. $f(x) = x^5 - x^3 + x^2 - 2x + 1$ and $p(x) = x + 1$
8. Show that we can carry out division in $R[x]$ under the less restrictive condition that the leading coefficient of $p(x)$ is a unit in R.
9. Find the remainder of $x^{15} + x^7 - 3x^2 + 1$ after division by $x - 1$ in $Q[x]$.
10. Find the remainder of $x^4 + 3x + 2$ after division by $x + 3$ in $Z_5[x]$.
11. Give an example of a polynomial of degree 2 in $Z_6[x]$ that has more than 2 roots. (Note that Z_6 is not a field and so this does not contradict Corollary 3.)
12. Determine which polynomials in $Z_2[x]$ have roots in Z_2.
13. Determine which quadratics in $Z_3[x]$ have no roots in Z_3.
14. Prove Corollary 2.
15. Prove Corollary 3. (Use induction on the degree of $g(x)$.)
16. Prove that $x^d - 1$ is a factor of $x^{p-1} - 1$ if and only if d is a factor of $p - 1$.

To the Teacher

The high school algebra curriculum typically combines both purely algebraic topics with what in college we usually call analysis, namely, the study of functions and their graphs. After studying lines as graphs of the relation $y = mx + b$, students graduate to studying the graphs of quadratic functions. The emphasis is on shape and translation. Is the parabolic graph turned up or down, narrow or wide, moved to the right or left? Does it cross the x-axis? Often, this is how polynomials enter the curriculum. These considerations are vitally important precalculus topics. However, they tend to obscure the purely algebraic aspects of polynomials that we also ask students to master such as the multiplication, division and factoring of polynomials. In the next few sections, you will be relearning many of these topics (albeit in an expanded context) and perhaps reflecting on the challenges this material poses to the first time learner. The *In the Classroom* essay, Section 3.9 at the end of this chapter, engages in a full discussion of the hurdles to teaching and learning this material. In this endnote, we look at some ways to connect polynomial arithmetic to numerical arithmetic.

A quick check that you have multiplied two polynomials correctly is to evaluate factors before and after you multiply. For instance, $(x + 1)(x + 3) \neq x^2 + 3$ because, letting $x = 3, (3 + 1)(3 + 3) \neq 3^2 + 3$. This is a reminder that the distributive law holds for polynomials because it holds for numbers. So $(x + 1)(x + 3) = x^2 + 4x + 3$. An interesting connection to arithmetic is the fact that the value of $p(x) = x^2 + 4x + 3$ cannot be a prime number for *any* strictly positive integer x. The infinite list $\{p(1), p(2), p(3), \ldots\} = \{8, 15, 24, \ldots\}$ is a list of only composite numbers. (How Fermat used similar reasoning in his work on perfect numbers is discussed in Section 1.9.)

Polynomial division does not always evaluate to what we expect from the Division Algorithm for integers, even when the coefficients are all integers. For instance, dividing $x^5 + 2x^4 + 2x^3 - 4x - 4$ by $x^3 + x - 2$, we get

$$x^5 + 2x^4 + 2x^3 - 4x - 4 = (x^2 + 2x + 1)(x^3 + x - 2) + (-x - 2).$$

Evaluating at $x = 5$, we have $4601 = 36 \cdot 128 - 7$. This is a true statement but it is not in the format of the Division Algorithm, which is $4601 = 35 \cdot 128 + 121$. Still, polynomial division has interesting numerical aspects. In the following carefully chosen (rigged?) polynomial, all coefficients are positive integers between 0 and 9.

$$x^5 + 4x^4 + 8x^3 + 9x^2 + 7x + 3 = (x^2 + 2x + 2)(x^3 + 2x^2 + 2x + 1) + (x + 1).$$

If we simply pull off the coefficients in order, we have the true statement that $148973 = 122 \cdot 1221 + 11$. The same result is obtained by substituting $x = 10$ into the polynomial equation. It's how our place system works! A bit more interesting is that the number fact is true in base 11, base 12, etc. For instance, in base 11, the expression $148973 = 122 \cdot 1221 + 11$ means

$$1 \cdot 11^5 + 4 \cdot 11^4 + 8 \cdot 11^3 + 9 \cdot 11^2 + 7 \cdot 11 + 3$$
$$= (1 \cdot 11^2 + 2 \cdot 11 + 2)(1 \cdot 11^3 + 2 \cdot 11^2 + 2 \cdot 11 + 1)$$
$$+ (1 \cdot 11 + 1),$$

which is equivalent to the base 10 fact that

$$231432 = 145 \cdot 1596 + 12.$$

Our particular example obviously works for any base greater than 10. But it is also valid in base 5 if we think of the coefficients as quantities of the powers of 5. (It's as if we have used base 5 inefficiently.)

$$1 \cdot 5^5 + 4 \cdot 5^4 + 8 \cdot 5^3 + 9 \cdot 5^2 + 7 \cdot 5 + 3$$
$$= (1 \cdot 5^2 + 2 \cdot 5 + 2)(1 \cdot 5^3 + 2 \cdot 5^2 + 2 \cdot 5 + 1)$$
$$+ (1 \cdot 5 + 1).$$

Let's clean up the left side. In base 5, $1 \cdot 5^5 + 4 \cdot 5^4 + 8 \cdot 5^3 + 9 \cdot 5^2 + 7 \cdot 5 + 3 = 2 \cdot 5^5 + 1 \cdot 5^4 + 0 \cdot 5^3 + 0 \cdot 5^2 + 2 \cdot 5 + 3$. The base 5 arithmetic fact is thus

$$210023 = 122 \cdot 1221 + 11.$$

The equivalent base 10 expression is

$$6888 = 37 \cdot 186 + 6.$$

Polynomials hold a wealth of arithmetic facts in very small packages.

Tasks

1. Determine if an irreducible polynomial in $\mathbf{Z}[x]$, with integer coefficients can assume composite values. (Note: A polynomial $p(x)$ in $\mathbf{Z}[x]$ is irreducible if its only factors in $\mathbf{Z}[x]$ are ± 1 and $\pm p(x)$.)
2. Interpret the fact that $(x^2 + x + 1)(x^2 + 2) + (x + 3) = x^4 + x^3 + 3x^2 + 3x + 5$ in base 10, in base 13, and in base 7.

3. Prove that if $p(x)$ in $\mathbf{Z}[x]$ is a reducible polynomial, then $p(x)$ is a prime number for at most finitely many integer values of x. (Note: A polynomial in $\mathbf{Z}[x]$ is reducible if it is not irreducible. See Task 1.)

3.2 THE EUCLIDEAN ALGORITHM FOR POLYNOMIALS

In number theory we use the Euclidean Algorithm to find the greatest common factor of two integers a and b without factoring either a or b. From that, we can determine if two numbers are relatively prime—a very important determination for procedures based on modular arithmetic. In this section, we develop its analog for polynomials. As with integers, we will be able to determine if two polynomials share a common factor through the division process rather than through factorization. This is good news because factoring polynomials can be even trickier than factoring numbers.

> **DEFINITION 1.** Let F be a field and let $g(x)$ and $f(x)$ be nonzero polynomials in $F[x]$. A **greatest common divisor** of $g(x)$ and $f(x)$ is a polynomial $d(x)$ of maximal degree that is a factor of both $f(x)$ and $g(x)$. We write $d(x) = \gcd(f(x), g(x))$.

As we shall see in the following example, we must say "a" greatest common divisor because a polynomial $d(x)$ that satisfies the definition is not unique.

EXAMPLE 1

Let $f(x) = x^5 + x^4 + x^3 - 2x^2 - 2x - 2$ and let $g(x) = x^4 + x^3 - x^2 - 2x - 2$ in $\mathbf{Q}[x]$. In the first exercise you will use Euclid's Algorithm for Polynomials to show that $d(x) = x^2 + x + 1$ is "a" greatest common divisor of $f(x)$ and $g(x)$. We can factor $f(x)$ and $g(x)$ as $f(x) = (x^3 - 2)(x^2 + x + 1)$ and $g(x) = (x^2 - 2)(x^2 + x + 1)$. But we can also factor $f(x)$ and $g(x)$ as

$$f(x) = \left(\frac{x^3}{2} - 1\right)(2x^2 + 2x + 2) \text{ and}$$

$$g(x) = \left(\frac{x^2}{2} - 1\right)(2x^2 + 2x + 2).$$

So $d_1(x) = (2x^2 + 2x + 2)$ is also a factor of both $f(x)$ and $g(x)$ and $d_1(x)$ has the same (maximal) degree as $d(x)$. ∎

If $f(x) = d(x)q(x)$, then $f(x) = (c \cdot d(x))(\frac{1}{c}q(x))$ for any nonzero constant c in F and similarly for $g(x)$. Thus it is easy to see that if $d(x)$ is a greatest common divisor of $f(x)$ and $g(x)$, then $c \cdot d(x)$ is also a greatest common divisor for any nonzero constant $c \in F$.

Euclid's Algorithm for Polynomials finds a greatest common divisor for two nonzero polynomials $f(x)$ and $g(x)$ in $F[x]$. It is based on the following three lemmas.

Section 3.2   The Euclidean Algorithm for Polynomials

LEMMA 1   Suppose $g(x) \in F[x]$ is a nonzero polynomial and that $f(x)$ is a factor of $g(x)$. Then $f(x)$ is a greatest common divisor of $g(x)$ and $f(x)$.

Proof. Clearly $f(x)$ is a factor of both $f(x)$ and $g(x)$. Since no polynomial of higher degree than $f(x)$ can divide $f(x)$, a greatest common divisor of $f(x)$ and $g(x)$ must be $f(x)$. ▲

LEMMA 2   Suppose that $g(x) \neq 0$ and, after applying the Division Algorithm to $f(x)$ and $g(x)$, we have $f(x) = g(x)q(x) + r(x)$. Let $h(x) \in F[x], h(x) \neq 0$. Then $h(x)$ divides both $f(x)$ and $g(x)$ if and only if $h(x)$ divides both $g(x)$ and $r(x)$.

Proof. If $f(x) = s(x)h(x)$ and $g(x) = t(x)h(x)$, then $r(x) = h(x)(s(x) - t(x)q(x))$. Conversely, if $g(x) = t(x)h(x)$ and $r(x) = w(x)h(x)$, then $f(x) = h(x)(t(x)q(x) + w(x))$. ▲

The next lemma is the critical fact behind the Euclidean Algorithm for Polynomials. It is an immediate corollary to Lemma 2. We leave its simple proof as an exercise.

LEMMA 3   Suppose that $f(x) = g(x)q(x) + r(x)$. Then $d(x)$ is a greatest common divisor of $f(x)$ and $g(x)$ if and only if $d(x)$ is a greatest common divisor of $g(x)$ and $r(x)$.

Proof. Exercise 2.

Lemma 3 tells us that if $f(x)$ is not a factor of $g(x)$, we can reduce the problem of finding a greatest common divisor of $f(x)$ and $g(x)$ to the easier problem of finding a gcd of $g(x)$ and $r(x)$, an easier problem because the degrees are lower. We can then iterate the process by dividing $g(x)$ by $r(x)$, etc., each time reducing degrees. The process must halt because at each step we replace $r(x)$ with 0 or with a polynomial of lower degree. After a finite number of steps, the remainder must be 0. In that case, Lemma 1 tells us that the last nonzero remainder is a greatest common divisor of $f(x)$ and $g(x)$.

Euclid's Algorithm for Polynomials in $F[x]$:

Let f and g be nonzero polynomials in $F[x]$. Apply the Division Algorithm a sufficient number of times, to obtain

$$f = gq_1 + r_1$$
$$g = r_1q_2 + r_2$$
$$r_1 = r_2q_3 + r_3$$
$$\vdots$$
$$r_{n-2} = r_{n-1}q_n + r_n$$
$$r_{n-1} = r_n q_{n+1} + 0.$$

Then $\gcd(f, g) = r_n$.

Notice that the steps are exactly analogous to Euclid's Algorithm for integers. To convince ourselves that the algorithm works, look at the last line of the algorithm. We see that r_n divides r_{n-1} and thus r_n is a greatest common divisor of r_n and r_{n-1}. (Lemma 1.) Go backward. From the next to the last line and from Lemma 3, we see that r_n is a greatest common divisor of r_{n-1} and r_{n-2}. Continuing backward, we see that r_n is a greatest common divisor of r_{n-2} and r_{n-3}, and so forth. When we reach the top line, we arrive at the conclusion that r_n is a greatest common divisor of f and g.

We previously observed that if $r_n(x)$ is a gcd of $f(x)$ and $g(x)$, then any constant multiple of $r_n(x)$ is also a gcd of $f(x)$ and $g(x)$. We show the converse, namely, that any two greatest common divisors differ from each other by a multiplicative constant as follows. Suppose that the polynomials $d(x)$ and $r_n(x)$ are both greatest common divisors of the polynomials $f(x)$ and $g(x)$. Lemma 2 tells us that any common divisor of $f(x)$ and $g(x)$ also divides r_n. Therefore, $d(x)$ divides $r_n(x)$ and $\deg(r_n) \geq \deg(d)$. By the definition of greatest common divisor, the degree of $d(x)$ is maximal; no common divisor has a higher degree. So $d(x)$ and $r_n(x)$ have the same degree. Therefore $d(x) = cr_n(x)$ where c is a nonzero constant in F. Thus the set of greatest common divisors of f and g is the set of all nonzero constant multiples of r_n.

It is more natural to speak of "the greatest common divisor" of f and g. Dividing any greatest common divisor of f and g through by its leading coefficient, we obtain a monic polynomial. (A monic polynomial has leading coefficient 1.) We define "**the greatest common divisor of f and g**" to be the unique monic greatest common divisor of f and g.

EXAMPLE 2

We find "the" greatest common divisor of $f(x) = x^5 + x^3 - x^2 + x^3 - 1$ and $g(x) = x^3 + 3x^2 + x + 3$ in $Q[x]$ with Euclid's Algorithm.

Step 1. $x^5 + x^3 - x^2 - 1 = (x^3 + 3x^2 + x + 3)(x^2 - 3x + 9) + (-28 - 28x^2)$

Step 2. $x^3 + 3x^2 + x + 3 = (-28 - 28x^2)(-x/28 - 3/28) + 0$

Thus "a" greatest common divisor of $f(x)$ and $g(x)$ is $(-28 - 28x^2)$. Dividing through by -28, we find that "the" greatest common divisor is $x^2 + 1$. ∎

An important corollary to the process described by Euclid's Algorithm is the fact that the greatest common divisor of two polynomials $f(x)$ and $g(x)$ can be expressed in terms of $f(x)$ and $g(x)$.

THEOREM 4 Let $f(x)$ and $g(x)$ be polynomials in $F[x]$. Let $d(x)$ be a greatest common divisor of $f(x)$ and $g(x)$. There are polynomials $s(x)$ and $t(x)$ such that

$$d(x) = s(x)f(x) + t(x)g(x).$$

Proof. To express $d(x)$ as in the theorem, we use Euclid's Algorithm in reverse, just as we did for integers. ▲

EXAMPLE 3

In Example 2, we found that the greatest common divisor of $f(x) = x^5 + x^3 - x^2 - 1$ and $g(x) = x^3 + 3x^2 + x + 3$ is $x^2 + 1$. From Step 1 of the algorithm we have

$$x^5 + x^3 - x^2 - 1 = (x^3 + 3x^2 + x + 3)(x^2 - 3x + 9) + (-28 - 28x^2).$$

Solving for $-28 - 28x^2$, we have

$$(x^5 + x^3 - x^2 - 1) - (x^3 + 3x^2 + x + 3)(x^2 - 3x + 9) = (-28 - 28x^2).$$

Multiplying through by $-1/28$ and regrouping we have

$$\frac{-1}{28}(x^5 + x^3 - x^2 - 1) + \left(\frac{1}{28}x^2 - \frac{3}{28}x + \frac{9}{28}\right)(x^3 + 3x^2 + x + 3) = (x^2 + 1).$$

Setting $s(x) = \frac{-1}{28}$ and $t(x) = \frac{1}{28}x^2 - \frac{3}{28}x + \frac{9}{28}$, and $d(x) = x^2 + 1$, we have

$$s(x)f(x) + t(x)g(x) = d(x).$$
∎

For polynomials, the concept that is analogous to "prime number" is that of an "irreducible polynomial." An irreducible polynomial cannot be nontrivially factored.

> **DEFINITION 2.** Let R be an integral domain and let $f(x)$ be a polynomial in $R[x]$. We say that $f(x)$ is **irreducible** if, given any factorization of $f(x)$ in $R[x]$ as $f(x) = h(x)g(x)$, either $g(x)$ or $h(x)$ is a unit in R. Otherwise, $f(x)$ is said to be **reducible**.

Since every nonzero constant in a *field* F is a unit, a polynomial $f(x)$ is irreducible in $F[x]$ if and only if we cannot factor $f(x)$ into two polynomials g and h, both of **lower** degree than f. For example, in $Q[x]$, the polynomial $p(x) = x^2 + 1$ factors as $(1/2)(2x^2 + 2)$. However, $1/2$ is a unit in Q. Since there is no factorization of $p(x)$ into polynomials **both** of degree less than 2, $p(x)$ is irreducible in $Q[x]$.

EXAMPLES

4. A polynomial $f(x)$ in $F[x]$ of degree 2 or 3 is reducible if and only if one of its factors is of degree 1, and hence has a root in F. Thus $x^2 - 2$ is irreducible in $Q[x]$ since it has no rational root. However, $x^2 - 2$ is reducible as $(x - \sqrt{2})(x + \sqrt{2})$ in $R[x]$.
5. The polynomial $p(x) = x^3 + x + 1$ is irreducible in $Z_2[x]$ since neither 0 nor 1 is a root. But $p(x)$ is reducible in $Z_3[x]$ as $p(x) = (x + 2)(x^2 + x + 2)$. ∎

Factoring polynomials over rings with zero divisors can lead to some surprises. Recall that 3 is a zero divisor, not a unit, in Z_6. If we consider the polynomial $3x + 3$ in $Z_6[x]$, then we can factor it in several ways:

$$3x + 3 = 3(x + 1) = (2x + 1)(3x + 3) = (2x^2 + 1)(3x + 3).$$

Unique factorization fails in $\mathbf{Z}_6[x]$. However, if F is a field, irreducible polynomials have many of the same properties that prime numbers in $\mathbf{Z}$ have. For instance, we shall see that polynomials in $F[x]$ can be factored uniquely into irreducible polynomials, up to multiplication by units and the order of the factors. The following set of propositions leads us to the **Unique Factorization of Polynomials in $F[x]$**.

PROPOSITION 5 Suppose that $p(x)$ and $f(x)$ are polynomials in $F[x]$. Suppose that $p(x)$ is irreducible and that $p(x)$ does **not** divide $f(x)$. Then the greatest common divisor of $f(x)$ and $p(x)$ is 1.

Proof. Since $p(x)$ does not divide $f(x)$, the greatest common divisor of $p(x)$ and $f(x)$ is a polynomial $d(x)$ of degree less than $p(x)$ that divides $p(x)$. The only polynomials of degree less than $p(x)$ that divide $p(x)$ are the nonzero constants in F. Hence $d(x)$ is a constant. "The" greatest common divisor of $p(x)$ and $f(x)$ is thus 1. ▲

PROPOSITION 6 Let $p(x)$ be a polynomial of degree > 0 in $F[x]$. Then either $p(x)$ is irreducible or it can be factored into the product of irreducible polynomials.

The proof is by induction on the degree of $p(x)$. It is left as an exercise. The following proposition echoes Euclid's Lemma of Chapter 1.

PROPOSITION 7 Let $f(x)$ and $g(x)$ be polynomials in $F[x]$. Suppose that $p(x)$ is irreducible in $F[x]$ and suppose that $p(x)$ divides the product $f(x)g(x)$. Then $p(x)$ divides $f(x)$ or $p(x)$ divides $g(x)$.

Proof. Suppose that p does not divide f. By Proposition 5, we know that $\gcd(f, p) = 1$. Thus we can find polynomials s and t such that $1 = sf + tp$. Multiplying through by g, we obtain $g = sfg + tpg$. Since p divides p and p divides fg, it must be the case that p divides g. ▲

THEOREM 8 (*Unique Factorization in $F[x]$*) Let F be a field and let $f(x)$ be a nonzero polynomial in $F[x]$. Suppose that $f(x) = p_1(x)p_2(x) \cdots p_k(x) = q_1(x)q_2(x) \cdots q_m(x)$, where each $p_i(x)$ and each $q_j(x)$ is an irreducible polynomial. Then $m = k$ and, after renumbering if necessary, $p_i(x) = c_i q_i(x)$ for each $i = 1, 2, \ldots, m$ where c_i is a nonzero element of F.

Proof. The proof is by induction on the degree of $f(x)$. If $\deg(f) = 0$, then $f(x)$ is a constant. In that case $m = k = 1$ and $p_1(x) = q_1(x) = f(x)$.

Now suppose that $n > 0$, that $\deg(f) = n$, and that the theorem holds for all polynomials of degree less than n. Suppose also that $f(x)$ factors as in expression (1), where

(1) $$f(x) = p_1(x)p_2(x) \cdots p_k(x) = q_1(x)q_2(x) \cdots q_m(x)$$

where each $p_i(x)$ and each $q_j(x)$ is an irreducible polynomial. Since $p_1(x)$ divides $q_1(x)q_2(x) \cdots q_m(x)$ and $p_1(x)$ is an irreducible polynomial, we know from Proposition 7 that $p_1(x)$ divides $q_j(x)$ for some $j = 1, \ldots, m$. Renumbering if necessary, we may suppose that $j = 1$. Since $q_1(x)$ is irreducible, we have $p_1(x) = c_1 q_1(x)$ for some constant c_1 in F. Because $F[x]$ is an integral domain, we can cancel $q_1(x)$ from both sides of equation (1) to obtain

(2) $$c_1 p_2(x) \cdots p_k(x) = q_2(x) \cdots q_m(x).$$

The polynomial in expression (2) has degree less than n. We may apply our induction hypotheses to it to conclude that $k = n$ and that $p_i(x) = c_i q_i(x)$ for each $i = 1, 2, \ldots, n$. ▲

We know that any positive integer can be factored uniquely into the product of primes, but to find a factorization of a given large integer can be difficult. Similarly, simply knowing that a polynomial factors uniquely into irreducible factors does not help us find a factorization. The rest of this chapter can be regarded as an investigation of issues involving factors and roots of polynomials.

EXAMPLE 6

In Section 3.4, we shall give a method by which we can determine that the polynomial $x^5 - 6x + 3$ is irreducible in $Q[x]$ and, by the way, necessarily irreducible in $Z[x]$. An appeal to a graphing calculator will show that the function $f(x) = x^5 - 6x + 3$ has three real roots. So $x^5 - 6x + 3$ must factor in $R[x]$. In Section 3.5 we shall show that it must have two complex roots that are complex conjugates, and that $x^5 - 6x + 3$ factors into irreducible linear terms in $C[x]$, as must all polynomials in $C[x]$. In Section 6.5, we shall show that even though $x^5 - 6x + 3$ has five distinct roots in C, it is not "solvable by radicals." We cannot find the roots of $x^5 - 6x + 3$ by a formula involving sums, products and radicals of its coefficients. ■

SUMMARY

In this section, we defined the **greatest common divisor (gcd)** of two polynomials $f(x)$ and $g(x)$ with coefficients in a field F. To find the gcd of f and g without factoring, we developed **Euclid's Algorithm for Polynomials.** It finds $\gcd(f(x), g(x))$ through repeated division. The steps of the algorithm can be used to express $\gcd(f(x), g(x))$ as a sum of the form $s(x) f(x) + t(x) g(x)$. We defined what it means for a polynomial to be **irreducible**. Finally, we proved the **Unique Factorization Theorem for Polynomials** with coefficients in a field. The progression of ideas and results in this section parallels our work in Chapter 1. There, our results were all ultimately based on the Well-Ordering Principle. This section also relied on the Well-Ordering Principle, this time applied to the degrees (all nonnegative integers) of the polynomials under consideration.

3.2 Exercises

1. Use Euclid's Algorithm to show that in $Q[x]$, the gcd of $f(x) = x^5 + x^4 + x^3 - 2x^2 - 2x - 2$ and $g(x) = x^4 + x^3 - x^2 - 2x - 2$ is $d(x) = x^2 + x + 1$.
2. Prove Lemma 3.

3. Use Euclid's Algorithm to find the greatest common divisor of the following pairs of polynomials.
   i. $f(x) = x^3 + x^2 - 5x - 2$ and $g(x) = x^4 - 2x^3 - x + 2$ in $Q[x]$
   ii. $f(x) = x^3 + 2x + 2$ and $g(x) = x^4 + 3x^3 + 4x + 2$ in $Z_5[x]$
   iii. $f(x) = x^3 + x^2 + x + 1$ and $g(x) = x^4 + x^2 + 1$ in $Z_2[x]$
4. Express the greatest common divisor of the following pairs of polynomials as a combination of the given polynomials:
   i. $f(x) = x^3 + x^2 - 5x - 2$ and $g(x) = x^4 - 2x^3 - x + 2$ in $Q[x]$
   ii. $f(x) = x^3 + 2x + 2$ and $g(x) = x^4 + 3x^3 + 4x + 2$ in $Z_5[x]$
   iii. $f(x) = x^3 + x^2 + x + 1$ and $g(x) = x^4 + x^2 + 1$ in $Z_2[x]$
5. Factor each of the following polynomials completely into irreducible polynomials in $Z_2[x]$.
   i. $x^4 + x^2$
   ii. $x^4 + x^3 + x$
   iii. $x^4 + x^3 + x^2 + x$
   iv. $x^4 + x^3 + x^2 + x + 1$
6. Are there any irreducible polynomials of degree 3 in $Z_3[x]$? If so, find an example.
7. Apply the quadratic formula to the following polynomials in $Z_5[x]$ to determine whether or not they are irreducible:
   i. $x^2 + 3x + 1$
   ii. $x^2 + 2$
   iii. $x^2 + 2x + 4$
8. Factor $x^4 - 1$ completely in $R[x]$ and then factor it completely in $C[x]$.
9. Prove Proposition 6.

To the Teacher

The following are problem that are typical in an algebra 2 or precalculus class. However, the degrees of the polynomials are atypically high.

Problem 1

Factor each of the following polynomials completely:

i. $p(x) = x^7 - 5x^6 + 5x^5 + x^4 + 11x^3 - 11x^2 - 8x - 12$
ii. $g(x) = x^5 - 3x^4 + 2x^3 - x^2 + 3x - 2$

(Let's assume this means to factor over the complex numbers.)

Problem 2

Reduce the following algebraic fraction to lowest terms:

$$\frac{g(x)}{p(x)} = \frac{x^5 - 3x^4 + 2x^3 - x^2 + 3x - 2}{x^7 - 5x^6 + 5x^5 + x^4 + 11x^3 - 11x^2 - 8x - 12}$$

Problem 3

How many *distinct* roots does the following polynomial have in the complex numbers?
$$p(x) = x^7 - 5x^6 + 5x^5 + x^4 + 11x^3 - 11x^2 - 8x - 12$$
(Assume that students would know that it has seven roots, counting multiplicities.)

You might think that they are all variations on the same problem because if we can do the first, we can do the next two. Indeed, that is so. But we don't actually need the answer to the first (hard) problem to do the others. The last two problems are considerably easier than the first!

To do Problem 2, we need the gcd of both numerator and denominator, easily computed with Euclid's Algorithm. In a few steps, we can tell that $\gcd(p(x), g(x)) = x^3 - x^2 - x - 2$. We then divide both the numerator and denominator by the gcd, again an easy algorithmic task. The answer is
$$\frac{x^2 - 2x + 1}{x^4 - 4x^3 + 2x^2 + x + 6}.$$
So Euclid's Algorithm enables us to reduce an algebraic fraction to lowest terms without actually knowing what the factors of the numerator and the denominator are.

Problem 3 is equally easy. Perhaps you recall from calculus that if a polynomial has a repeated root, then that root is also a root of its derivative. In symbols, if $n > 1$ and $p(x) = (x-a)^n f(x)$, then $p'(x) = n(x-a)^{n-1} f(x) + (x-a)^n f'(x)$. Thus $p(a) = 0$ and $p'(a) = 0$. If you do the Worksheet 4 on derivatives that follows this section, you will discover the following stronger result:

> A polynomial $p(x)$ and its derivative $p'(x)$ share a root a if and only if a is a repeated root of $p(x)$.

You can prove that the greatest common divisor of a polynomial $p(x)$ and its derivative $p'(x)$ is the product of all repeated factors of $p(x)$, each to the degree one less than in $p(x)$.

EXAMPLE A

Let $q(x) = (x-2)^3(x-1)^2(x+4) = x^6 - 4x^5 - 7x^4 + 62x^3 - 124x^2 + 104x - 32$. Then
$$q'(x) = 6x^5 - 20x^4 - 28x^3 + 186x^2 - 248x + 104$$
and
$$\gcd(q(x), q'(x)) = x^3 - 5x^2 + 8x - 4 = (x-1)(x-2)^2.$$
Thus $\gcd(q(x), q'(x))$ is the product of the repeated factors of $q(x)$, namely $(x-2)$ and $(x-1)$, each to the degree one less than in $q(x)$. ∎

So if we simply subtract the *degree* of $\gcd(p(x), p'(x))$ from the degree of $p(x)$, we know how many distinct roots $p(x)$ has. In Example A, we can see that $q(x)$ has 3 distinct roots and has degree 6. The degree of $\gcd(q(x), q'(x))$ is 3 and indeed $3 = 6 - 3$. So to complete task 3, where $p(x) = x^7 - 5x^6 + 5x^5 + x^4 + 11x^3 - 11x^2 - 8x - 12$, we note that $\gcd(p(x), p'(x)) = x^3 - x^2 - x - 2$, which has degree 3. Thus $p(x)$ has four distinct roots in the complex numbers.

Problem 3 took us a bit outside the purview of high school math and a bit beyond the current section into the next. The point is to show how powerful the Euclidean Algorithm is. It is simple to do—just a bit of high school algebra—yet it helps us analyze polynomials in the dark, without total knowledge of its roots.

Task 1

With the knowledge that $(x - 2)$ is a factor of $x^3 - x^2 - x - 2$, factor $p(x) = x^7 - 5x^6 + 5x^5 + x^4 + 11x^3 - 11x^2 - 8x - 12$ completely. (There are enough hints to do this within the text!)

Task 2

Factor the polynomial $x^5 - 19x^4 + 139x^3 - 485x^2 + 800x - 500$ in $C[x]$ using the techniques of this essay. (**Hint.** You might want to consider the second derivative.)

3.3 WORKSHEET 4: DERIVATIVES

The derivative of a polynomial can be defined strictly algebraically, without using limits. In this worksheet, you will rediscover basic facts about the derivative from a strictly algebraic point of view and you will see how to use the derivative to detect multiple roots of polynomials.

> **DEFINITION 1.** Let F be a field and $p(x) = a_n x^n + a_{n-1} x^{n-1} + \ldots + a_0$ be a polynomial in $F[x]$. The **derivative** of $p(x)$ is the polynomial
> $$D(p(x)) = n a_n x^{n-1} + (n-1) a_{n-1} x^{n-2} + \ldots + a_1.$$

$D(p(x))$ is the usual derivative from calculus when $p(x) \in \mathbf{R}[x]$. But it can look a bit different over other fields. For instance, consider $p(x) = x^5 + 2x^3 + 3x + 1$ as a polynomial in $\mathbf{Z}_5[x]$. Then $D(p(x)) = 5x^4 + 6x^2 + 3 = 0x^4 + x^2 + 3 = x^2 + 3$. In what follows, F will denote a field and all polynomials are members of $F[x]$.

Task 1

Prove that $D(sf(x) + tg(x)) = sD(f(x)) + tD(g(x))$ for all polynomials $f(x)$ and $g(x)$ in $F[x]$ and constants s and t in F. (This shows that D is a linear transformation.)

Task 2

Prove that the usual product rule for derivatives holds.
$$D(f(x)g(x)) = D(f(x))g(x) + D(g(x))f(x).$$
No limits of course!

Task 3

Suppose that F is a field of characteristic 0. Let $f(x) \in F[x]$, $\deg(f(x)) > 1$. Show that the $\deg(D(f(x))) = \deg(f(x)) - 1$. (We have already shown that this is not the case for fields like $\mathbf{Z}_5$.)

Let $p(x)$ and $f(x)$ be polynomials in $F[x]$. Suppose that $f(x) \neq 0$ and $\deg(p(x)) \geq 1$. Then $p(x)$ is said to be a **multiple factor** of $f(x)$ if $f(x) = p(x)^k q(x)$, where $k > 1$. We say a is a **multiple root** of $f(x)$ if $(x - a)$ is a multiple factor of $f(x)$. If $f(x) = (x - a)^k q(x)$ and a is not a root of $q(x)$, then k is called the **multiplicity** of the root a.

Task 4

Prove the following propositions and corollaries. Through them, you will discover how the derivative and the Euclidean Algorithm for Polynomials together can yield important information about the roots of a polynomial $p(x)$, bypassing the often elusive process of actually finding those roots.

Note: From here on, assume that F has characteristic 0.

PROPOSITION A Suppose that $p(x)$ has degree ≥ 1. If $f(x) = p(x)^k q(x)$ and $k > 1$, then $p(x)$ is a factor of $D(f(x))$.

COROLLARY 1 If $f(x)$ has a multiple factor $p(x)$, then $f(x)$ and $D(f(x))$ are not relatively prime.

COROLLARY 2 If $f(x)$ has a multiple root at $x = a$, then a is a root of $D(f(x))$.

COROLLARY 3 If $f(x)$ and $D(f(x))$ are relatively prime, then $f(x)$ does not have a multiple root.

(It's here in Corollary 3 that the Euclidean Algorithm and the derivative come together to give us important information about the root structure of a polynomial. There's more to come.)

COROLLARY 4 Suppose that $f(x)$ has a root at $x = a$ of multiplicity k. Then $D(f(x))$ has a root at a of multiplicity at most $k - 1$.

Hint. Express $f(x)$ as $f(x) = (x - a)^k g(x)$ and note that a is not a root of $g(x)$. Use the product rule to express $D(f(x))$ as $(x - a)^{k-1} h(x)$. Evaluate $h(x)$ at a.

The converse to Corollary 1 is also true, namely,

PROPOSITION B If $f(x)$ and $D(f(x))$ are not relatively prime, then $f(x)$ has a multiple factor.

Here's a sketch of the proof. You fill in the details.

1. Let $g(x)$ be the greatest common divisor of $f(x)$ and $D(f(x))$ and let $p(x)$ be an irreducible factor of $g(x)$ of degree greater than or equal to 1. Then $p(x)$ divides both $f(x)$ and $D(f(x))$.
2. Write $f(x) = p(x)q(x)$. Use the product rule to deduce that $p(x)$ must divide $q(x)$. Conclude that $p(x)$ is a multiple factor of $f(x)$.

As we know, a polynomial $f(x)$ may have coefficients in a field F but no roots in F. Its roots may be found in a field E that contains F. A familiar example would be $f(x) = x^2 - 2$ in $\mathbf{Q}[x]$, which has its roots in $\mathbf{R}$ where $\mathbf{Q} \subseteq \mathbf{R}$. The next proposition says that using the derivative and Euclidean Algorithm, we can determine when $f(x) \in F[x]$ has no multiple roots in a field E, where $F \subseteq E$.

PROPOSITION C Suppose that E is a field such that $F \subseteq E$. A polynomial $f(x) \in F[x]$ has multiple irreducible factors in $E[x]$ if and only if $f(x)$ and $D(f(x))$ are not relatively prime in $F[x]$.

Hint. The polynomials $f(x)$ and $D(f(x))$ are relatively prime if and only if we can find polynomials $s(x)$ and $t(x)$ such that $1 = s(x)f(x) + t(x)D(f(x))$. To do this, we use the Euclidean Algorithm, which uses only addition and multiplication of the coefficients of $f(x)$ and $D(f(x))$. Since $F \subseteq E$, the arithmetic is identical whether we consider the polynomials as members of $F[x]$ or members of $E[x]$.

COROLLARY 5 If $f(x) \in F[x]$ is irreducible, then $f(x)$ has no multiple zeros in any field E such that $F \subseteq E$.

Hint. Since $f(x)$ is irreducible, the only possible common divisors of $f(x)$ and $D(f(x))$ are c and $f(x)$, where c is a nonzero constant in F.

Additional Tasks

1. Find a nonconstant polynomial in $\mathbf{Z}_3[x]$ that has 0 as its derivative.
2. Let p be a prime. Characterize all nonconstant polynomials in $\mathbf{Z}_p[x]$ that have derivative 0.
3. Find the greatest common divisor of each polynomial $f(x) \in \mathbf{Q}[x]$ and its derivative and determine whether $f(x)$ has multiple factors, and, in particular, multiple roots.

   i. $f(x) = x^5 - 5x^4 + 7x^3 - 2x^2 + 4x - 8$
   ii. $f(x) = x^6 + x^5 + 5x^4 + 4x^3 + 8x^2 + 4x + 4$
   iii. $f(x) = x^4 + x^3 + 3x^2 + 2x + 2$

3.4 FACTORING IN Z[x] AND Q[x]

In this section, we look at how factoring polynomials in $\mathbf{Z}[x]$ is related to factoring polynomials in $\mathbf{Q}[x]$. A polynomial in $\mathbf{Z}[x]$ can be regarded as a polynomial in $\mathbf{Q}[x]$ because $\mathbf{Z} \subseteq \mathbf{Q}$. Suppose $p(x)$ is a polynomial with *integer* coefficients that is reducible in $\mathbf{Q}[x]$. We can find $q(x)$ and $t(x)$ in $\mathbf{Q}[x]$ such that $p(x) = t(x)q(x)$, meaning $p(x)$ can be factored into two polynomials with *rational* coefficients. But we can do better. In this section, we will show that if $p(x)$ is a polynomial with integer coefficients that factors in $\mathbf{Q}[x]$, then we can factor $p(x)$ into polynomials with *integer* coefficients.

EXAMPLE 1

Let $p(x) = 6x^2 + 7x + 2$ in $Z[x]$. We can factor $p(x)$ as $(6x + 3)(x + \frac{2}{3})$ in $Q[x]$. Since $p(x)$ factors in $Q[x]$, it must factor in $Z[x]$, as we shall show. In $Z[x]$, its factorization is $(2x + 1)(3x + 2)$. ∎

DEFINITION 1. Let $p(x) = a_n x^n + a_{n-1} x^{n-1} + \ldots + a_0$ be a polynomial in $Z[x]$. The greatest common divisor of the coefficients of $p(x)$ is the **content** of $p(x)$. We say that $p(x)$ is **primitive** if its content is 1, that is, if the greatest common divisor of its coefficients is 1.

EXAMPLE 2

The content of the polynomial $p(x) = 6x^3 + 12x^2 + 8x + 14$ is 2. The polynomial $q(x) = 3x^3 + 6x^2 + 4x + 7$ is primitive because the only common positive integer that divides all the coefficients of $q(x)$ is 1. ∎

THEOREM 1 (*Gauss's Lemma*) Let $f(x)$ and $g(x)$ be primitive polynomials in $Z[x]$. The product $h(x) = f(x)g(x)$ is also primitive.

Proof. Let $f(x) = a_n x^n + a_{n-1} x^{n-1} + \ldots + a_0$ and $g(x) = b_m x^m + b_{m-1} x^{m-1} + \ldots + b_0$. Suppose that $f(x)$ and $g(x)$ are both primitive. Let $h(x) = f(x)g(x)$ and let p be any prime in Z. We shall show that $h(x)$ is primitive by showing that p cannot divide all the coefficients of $h(x)$.

Since $f(x)$ is primitive we can find a position i_0 such that p divides the coefficients a_n through a_{i_0+1} but such that p does not divide a_{i_0}. So a_{i_0} is the first coefficient, starting from the leading coefficient of $f(x)$, that is not divisible by p. Similarly, we can find j_0 such that p divides the coefficients b_m through b_{j_0+1}, but such that p does not divide b_{j_0}. Let $k = i_0 + j_0$. The kth coefficient in the product $h(x)$ is

$$c_k = \sum_{i+j=k} a_i b_j = (a_k b_0 + a_{k-1} b_1 + \ldots + a_{i_0+1} b_{j_0-1}) + a_{i_0} b_{j_0} + (a_{i_0-1} b_{j_0+1} + \ldots + a_0 b_k).$$

(Coefficients indexed higher than the leading coefficient in either $f(x)$ or $g(x)$ are 0.)

The sums in each set of parentheses are divisible by p. (In the first set, each of the a_i's is divisible by p and in the second set, each of the b_j's is divisible by p.) If c_k is divisible by p, then $a_{i_0} b_{j_0}$ must also be divisible by p. In turn, one of a_{i_0} or b_{j_0} must be divisible by p, contradicting our choice of a_{i_0} and b_{j_0}. Thus for each prime p, at least one coefficient of $h(x)$ is not divisible by p, proving that $h(x)$ is primitive. ▲

The following theorem summarizes the assertions made in the opening paragraph of this section about factoring in $Z[x]$ and $Q[x]$.

THEOREM 2 Let $f(x)$ be a polynomial in $Z[x]$. Suppose that $f(x)$ factors in $Q[x]$ so that $f(x) = p(x)q(x)$ where $p(x)$ and $q(x)$ have rational coefficients. Then we can find polynomials $p_1(x)$ and $q_1(x)$ in $\mathbf{Z}[x]$ such that $f(x) = q_1(x)p_1(x)$ where $\deg(p) = \deg(p_1)$ and $\deg(q) = \deg(q_1)$.

Proof. First we prove that the theorem holds for primitive polynomials in $\mathbf{Z}[x]$ and then we show that it holds for all polynomials in $\mathbf{Z}[x]$. Assume that $f(x)$ is a primitive polynomial in $\mathbf{Z}[x]$ and that $f(x) = p(x)q(x)$ in $\mathbf{Q}[x]$. Assume also that the coefficients of $p(x)$ and $q(x)$ are rational numbers expressed as fractions in lowest terms. Let s and t be the least common multiples of the denominators of the coefficients of $p(x)$ and $q(x)$ respectively. Then $sp(x)$, and $tq(x)$, and $stf(x)$ are all in $\mathbf{Z}[x]$. Let c_p and c_q be the contents of $sp(x)$ and $tq(x)$ respectively. Then $sp(x) = c_p p_1(x)$ and $tq(x) = c_q q_1(x)$ where both p_1 and q_1 are primitive. Thus we have

$$(1) \qquad stf(x) = c_p c_q p_1(x) q_1(x).$$

Since $f(x)$ is primitive, st is the content of the left side of equation (1). By Gauss's Lemma, $p_1(x)q_1(x)$ is primitive so that $c_p c_q$ is the content of the right side. Thus $st = c_p c_q$ and we may cancel it from both sides of the equation to obtain

$$f(x) = p_1(x)q_1(x) \text{ in } \mathbf{Z}[x].$$

Since $p_1(x)$ and $p(x)$ differ by a constant multiple, they have the same degree. Similarly, $q_1(x)$ and $q(x)$ have the same degree.

Suppose that $f(x) \in \mathbf{Z}[x]$ and that c is the content of $f(x)$. Suppose that $f(x) = p(x)q(x)$ in $\mathbf{Q}[x]$. Set $f_1(x) = \frac{f(x)}{c}$. Then $f_1(x) \in \mathbf{Z}[x]$ and $f_1(x)$ is primitive. We can factor $f_1(x)$ in $\mathbf{Q}[x]$ as

$$f_1(x) = \left(\frac{1}{c}p(x)\right)q(x).$$

Since the theorem holds for primitive polynomials, we can factor $f_1(x)$ as $f_1(x) = p_1(x)q_1(x)$ in $\mathbf{Z}[x]$. Thus $f(x)$ in $\mathbf{Z}[x]$ factors as $f(x) = (cp_1(x))q_1(x)$. Clearly, $\deg(cp_1(x)) = \deg(p(x))$ and $\text{degree}(q_1(x)) = \deg(q(x))$. ▲

EXAMPLE 3

The polynomial $p(x) = \frac{3x^3}{2} + 3x - \frac{3}{2}$ factors in $\mathbf{Q}[x]$ if and only if $2p(x) = 3x^3 + 6x - 3$ factors in $\mathbf{Z}[x]$. Now $2p(x) = 3(x^3 + 2x - 1)$ factors in $\mathbf{Z}[x]$ if and only if the primitive polynomial $x^3 + 2x - 1$ factors in $\mathbf{Z}[x]$. So to determine whether or not $p(x)$ factors in $\mathbf{Q}[x]$, we check whether or not $x^3 + 2x - 1$ factors in $\mathbf{Z}[x]$. Suppose that $x^3 + 2x - 1 = (x + a)(x^2 + bx + c) = x^3 + (a + b)x^2 + (ab + c)x + ac$ with a, b, and c in $\mathbf{Z}$. Equating coefficients we find that $a + b = 0$ and $ac = -1$. Thus $a = +1$ or -1. Assume first that $a = 1$. Then $b = -1$ and, since $ab + c = 2$, we have $c = 3$. But if $c = 3$, $ac \neq -1$. The assumption that $a = -1$ leads to the same conclusion. So $p(x)$ is irreducible in $\mathbf{Q}[x]$. ∎

The following two results can help us determine whether or not a polynomial in $\mathbf{Z}[x]$ is reducible.

PROPOSITION 3 Suppose that $f(x)$ is in $\mathbf{Z}[x]$ and p is a prime. Let $f_p(x)$ be the associated polynomial in $\mathbf{Z}_p[x]$ obtained by reducing the coefficients of $f(x)$ mod p. If $\deg(f(x)) = \deg(f_p(x))$ and if $f_p(x)$ is irreducible in $\mathbf{Z}_p[x]$, then $f(x)$ is irreducible in $\mathbf{Z}[x]$.

Proof. We prove the contrapositive. Suppose that $f(x)$ is not irreducible in $Z[x]$ and that $f(x) = g(x)h(x)$ in $Z[x]$, where the degrees of g and h are both less than the degree of f. Let p be a prime such that $\deg(f(x)) = \deg(f_p(x))$. Then $f_p(x) = g_p(x)h_p(x)$ in $Z_p[x]$. Since $\deg(f_p(x)) = \deg(f(x))$ and the degrees of both $g_p(x)$ and $h_p(x)$ are less than $\deg(f(x))$, we see that $f_p(x)$ is reducible in $Z_p[x]$. ▲

EXAMPLES

4. In Example 3, we determined that the polynomial $p(x) = x^3 + 2x - 1$ was irreducible in $Z[x]$ by a somewhat complicated argument. Now let's consider $p_3(x)$ in $Z_3[x]$. Since it is a cubic polynomial, we can check to see if it has a nontrivial factorization by checking to see if it has a root in Z_3. Substituting in 0, 1 and 2, we see that it does not have a root and is therefore irreducible in both $Z_3[x]$ and in $Z[x]$.

5. Let $f(x) = x^5 + 2x^4 + 10x^3 - x^2 + 12x + 5$. Then $f_2(x) = x^5 + x^2 + 1$. By checking all possible factorizations of $f_2(x)$ in $Z_2[x]$, a task with a finite number of steps, we find that $x^5 + x^2 + 1$ is irreducible in $Z_2[x]$ and conclude that $f(x) = x^5 + 2x^4 + 10x^3 - x^2 + 12x + 5$ is irreducible in $Z[x]$. ■

THEOREM 4 (*Eisenstein's Criterion*) Let $f(x) = a_n x^n + a_{n-1} x^{n-1} + \ldots + a_0$ be a polynomial in $Z[x]$. Suppose p is a prime such that

i. p does **not** divide a_n,
ii. p does divide $a_{n-1}, a_{n-2}, \ldots, a_0$, but p^2 does **not** divide a_0.

Then $f(x)$ is irreducible in $Q[x]$.

Here's an example before the proof. Let $f(x) = 2x^4 + 9x^3 + 6x^2 + 15x + 21$. The prime $p = 3$ does not divide the leading coefficient, it does divide all other coefficients, but $3^2 = 9$ does not divide $a_0 = 21$. By Eisenstein's Criterion, $f(x)$ is irreducible. However, the converse of the theorem does not hold. If we cannot find a prime satisfying Eisenstein's conditions for a polynomial $p(x)$, it does not mean that $p(x)$ is reducible. For instance, there is no prime that works for either $g(x) = x^2 + x + 1$ or $h(x) = x^4 - 1$ but $g(x)$ is irreducible in $Q[x]$ whereas $h(x)$ is not.

Proof. Suppose that p is a prime satisfying Eisenstein's Criterion for $f(x)$. But suppose that $f(x) = g(x)h(x)$ where $\deg(h(x)) < n$ and $\deg(g(x)) \geq 1$. Let $g(x) = b_k x^k + b_{k-1} x^{k-1} + \ldots + b_0$ and $h(x) = c_m x^m + c_{m-1} x^{m-1} + \ldots + c_0$. Since p does not divide a_n, and since $a_n = c_m b_k$, p does not divide either c_m or b_k. Since p, but not p^2, divides a_0, p fails to divide exactly one of b_0 or c_0. Let us assume that p divides b_0 and does not divide c_0. Since p does not divide the leading coefficient b_k, we can find the smallest index i such that p does not divide b_i but does divide $b_0, b_1, \ldots, b_{i-1}$. Note that $i \neq n$ and $i \neq 0$ and that $a_i = b_0 c_i + b_1 c_{i-1} + b_i c_0$. Thus $a_i - (b_0 c_i + b_1 c_{i-1} + \ldots + b_{i-1} c_1) = b_i c_0$. Since p divides a_i and each b_j in the parenthesized term, p divides $b_i c_0$. But that is impossible since the prime p divides neither b_i nor c_0. ▲

(You might have noticed that the proof was similar to that of Gauss's Lemma. Eisenstein was a student of Gauss.)

EXAMPLE 6

Let's apply the Eisenstein Criterion to $x^5 - 6x + 3$ with $p = 3$. We see that 3 does not divide the leading coefficient $a_5 = 1$ and 3 does divide all other coefficients. But 3^2 does not divide $a_0 = 3$. Thus $x^5 - 6x + 3$ is irreducible in $Q[x]$ and hence, irreducible in $Z[x]$. ∎

EXAMPLE 7

Let p be a prime. The **pth roots of unity** are the complex solutions to the equation $x^p - 1 = 0$. Of course, $x = 1$ is a pth root of unity for any p. There are $p - 1$ other roots in the complex numbers C. They are the roots of $\Phi_p(x) = \frac{x^p-1}{x-1} = x^{p-1} + x^{p-2} + \ldots + 1$. The polynomial $\Phi_p(x)$ is called the **pth cyclotomic polynomial**. It is irreducible in $Q[x]$ but Eisenstein's Criterion is not directly applicable. To prove its irreducibility, we consider the polynomial $f(x) = \Phi_p(x+1) = \frac{(x+1)^p - 1}{(x+1) - 1} = x^{p-1} + \binom{p}{1}x^{p-2} + \binom{p}{2}x^{p-3} + \ldots + \binom{p}{p-1}$. If $\Phi_p(x)$ is reducible, so is $f(x)$ because if $\Phi_p(x) = g(x)h(x)$, then $f(x) = \Phi_p(x+1) = g(x+1)h(x+1)$. So we will check the irreducibility of $f(x)$ with Eisenstein's Criterion: p does not divide the leading coefficient 1 but p does divide each of the binomial coefficients $\binom{p}{i}$ for $i = 1$ to $p - 1$. Since $\binom{p}{p-1} = p$, we see that p^2 does not divide the constant term. Since $f(x)$ is irreducible, so is $\Phi_p(x)$. ∎

SUMMARY

In this section, we proved that if a polynomial $f(x)$ with integer coefficients factors in $Q[x]$ as $f(x) = p(x)q(x)$ where p and q have rational coefficients, then it factors in $Z[x]$ as $f(x) = p_1(x)q_1(x)$, where p_1 and q_1 have integer coefficients. Also, $\deg(p) = \deg(p_1)$ and $\deg(q) = \deg(q_1)$. **Gauss's Lemma** is key to proving this fact. It states that the product of two **primitive** polynomials (each with **content** 1) is also primitive. Still, factorization remains a difficult task. **Eisenstein's Criterion** gives us a condition that, if it holds, tells us that a polynomial is irreducible (but not conversely). The reader who does Worksheet 5 in Section 3.8 of this chapter will discover an algorithm for factoring polynomials in $Z[x]$ that can be extended to $Q[x]$. (It is rather tedious to execute by hand.) The existence of such an algorithm for $Q[x]$ sets it apart from $C[x]$ because no such algorithm exists for $C[x]$.

3.4 Exercises

1. Determine which of the following polynomials are primitive.
   i. $3x^3 + 2x + 5$
   ii. $6x^4 + 3x + 9$

2. Show that any polynomial $p(x)$ in $Q[x]$ can be written as $p(x) = tq(x)$, where $t \in Q$ and $q(x) \in Z[x]$ is primitive.
3. Prove that if $f(x)g(x)$ is primitive, then $f(x)$ and $g(x)$ must also be primitive.
4. Determine the irreducibility of the following polynomials in $Q[x]$.
   i. $x^4 + x + 1$
   ii. $x^3 - \frac{x^2}{2} - \frac{1}{2}$
5. Use Proposition 3 to determine the irreducibility of the following polynomials in $Z[x]$.
   i. $x^3 + x + 1$
   ii. $x^4 + x^2 + x + 1$
6. Here is **Descartes' Criterion** for determining the **rational** roots of a polynomial in $Z[x]$:

   If $a_n x^n + a_{n-1} x^{n-1} + \ldots + a_0$ has a rational root $x = \frac{s}{t}$, where s and t are relatively prime, then t divides a_n and s divides a_0.

   Prove that Descartes' assertion is correct and use it to find the rational roots of the following polynomials. (**Hint.** By factoring a_0 and a_n, you can make a list of all possible values $\frac{s}{t}$ and substitute to find which, if any, is a root.)
   i. $x^3 - x + 1$
   ii. $2x^3 + x - 1$
   iii. $2x^3 - x^2 + 2x - 1$
   iv. $6x^4 + x^3 + 4x^2 + x - 2$
7. Determine the irreducibility of the following polynomials in $Q[x]$.
   i. $f(x) = 2x^7 + 5x^3 - 25x + 15$
   ii. $g(x) = x^4 + 3x^2 + 2$
   iii. $h(x) = x^5 + 2x^3 + 2x^2 + 2$
   iv. $k(x) = x^4 + 1$

To the Teacher

The following looks to be a simple task for the high school algebra student:

Factor $p(x) = 6x^2 + 16x + 8$.

Two possible answers: $(3x + 2)(2x + 4)$ and $(6x + 4)(x + 2)$.

Let's make sure that we understand the two given answers and the questions they should prompt. The answers are both correct factorizations. This raises the following questions:

Is unique factorization violated? (No)

Are the factorizations complete? (Maybe)

If $p(x)$ is considered to be a polynomial in $Q[x]$, then both answers factor $p(x)$ into irreducible factors which differ only by units in $Q[x]$. Namely, $(3x + 2) = \frac{1}{2} \cdot (6x + 4)$ and $(2x + 4) = 2 \cdot (x + 2)$. So unique factorization is not violated in $Q[x]$. But 2 and $1/2$ are not units in $Z[x]$. So we have two distinct factorizations. However, in $Z[x]$, neither factorization is complete because the factors are not irreducible. The number 2 is not a unit in $Z[x]$ and $p(x) = 2$ is an irreducible polynomial in $Z[x]$.

A complete factorization of $p(x)$ in $\mathbf{Z}[x]$ into irreducible factors has three factors: $2(x+2)(3x+2)$. Similarly, we can express $10x^2 + 20$ as the product of three irreducible factors: $2 \cdot 5 \cdot (x^2 + 2)$.

This raises an additional question: Are polynomials in $\mathbf{Z}[x]$ uniquely factorable? After all, we proved unique factorization for polynomials with coefficients in a field but $\mathbf{Z}$ is not a field. The answer is yes. (A good thing for test correctors!). Let's see how. If a polynomial $f(x)$ has content c, first factor it as $f(x) = c \cdot p(x)$ in $\mathbf{Z}[x]$. Then $p(x)$ has content 1 and can only be factored in one way (up to sign) in $\mathbf{Q}[x]$ into irreducible factors with *integer* coefficients. So to obtain the irreducible factors of $f(x)$ uniquely (up to sign) in $\mathbf{Z}[x]$, factor c into primes and then factor $p(x)$.

EXAMPLE

The polynomial $14x^3 - 14x^2 - 42x - 126$ has content 14. It factors into irreducible factors as $(2x - 6) \cdot (7x^2 + 14x + 21)$ in $\mathbf{Q}[x]$. But in $\mathbf{Z}[x]$, its factorization is $2 \cdot 7 \cdot (x - 3)(x^2 + 2x + 3)$. ∎

The word "content" is not part of the standard vocabulary of high school mathematics (and need not be) but the concept of *content* is key to understanding the basic high school task of factoring over the integers. Issues that arise around factoring are addressed at greater length in the *In the Classroom* essay in Section 3.9.

Task

Investigate one or more high school texts in which the factoring of polynomials is addressed. Write a critique of its treatment of this issue in the light of what you have learned in this section.

3.5 THE FUNDAMENTAL THEOREM OF ALGEBRA

To Descartes, a "root" of a polynomial had to be a nonnegative real number. For him and his contemporaries, the polynomial $x^2 - x - 2$ has one root at $x = 2$ and one "false" root at $x = -1$. While we no longer call negative roots "false," we do call the roots of the polynomial $x^2 + 1$ "imaginary." This terminology hints at the history of the difficult intellectual process to determine the natural number system in which to do mathematics and, in particular, solve polynomial equations. The Fundamental Theorem of Algebra helped to settle the question. It states that every nonconstant polynomial with complex coefficients has a complex root. There are no irreducible polynomials in $\mathbf{C}[x]$ of degree greater than 1. The proof we give below relies on a few simple facts from the calculus of functions of two variables. Gauss first proved the theorem in 1799. The proof given below is very different from his. A glimpse into Gauss's method is given at the end of this section. For an in-depth look at six of the hundreds of different proofs of this theorem, see the book by Fine and Rosenberg, entitled *The Fundamental Theorem of Algebra*.[1]

---

[1] Fine, Benjamin and Rosenberger, Gerhard. *The Fundamental Theorem of Algebra*. UTM, New York: Springer-Verlag, 1997.

Preliminaries

Any polynomial $p(z)$ in $\mathbf{C}[z]$ can be written as the sum of its real and imaginary parts, namely, $p(z) = p(x+iy) = a(x, y) + ib(x, y)$. Both $a(x, y)$ and $b(x, y)$ are polynomials with real coefficients in the two real variables x and y. Thus both $a(x, y)$ and $b(x, y)$ are continuous functions in the xy-plane. The composite function $|p(z)| = \sqrt{a(x, y)^2 + b(x, y)^2}$ is also continuous in the xy-plane.

EXAMPLE 1

Let $p(z) = z^4 + 3iz^2 + 2 + i = (x+iy)^4 + 3i(x+iy)^2 + (2+i)$. Then the real part of $p(z)$ is the function $a(x, y) = x^4 - 6x^2y^2 + y^4 - 6xy + 2$ and the imaginary part is $b(x, y) = 4x^3y - 4xy^3 + 3x^2 - 3y^2 + 1$. ∎

Let $r > 0$. The closed disk in the xy-plane centered at a point (x_0, y_0) of radius r is the set of points $\{(x, y): (x - x_0)^2 + (y - y_0)^2 \leq r^2\}$. Just as any continuous function of one variable has a minimum value on a closed interval, any continuous function of two variables has a minimum value on a closed disk. The strategy of the proof of the Fundamental Theorem of Algebra presented below is to show that, given a polynomial $p(z)$, the continuous function $|p(z)|$ assumes the value 0 (a global minimum) at some point z_0, which is found in a well-chosen closed disk. If $|p(z_0)| = 0$, then $p(z_0) = 0$ and so z_0 is the root postulated by the theorem.

Not all continuous functions on $\mathbf{C}$ assume a global minimum. For instance, the function $f(z) = \frac{1}{1+|z|^2}$ does not. The insight as to why the absolute value of a nonconstant polynomial must have a global minimum of 0 comes from investigating polynomials of the form $p(z) = 1 - z^k + z^{k+1}g(z)$ where $g(z)$ is also a polynomial. For such a polynomial, $|p(z)|$ assumes values less than 1 near 0. To see this, first assume that $g(z) = 0$. Restricting z to real values $x + i0$, the graph of $p(x) = 1 - x^k$ looks like either Graph 1 or Graph 2 in Figure 1, depending on whether k is even or odd.

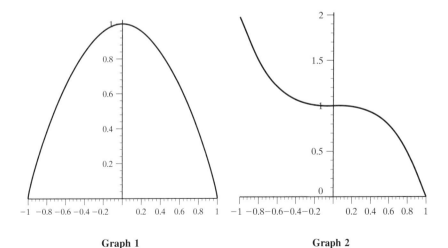

Graph 1   Graph 2

FIGURE 1

In both cases, there are real values of x to the right of 0 for which $p(x)$ is smaller in absolute value than 1. Even if $g(x)$ is not identically zero, the term x^{k+1} is so small near 0 that the term $x^{k+1}g(x)$ is negligible and the respective graphs are similar. (A rigorous argument for this will appear within the proof of the Fundamental Theorem.) When x is replaced by the complex variable z, the function $|f(z)| = |1 - z^k + z^{k+1}g(z)|$ does not have a minimum value of 1 at $z = 0$ since the function restricted to real values of z does not. This observation will lead to a contradiction of the (wrong) assertion that a polynomial in $\mathbf{C}[x]$ could fail to have a root.

Of course some *real-valued* functions (e.g., $f(x) = 1 + x^2$), do assume a minimum value of 1 at 0. This is never the case for nonconstant polynomials of a *complex variable*. The key is that we can rotate and scale the complex plane so that a given nonconstant polynomial $p(z)$ for which $p(0) = 1$ has the form $f(w) = 1 - w^k + w^{k+1}t(z)$ on the rotated plane. Here's how.

Suppose that $p(z) = 1 + a_1 z + \ldots + a_k z^k + \ldots + a_n z^n$ and suppose that $a_1 = a_2 = \ldots = a_{k-1} = 0$ and that $a_k \neq 0$. Let $r = \sqrt[k]{-1/a_k}$ and make the substitution $rw = z$. Then

$$p(z) = p(rw) = 1 + a_k(w\sqrt[k]{-1/a_k})^k + (rw)^{k+1}g(rw) = 1 - w^k + w^{k+1}G(w),$$

where $G(w) = r^{k+1}g(rw)$. Let $f(w) = 1 - w^k + w^{k+1}G(w)$. By looking at the values that $f(w)$ assumes when w is restricted to the real line, we are looking at the values that $p(z)$ assumes on the line $z = rt, -\infty < t < \infty$, in the complex plane. (In essence we are rotating the plane.) Since $|f(w)|$ does not have a minimal value of 1 when w is real, $|p(z)|$ does not have a minimal value of 1 on the line $z = rt$.

EXAMPLE 2

Let $p(z) = 1 + z^2$. Restricting z to the real line, the graph of $y = p(x)$ is as shown in Figure 2.

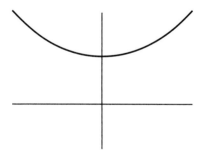

FIGURE 2

Let $r = \sqrt[2]{-1/1} = \sqrt[2]{-1} = i$. (Notice that taking the square root of -1 or, in general, finding $\sqrt[k]{-1/a_k}$, is not always possible in the case of real variables. This is what makes finding roots in $\mathbf{C}$ different from $\mathbf{R}$.) Along the line $z = rt, p(z) = 1 + (it)^2 = 1 - t^2$, the graph of which is shown in Figure 3. ∎

Section 3.5   The Fundamental Theorem of Algebra

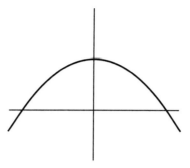

FIGURE 3

With these insights, the proof of the Fundamental Theorem of Algebra unwraps readily.

The Fundamental Theorem of Algebra

Every polynomial $p(z) = a_n z^n + a_{n-1} z^{n-1} + \ldots + a_0$ in $C[x]$ of degree $n \geq 1$ has a root in C.

Proof. Let $p(z) = a_n z^n + a_{n-1} z^{n-1} + \ldots + a_0$. Since $p(z)$ has a root if and only if $p(z)/a_n$ has a root, we shall assume that $a_n = 1$. If $a_0 = 0$, then p has a root at 0. So assume that $a_0 \neq 0$. First we find a closed disk D outside of which p has no root and then we show that the function $|p(z)|$ achieves an absolute (global) minimum value at some point z_0 in D. Later in the proof, we will prove that $|p(z_0)| = 0$.

Let $p(z) = z^n + a_{n-1} z^{n-1} + \ldots + a_0$ be a polynomial in $C[x]$. By the form of the triangle inequality that states $|s + t| \geq |s| - |t|$, we have

$$|p(z)| = |z^n + (a_{n-1} z^{n-1} + \ldots + a_0)| \geq |z^n| - |a_{n-1} z^{n-1} + \ldots + a_0|.$$

Since

$$\lim_{|z| \to \infty} \frac{|z^n|}{|z^n|} - \left| a_{n-1} \frac{z^{n-1}}{|z^n|} + \ldots + \frac{a_0}{|z^n|} \right| = 1,$$

we have

$$\lim_{|z| \to \infty} |z^n| - |a_{n-1} z^{n-1} + \ldots + a_0| = \infty.$$

Thus $\lim_{|z| \to \infty} |p(z)| = \infty$. This means that for any real number $Q > 0$, we can find $R > 0$ such that if $|z| > R$, then $|p(z)| > Q$. In particular, let $Q = 1 + |a_0|$ and choose R so that if $|z| > R$, then $|p(z)| > 1 + |a_0|$. The continuous function $|p(z)|$ assumes a minimum value M at some point z_0 in the disk $D_R = \{(x, y) : |z| \leq R\}$. Since $0 \in D_R$ and $|p(0)| = |a_0|$, we have $M \leq |a_0|$. Because of our choice of R, we know that $|p(z)| > 1 + |a_0|$ for $z \notin D_R$. So $|p(z_0)| \leq |p(z)|$ for all z in C, which is to say that $|p(z)|$ has a global minimum value at z_0.

Next we will show that $|p(z_0)| = 0$ and hence $p(z_0) = 0$. First, we translate our function so that the absolute minimum value occurs at $z = 0$. Let $f(z) = p(z + z_0) = b_0 + b_1 z + \ldots + z^n$. Since $|p(z)|$ has a minimum value at $z = z_0$, $|f(z)|$ has the same minimum at $z = 0$. Our task is to show that $|f(0)| = 0$ because, if this is the case, then $f(0) = 0$ and hence $p(z_0) = 0$, giving us our desired root. To that end, we will assume that $f(0) \neq 0$ and obtain a contradiction. On the way, a few substitutions and variable changes will be made so that ultimately, the value of $f(0)$ becomes clear.

Suppose that $f(0) = b_0$, and $b_0 \neq 0$. Let

$$g(z) = \frac{1}{b_0} f(z) = 1 + c_1 z + \ldots + c_n z^n,$$

where $c_i = \frac{b_i}{b_0}$. If $|f(z)|$ has a minimum value of $|b_0|$ at $z = 0$, then $|g(z)|$ has a minimum value of 1 at $z = 0$. Suppose that $c_1 = c_2 = \ldots = c_{k-1} = 0$ and that $c_k \neq 0$. Next we make a change of variable and gather terms. Let $r = \sqrt[m]{-1/c_k}$ and let

$$h(w) = g(rw) = 1 - w^k + w^{k+1} m(w).$$

The function $|h(w)|$ has a minimum value at $w = 0$ if and only if $|g(z)|$ has a minimum at $z = 0$. The next part of the proof will show that $|h(w)|$ assumes values less than 1, thus obtaining a contradiction to the original assertion that $f(0) = b_0$ and $b_0 \neq 0$.

By the triangle inequality, $|h(w)| \leq |1 - w^k| + |w^{k+1}| \cdot |m(w)|$. Restrict w to real values such that $0 < w < 1$. Then $|1 - w^k| = 1 - w^k$ and $|w^{k+1}| \cdot |m(w)| = w^{k+1} |m(w)|$. Thus

$$|h(w)| \leq 1 - w^k + w^{k+1} |m(w)| = 1 - w^k (1 - w |m(w)|).$$

Since $\lim_{w \to 0} w |m(w)| = 0$, we can choose w_0 so small that $1 - w_0 |m(w_0)| > 0$ and

$$|h(w_0)| \leq 1 - w_0^k (1 - w_0 |m(w_0)|) < 1.$$

We thus have a contradiction to the assertion that the minimum value of $h(w)$, and hence of $g(w)$, is 1. Therefore, the minimum value of $|p(z)|$ is zero at z_0 and we have proven that $p(z)$ has a root z_0 in C. ▲

The situation for polynomials in $R[x]$ is almost as good. Every polynomial with real coefficients factors into irreducible linear and quadratic factors. (But actually finding a factorization for a given polynomial may be very problematic.)

COROLLARY 1 All polynomials in $R[x]$ of degree > 2 are reducible.

Proof. Suppose that $f(x)$ is a polynomial with real coefficients and that $\deg(f(x)) > 2$. If $f(x)$ has a real root, it is reducible and we are done. Assume that $f(x)$ has no real root. By the Fundamental Theorem of Algebra, $f(x)$ must have a complex root $z = a + bi$. Let $z_1 = a - bi$. Then the polynomial $g(x) = (x - z)(x - z_1) = x^2 - 2ax + (a^2 + b^2)$ has real coefficients and it is irreducible in $R[x]$. We shall show that $g(x)$ divides $f(x)$ thus showing that $f(x)$ is reducible. Applying the Division Algorithm to $f(x)$ and $g(x)$, we have

$$f(x) = q(x) g(x) + (sx + t),$$

where s and t are real numbers. Since z is a root of $f(x)$ and of $g(x)$, we have $0 = f(z) = q(z)g(z) + (sz + t)$. Thus $sz + t = 0$. If $s \neq 0$, then $z = -t/s$, and z is real contradicting the assumption on z. So $s = 0$ and $t = 0$. ▲

EXAMPLE 3

Let $p(x) = x^4 - 6x^3 + 26x^2 - 46x + 65$. An arithmetic check will verify that $p(1 + 2i) = 0$. Since p has real coefficients, we know that its complex conjugate $1 - 2i$ is also a root. Thus $(x - (1 + 2i))(x - (1 - 2i)) = x^2 - 2x + 5$ is a factor of $x^4 - 6x^3 + 26x^2 - 46x + 65$. Dividing p by $x^2 - 2x + 5$, we find that $x^4 - 6x^3 + 26x^2 - 46x + 65 = (x^2 - 2x + 5)(x^2 - 4x + 13)$. An application of the quadratic formula to $x^2 - 4x + 13$ yields roots of $2 + 3i$ and $2 - 3i$. Thus $p(x)$ factors completely as $(x^2 - 2x + 5)(x^2 - 4x + 13)$ in $\mathbf{R}[x]$ and as $(x - (1 + 2i))(x - (1 - 2i)) \cdot (x - (2 + 3i))(x - (2 - 3i))$ in $\mathbf{C}[x]$. ■

A polynomial $p(z) = p(x + iy) = a(x, y) + ib(x, y)$ has a root at $z_0 = x_0 + iy_0$ if and only if its real and imaginary parts are simultaneously zero at (x_0, y_0). If we can solve the system of equations $a(x, y) = 0$ and $b(x, y) = 0$, we can find the roots of $p(z)$.

EXAMPLE 4

In this example we will find the roots of $p(z) = z^3 + (-3i - 2)z^2 + (5 + 6i)z - 15i = (x + iy)^3 + (-3i - 2)(x + iy)^2 + (5 + 6i)(x + iy) - 15i$ by finding out where its real and imaginary parts are simultaneously equal to 0. Its real part is $a(x, y) = x^3 - 3xy^2 + 6xy - 2x^2 + 2y^2 - 6y + 5x$. Its imaginary part is $b(x, y) = -y^3 - 3x^2 + 3x^2y + 5y + 3y^2 + 6x - 4xy - 15$. The solutions to the simultaneous equations $a(x, y) = 0$ and $b(x, y) = 0$ are $(x, y) = (1, 2), (1, -2)$ and $(0, 3)$. The roots of $p(z)$ are thus $z = 1 + 2i, z = 1 - 2i$, and $z = 3i$. The graph in Figure 4 shows the level curves $a(x, y) = 0$ and $b(x, y) = 0$ and their points of intersection. (The graph of $a(x, y) = 0$ is darker.) ■

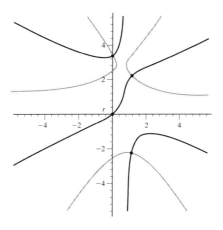

FIGURE 4

EXAMPLE 5

In this example, we find the roots of $p(z) = z^n - 1$ by using the same approach as in Example 4 but in polar form. Let $z = r(\cos(\theta) + i\sin(\theta))$, $r \geq 0$. By de Moivre's formulas, $z^n = r^n(\cos(n\theta) + i\sin(n\theta))$. The real part of $p(z)$ is $T = r^n \cos(n\theta) - 1$ and the imaginary part is $U = r^n \sin(n\theta)$. Now U is 0 when $\theta = m\left(\frac{\pi}{n}\right)$, $m = 0, \pm 1, \pm 2, \ldots$. For T to be simultaneously equal to 0, we must have $r = 1$ and $m = 0, \pm 2, \pm 4, \ldots$. For $n = 4$, we find that the solutions to $z^4 - 1 = 0$ are found when $r = 1$ and $\theta = m\left(\frac{\pi}{4}\right)$, $m = 0, \pm 2, \ldots$ or, nonredundantly, $\theta = 0, \pm\frac{\pi}{2}$, and π. As complex numbers, the roots are $z = 1, i, -i$, and -1. ∎

In 1799, Gauss proved that the simultaneous equations $a(x, y) = 0$ and $b(x, y) = 0$ **must** have a solution in the xy-plane. His proof is rather old-fashioned in that it makes fairly intensive use of trigonometry. But it also anticipates methods now common to topological thinking but were then many years in the future. What follows is the thrust of Gauss's argument as presented in J. V. Uspenski's book *Theory of Equations* (1948, McGraw Hill). We paraphrase freely.

Let $f(z) = z^n + az^{n-1} + bz^{n-2} + \ldots + k$ where $a, b, \ldots, k$ are complex numbers. In polar form, we have

$$a = A(\cos(\alpha) + i\sin(\alpha)), b = B(\cos(\beta) + i\sin(\beta)), \text{etc.}$$

Let $z = r(\cos(\theta) + i\sin(\theta))$, $r \geq 0$. Using de Moivre's formula and the identities for the sine and cosine of the sum of two angles, we have $f(z) = T + iU$, where

$$T = r^n \cos(n\theta) + r^{n-1}A\cos((n-1)\theta + \alpha) + r^{n-2}B\cos((n-2)\theta + \beta) + \ldots$$

and

$$U = r^n \sin(n\theta) + r^{n-1}A\sin((n-1)\theta + \alpha) + r^{n-2}B\sin((n-2)\theta + \beta) + \ldots.$$

For r sufficiently large, r^n is very much greater than r^{n-1} and the functions T and U behave very much like $r^n \cos(n\theta)$ and $r^n \sin(n\theta)$, respectively. Gauss articulates this assertion and supports it rigorously. (Here "r sufficiently large" means $r > 1 + \sqrt{2}C$, where C is any number greater than the maximum of $A, B, \ldots$.)

EXAMPLE 6

Consider the polynomial $f(z) = z^5 + z + 1$. In polar coordinates $f(z) = r^5(\cos(\theta) + i\sin(\theta))^5 + r(\cos(\theta) + i\sin(\theta)) + 1$. Thus $T = r^5 \cos(5\theta) + r\cos(\theta) + 1$. The value of C is 1. When r is less than $1 + \sqrt{2}$ (e.g., $r = 0.5$), the graphs of T and $r^5(\cos(5\theta))$ can behave very differently, as in Graph 1 in Figure 5. The graphs become closer as r gets larger, as shown in Graph 2 for $r = 1.5$. But for $r > 1 + \sqrt{2}$ (e.g., $r = 2.5$), their roots, high points and low points coincide very closely as in Graph 3, in which they are barely distinguishable. ∎

Section 3.5 The Fundamental Theorem of Algebra 143

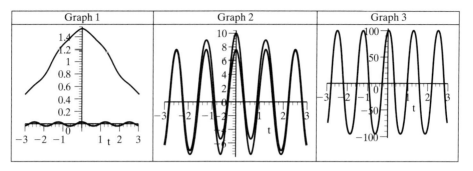

FIGURE 5

It is the interplay of the highs and lows of $\cos(n\theta)$ and $\sin(n\theta)$ and hence T and U that Gauss uses to show that $f(z)$ must have a root. On a circle of radius $r > 0$, $\cos(n\theta)$ changes sign exactly $2n$ times, at each of its $2n$ roots. At each of the $2n$ roots of $\cos(n\theta)$, $\sin(n\theta)$ has a nonzero value and these values alternate signs as we travel around the circle. Similarly, on a circle of radius $r > 1 + \sqrt{2}C$, T changes sign exactly $2n$ times at its $2n$ roots. At these roots, the value of U is nonzero and these nonzero values alternate sign as we go around the circle. As r grows, the roots of T on the circle of radius r trace out the level curve $T = 0$. For $p(z) = z^3 + (-3i - 2)z^2 + (5 + 6i)z - 15i$, which we investigated in Example 4, the graph of level curve $T(x, y) = 0$ and of a circle $x^2 + y^2 = r^2$ are given in Figure 6.

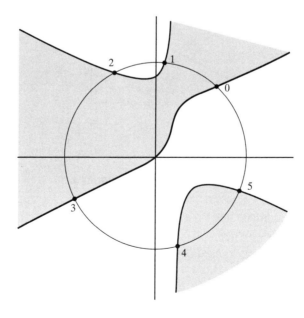

FIGURE 6

The shaded parts are where $T < 0$ and the unshaded parts are where $T > 0$. The level curve $T = 0$ intersects the circle at $2n$ points. (In our case $n = 3$.) As Gauss suggested, consider the shaded parts to be land, and the unshaded parts to be seas. Imagine the level curve $T = 0$ to be the seashore. Starting at a point outside the circle, suppose we walk along the seashore toward the circle, always keeping land on our right. Label the point at which we enter 0. Label the other points of intersections of the circle with $T = 0$ sequentially, going counterclockwise. With the land still on our right, we must eventually exit the circle. To keep the land on our right, we must exit at a point labeled 1, or 3, or 5, etc. (In our diagram, we exit at 3.) Since the sign of U is alternately positive and negative at the labeled intersections, U must have opposite signs at the two points where we enter and leave. Thus if we track the values of U along the seashore on which we have walked, U will have changed signs and our route along which $T = 0$ will have intersected a point where $U = 0$. This common root is what we sought.

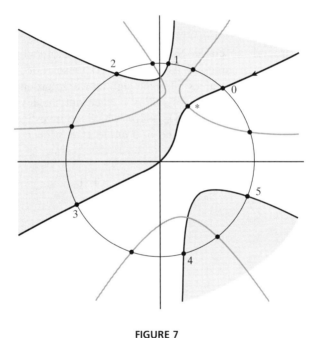

FIGURE 7

In Figure 7, level curves of both $U = 0$ and $T = 0$ are drawn and the root of $p(z)$ that occurs on our path along the "seashore" that commences at intersection 0 is indicated by an asterisk.

SUMMARY

In this section, we gave a proof of the **Fundamental Theorem of Algebra**. Our proof is not algebraic but grounded in the analysis of multivariate calculus. We also indicated the strategy of Gauss's proof of 1799. One of the first rigorous proofs of this important theorem, Gauss's proof is based on trigonometry. Understandably,

Section 3.5 The Fundamental Theorem of Algebra 145

not every reader of this book will choose to negotiate the challenging proof of the Fundamental Theorem of Algebra. But every student and teacher of mathematics should understand the statement of this theorem and its corollary (Corollary 1) for polynomials in $R[x]$. The simple fact the every polynomial $p(x)$ in $R[x]$ has a root in C is counterbalanced by the fact that such a root cannot always be expressed as the sum and product of roots of its coefficients. In short, we know its roots exist, but finding them is another matter.

3.5 Exercises

1. Let $z = x + iy$. Find the real and imaginary parts of each of the following polynomials.
   i. $z^2 + 2z + 1 - 2i$
   ii. $z^4 + 5z^2 + 6$
   iii. $z^3 + i$
2. With the help of the quadratic formula, find all the complex roots of each of the polynomials given in Exercise 1.
3. Prove that for any complex numbers s and t, $|s+t| \geq |s| - |t|$. **Hint.** Apply the usual triangle inequality to s expressed as $(s+t) - t$.
4. Use the fact that i is a root of each of the following polynomials to completely factor each as the product of linear terms.
   i. $z^3 - iz^2 + 3iz + 3$
   ii. $z^3 + (2-i)z^2 + (1-4i)z - i - 2$
   iii. $z^4 + 5z^3 + 7z^2 + 5z + 6$
5. Let $p(z) = 1 - z^4$. Show that $|p(z)|$ does not have a maximum value at $z=0$.
6. Prove by induction that every complex polynomial can be factored as the product of linear terms.
7. Prove that every polynomial in $R[x]$ can be factored as the product of linear and quadratic factors.
8. Express $p(z) = z^2 - (1+2i)z - 1 + i$ as $a(x, y) + ib(x, y)$, the sum of its real and imaginary parts. Then find the roots of p by finding where a and b are simultaneously equal to 0. (Use the graphing method as illustrated in Example 4.)
9. Express the real and imaginary parts of the polynomial p given in Exercise 8 in polar coordinates. By finding where the polar expressions are simultaneously equal to zero, find the polar coordinates of the roots of p.

To the Teacher

It's okay to skip the proof of the Fundamental Theorem of Algebra. Tough going for you, it is well beyond the reach of the high school curriculum. But students coming out of Algebra 2 are expected to know that

1. Counting repetitions, a polynomial of degree n with real coefficients has n complex roots.
2. The complex roots come in conjugate pairs $a \pm bi$.

The proof of the latter fact as found in Corollary 1 in this section is well within the reach of a student in Algebra 2. It is yet another application of the Remainder Theorem.

Here are some exercises for high school students that can reinforce the message of the Fundamental Theorem.

1. Find a monic polynomial of degree 4 that has the following list of roots: 1, 3, 4. (There are 3 answers depending on which root you repeat.)
2. Find a monic polynomial of degree 4 with real coefficients if you know that 1, 2 and $3 + 4i$ are roots. (Only one answer since complex roots come in conjugate pairs.)
3. If a polynomial with real coefficients has no repeated roots and 4 real roots, how do you know that its degree is even?
4. Counting repetitions, a polynomial of odd degree with real coefficients has an odd number of real roots. (Give a graphical demonstration for degree 5.)
5. The graph of a polynomial of even degree with real coefficients has an odd number of "bumps" or "turning points" (local maximums and minimums).
6. Find a polynomial of degree three with the following roots: 1, and $1 + 5i$, and $3i$. (The resulting polynomial must have complex coefficients. Often, students have had no sighting of such polynomials.)

Task

Continue the above list with exercises suitable for high school students.

3.6 SOLVING CUBIC AND QUARTIC EQUATIONS

> "In this book, learned reader, you have the rules of algebra. It is so replete with new discoveries and demonstrations by the author—more than seventy of them—that its forerunners are of little account or, in the vernacular, are washed out."

This is how Girolomo Cardano began his own book, *The Great Art or The Rules of Algebra*, first published in Milan in 1545. In his book, he presents methods for solving all cubic and quartic ("biquadratic") equations, the culmination of the algebra of his times. Negative and indeed complex roots were accounted for, a breakthrough in mathematical thought. The results are not all original to Cardano. But his book is comprehensive and it forwarded the quest for methods to solve all polynomial equations by similar methods—a quest doomed to failure as Abel and Galois showed almost 300 years later.

What follows is a reworking of Cardano's methods for solving the general cubic equation

(1) $$x^3 + ax^2 + bx + c = 0.$$

in terms of the coefficients a, b, and c. (Cardano himself would have proceeded less generally. He presented solutions to equations with specified coefficients like

$x^3 + 2x^2 + 3x - 5 = 0$ that served as models for his methods.) First, we introduce a change of variable $x = y - \frac{a}{3}$ to eliminate the square term. Substituting for x, we obtain

$$\left(y - \frac{a}{3}\right)^3 + a\left(y - \frac{a}{3}\right)^2 + b\left(y - \frac{a}{3}\right) + c$$

$$= y^3 + \left(b - \frac{a^2}{3}\right)y + \left(c - \frac{ba}{3} + \frac{2a^3}{27}\right).$$

Set $p = \left(b - \frac{a^2}{3}\right)$ and $q = \left(c - \frac{ba}{3} + \frac{2a^3}{27}\right)$ and consider

(2) $$y^3 + py + q = 0.$$

If r is a root of (2), then $r - \frac{a}{3}$ is a root of (1). So we look for a solution to (2) in terms of p and q. If either p or q is 0, then the solutions can be found with de Moivre's formula or with the quadratic formula. So assume neither is 0. Our goal is to reduce the process of finding the solution to (2) to taking certain square roots and cube roots. To do this we introduce new variables u and v and set $y = u + v$ and $3uv = -p$. Substituting $y = u + v$ in (2), expanding and regrouping, we obtain

$$(u+v)^3 + p(u+v) + q = u^3 + v^3 + (p + 3uv)(u+v) + q = 0.$$

Since $p + 3uv = 0$, we have $u^3 + v^3 = -q$. Notice that u^3 and v^3 are the roots of the quadratic equation

$$(t - u^3)(t - v^3) = 0$$

or, equivalently,

$$t^2 - (u^3 + v^3)t + u^3v^3 = t^2 + qt - \frac{p^3}{27} = 0.$$

Using the quadratic formula, we find that the roots of $t^2 + qt - \frac{p^3}{27}$ are A and B, where

$$A = -\frac{q}{2} + \sqrt{\frac{q^2}{4} + \frac{p^3}{27}} \quad \text{and} \quad B = -\frac{q}{2} - \sqrt{\frac{q^2}{4} + \frac{p^3}{27}}.$$

Set $u^3 = A$ and $v^3 = B$. Now we can solve for u and v by taking cube roots. For each cube root we have three possibilities. Let $\sqrt[3]{A}$ denote any one of the three cube roots of A and let $\omega = \frac{-1+\sqrt{-3}}{2}$ so that $\omega^3 = 1$. Then the other two roots are $\omega\sqrt[3]{A}$ and $\omega^2\sqrt[3]{A}$. Set u equal to $\sqrt[3]{A}$. Then from among the three cube roots of B, choose the value of $\sqrt[3]{B}$ that makes the following equation true:

$$\sqrt[3]{A}\sqrt[3]{B} = uv = \frac{-p}{3}.$$

Then $y_1 = \sqrt[3]{A} + \sqrt[3]{B}$ is a root of (2). The other two roots are $y_2 = \omega\sqrt[3]{A} + \omega^2\sqrt[3]{B}$, and $y_3 = \omega^2\sqrt[3]{A} + \omega\sqrt[3]{B}$.

EXAMPLE 1

We will solve $x^3 + 3x^2 + 9x + 14 = 0$ with Cardano's method. Since $a = 3$, we substitute $y - 1$ in for x so that we must solve $y^3 + 6y + 7 = 0$. Let $p = 6$ and $q = 7$. Then

$$A = -\frac{q}{2} + \sqrt{\frac{q^2}{4} + \frac{p^3}{27}} = \frac{-7}{2} + \sqrt{\frac{81}{4}} = 1$$

$$B = -\frac{q}{2} - \sqrt{\frac{q^2}{4} + \frac{p^3}{27}} = \frac{-7}{2} - \sqrt{\frac{81}{4}} = -8.$$

We set $u = 1$ and $v = -2$ to find that $y_1 = -1$ is a root of $y^3 + 6y + 7$. Then $y_2 = \omega u + \omega^2 v = \left(\frac{-1+\sqrt{-3}}{2}\right) + \left(\frac{-1-\sqrt{-3}}{2}\right)(-2) = \frac{1}{2} + \frac{3\sqrt{-3}}{2}$. Similarly, $y_3 = \omega^2 u + \omega v = \left(\frac{-1-\sqrt{-3}}{2}\right) + \left(\frac{-1+\sqrt{-3}}{2}\right)(-2) = \frac{1}{2} - \frac{3\sqrt{-3}}{2}$. Now since $x = y - 1$, the roots of $x^3 + 3x^2 + 9x + 14$ are $x_1 = -2, x_2 = -\frac{1}{2} + \frac{3\sqrt{-3}}{2}$, and $x_3 = -\frac{1}{2} - \frac{3\sqrt{-3}}{2}$. ∎

Now we show, as Cardano's student Ferrari did, how to solve a "biquadratic" equation or fourth degree polynomial by solving a cubic and two quadratic equations. Suppose that the equation to be solved is $f(x) = x^4 + ax^3 + bx^2 + cx + d$. If we substitute $y - \frac{a}{4}$ for x, the coefficient on the cubic term in the resulting expression is 0. So we shall assume that $a = 0$. What we want to solve is

(3) $$x^4 = -bx^2 - cx - d.$$

To do this we introduce a new variable y and add $yx^2 + \frac{y^2}{4}$ to both sides of (3) to obtain

(4) $$x^4 + yx^2 + \frac{y^2}{4} = (-b+y)x^2 - cx + \left(\frac{y^2}{4} - d\right)$$

or

(5) $$\left(x^2 + \frac{y}{2}\right)^2 = (-b+y)x^2 - cx + \left(\frac{y^2}{4} - d\right).$$

The left side of (5) is a perfect square. So we now look for y such that the right side also becomes a perfect square. Recall that a quadratic $Ax^2 + Bx + C$ is a perfect square if and only if its discriminant $B^2 - 4AC$ is 0. Let $A = (-b+y)$, $B = -c$, and $C = \frac{y^2}{4} - d$. The discriminant of the right side of (5), set equal to 0, is

(6) $$y^3 - by^2 - 4dy + (4bd - c^2) = 0.$$

Equation (6) is a cubic that can be solved by Cardano's method. Set y_0 equal to any root of (6). Since $(-b+y_0)x^2 - cx + \left(\frac{y_0^2}{4} - d\right)$ is a perfect square, it is equal to $(ex+f)^2 = e^2x^2 + 2efx + f^2$. Equating coefficients, we find that $e = \pm\sqrt{-b+y_0}$ and $f = \pm\sqrt{\frac{y_0^2}{4} - d}$ with signs chosen so that $2ef = -c$. To solve the quartic equation for x, we solve two quadratic equations for x:

$$\left(x^2 + \frac{y_0}{2}\right) = +(ex+f) \text{ and } \left(x^2 + \frac{y_0}{2}\right) = -(ex+f).$$

EXAMPLE 2

In this example, we solve $x^4 - 4x^2 + x + 2 = 0$. Let $b = -4, c = 1$ and $d = 2$. Then Equation (6) becomes $y^3 + 4y^2 - 8y - 33 = 0$, which has a root at $y = -3$. (Any root of (5) will work, so we chose to use the rational root.) Setting $e = 1$ and $f = -\frac{1}{2}$, we solve $x^2 - \frac{3}{2} = x - \frac{1}{2}$ and $x^2 - \frac{3}{2} = -x + \frac{1}{2}$. From the first equation we obtain the roots $\frac{1}{2} + \frac{\sqrt{5}}{2}$ and $\frac{1}{2} - \frac{\sqrt{5}}{2}$. From the second we obtain the roots -2 and 1. ∎

SUMMARY

Solving cubic and quartic equations with **Cardano's method** presents us with some serious hurdles—the computations are indeed complicated. The very fact that the complexity and subtlety needed to solve cubics and quartics is so much greater than what is required for quadratics strongly suggests that the higher order equations might not be solvable by similar techniques. So why study them when computer algebra systems are at our service? Understanding how Cardano solved the cubic enables us to understand the question of solvability. The goal is to express the roots of a polynomial by taking simple roots of various prescribed expressions involving its coefficients, which are themselves regarded as abstract variables. While executing such a program for the cubic equation is difficult, to answer that such a program is not available for the quintic equation is far more difficult.

3.6 Exercises

1. Use Cardano's method to solve the following cubic equations.
   i. $x^3 - 6x - 6 = 0$
   ii. $x^3 + 6x^2 + 9x + 8 = 0$
   iii. $x^3 + 6x^2 - 36 = 0$

2. Use the method of this section to solve the following equations.
   i. $x^4 + 8x^2 + 8x + 2 = 0$
   ii. $x^4 - x^2 - 2x - 1 = 0$

To the Teacher

In this section, we looked at techniques for solving cubics and quartics with an historical eye. We can understand the validity of the techniques and, either under duress or with a mathematician's perverse joy of process, carry out the techniques on a few problems. (Shall we praise technology?) But we don't need to go so far back to find those techniques as part of the standard high school curriculum. What follows are a few excerpts from old New York State Regents exams.

1. From the January, 1964 Algebra Regents Exam:

    Solve the equation $2x^3 - 5x^2 - x + 6 = 0$. [10]

2. From the June, 1952 Advanced Algebra Regents Exam:

    Solve the equation $x^4 - 2x^3 - 3x^2 + 4x - 12 = 0$.

The NYS Regents Exams were standardized exams given to most public school and many private school students in New York State. The first is from the Intermediate Algebra exam, a course routinely taken by sophomores and juniors. It was a half-year course, with the other half-year devoted to trigonometry. The last two questions are drawn from an Advanced Algebra exam. That course was also one-half year in length, taken by juniors or seniors. It has since disappeared, replaced by precalculus. The examples are typical with variations found on all other exams of that era. (Regents exams are still given in New York, but the courses of study have changed.) More examples can be found in the New York State Archives available at

http://www.nysl.nysed.gov/regentsexams.htm

Tasks

1. Solve the given Regent's problems.
2. Write an essay that discusses the role of technology versus the role of mathematical technique in the high school classroom. Reflect on how you might bring these into good balance.

3.7 POLYNOMIAL CONGRUENCES

In Chapter 1, we developed the theory of congruences for integers and, based on that theory, we constructed the fields Z_p, for prime numbers p. In this section we develop a parallel theory for polynomials with coefficients in a field F. The role of the prime number p is played by an irreducible polynomial $p(x)$ of degree >1, which, since it is irreducible, has no root in F. Our goal is not simply to construct a new field but to construct a field that both contains F and contains a root of $p(x)$. We can then guarantee if a polynomial does not have a root in the field F from which its coefficients come, it has a root in some field that contains F. No polynomial shall remain rootless!

> **DEFINITION 1.** Suppose that F is a field and that $p(x)$ is a polynomial in $F[x]$. Let $f(x)$ and $g(x)$ be in $F[x]$. We say that **$f(x)$ is congruent to $g(x)$ modulo $p(x)$** if $f(x) - g(x)$ is divisible by $p(x)$. We write $f(x) \equiv g(x) \bmod p(x)$.

EXAMPLES

1. Let $f(x) = x^2 - 4$, $g(x) = x^2 - x - 2$, and $p(x) = x - 2$ be polynomials in $\mathbf{Q}[x]$. Then $f(x) \equiv g(x) \bmod p(x)$ because $f(x) - g(x) = x - 2$, which is indeed divisible by $(x - 2)$.
2. Let $p(x) = x^2 + x + 1$ be a polynomial in $\mathbf{Z}_2[x]$. To check that $x^4 + x^2 + 1 \equiv x^3 + x^2 + x \bmod (x^2 + x + 1)$, first take the difference $(x^4 + x^2 + 1) - (x^3 + x^2 + x) = x^4 + x^3 + x + 1$. Then factor that difference to see that $x^4 + x^3 + x + 1 = (x^2 + x + 1)(x^2 + 1)$ in $\mathbf{Z}_2[x]$. ∎

The essential properties of polynomial congruences are stated in the following theorem. The proofs of the analogous properties for integer congruences apply almost verbatim. Most are left as exercises.

THEOREM 1 Let F be a field and suppose that $f(x) \equiv g(x) \bmod p(x)$ and $a(x) \equiv b(x) \bmod p(x)$ for polynomials $f(x), g(x), a(x), b(x)$, and $p(x)$ in $F[x]$. Then

i. $f(x) \equiv f(x) \bmod p(x)$.
ii. If $f(x) \equiv g(x) \bmod p(x)$, then $g(x) \equiv f(x) \bmod p(x)$.
iii. If $f(x) \equiv g(x) \bmod p(x)$ and $g(x) \equiv h(x) \bmod p(x)$, then $f(x) \equiv h(x) \bmod p(x)$ for any $h(x)$ in $F[x]$.
iv. $(f(x) + a(x)) \equiv (g(x) + b(x)) \bmod p(x)$.
v. $f(x)a(x) \equiv g(x)b(x) \bmod p(x)$.
vi. If $h(x)$ is any polynomial in $F[x]$, then $h(x)f(x) \equiv h(x)g(x) \bmod p(x)$.
vii. $f(x) \equiv g(x) \bmod p(x)$ if and only if $f(x)$ and $g(x)$ have the same remainder after division by $p(x)$.

Parts i, ii, and iii establish that the relation $f(x) \equiv g(x) \bmod p(x)$ is an **equivalence relation** on $F[x]$. Parts iv, v, and vi establish the arithmetic properties of congruence.

Proof of v. Since $f(x) \equiv g(x) \bmod p(x)$, we can find $q(x)$ in $F[x]$ such that $f(x) - g(x) = q(x)p(x)$. Thus $f(x) = g(x) + q(x)p(x)$. Similarly, we can find $c(x)$ such that $a(x) = b(x) + c(x)p(x)$. Thus $a(x)f(x) = g(x)b(x) + p(x)[c(x)g(x) + b(x)q(x) + c(x)q(x)p(x)]$. Setting $t(x) = [c(x)g(x) + b(x)q(x) + c(x)q(x)p(x)]$, we have $a(x)f(x) - g(x)b(x) = p(x)t(x)$. Thus $f(x)a(x) \equiv g(x)b(x) \bmod p(x)$. ▲

Notice how similar the above proof is to that of part iii of Theorem 2 in Section 2.1.

Proof of vii. Let $f(x) = q(x)p(x) + r_f(x)$ and $g(x) = s(x)p(x) + r_g(x)$, so that r_f and r_g are the remainders after division by p. First suppose that $r_f(x) = r_g(x)$. Then $f(x) - g(x) = p(x)(q(x) - s(x))$ and so $f(x) \equiv g(x) \bmod p(x)$. To prove the converse, assume that $f(x) \equiv g(x) \bmod p(x)$. Then $q(x)p(x) + r_f(x) \equiv s(x)p(x) + r_g(x) \bmod p(x)$ and $r_f(x) \equiv r_g(x) \bmod p(x)$. Thus the difference $r_f(x) - r_g(x)$ is divisible by $p(x)$. Since the degree of $r_f(x) - r_g(x)$ is less than that of $p(x)$, the difference $r_f(x) - r_g(x)$ must be zero. ▲

Part vii says that any polynomial in the equivalence class of $f(x)$ mod $p(x)$ is congruent to exactly one polynomial of degree less than $\deg(p(x))$, namely $r_f(x)$, the remainder of f after division by p. The analogous statement for integers is that every integer n modulo m is congruent to exactly one integer r such that $0 \leq r < m$. As we conveniently represented the entire equivalence class of n mod m by r, we can represent the entire equivalence class of $f(x)$ mod $p(x)$ by its remainder after division by $p(x)$. That remainder is the **principal representative** of the equivalence class of $f(x)$.

EXAMPLE 3

In $Q[x]$, let $p(x) = x^3 + x + 1$ and let $f(x) = x^7 + 3x + 2$. The remainder of $f(x)$ after division by $p(x)$ is $r(x) = 2x^2 + 3x + 1$. (Check it out with long division!) All polynomials in the equivalence class of $f(x)$ mod $p(x)$ are congruent to $2x^2 + 3x + 1$. The Division Algorithm for Polynomials guarantees that every polynomial in $Q[x]$ is congruent mod $p(x)$ to a polynomial of degree 2 or less. ∎

The division of polynomials can be tedious without the help of technology. The next example shows that, when looking for the principal representative of $f(x)$ mod $p(x)$, we can sometimes bypass the division process by using the arithmetic properties given in Theorem 1.

EXAMPLE 4

In $Q[x]$, let $p(x) = x^2 + 2$. Then $x^2 \equiv -2$ mod $p(x)$. By using congruence properties, we can find out what $x^4 + 3x^3 + x + 2$ is congruent to without actually dividing by $p(x)$. First, rewrite $x^4 + 3x^3 + x + 2$ as $(x^2)^2 + 3(x^2)x + x + 2$, and then substitute -2 for x^2 to obtain the following congruence:

$$x^4 + 3x^3 + x + 2 \equiv (-2)^2 + 3(-2)x + x + 2 \text{ mod } p(x).$$

Gathering terms on the right side of the congruence, we have

$$x^4 + 3x^3 + x + 2 \equiv -5x + 6 \text{ mod } p(x).$$

Since $-5x + 6$ has degree less than 2, we are done! ∎

EXAMPLE 5

Let $p(x) = x^2 + x + 1$ in $Z_2[x]$. In this example, we find the principal representative of x^4 mod $p(x)$. Since $+1 \equiv -1$ mod 2, we have

$$x^2 \equiv x + 1 \text{ mod } (x^2 + x + 1).$$

Thus

$$x^4 \equiv (x+1)^2 \equiv x^2 + 2x + 1 \equiv x^2 + 1 \text{ mod } (x^2 + x + 1).$$

In turn, we have

$$x^2 + 1 \equiv (x+1) + 1 \equiv x + 2 \equiv x \bmod (x^2 + x + 1).$$

Thus $x^4 \equiv x \bmod (x^2 + x + 1)$ in $\mathbf{Z}_2[x]$. ∎

> **DEFINITION 2.** We denote the set of equivalence classes defined by the mod $p(x)$ equivalence relation on $F[x]$ by $\boldsymbol{F[x]/\langle p(x)\rangle}$.

The equivalence class of a polynomial $f(x) \in F[x]$ is denoted by $[f(x)]$ or, more simply, by the principal representative of $f(x)$ of degree less than $\deg(p(x))$.

EXAMPLE 6

Let $p(x) = x^2 + 2x + 1$ in $\mathbf{Z}_3[x]$. Every polynomial in $\mathbf{Z}_3[x]$ is congruent to exactly one of the nine polynomials in the following set.

$$F[x]/\langle p(x)\rangle = \{0, 1, 2, x, x+1, x+2, 2x, 2x+1, 2x+2\}$$

To find which of the nine elements of $F[x]/\langle p(x)\rangle$ the polynomial x^4 is congruent to, note $x^2 \equiv -2x - 1 \bmod p(x)$ and that $-2x - 1 \equiv x + 2 \bmod p(x)$ since $-2 \equiv 1 \bmod 3$ and $-1 \equiv 2 \bmod 3$. Since $x^4 = (x^2)^2$, we have $x^4 \equiv (x+2)^2 \bmod p(x)$. Expanding and substituting again, we obtain $(x+2)^2 \equiv x^2 + x + 1 \equiv x + 2 + x + 1 \equiv 2x \bmod p(x)$. Thus $[x^4] = [2x]$ in $F[x]/\langle p(x)\rangle$. ∎

EXAMPLE 7

Let $p(x) = x^2 - 3$ in $\mathbf{Q}[x]$. Then $F[x]/\langle p(x)\rangle = \{ax + b : a \in \mathbf{Q} \text{ and } b \in \mathbf{Q}\}$. The set $F[x]/\langle p(x)\rangle$ is an infinite set because a and b can be any rational numbers. ∎

The set $F[x]/\langle p(x)\rangle$ becomes a ring when the operations of addition and multiplication are defined on it as follows. Let $[f(x)]$ and $[g(x)]$ be the equivalence classes of $f(x)$ and $g(x)$ in $F[x]/\langle p(x)\rangle$.

Addition: $\qquad [f(x)] + [g(x)] = [g(x) + f(x)]$

Multiplication: $\qquad [f(x)] \cdot [g(x)] = [f(x)g(x)]$

Thus, to add or multiply equivalence classes, we find the equivalence classes of the sum or product of their representatives. Parts iv through vi of Theorem 1 guarantee that addition and multiplication are well-defined.

EXAMPLE 8

Let $p(x) = x^3 + 2x + 1$ in $\mathbf{Q}[x]$. Then in $\mathbf{Q}[x]/\langle p(x)\rangle$ we have

$$[x^2 + x + 1] + [2x + 3] = [x^2 + 3x + 4],$$

and
$$[x^2 + x + 1][2x + 3] = [2x^3 + 5x^2 + 5x + 3]$$
$$= [2(-2x - 1) + 5x^2 + 5x + 3]$$
$$= [5x^2 + x + 1].$$ ∎

EXAMPLE 9

Let $p(x) = x^3 + 2$ in $Q[x]$. Every polynomial in $Q[x]$ is congruent to exactly one polynomial of degree 2 or less. Let $[f(x)] = [2x^2 + 3x + 1]$ and $[g(x)] = [x + 3]$. Then $[f(x)] + [g(x)] = [(2x^2 + 3x + 1) + (x + 3)] = [2x^2 + 4x + 4]$ and $[f(x)][g(x)] = [(2x^2 + 3x + 1)(x + 3)] = [2x^3 + 9x^2 + 10x + 3]$. To find a representative for $[2x^3 + 9x^2 + 10x + 3]$ of degree less than 3, we can either divide to find the remainder, or replace x^3 by -2. Replacing, $[f(x)][g(x)] = [2(-2) + 9x^2 + 10x + 3] = [9x^2 + 10x - 1]$. ∎

The operations of addition and multiplication in $F[x]/\langle p(x) \rangle$ are easily seen to be associative and commutative and the operations distribute. These properties all carry over from $F[x]$. The identity element for addition is $[0]$. (Note that $[0] = [p(x)]$.) The identity element for multiplication is $[1]$. If we identify each element $a \in F$ with its equivalence class $[a] \in F[x]/\langle p(x) \rangle$, we see that $F \subseteq F[x]/\langle p(x) \rangle$. Our results are summarized in the following theorem.

THEOREM 2 $F[x]/\langle p(x) \rangle$ is a commutative ring that contains F.

Note: In what follows we will omit the square brackets that denote equivalence classes.

Let $p_1(x) = x^2 + 2x + 2$ and $p_2(x) = x^2 + x + 1$ in $Z_3[x]$. Set $F_1 = Z_3[x]/\langle p_1(x) \rangle$ and set $F_2 = Z_3[x]/\langle p_2(x) \rangle$. Each of these rings has nine elements of the form $a + bx$, where a and b can be 0, 1, or 2. The sum of $(2x + 1)$ and $(2x + 2)$ is the same in F_1 or F_2, namely $(2x + 1) + (2x + 2) = 4x + 3 = x$. In fact, the addition tables for both rings are identical. (We never increase the degree when we add.) However, products differ. In F_1, we first note that $x^2 = -2x - 2 = x + 1$. Thus $(2x + 1)(2x + 2) = 4x^2 + 6x + 2 = x^2 + 2 = (x + 1) + 2 = x$. In F_2, notice that $x^2 = 2x + 2$ and so $(2x + 1)(2x + 2) = x^2 + 2 = (2x + 2) + 2 = 2x + 1$. The multiplication tables for F_1 and F_2 are given in the following charts.

F_1,*	0	1	2	x	$x + 1$	$x + 2$	$2x$	$2x + 1$	$2x + 2$
0	0	0	0	0	0	0	0	0	0
1	0	1	2	x	$x + 1$	$x + 2$	$2x$	$2x + 1$	$2x + 2$
2	0	2	1	$2x$	$2x + 2$	$2x + 1$	x	$x + 2$	$x + 1$
x	0	x	$2x$	$x + 1$	$2x + 1$	1	$2x + 2$	2	$x + 2$
$x + 1$	0	$x + 1$	$2x + 2$	$2x + 1$	2	x	$x + 2$	$2x$	1
$x + 2$	0	$x + 2$	$2x + 1$	1	x	$2x + 2$	2	$x + 1$	$2x$
$2x$	0	$2x$	x	$2x + 2$	$x + 2$	2	$x + 1$	1	$2x + 1$
$2x + 1$	0	$2x + 1$	$x + 2$	2	$2x$	$x + 1$	1	$2x + 2$	x
$2x + 2$	0	$2x + 2$	$x + 1$	$x + 2$	1	$2x$	$2x + 1$	x	2

Chart 1. Multiplication in F_1

$F_2, *$	0	1	2	x	$x+1$	$x+2$	$2x$	$2x+1$	$2x+2$
0	0	0	0	0	0	0	0	0	0
1	0	1	2	x	$x+1$	$x+2$	$2x$	$2x+1$	$2x+2$
2	0	2	1	$2x$	$2x+2$	$2x+1$	x	$x+2$	$x+1$
x	0	x	$2x$	$2x+2$	2	$x+2$	$x+1$	$2x+1$	1
$x+1$	0	$x+1$	$2x+2$	2	x	$2x+1$	1	$x+2$	$2x$
$x+2$	0	$x+2$	$2x+1$	$x+2$	$2x+1$	0	$2x+1$	0	$x+2$
$2x$	0	$2x$	x	$x+1$	1	$2x+1$	$2x+2$	$x+2$	2
$2x+1$	0	$2x+1$	$x+2$	$2x+1$	$x+2$	0	$x+2$	0	$2x+1$
$2x+2$	0	$2x+2$	$x+1$	1	$2x$	$x+2$	2	$2x+1$	x

Chart 2. Multiplication in F_2

Not only are the multiplication tables different, their essential properties are different. In the ring F_1 each nonzero element has a multiplicative inverse so that F_1 is a field. In F_2, the elements $x+2$ and $2x+1$ are zero divisors and have no multiplicative inverses. The difference comes from the fact that $p_1(x)$ is irreducible in $Z_3[x]$ while $p_2(x) = x^2 + x + 1 = (x+2)^2$. These differences parallel the differences between the rings Z_n and Z_p, where n is a composite integer and p is a prime integer.

THEOREM 3 $F[x]/\langle p(x) \rangle$ is a field if and only is $p(x)$ is irreducible in $F[x]$.

Proof. If $p(x)$ is reducible, then $p(x) = f(x)g(x)$, where $f(x)$ and $g(x)$ are non-constant polynomials with degrees less than that of $p(x)$. Their representatives in $F[x]/\langle p(x) \rangle$ are zero divisors since $f(x)g(x) = p(x) = 0$. Thus $F[x]/\langle p(x) \rangle$ cannot be a field.

For the converse, assume that $p(x)$ is irreducible. We must show that every nonzero element $f(x)$ in $F[x]/\langle p(x) \rangle$ has a multiplicative inverse. If $f(x)$ is not 0 in $F[x]/\langle p(x) \rangle$, then it is not divisible by $p(x)$ and hence it is relatively prime to $p(x)$. So we can find polynomials $s(x)$ and $t(x)$ such that $1 = s(x)f(x) + t(x)p(x)$ in $F[x]$. Reducing mod $p(x)$, we have $s(x)f(x) \equiv 1 \bmod p(x)$. Thus $s(x)$ or its remainder after division by $p(x)$ is the multiplicative inverse of $f(x)$ in $F[x]/\langle p(x) \rangle$. ▲

EXAMPLE 10

The polynomial $p(x) = x^3 - 2$ is irreducible in $Q[x]$ and so $Q[x]/\langle p(x) \rangle$ is a field. We'll use Euclid's Algorithm to find the multiplicative inverse of $f(x) = x^2 + 2x + 1$.

$$x^3 - 2 = (x-2)(x^2 + 2x + 1) + 3x$$

$$x^2 + 2x + 1 = \left(\frac{x}{3} + \frac{2}{3}\right)(3x) + 1$$

Now we solve for 1:

$$1 = (x^2 + 2x + 1) - (3x)\left(\frac{x}{3} + \frac{2}{3}\right)$$

$$1 = (x^2 + 2x + 1) - [(x^3 - 2) - (x-2)(x^2 + 2x + 1)]\left(\frac{x}{3} + \frac{2}{3}\right).$$

Regrouping, we obtain

$$1 = \left[(x-2)\left(\frac{x}{3}+\frac{2}{3}\right)+1\right](x^2+2x+1) - \left(\frac{x}{3}+\frac{2}{3}\right)(x^3-2)$$

$$1 = \left(\frac{x^2}{3}-\frac{1}{3}\right)(x^2+2x+1) - \left(\frac{x}{3}+\frac{2}{3}\right)(x^3-2).$$

Thus $1 \equiv \left(\frac{x^2}{3}-\frac{1}{3}\right)(x^2+2x+1) \bmod (x^3-2)$ and the multiplicative inverse of $f(x)$ is $s(x) = \frac{x^2}{3} - \frac{1}{3}$. ∎

Let's look again at the field F_1 with 9 elements given in Chart 1. As a visual trick, replace the symbol x by another symbol, say α. Then the list of elements in F_1 is $\{0, 1, 2, \alpha, \alpha+1, \alpha+2, 2\alpha, 2\alpha+1, 2\alpha+2\}$. Its multiplication table looks like this:

*	0	1	2	α	$\alpha+1$	$\alpha+2$	2α	$2\alpha+1$	$2\alpha+2$
0	0	0	0	0	0	0	0	0	0
1	0	1	2	α	$\alpha+1$	$\alpha+2$	2α	$2\alpha+1$	$2\alpha+2$
2	0	2	1	2α	$2\alpha+2$	$2\alpha+1$	α	$\alpha+2$	$\alpha+1$
α	0	α	2α	$\alpha+1$	$2\alpha+1$	1	$2\alpha+2$	2	$\alpha+2$
$\alpha+1$	0	$\alpha+1$	$2\alpha+2$	$2\alpha+1$	2	α	$\alpha+2$	2α	1
$\alpha+2$	0	$\alpha+2$	$2\alpha+1$	1	α	$2\alpha+2$	2	$\alpha+1$	2α
2α	0	2α	α	$2\alpha+2$	$\alpha+2$	2	$\alpha+1$	1	$2\alpha+1$
$2\alpha+1$	0	$2\alpha+1$	$\alpha+2$	2	2α	$\alpha+1$	1	$2\alpha+2$	α
$2\alpha+2$	0	$2\alpha+2$	$\alpha+1$	$\alpha+2$	1	2α	$2\alpha+1$	α	2

Chart 3.

Now think of the elements of F_1 not as polynomials but simply as elements in a field with 9 elements. We can again form polynomials, this time with coefficients in F_1. They are elements in the set $F_1[x]$. Some typical elements in $F_1[x]$ would be $f(x) = x^2 + 2x + 1, g(x) = \alpha x^3 - (\alpha+1)x - \alpha, h(x) = x^2 + 2\alpha x + 2 + \alpha$, etc. We can evaluate these polynomials at various elements in F_1. For instance, $f(\alpha+1) = (\alpha+1)^2 + 2(\alpha+1) + 1 = 2\alpha + 2$. Let us evaluate the modulus polynomial $p(x)$ at α:

$$p(\alpha) = \alpha^2 + 2\alpha + 2 = (\alpha+1) + 2\alpha + 2 = 3\alpha + 3 = 0.$$

Whereas $p(x)$ is irreducible in $Z_3[x]$ and has no root in Z_3, $p(x)$ has a root, namely α, in the bigger field F_1. Therefore, $p(x)$ must factor in $F_1[x]$. We can use long division to find a factorization.

$$\begin{array}{r}
x + (2+\alpha) \\
x - \alpha \,\overline{\smash{\big)}\, x^2 + 2x + 2} \\
\underline{x^2 - \alpha x} \\
(2+\alpha)x + 2 \\
\underline{(2+\alpha)x - \alpha(2+\alpha)} \\
2 + 2\alpha + \alpha^2
\end{array}$$

Since $2 + 2\alpha + \alpha^2 = 0$, we see that the polynomial $x^2 + 2x + 2$ factors as

$$x^2 + 2x + 2 = (x - \alpha)(x + (2+\alpha)).$$

Since the field F_1 contains Z_3, we have constructed a field that both contains the coefficient field of $p(x)$, namely Z_3, and has a root of $p(x)$, namely α. More generally, when $p(x)$ is irreducible in $F[x]$ and hence without a root in F, then $F[x]/\langle p(x)\rangle$ is an "extension" field of F in which $p(x)$ naturally has a root.

EXAMPLE 11

The polynomial $p(x) = x^3 - 2$ is irreducible in $Q[x]$ and hence rootless in Q. But $p(x)$ has a root in the larger field $F = Q[x]/\langle p(x)\rangle$. Let's denote its root $[x]$ by the symbol β. Since β is a root of $p(x)$, we know that $x - \beta$ is a factor of $p(x)$ in $F[x]$. We can apply the Division Algorithm to factor $p(x)$ as

$$p(x) = (x - \beta)(x^2 + \beta x + \beta^2)$$

in $F[x]$. The quadratic polynomial $q(x) = x^2 + \beta x + \beta^2$ is irreducible in $F[x]$ because its discriminant $\Delta = \beta^2 - 4\beta^2 = -3\beta^2$ has no square root in F. (We leave the details as an exercise.) But we can repeat our construction. Let $E = F[x]/\langle x^2 + \beta x + \beta^2\rangle$ and let α denote $[x]$ in E so that α is a root of $x^2 + \beta x + \beta^2$ in E. Now we can divide $x^2 + \beta x + \beta^2$ by $x - \alpha$ as follows.

$$
\begin{array}{r}
x + (\alpha + \beta) \\
x - \alpha \overline{)\,x^2 + \beta x + \beta^2} \\
\underline{x^2 - \alpha x } \\
(\alpha + \beta)x + \beta^2 \\
\underline{(\alpha + \beta)x - (\alpha^2 + \alpha\beta)} \\
\beta^2 + \alpha\beta + \alpha^2
\end{array}
$$

Since α is a root of $x^2 + \beta x + \beta^2$, the remainder is 0. Thus $x^2 + \beta x + \beta^2 = (x - \alpha)(x + (\alpha + \beta))$ in E. Better, $x^3 - 2 = (x - \beta)(x - \alpha)(x + (\alpha + \beta))$ in E. So we have constructed a field E in which the polynomial $x^3 - 2$ factors into three linear terms and has three roots. ■

As we can see from Example 11, we can iterate the process of constructing a field in which $p(x)$ has at least one root to obtain a field K over which $p(x)$ factors completely into linear terms. Thus every polynomial factors completely into linear terms in some field that contains its coefficients. That is the message of the following theorem.

THEOREM 4 Let F be a field and suppose that $p(x)$ is a polynomial in $F[x]$ of degree $n > 0$. There is a field K such that $F \subseteq K$, over which $p(x)$ factors into linear terms.

Proof. The proof is by induction on the degree of $p(x)$. If $p(x)$ is of degree 1, then $K = F$. Suppose $\deg(p(x)) > 1$ and that $p(x) = p_1(x)p_2(x)\cdots p_k(x)$, where each factor is irreducible. Suppose that the assertion holds for polynomials of degree $0 < m < n$. Let $F_1 = F[x]/\langle p_1(x)\rangle$. Since $p_1(x)$, and therefore $p(x)$, has at least one root α_1 in F_1, $p(x)$ can be factored as $(x - \alpha_1)q(x)$ in $F_1[x]$ where $\deg(q(x)) < n$. By the induction hypothesis, we can find a field K such that $F \subseteq F_1 \subseteq K$ over which $q(x)$, and hence $p(x)$, factors into linear terms. ▲

EXAMPLE 12

The polynomial $p(x) = x^3 + 2x + 1$ is irreducible in $\mathbf{Z}_3[x]$. Let β be a root of $p(x)$ in $F = \mathbf{Z}_3[x]/\langle p(x) \rangle$. Dividing $p(x)$ by $x - \beta$, we have $p(x) = (x - \beta)(x^2 + \beta x + (2 + \beta^2))$. The quadratic factor is *not* irreducible in $F[x]$ because $(x^2 + \beta x + (2 + \beta^2))$ factors as $(x - (-1 - \beta)/2)(x - (1 - \beta)/2) = (x + 2\beta + 2)(x + 2\beta - 2)$ in $F[x]$. (To factor the quadratic, use the quadratic formula mod 3.) ∎

If $p(x)$ is a polynomial of degree n, then we need to apply our construction procedure at most $n - 1$ times to find a field in which $p(x)$ has n roots (counting multiplicities) and hence factors into linear terms. However, $n - 1$ is an upper bound as Example 12 shows.

If, in Example 11, we use the *symbol* $\sqrt[3]{2}$ instead of β—just a change of graphics, really—then $p(x) = x^3 - 2$ factors as

$$(x - \sqrt[3]{2})(x^2 + \sqrt[3]{2}x + (\sqrt[3]{2})^2)$$

as it would in $\mathbf{R}[x]$ or $\mathbf{C}[x]$. The field $\mathbf{R}$ of real numbers contains the field $\mathbf{Q}$ of rational numbers. The polynomial $p(x) = x^3 - 2$ has no root in $\mathbf{Q}$ but it does have a root in $\mathbf{R}$, namely, the real number $2^{1/3}$. However, the set of real numbers contains all sorts of numbers like π and $\sqrt{17}$ that have nothing to do with finding the roots of the polynomial $x^3 - 3$. So next we look for the smallest subset of $\mathbf{R}$ that contains $2^{1/3}$ and that is itself a field when we add and multiply its elements as real numbers. (A subset E of a field F that is a field using the same operations as in F is called a **subfield** of F.) We are looking for a subfield $\mathbf{Q}_1$ of $\mathbf{R}$ that contains $\mathbf{Q}$ and $2^{1/3}$ and is minimal in the sense that its elements must be contained in any other subfield of $\mathbf{R}$ that contains both $\mathbf{Q}$ and $2^{1/3}$.

Let $\mathbf{Q}_1 = \{a_0 + a_1 2^{1/3} + a_2 2^{2/3} : a_i \in \mathbf{Q} \text{ for } i = 0, 1, 2\}$. Typical elements of $\mathbf{Q}_1$ are the real numbers $2 - 2^{1/3} + 2^{2/3}$ and $5 + \frac{7}{3} \cdot 2^{1/3} + \frac{19}{7} \cdot 2^{2/3}$. Since $(2^{1/3})^3 = 2$, powers of $2^{1/3}$ higher than $2^{2/3} = (2^{1/3})^2$ are not needed in sums or products of members of $\mathbf{Q}_1$. For instance, $(1 + 2^{2/3})^2 = 1 + 2 \cdot 2^{2/3} + 2^{4/3} = 1 + 2 \cdot 2^{1/3} + 2 \cdot 2^{2/3}$. Thus it is not difficult to see that $\mathbf{Q}_1$ is closed under addition and multiplication, and that it contains 0 and 1. What may be surprising is that the multiplicative inverses of nonzero numbers of the form $a_0 + a_1 2^{1/3} + a_2 2^{2/3}$ are also of that form, as we now show. Since the polynomial $p(x) = x^3 - 2$ is irreducible in $\mathbf{Q}[x]$, any nonzero polynomial $g(x) = a_2 x^2 + a_1 x + a_0$ in $\mathbf{Q}[x]$ is relatively prime to $p(x)$. So we can find polynomials $s(x)$ and $t(x)$ in $\mathbf{Q}[x]$ such that $1 = s(x)p(x) + t(x)g(x)$. Substituting $2^{1/3}$ for x and noting that $p(2^{1/3}) = 0$, we see that

$$1 = t(2^{1/3}) \cdot g(2^{1/3}).$$

The expression $t(2^{1/3})$ can be reduced to the form $a_0 + a_1 2^{1/3} + a_2 2^{2/3}$ with $a_i \in \mathbf{Q}$ by replacing $2^{n/3}$ by $2^q 2^{r/3}$ where $n = 3q + r$ and $r = 0, 1,$ or 2. Thus the multiplicative inverses of expressions of the form $a_0 x^2 + a_1 x + a_2$ are also of that form, as asserted. So $\mathbf{Q}_1 = \{a + b 2^{1/3} + c 2^{2/3} : a, b \text{ and } c \in \mathbf{Q}\}$ is a field that is a subfield of $\mathbf{R}$. Any subfield of $\mathbf{R}$ that contains $\mathbf{Q}$ and the irrational number $2^{1/3}$

must contain all the members of Q_1 as well. In that sense, Q_1 is the smallest such subfield.

It is easy to verify that the map $f\colon Q_1 \to Q[x]/\langle x^3 - 2\rangle$ given by $f((a + b2^{1/3} + c2^{2/3})) = (a + b[x] + c[x^2])$ is one-to-one correspondence. We just match up the elements of Q_1 and $Q[x]/\langle x^3 - 2\rangle$ by matching up their coefficients. This correspondence holds up under addition and multiplication. For instance, $(2 + 3[x]) \cdot (3 - [x^2]) = 6 + 9[x] - 2[x^2] - 3[x^3] = 9[x] - 2[x^2]$ and similarly $(2 + 3 \cdot 2^{1/3}) \cdot (3 - 2^{2/3}) = 6 + 9 \cdot 2^{1/3} - 2 \cdot 2^{2/3} - 3 \cdot 2^{3/3} = 9 \cdot 2^{1/3} - 2 \cdot 2^{2/3}$. So, whether we construct the field $Q[x]/\langle x^3 - 2\rangle$ to obtain a root of $x^3 - 2$ or find the smallest subfield of R that has a root of $x^3 - 2$, we end up with essentially the same field.

SUMMARY

In this section, we defined what it means for two polynomials in $F[x]$ to be **congruent mod $p(x)$**. Through this equivalence relation, we constructed the ring $F[x]/\langle p(x)\rangle$ in much the same way as we constructed the rings Z_m. When $p(x)$ is irreducible, $F[x]/\langle p(x)\rangle$ is a field that contains F. The importance of this construction is that $p(x)$ has a root, and hence factors, in $F[x]/\langle p(x)\rangle$. In conclusion we found that, given any polynomial $p(x)$ with coefficients in a field F, we can find a field K in which $p(x)$ factors completely into linear terms. In the last chapter of this book, we shall look into the question of when a polynomial is "solvable by radicals." That question asks if we can find all the roots of a given polynomial $p(x)$ by repeatedly constructing fields of the form $F[x]/\langle p(x)\rangle$, where $p(x)$ has the particularly simple form $x^p - a$. But to get to the answer, we need a trip into the more abstract aspects of algebra. We begin that journey in the next chapter.

3.7 Exercises

1. Determine whether or not $x^3 + 2x + 1$ and $x^4 + 3x - 2$ are congruent mod $(x^2 + 2x + 2)$ in $Q[x]$.
2. Determine whether $x^4 + x^3 + x^2 + 2$ and $x^3 + 1$ are congruent mod $(x^2 + 2)$ in $Z_3[x]$.
3. Prove that if c is any element in a field F, and $f(x)$ is any polynomial in $F[x]$, then $f(x) \equiv f(c)$ mod $(x - c)$.
4. Find the principal representative of each of the following:
   i. $x^3 - x + 1$ mod $x + 2$ in $Q[x]$
   ii. $x^7 + x + 1$ mod $x^3 + x + 1$ in $Z_3[x]$
   iii. $x^4 + 2x + 4$ mod $x^2 + 1$ in $Z_5[x]$
5. Use the technique of Example 5 to find the principal representative of the indicated polynomial:
   i. $x^3 - x + 1$ mod $(x + 2)$ in $Q[x]$
   ii. $x^7 + x + 1$ mod $(x^3 + x + 1)$ in $Z_3[x]$
   iii. $x^4 + 2x$ mod $(x^2 + 1)$ in $Z_5[x]$
6. Let $p(x) = x^3 + x + 1$ in $Q[x]$. For each of the following, find the polynomial of degree less than 3 congruent to it mod $p(x)$ without dividing.
   i. $x^3 + 3x^2 - 2x + 3$

      ii. x^5
      iii. $x^6 + x^3 + 1$
7. Repeat Exercise 6, regarding each polynomial as a polynomial in $\mathbf{Z}_5[x]$.
8. List a complete set of representatives of $F[x]/\langle p(x)\rangle$ where $F = \mathbf{Z}_2$ and $p(x) = x^4 + x^2 + 1$.
9. Let $p(x) = x^2 + x + 1$ in $\mathbf{Z}_2[x]$. Write down the multiplication table for $\mathbf{Z}_2[x]/\langle p(x)\rangle$.
10. Find the multiplicative inverse of $x^2 + 1$ in $\mathbf{Q}[x]/\langle x^4 - 2\rangle$.
11. Let $p(x) = x^2 + 1$ in $\mathbf{Z}_3[x]$. Write down the multiplication table for $F_3 = \mathbf{Z}_3[x]/\langle p(x)\rangle$.
12. Find a field with 32 elements and another with 27 elements.
13. Let $p(x) = x^3 + x + 1$ and $F = \mathbf{Z}_2[x]/\langle p(x)\rangle$. Factor $p(x)$ in $F[x]$. Determine if $p(x)$ factors into linear (degree 1) terms.
14. Let $p(x) = x^3 + 2x + 2$ and $F = \mathbf{Z}_3[x]/\langle p(x)\rangle$. Factor $p(x)$ into linear factors in $F[x]$.
15. Express the multiplicative inverse of $1 + 2^{1/3} - 3(2^{2/3})$ as $a_0 + a_1 2^{1/3} + a_2 2^{2/3}$.
16. Let β be as in Example 11. Show that $\Delta = \beta^2 - 4\beta^2 = -3\beta^2$ has no square root in F.
17. Let β be as in Example 11. Thus β is a root of the polynomial $x^3 - 2$ in $F = \mathbf{Q}[x]/\langle p(x)\rangle$. Also let α be a root of $q(x) = x^2 + \beta x + \beta^2$ in $F_2 = F[x]/\langle x^2 + \beta x + \beta^2\rangle$. Then $\xi = \frac{2\alpha}{\beta} + 1$ is an element of F_2.
   i. Show that $\xi^2 = -3$.
   ii. Use the quadratic formula to show that the roots of $q(x)$ are $\alpha_{1,2} = \left(\frac{-1 \pm \xi}{2}\right)\beta$ or, renaming ξ as $\xi = \sqrt{-3}$, obtain $\alpha_{1,2} = \left(\frac{-1 \pm \sqrt{-3}}{2}\right)\beta$.

To the Teacher

In this section, we studied remainders. To get remainders, we had to divide polynomials. Without technology, this is a tedious task. So let's upgrade **synthetic division** to help us out. Synthetic division is a technique very familiar to and appreciated by the high school algebra student. It provides a way of dividing any monic polynomial $p(x)$ by a linear factor $(x - a)$. Synthetic division feels like a short cut because we skip the writing and rewriting of monomials like x^i. In this note, we find a similar scheme for dividing any polynomial $g(x) = a_n x^n + \ldots + a_0$ by a monic polynomial $f(x) = x^m + b_{m-1} x^{m-1} + \ldots + b_0$. The procedure is given below. It is similar to row reduction for matrices. Alongside of the general procedure, we will do the specific example of dividing $g(x) = x^4 + 5x^3 + 2x + 3$ by $f(x) = x^2 - 1$.

Suppose that the difference of the degrees of $g(x)$ and $f(x)$ is $n - m = q$. Write down $q + 1$ vectors, each of length $n + 1$. The first is $[1, b_{m-1}, \ldots, b_0, 0, \ldots, 0]$, which lists the coefficients of $f(x)$ followed by q trailing 0s. The rest are similar but with the 0s shifted one by one until the last vector has q leading 0s. We obtain the

following list:
$$v_1 = [1, b_{m-1}, \ldots, b_0, 0, \ldots, 0]$$
$$v_2 = [0, 1, b_{m-1}, \ldots, b_0, \ldots, 0]$$
.
.
.
$$v_{q+1} = [0, 0, \ldots, 1, b_{m-1}, \ldots, b_0]$$

Our example: $q = 2$ and the three vectors are
$$v_1 = [1, 0, -1, 0, 0]$$
$$v_2 = [0, 1, 0, -1, 0]$$
$$v_3 = [0, 0, 1, 0, -1].$$

Now we build a matrix with $q + 2$ rows. The top row just lists the coefficients of $g(x)$.

$$r_1 = [a_n, a_{n-1}, \ldots, a_0]$$ (Don't forget to list the coefficients that are 0.)

Our example: $r_1 = [1, 5, 0, 2, 3]$.

For the second row, we multiply v_1 by a_n and subtract it from r_1 so that
$$r_2 = r_1 - a_n \cdot v_1 = [0, c_2, \ldots, c_m, 0, \ldots, 0].$$

Our example: $r_2 = [1, 5, 0, 2, 3] - 1 \cdot [1, 0, -1, 0, 0] = [0, 5, 1, 2, 3]$.

The third row is $r_3 = r_2 - c_2 \cdot v_2$.

Our example: $[0, 5, 1, 2, 3] - 5 \cdot [0, 1, 0, -1, 0] = [0, 0, 1, 7, 3]$.

In general, if $r_i = [0, 0, \ldots, c_i, \ldots]$, then $r_{i+1} = r_i - c_i \cdot v_i$. (We do this even if $c_i = 0$.)

Our example: (We have only one more row to compute.)
$$r_4 = [0, 0, 1, 7, 3] - 1 \cdot [0, 0, 1, 0, -1] = [0, 0, 0, 7, 4].$$

Putting all our rows into a matrix, we end up with a $(q + 2) \times (n + 1)$ matrix.

Our example:
$$\begin{bmatrix} 1 & 5 & 0 & 2 & 3 \\ 0 & 5 & 1 & 2 & 3 \\ 0 & 0 & 1 & 7 & 3 \\ 0 & 0 & 0 & 7 & 4 \end{bmatrix}$$

To read off the answer, circle the upper $q + 1$ diagonal entries. They are the coefficients of the quotient, starting with the leading coefficient of x^q. Box in the last m terms in the last row. They are the coefficients of the remainder, starting with the leading coefficient of x^{m-1}. If the last row is a row of zeros, $f(x)$ divides $g(x)$.

Our example: The quotient is $x^2 + 5x + 1$ and the remainder is $7x + 4$.

It's much shorter without the explanation. Here's another example.

EXAMPLE 1

Divide $x^5 + x^4 + 3x^3 + 2x^2 + 3x + 2$ by $x^2 + x + 2$.

Vectors:
$$v_1 = [1, 1, 2, 0, 0, 0]$$
$$v_2 = [0, 1, 1, 2, 0, 0]$$
$$v_3 = [0, 0, 1, 1, 2, 0]$$
$$v_4 = [0, 0, 0, 1, 1, 2]$$
$$r_1 = [1, 1, 3, 2, 3, 2]$$

Computations:
$$r_1 = [1, 1, 3, 2, 3, 2]$$
$$r_2 = r_1 - v_1 = [0, \mathbf{0}, 1, 2, 3, 2]$$
$$r_3 = r_2 - 0 \cdot v_2 = [0, 0, \mathbf{1}, 2, 3, 2]$$
$$r_4 = r_3 - 1 \cdot v_3 = [0, 0, 0, \mathbf{1}, 1, 2]$$
$$r_5 = r_4 - 1 \cdot v_4 = [0, 0, 0, 0, \underline{0}, \underline{0}]$$

Matrix:
$$\begin{bmatrix} 1 & 1 & 3 & 2 & 3 & 2 \\ 0 & 0 & 1 & 2 & 3 & 2 \\ 0 & 0 & 1 & 2 & 3 & 2 \\ 0 & 0 & 0 & 1 & 1 & 2 \\ 0 & 0 & 0 & 0 & 0 & 0 \end{bmatrix}$$

The quotient is $x^3 + 0x^2 + x + 1$ and the remainder is 0. ■

With a bit of practice, the process can be streamlined further, to one's own taste. For instance, we can stack the computations, omit the matrix, and extract the bold and underlined numbers for the coefficients of the quotient and remainder, respectively.

EXAMPLE 2

Divide $x^4 + 4x^3 + 6x^2 + 5x + 2$ by $x^3 + x + 2$.

Computations:

	1, 4, 6, 5, 2
	1, 0, 1, 2, 0 (reference row)
(subtract)	0, **4**, 5, 3, 2
	0, 4, 0, 4, 8 (reference row shifted and multiplied by 4)
(subtract)	0, 0, $\underline{5}$, $\underline{-1}$, $\underline{-6}$

Answer: The quotient is $x + 4$ and the remainder is $5x^2 - x - 6$. ■

It's not magic; it's just a bit speedier. It makes the computations involved in polynomial division quite transparent and it points to more advanced computational techniques that are behind computer algebra systems.

Tasks

1. Use synthetic division to divide $g(x)$ by $f(x)$ for each of the given pairs of polynomials.

   i. $f(x) = x^2 - 2x + 5$ and $g(x) = 3x^5 + 4x^3 + 2x + 2$
   ii. $f(x) = x^2 - 1$ and $g(x) = 5x^4 + 3x^3 + 2x - 7$
   iii. $f(x) = x^2 + 4x + 4$ and $g(x) = 3x^5 + 4x^3 + 2x + 2$ in $\mathbf{Z}_5[x]$

3.8 WORKSHEET 5: LAGRANGE INTERPOLATION

In this worksheet, we turn our attention to an algorithm for factoring polynomials in $\mathbf{Z}[x]$. If a polynomial in $\mathbf{Z}[x]$ of degree greater than one can be factored into two polynomials of lower degree (not always the case!), then the algorithm will find that factorization in a finite number of steps. The situation is very different for polynomials in $\mathbf{C}[x]$. There, polynomials always have factorizations into linear terms, but there is no algorithm to accomplish that factorization.

Our algorithm, called the **Lagrange Interpolation**, is based on a corollary to the Chinese Remainder Theorem for Polynomials. The statement of the Chinese Remainder Theorem for Polynomials is analogous to the Chinese Remainder Theorem for integers, with numbers replaced by polynomials. Its proof is also directly analogous. (See Theorem 10 of Section 2.1.)

The Chinese Remainder Theorem for Polynomials: Let F be a field. Let $a_1(x), a_2(x), \ldots, a_n(x)$ be polynomials in $F[x]$ and let $b_1(x), b_2(x), \ldots, b_n(x)$ be a set of polynomials in $F[x]$ that are mutually relatively prime. We can find a polynomial $f(x)$ in $F[x]$ that simultaneously satisfies each of the congruences:

(1) $\qquad f(x) \equiv a_i(x) \bmod b_i(x)$ for $i = 1, 2, \ldots, n$.

Furthermore, $g(x)$ is also a solution to (1) if and only if

$$f(x) \equiv g(x) \bmod b_1(x)b_2(x)\cdots b_n(x).$$

The following exercise will be far less tedious if a CAS such as Maple is used to carry out polynomial division.

Task 1

Prove the Chinese Remainder Theorem for Polynomials.

Task 2

Use the procedure outlined in the proof of the Chinese Remainder Theorem to solve the following pairs of congruences.

   i. $p(x) \equiv 9 \bmod (x - 1)$
      $p(x) \equiv 15 \bmod (x - 2)$ in $\mathbf{Q}[x]$

ii. $p(x) \equiv 5 + 2x \mod (x^2 + 2)$
   $p(x) \equiv 2 + 2x \mod (x^2 - 1)$ in $Q[x]$
iii. $p(x) \equiv 1 + x \mod (x^2)$
   $p(x) \equiv x \mod (x^2 + x + 1)$ in $Z_2[x]$

Let $h(x)$ be in $F[x]$ and $b \in F$. The corollary to the Remainder Theorem for Polynomials can be restated as follows: $h(x) \equiv b \mod (x - a)$ if and only if $h(a) = b$. We can use this fact to prove the following corollary to the Chinese Remainder Theorem.

COROLLARY 1 Let $r_1, \ldots, r_n$ be distinct elements of F and let $s_1, \ldots, s_n$ be arbitrary elements of F. Then we can find a unique polynomial $q(x)$ in $F[x]$ of degree less than n such that $q(r_i) = s_i$ for each $i = 1, \ldots, n$.

Task 3

i. Prove Corollary 1.
ii. Find a polynomial $g(x)$ in $Q[x]$ of degree < 4 such that $g(1) = 2$, $g(2) = 3$, $g(3) = -2$, $g(-1) = 3$.

Note: Given a set of $d + 1$ pairs (x_i, y_i), there is a unique polynomial $p(x)$ of degree less than or equal to d such that $p(x_i) = y_i$ for $i = 0, 1, 2, \ldots, d$. The polynomial can be obtained by simple linear algebra techniques. In the following, those techniques can replace the Chinese Remainder Theorem.

Lagrange Interpolation

Let $p(x)$ be a polynomial with integer coefficients. If $p(x)$ factors in $Q[x]$, then it factors in $Z[x]$. Our goal is to find a factorization in $Z[x]$. Suppose that $n > 1$ and that $\deg(p(x)) = n$. Set $d = n/2$ if n is even or $d = (n - 1)/2$ if n is odd. If $p(x)$ has a nontrivial factorization, then it has a factor of degree less than or equal to d.

Step 1. Pick $d + 1$ distinct integers $n_0, n_1, \ldots, n_d$. Evaluate $p(x)$ at each of these integers, to produce the vector $s = [p(n_0), p(n_1), \ldots, p(n_d)]$. Let $r_i = p(n_i)$ for $i = 0, \ldots, d$.

Step 2. List **all** the vectors of the form $v = [s_0, s_1, \ldots, s_d]$, where s_i is an integer divisor of r_i. (For instance, if $r_0 = 6$, the first entry s_0 could assume any of the values, $-6, -3, -2, -1, 1, 2, 3,$ and 6).

Step 3. For each vector $v = [v_0, v_1, \ldots, v_d]$ listed in Step 2, solve the Chinese Remainder Theorem as in it appears in Corollary 1 (or use linear algebra techniques) to find a polynomial $a_v(x)$ of degree d or less such that

$$a_v(n_i) = v_i \text{ for } i = 1, \ldots, d.$$

Step 4. Divide $p(x)$ by each of the polynomials $a_v(x)$ to see whether it is a factor of $p(x)$.

Any factor of $p(x)$ of degree d or less will be listed among the $a_{v(x)}$. So $p(x)$ is irreducible if and only if none of the $a_v(x)$ divide $p(x)$.

Task 4

These require lots of patience!
  i. Carry out Steps 1 through 4 for the polynomial $p(x) = x^4 - x^2 - 2x - 1$ to factor it completely. Use $n_0 = 0, n_1 = 1$, and $n_2 = -1$.
  ii. Carry out Steps 1 through 4 for $p(x) = x^5 + x + 1$.
  iii. Carry out Steps 1 through 4 for $p(x) = x^6 + 3x^4 + 2x^3 + 2x^2 + 3x + 1$.

Task 5

Assume that $p(x)$ has a factor $a(x)$ of degree less than or equal to d. Prove that $a(x) = a_v(x)$ for some v. **Hint.** To do this note that $p(x) = q(x)a(x)$ and therefore, for each n_i, we have that $p(n_i) = r_i = a(n_i)q(n_i)$.

3.9 IN THE CLASSROOM: ROOTS AND FACTORING

As a high school teacher, you will teach the algebra that you learned in high school. But you go into the high school classroom with a much deeper understanding of the material at the core of the high school algebra curriculum: factoring and finding roots of polynomial functions. This section will outline the curriculum commonly taught in elementary algebra and intermediate algebra and it will discuss questions that arise for the high school mathematics teacher as a result of a deeper understanding of such theory. Problems that teachers can use in their high school courses to bring these issues into the classroom will be also presented.

The High School Curriculum

Students typically take an elementary algebra course, an intermediate algebra course, and then perhaps a precalculus/advanced algebra course during high school. We begin with a discussion of the content covered in such courses and with some examples of problems from current popular textbooks. The goal of this outline is to provide a vehicle to discuss the evolution of the material as it is learned by the high school student and the difficulties or questions that arise for the teacher who has an advanced perspective on the material.

Elementary Algebra

In this course, students are introduced to polynomials as functions and as expressions of the form $a_n x^n + a_{n-1} x^{n-1} + \ldots + a_1 x + a_0$, and to polynomial addition, subtraction, and multiplication. At this level, students usually work with polynomials that generally have integer coefficients and only sometimes see rational coefficients. Students then learn to factor polynomials. In general, factoring is done over the ring of integers. The emphasis is on finding the greatest common factor of the terms of the polynomial and recognizing special forms, such as the difference of perfect squares, $a^2 - b^2 = (a+b)(a-b)$, or perfect-square trinomials, $a^2 + 2ab + b^2 = (a+b)^2$. Other methods include factoring by grouping, guessing and checking, using the quadratic formula to find rational zeros if they exist, and using the AC method. (A description and example of the AC method is found in the exercises at the end of this section.)

After factoring techniques, the course addresses finding real solutions to quadratic equations. The following methods are typically used: using square roots, factoring, completing the square, applying the quadratic formula, and using technology. The discriminant $b^2 - 4ac$ of a quadratic expression $ax^2 + bx + c$ is used to give information about the number and nature of its roots. If the discriminant is positive, the quadratic expression has two real roots; if it is zero, exactly one real root; and if it is negative, no real roots. In the case where the real roots are actually rational (the discriminant is a perfect square), the quadratic expression factors over the integers. At this time, students do not find complex solutions and simply state "no real roots" when the discriminant is negative. Connections are made between the number of x-intercepts of the graph and the number of real roots. Students often use a graphing calculator to graph a polynomial function and then examine a table of function values to find roots or they use the *zoom* and *trace* features of the calculator to approximate the roots. They may also use special menu commands, such as *zero* or *root* to locate the zeros or they can also graph $y = 0$ and use the *intersect* command.

One of the principal tasks of the teacher at this level is to make the connection between the arithmetic of numbers and the arithmetic of polynomials concrete—how dealing with x's and y's relates to dealing with 2s and 3s. Manipulatives such as algebra tiles can help and many current popular textbooks make use of such representations. Algebra tiles have the following pieces: a 1 unit by 1 unit square; a rectangle with dimensions x units by 1 unit and so area x square units; and a square with dimensions x units by x units, and so area x^2 square units. The pieces also typically come in two colors to represent positive and negative quantities or are somehow labeled to designate the sign. The blocks are made so that the length representing x units is not a multiple of the length representing 1 unit. This prevents students from "trading" a certain number of unit squares for the other pieces as they often do when using base-10 blocks in elementary school. The following examples illustrate the use of these tiles.

EXAMPLE 1

Model $(2x^2 + 3x - 4) + (x^2 - 4x + 2)$ with algebra tiles.

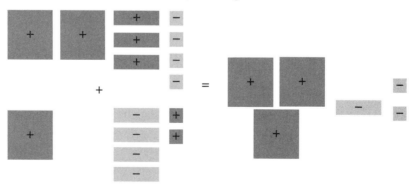

First represent each polynomial with the tiles. Then for addition, combine the tiles and recognize additive inverses to represent the sum using the least amount of blocks, $3x^2 - x - 2$.

EXAMPLE 2

Model $(x + 3)(x + 2)$ with algebra tiles. Give the simplified product. Then use the distributive property (FOIL) to verify the product found.

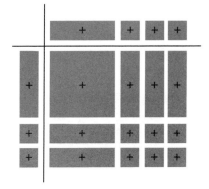

Just as in elementary school, where $2 \cdot 3$ can be modeled by the area of a rectangle with dimensions 2 units by 3 units for a total of 6 square units, polynomial products can be modeled in a similar fashion. The grid and the algebra tiles along the outside define the dimensions of the rectangle. Use tiles to create a rectangle with those dimensions. Total up the blocks that represent the area and get $x^2 + 5x + 6$ for the product. This area model illustrates nicely the four partial products, $x \cdot x, 2 \cdot x, 3 \cdot x$, and $3 \cdot 2$, computed when using the distributive property (FOIL). Find them!

It should be noted that once negative quantities are involved, the use of the tiles becomes less clear and students need to make decisions about the signs of the tiles used to fill in the rectangle based on their knowledge of multiplying integers. The next example illustrates this.

EXAMPLE 3

Model $(x - 3)(x - 2)$.

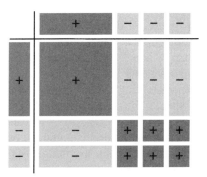

Note the choice of signs inside the grid. The area represented by the rectangle is $x^2 - 5x + 6$.

EXAMPLE 4

Represent the polynomial $x^2 + 6x + 8$ with algebra tiles. Arrange the tiles into a rectangle. Using the tiles, what is the factorization of this polynomial?

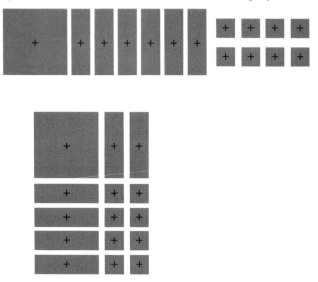

Since the lengths of the sides of the rectangle formed are $x + 2$ and $x + 4$, we know that $x^2 + 6x + 8 = (x + 2)(x + 4)$.

We note that if we represent a quadratic in terms of the algebra tiles and the tiles cannot be arranged into a rectangle, then the quadratic is not factorable over the integers. Sometimes, however, we can add in pairs of tiles that represent additive inverses (and therefore have an overall value of zero) in order to form a rectangle. An example is explored in the exercises.

Intermediate Algebra

In Intermediate Algebra, the curriculum is extended to the study of more general polynomials. At this level, definitions and theorems are investigated and applied to polynomials of degree n. But examples encountered generally involve polynomials of degree less than or equal to 5. In addition to adding, subtracting, and multiplying such polynomials, students learn to divide polynomials by using long division and synthetic division. Factoring techniques continue to emphasize common monomials and recognizing special forms that now include forms like the sum and difference of two cubes: $x^3 + a^3 = (x + a)(x^2 - ax + a^2)$ and $x^3 - a^3 = (x - a)(x^2 + ax + a^2)$. The techniques of solving quadratic equations by taking the square root of both sides of an equation, by factoring, by graphing, by completing the square, and by using the quadratic formula are revisited.

New to this course is the use of imaginary and complex numbers in connection with solving quadratics. Coverage typically includes the definition of i, imaginary and complex numbers and their conjugates, the arithmetic of complex numbers, the relationship among complex, real, irrational, rational, integer, and whole numbers. Students can now use the quadratic formula to find all solutions, real or complex. Material relating the graphs of polynomial functions, the zeros of those functions and the solutions of corresponding polynomial equations is done at a higher level than the previous course because irrational and complex zeros are considered.

New results found in this course include the Factor (Root) Theorem, the Remainder Theorem, the Rational Root Theorem (our Descartes' Criterion), and the Complex Conjugate Root Theorem. Typically this course culminates with a statement of the Fundamental Theorem of Algebra with no proof.

The next example is for those who have not seen synthetic division in a while (or at all). This method can be used to divide a polynomial only by a linear binomial of the form $x - r$. When dividing by nonlinear divisors, long division must be used. In synthetic division, you do not write the variables.

EXAMPLE 5

Find the quotient $(x^3 + 3x^2 - 4x - 12) \div (x - 2)$ using synthetic division.

Write down the coefficients of the polynomial $x^3 + 3x^2 - 4x - 12$ and then write the r-value, 2, of the divisor $x - 2$, on the left. Leave a space below the coefficients, draw a line, and write the first coefficient, 1, of $x^3 + 3x^2 - 4x - 12$ below the line.

$$
\begin{array}{r|rrrr}
2\rfloor & 1 & 3 & -4 & -12 \\
\hline
 & 1 &
\end{array}
$$

Multiply the r-value, 2, by the number 1 that is below the line to get 2 and record the 2 below the next coefficient which is 3. Then add the 3 and the product 2 to get 5 and record the 5 below the line.

$$
\begin{array}{r|rrrr}
2\rfloor & 1 & 3 & -4 & -12 \\
 & & 2 & & \\
\hline
 & 1 & 5 &
\end{array}
$$

Now we use the new number 5 below the line. We multiply 2 by 5 to get 10 and add -4 and 10 to get 6. Continue this process. We multiply 2 by 6 to get 12 and add -12 to 12 to get 0. The final number on the right below the line is the remainder. The numbers below the line before the remainder are the coefficients of the quotient.

$$
\begin{array}{r|rrrr}
2\rfloor & 1 & 3 & -4 & -12 \\
 & & 2 & 10 & 12 \\
\hline
 & 1 & 5 & 6 & \underline{|0}
\end{array}
$$

In this problem, 0 is the remainder and 1, 5, 6 are the coefficients of the quotient, $x^2 + 5x + 6$. ∎

170　Chapter 3　Polynomials

There is a significant role for technology to assist students in finding roots of polynomials. The next example illustrates the integration of the theorems learned in this course and the use of technology.

EXAMPLE 6

Find all of the zeros of $f(x) = 2x^3 - 3x^2 - 3x - 5$. Factor the function over the set of complex numbers.

From the Rational Root Theorem, students know that any rational root $\frac{a}{b}$ is such that a is a factor of -5 and b is a factor of 2. Such possible quotients include: $\pm 1, \pm 5, \pm \frac{1}{2}, \pm \frac{5}{2}$. Rather than testing all of these to see which are really rational roots, students consider the graph of the function on a window that includes all these rational values. See Figure 1.

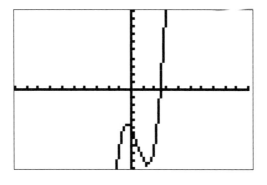

FIGURE 1

The graph indicates a zero between 2 and 3. To see if this zero is rational, students test $\frac{5}{2}$, the only possible rational zero in that range. Evaluating the function, they verify that $f(\frac{5}{2}) = 0$ and recognize that $x - \frac{5}{2}$ is a factor by the Factor Theorem. They can now use long division (or synthetic division) to find the resulting quadratic factor $2x^2 + 2x + 2$. Then they can use the quadratic formula to find the zeros of $2x^2 + 2x + 2$ (first likely factoring out the 2 common to all terms) to get $\frac{-1 \pm i\sqrt{3}}{2}$. Therefore, the zeros of $f(x) = 2x^3 - 3x^2 - 3x - 5$ are $x = \frac{5}{2}, \frac{-1 \pm i\sqrt{3}}{2}$ and $f(x) = 2(x - \frac{5}{2})(x - \frac{-1+i\sqrt{3}}{2})(x - \frac{-1-i\sqrt{3}}{2}) = (2x - 5)(x + \frac{1}{2} - \frac{\sqrt{3}}{2}i)(x + \frac{1}{2} + \frac{\sqrt{3}}{2}i)$. Note that the graph narrows down the testing of possible rational zeros. If the graph happened to have an x-intercept and none of the possible rational zeros were really zeros, then the value of this x-intercept must be irrational. Students could find decimal approximations to such roots by using the Zoom and Trace features on their calculator. Students need to be reminded that complex zeros cannot be determined by the graph.　■

From Your Vantage Point

As algebra teachers, you will teach material that you learned in high school. But now you have a deeper understanding of the core issues that surround the theory of polynomials. What questions or issues does this depth of knowledge raise for the classroom? What tools and responsibilities does the high school teacher have now as a result of his or her greater understanding of this material? In the following paragraphs, one issue will be explored:

- What does it mean to "factor completely"?

After an examination of current popular algebra textbooks, it appears that they are not always consistent about stating what ring a given polynomial is to be factored over or in what ring admissible roots are to be found. It is often assumed that students will be factoring over the integers and some teacher editions do point that out. But this information is not usually found in the student version of the text. The directions "factor completely" are often not clear and seem to mean different things at different times. Various textbooks define prime or irreducible polynomials but others rely on students just knowing that they cannot factor a particular polynomial any further (using methods discussed so far in their course). Even when a definition is given, it might be given as, "a prime polynomial is one that cannot be factored." How does a student know if a given polynomial can be factored further or not? Is it their shortcoming if they can't factor a given polynomial over the integers or is the polynomial really irreducible?

You know that irreducibility depends on what ring we are working over. For example, $x^4 - 25$ is $(x^2 + 5)(x^2 - 5)$ when factored over the integers; it is $(x^2 + 5)(x + \sqrt{5})(x - \sqrt{5})$ when factored over the set of real numbers; and it is $(x + \sqrt{5}i)(x - \sqrt{5}i)(x + \sqrt{5})(x - \sqrt{5})$ when factored over the set of complex numbers. When will students know to say that $x^2 - 5$ is prime and when do we want them to say $x^2 - 5 = (x + \sqrt{5})(x - \sqrt{5})$, especially if we have not been clear about our ring? We have found texts that have all other examples in a particular section factored only over the integers but that give a single example like the one above to show students what can be done over the real numbers. Will your students know what you want and when? As a teacher, you will have to be very clear to your students what ring you are working with at any given time. You have to think carefully about how you will explain to students at different levels whether polynomials that arise as examples and in homework are irreducible over the given ring or not. The tools you will be able to give them that will enable them to make these decisions on their own will be limited.

Elementary algebra students tackle the question of factoring quadratic polynomials. To factor over the integers, they can use a system of guessing and checking factors such as the AC method. They can use the discriminant to determine the nature of the roots. To factor over the ring of real numbers, they can graph the quadratic function to see if it has real roots and they can use the quadratic formula to find the real roots, if there are any. Even in this simple situation, the teacher must establish context because $x^2 - 5$ is prime in $Z[x]$ but reducible in $R[x]$.

EXAMPLE 7

Factor $6x^2 + 23x + 20$.

Students can use the quadratic formula to solve $6x^2 + 23x + 20 = 0$ and get $x = -5/2$ or $x = -4/3$. Students then work backwards to find the factors.

$$x = -5/2 \qquad x = -4/3$$
$$2x = -5 \qquad 3x = -4$$
$$2x + 5 = 0 \qquad 3x + 4 = 0$$

The factored form of $6x^2 + 23x + 20$ is $(2x + 5)(3x + 4)$. ∎

Consider the following examples found in current popular elementary algebra textbooks. The assumption is that we are working over the integers.

EXAMPLE 8

With the directions "Factor each polynomial completely," the following problem and solution were given: $81a^4 - 9b^2 = (9a^2 + 3b)(9a^2 - 3b)$. In this problem, the 9 that was common to both original terms was not factored out first. But on a practice worksheet for the same section with the exact directions, we find the solution $4y^2 - 16 = 4(y + 2)(y - 2)$ with the common factor of 4 factored out first. Is either factored completely over the integers? ∎

EXAMPLE 9

The following is even more puzzling. The solution to the problem "Factor $7y^3 - 21y^2 + 14y$" is given as $7y(y^2 - 3y + 2)$ in the teacher's edition. If you are bothered by the answer given, it should be noted that this was in an elementary algebra text in a section on factoring out common monomials that was prior to sections on factoring general quadratic trinomials. But does the student in this section of the textbook think they have factored it properly if they get this answer? Would you accept this as a solution? Should we change the directions to indicate that we only want them to practice factoring out the common monomial at this time? We could then use this same problem in the next section on factoring trinomials to factor it completely. ∎

Significantly more troublesome was the discovery of the following problem in an elementary algebra textbook.

EXAMPLE 10

"Factor $x^9 - x^2$." The solution given was $x^2(x^7 - 1)$. None of us would consider this factored over the integers and nor should the students! Even at the elementary level, a high school student with some understanding of the connection between factors and roots (even without a formal statement of the Factor Theorem) could quickly decide that 1 is a root of $x^7 - 1$ and that $(x - 1)$ is also a factor. What would this elementary algebra student then do? This question begs for the use of long division (or synthetic division) of polynomials to find what sixth degree polynomial factor we are missing. However, this is often part of the intermediate algebra course. So let's pick it up there.

An intermediate algebra student can use division to obtain $x^7 - 1 = (x - 1)(x^6 + x^5 + \ldots + 1)$. However, once a student obtains $x^6 + x^5 + x^4 + x^3 + x^2 + x + 1$, the tools available to determine that it is irreducible over the integers are very limited. (In Section 3.4, Example 7, we applied the Eisenstein's Criterion to a related polynomial to prove that this cyclotomic polynomial is irreducible in $Q[x]$.)

What can students in a high school algebra class do with $x^6 + x^5 + x^4 + x^3 + x^2 + x + 1$? We could just have a statement that when p is prime, $x^{p-1} + x^{p-2} + \ldots + 1$ is irreducible over the integers. But given the rise of handheld technology in the math classroom, it is likely that students will try to graph $y = x^6 + x^5 + x^4 + x^3 + x^2 + x + 1$ in the standard viewing window, $x = -10 \ldots 10, y = -10 \ldots 10$ as in Figure 2.

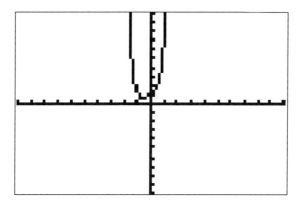

FIGURE 2

Since the graph does not appear to cross the x-axis, students will likely deduce that this polynomial has no real roots. But how can a student tell that the graph does not turn around and cross the x-axis beyond the window chosen so far? Even with some calculus to further investigate the graph of this function, this polynomial and its first and second derivatives are not easy for students to use. If their course covers Descartes' Rule of Signs, students can observe the sign changes in the coefficients of $f(x) = x^7 - 1$ and $f(-x) = -x^7 - 1$ and deduce that there is only one positive real root, $x = 1$, and no negative real roots. Therefore, $(x - 1)$ is the only linear factor of $x^7 - 1$ but $x^6 + x^5 + x^4 + x^3 + x^2 + x + 1$ might still be factorable as the product of three quadratics, or a quadratic and a fourth degree polynomial, but not the product of two cubic polynomials. (Why?) Students could investigate the coefficients as we did in Example 3 of Section 3.4. This is hard stuff!

If we use a TI-89 or some calculator that can do symbolic calculations, we have further surprises. For example, in EXACT mode with Complex Format set to REAL, the command "factor(x^7 − 1)" will return $(x - 1)(x^6 + x^5 + x^4 + x^3 + x^2 + x + 1)$, but "factor(x^7 − 1, x)" returns $(x - 1)(x^2 + 2\sin(\frac{\pi}{14})x + 1)(x^2 - 2\sin(\frac{3\pi}{14})x + 1)(x^2 + 2\cos(\frac{\pi}{7})x + 1)$. Here the student can observe that over the ring of real numbers, $x^6 + x^5 + x^4 + x^3 + x^2 + x + 1$ factors as the product of three irreducible quadratics. The quadratic formula can now be used to find the six complex roots.

The TI-89 can factor and solve over the complexes. The command "cFactor(x^7 − 1, x)" in EXACT mode with Complex Format set to REAL returns:

$$(x-1)\left(x+-e^{\frac{-4i\pi}{7}}\right)\left(x+-e^{\frac{4i\pi}{7}}\right)\left(x+e^{\frac{5i\pi}{7}}\right)$$

$$\left(x+e^{\frac{-5i\pi}{7}}\right)\left(x+e^{\frac{i\pi}{7}}\right)\left(x+e^{\frac{-i\pi}{7}}\right),$$

"cZeros(x^7 − 1, x)" returns:

$$\left\{1, e^{\frac{4i\pi}{7}}, e^{\frac{-4i\pi}{7}}, -e^{\frac{-5i\pi}{7}}, -e^{\frac{5i\pi}{7}}, -e^{\frac{-i\pi}{7}}, -e^{\frac{i\pi}{7}}\right\},$$

and "cSolve(x^7 − 1 = 0, x)" returns:

$$x = -e^{\frac{i\pi}{7}} \text{ or } x = -e^{\frac{-i\pi}{7}} \text{ or } x = -e^{\frac{5i\pi}{7}}$$

$$\text{or } x = -e^{\frac{-5i\pi}{7}} \text{ or } x = e^{\frac{-4i\pi}{7}} \text{ or } x = e^{\frac{4i\pi}{7}} \text{ or } x = 1.$$

However, "cSolve(x^7 − 1 = 0, x)" in EXACT mode with Complex Format set to RECTANGULAR returns

$$x = -\cos\left(\tfrac{\pi}{7}\right) - \sin\left(\tfrac{\pi}{7}\right)i \text{ or } x = -\cos\left(\tfrac{\pi}{7}\right) + \sin\left(\tfrac{\pi}{7}\right)i \text{ or}$$

$$x = \sin\left(\tfrac{3\pi}{14}\right) - \cos\left(\tfrac{3\pi}{14}\right)i \text{ or } x = \sin\left(\tfrac{3\pi}{14}\right) + \cos\left(\tfrac{3\pi}{14}\right)i \text{ or}$$

$$x = -\sin\left(\tfrac{\pi}{14}\right) - \cos\left(\tfrac{\pi}{14}\right)i \text{ or } x = -\sin\left(\tfrac{\pi}{14}\right) + \cos\left(\tfrac{\pi}{14}\right)i \text{ or } x = 1.$$

You can reconcile all these answers with your knowledge of the material in Section 2.6. But it is difficult territory for the high school student. Be sure to experiment with your technology to be ready for what output it provides.

There are different ways to explore this problem with high school students and revisit it from one course to the next. But the question remains whether students should be left to believe that $x^2(x^7 - 1)$ is the factored form of $x^9 - x^2$. We can do so much better than that! ■

There are several lessons to be learned here. Be careful of your directions to students. Inspect textbook problems. Carefully look over examples and solutions. Give students clear directions about what ring to work over. Stress what tools students can use to determine whether a certain polynomial is irreducible over the ring in which you are working. Be prepared to answer questions regarding the technology.

Here are some questions to ask yourself. Do we, and does our text, revisit factoring and solving equations once complex numbers are covered? Are we as teachers careful about our wording of instructions to students verbally and in writing as to what types of solutions we are looking for or over what ring we are factoring? As a teacher, do you notice the subtle differences in the directions given in the text—"find all zeros" versus "find all real zeros"? Is the text consistent and correct with what it asks for and what it says the solutions are in the teacher manual? Are the teachers consistent and correct in what they say and write to students? What role does technology play in finding answers to questions of factoring and finding roots? What is it capable of? What methods does it use? What limitations does it have?

Exercises for Future Teachers to Explore Ideas More Fully

1. **Recall:** A polynomial of degree n over a field has at most n zeros counting multiplicities. Show that this is not true for arbitrary polynomial rings. For example, show $x^2 + 3x + 2$ has four zeros in Z_6. Show that this statement is false for a ring that has a zero divisor.

2. Write pairs of polynomials such that one polynomial's coefficients are the second polynomial's coefficients, but in reverse order. How are the zeros of each polynomial related?

3. **How to Factor Using the AC Method.** If a quadratic trinomial $ax^2 + bx + c$ factors over the integers, students often attempt the factorization by guessing and checking. When the leading coefficient is 1, they use their knowledge of binomial multiplication and look for factors of c (including negative ones) that add to b. When the leading coefficient is not 1, students have a more difficult time. The following is a description of a method often referred to as the AC method (because you multiply the "a" and "c" coefficients) which eliminates the guesswork of deciding whether $ax^2 + bx + c$ factors over the integers. It finds those factors if they exist.

   The AC Method

   i. List all the factors of the product ac. Do not forget the negative factors.
   ii. Find the pair of factors that sums to b. Call these factors m and n.
   iii. Rewrite the original expression, replacing bx by the sum $mx + nx$ to get $ax^2 + mx + nx + c$. Be sure to let m be the value that has a common factor with a and n be the value that has a common factor with c.
   iv. Factor by grouping the first two terms and the second two terms.

   Here's how to factor $2x^2 - x - 6$ over the integers using the AC method:

   The factors of $(2)(-6) = -12$ that add to -1 are 3 and -4. Rewriting and factoring by grouping, we get

   $$2x^2 - x - 6 = 2x^2 + -4x + 3x - 6$$
   $$= (2x^2 - 4x) + (3x - 6)$$
   $$= 2x(x - 2) + 3(x - 2)$$
   $$= (x - 2)(2x + 3).$$

   a. Factor $3x^2 - 4x - 15$ over the integers using this method.
   b. Explain why the method always works if the quadratic factors over the integers.
   c. Explain why the order of mx and nx (and thus which terms are grouped for factoring) matters when you replace the bx term.

4. In an intermediate algebra text (assuming we are factoring over the integers), you find the problem "Factor $-x^6 - 64$ completely" with the solution given as $-1(x^2 + 4)(x^4 - 4x^2 + 16)$. Students were expected to recognize $x^6 + 64$ as the special "sum of two cubes" and use $a^3 + b^3 = (a+b)(a^2 - ab + b^2)$ to factor it. Now consider that you are the teacher of this course. You have found other errors in your textbook. How certain are you that $x^4 - 4x^2 + 16$ is factored completely over the integers?

   a. Suppose the question had been to find any real roots of $-x^6 - 64$. Brainstorm ways to explain to high students at either the elementary or intermediate algebra or precalculus level that this polynomial does not have any real zeros.
   b. Now factor $-x^6 - 64$ over the complexes.
   c. To factor $-x^6 - 64$ over the complexes, you had to factor $x^4 - 4x^2 + 16$ over the complexes. To do this, you may have started by using the quadratic formula on $x^4 - 4x^2 + 16 = 0$ to get that $x^2 = 2 \pm 2\sqrt{3}i$. You then had at your disposal de Moivre's formula to find the square roots of these complex numbers. Without calling on that result, compute these square roots algebraically as you could do with an intermediate algebra or precalculus class, which has some basic knowledge of complex numbers.

5. Consider the polynomial $x^7 - 1$ in Example 10.

   a. Use de Moivre's formula to find the 7th roots of unity.
   b. The command "factor(x^7 − 1, x)" in EXACT mode in REAL Complex Format on the TI-89 returns $(x-1)(x^2 + 2\sin(\frac{\pi}{14})x + 1)(x^2 - 2\sin(\frac{3\pi}{14})x + 1)(x^2 + 2\cos(\frac{\pi}{7})x + 1)$. We know $x = 1$ is a root. Make use of the quadratic formula to find the six complex roots of $x^7 - 1$. How do the roots you find compare to those given by the TI-89 in EXACT mode in RECTANGULAR Complex Format using the command "cSolve(x^7 − 1 = 0, x)"?
   c. The commands "cZeros(x^7 − 1, x)" and "cSolve(x^7 − 1 = 0, x)" on the TI-89 in EXACT mode in REAL Complex Format return the roots $x = 1$, $x = e^{\frac{4i\pi}{7}}$, $x = e^{\frac{-4i\pi}{7}}$, $x = -e^{\frac{-5i\pi}{7}}$, $x = -e^{\frac{5i\pi}{7}}$, $x = -e^{\frac{-i\pi}{7}}$, $x = -e^{\frac{i\pi}{7}}$. Reconcile these answers with the ones you got using de Moivre's formula in part a and with the answers you found in part b. The relationship $z = r(\cos\theta + i\sin\theta) = e^{\ln(r) + i\theta}$, for $r > 0$, which is found in the *To the Teacher* essay in Section 2.7, and some trigonometry will be helpful.

6. Determine whether the following polynomials are irreducible over the integers. Explain your results. Then explain how you might explore these polynomials with high school students. Explain ways to grow your discussion of each polynomial from course to course as students learn more material, particularly about irrational numbers, complex numbers, the Factor Theorem, the Rational Root

Theorem, Descartes' Rule of Signs, and the Fundamental Theorem of Algebra. What role might technology play in these explorations?

a. $x^5 - 7x^3 + 15x^2 - 7$
b. $4x^4 - 45x^2 + 20x + 21$

7. We gave an example of using algebra tiles to factor quadratic trinomials. The example given had all positive quantities. Recall that to factor with the tiles, we represented the polynomial with the tiles, arranged the tiles into a rectangle, and read off the dimensions of the rectangle to find the factors. How can we use the tiles to factor $x^2 - x - 6$? Actually work with the tiles and try to avoid finding the factors first and then modeling the product of those factors. What issues come up when using only the tiles?

Exercises for the High School Classroom

1. Factor the polynomial $4x^2 + 11x + 6$ over the integers using the following different methods:

    a. Algebra tiles
    b. Guess-and-check
    c. Quadratic formula
    d. AC method

    Which method do you prefer? Why?

2. Find all values of x for which $(x^2 - 5x + 5)^{x^2-9x+20} = 1$. (This problem comes from the 1988 NCTM Yearbook, *The Ideas of Algebra K–12* (p. 19). A discussion of its use with an actual high school class can be found in NCTM's *Professional Standards for Teaching Mathematics* (1991), p. 43.) Try this problem yourself. It is really quite clever. Even students who do not know where to start can get interested immediately by just having them try some values for x.

3. Write a polynomial function, $p(x)$, in factored form that satisfies all the following conditions:

    p is of degree 4; $p(0) = 100$; the zeros of p include: $1, 2, 5i$.

4. Use the Rational Root Theorem and the polynomial $x^2 - 2$ to show that $\sqrt{2}$ is irrational.

Bibliography

Leitze, Annette Ricks and Nancy A. Kitt. "Using Homemade Algebra Tiles to Develop Algebra and Prealgebra Concepts." *The Mathematics Teacher* 93, 6 (Sept 2000): 462.

National Council of Teachers of Mathematics. *Professional Standards for Teaching Mathematics*. Reston, Va: National Council of Teachers of Mathematics, 1991.

———. *NCTM 1988 Yearbook: The Ideas of Algebra K–12*. Reston, Va: National Council of Teachers of Mathematics. 1988.

Shultz, James E., et al. *Algebra 1: Annotated Teacher's Edition*. Austin: Holt, Rinehart, and Winston, 2003.
_____. *Algebra 2: Annotated Teacher's Edition*. Austin: Holt, Rinehart, and Winston, 2003.
The University of Chicago School Mathematics Project. *Algebra: Teacher's Edition Part 1 & 2*. Glenview, Ill.: Scott, Foresman and Company, 1998.
_____. *Advanced Algebra: Teacher's Edition Part 1 & 2*. Glenview, Ill.: Scott, Foresman and Company, 1998.
_____. *Precalculus and Discrete Mathematics: Teacher's Edition*. Glenview, Ill.: Scott, Foresman and Company, 1992.

3.10 FROM THE PAST: VIETA AND SYMBOLIC ALGEBRA

The work of the sixteenth-century French mathematician Francois Vieta had a profound influence on the development of algebra as we know it. Yet his name is not currently well known. Born in 1540, he earned a law degree in 1559, but he also studied mathematics and astronomy. Like many of his contemporaries whom we call mathematicians, Vieta pursued mathematics in his spare time. Most of his working hours were spent as an advisor to the rich and powerful. The affairs of sixteenth-century France were dominated by the religious conflict between Roman Catholics and the Huguenots, a conflict that affected all aspects of life. Vieta served as an advisor to the Catholic king, Henry III. He carried messages for the king and conducted delicate negotiations. He used his mathematical talents to decode secret messages sent between enemies of the king. He was very successful in this endeavor. Vieta remained a powerful political figure in France until his death in 1603. Most of his significant mathematics comes from the period 1584–1603, his busiest period politically as well as mathematically. The works of Vieta have been collected, translated into English, and published under the title *The Analytic Art*. These works were extremely influential. The mathematical lineage from Vieta through Descartes and Fermat to the development of the calculus and of analytic geometry is direct.

The key to Vieta's importance in the history of algebra is found in the title given to his collected works, *The Analytic Art*. The word "analytic" refers to one of the two forms of Greek mathematics—synthesis and analysis. These were forms of both doing and presenting mathematics. In synthetic mathematics, one started from definitions, axioms, and theorems that had already been proved, and proceeded to prove new theorems using rigorous logic to guarantee their correctness. Euclid's *Elements* is an example of this kind of mathematics. Aristotle put his authority behind this form. What it lacked was any indication of how the proofs themselves had been discovered. Analysis provided a path to a proof. Analysis began with what was to be proved and worked towards results already known. With luck one could reverse the steps and develop a synthetic proof—then the paradigm of rigor. Take, for example, Proposition 11, Book II of Euclid's *Elements*. Many of Euclid's propositions concern constructions and the statements are presented as infinitive clauses describing an activity.

> *To cut a given straight line so that the rectangle contained by the whole and one of the segments is equal to the square on the remaining segment.*

Section 3.10 From the past: Vieta and Symbolic Algebra

Euclid presents a synthetic proof, which is a construction that yields the two portions of the given segment. The analytic approach would begin with assuming that a segment of length a had been cut into two pieces, one of length x, the other of length $a - x$ and that $a(a - x) = x^2$. The assumption is that such an x exists before it is proved to exist. The equation is solved and the steps reversed to develop a synthetic proof. Thus analysis often went from an equation involving an unknown to a determination of the value of the unknown.

What we think of a cubic equation would often have been presented as a geometric problem:

> Given a volume Q and an area P find a length x such that the volume of the cube of side x together with the volume of a solid with base P and height x would equal the given volume Q.

The synthetic theorem would give conditions for the existence of such a length x based on P and Q. The analytic problem would be to solve the equation $x^3 + Px = Q$. When cubic equations were first considered, each equation was presented with specific numerical values for P and Q. The analysis would proceed from the equation to a value of x. With the specific numerical values it was very difficult to reverse the steps to develop a synthetic proof. Vieta's most important innovation was to express both variables (unknowns) and constants by letters of the alphabet. Before Vieta, equations had been expressed with particular numerical coefficients. One would solve equations like $x^2 + 4x = 8$, for instance. Vieta set up his notation to solve whole classes of equations at once. He would solve $x^2 + ax = b$. With this change, he was able to see the form underlying the classes of equations that he worked with and thus advance the methodology of solution. This made the reversal of the steps of the analysis much easier. In the actual evolution of algebra, the requirement of a synthetic proof disappeared as the steps of an analytic solution of an equation became an acceptable form of proof. This evolution began with Vieta. Before considering his notation in greater detail let us see what he inherited from earlier mathematicians.

The first influence on Vieta came from Diophantus of Alexandria (c. 250 CE). He was an anomaly among Greek mathematicians in that he used an algebraic notation to study solutions of equations rather than geometric arguments to study the relations between geometric quantities. His work has been so influential that polynomial equations with integer coefficients are still called Diophantine equations. He solved equations in one variable and, rather than holding to integers, he allowed rational numbers as solutions. Here is how Diophantus expressed his equations. He used the symbol ς for the unknown quantity. He used Δ^Y as the square of ς and K^Y for the cube of ς. Higher powers of the unknown were expressed as combinations of these symbols. For example, ς raised to the fourth power would be $\Delta^Y \Delta$ since it is a square times a square while the fifth power was ΔK^Y, a square times a cube. He used the symbol M for the units or pure numbers. All his equations had specific numerical coefficients and the coefficient followed the power of the unknown so that $3x^2$ would be written $\Delta^Y 3$. The Greeks didn't use Arabic numerals. For 3, Diophantus would use the letter γ. Thus $3x^2$ would appear as $\Delta^Y \gamma$. (In the numeration system α stood for 1, β for 2, up to θ for 9. Then the sequence of numerals $10, 20, \ldots$ were represented by $\iota, \kappa, \ldots$.) All the terms with positive coefficients came first and there were no plus signs between them! The terms that we would view as having negative coefficients were then subtracted as a group. Diophantus would regroup the polynomial $3x^3 - 4x^2 + 5x - 7$ as $(3x^3 + 5x) - (4x^2 + 7)$, which he would represent as $K^Y \gamma \varsigma \varepsilon - \Delta^Y \delta M \varphi$. With this notation, Diophantus was able to state complex problems and find solutions to them. Yet every problem was solved as a specific case. In a sense, he was giving a recipe for the solution by showing one example. A problem from Diophantus work that has become famous is Problem 8 from Book II. In this problem, Diophantus gives a method for writing a square number as the sum of two squares. He intended that the given square be a perfect square, such as 9 or 25, and that the two squares found be squares of rational numbers. His solution consisted of solving the problem for only a particular perfect square, namely 16. He shows that

$$16 = \left(\frac{12}{5}\right)^2 + \left(\frac{16}{5}\right)^2 = \frac{144}{25} + \frac{256}{25}.$$

The next influence on Vieta was the group of sixteenth-century Italian algebraists that solved the cubic and quartic equations for the first time. Girolamo Cardano (1501–1576) was the first of this group to publish these solutions. His book, *Ars Magna* or *The Great Art*, presented a nearly complete program for solution to all cubic equations with rational coefficients as well as a method for solving some quartic equations. Yet he presents his methods through specific examples. For example, under the heading "On the Cube and the First Power Equal to a Number," he solves $x^3 + 6x = 20$ rather than $x^3 + ax = b$. His notation reflects a geometric point of view. He posits a quantity GH, a length, and says that a cube (meaning its volume) GH^3 and 6 times its side GH is equal to 20. Polynomials and equations were expressed with abbreviations of Latin terms for the unknown, its powers, the constant term, and the numerical operations. Since the coefficients represented geometric quantities, Cardano did not use negative numbers. He recognized that some equations had solutions that were not positive numbers. He called these false solutions. For the variable x Cardano wrote pos. (We would translate this as "thing.") Thus the unknown was both a "thing" and a line segment. He used quad. for x^2 and cub. for x^3. In the same manner as Diophantus, he used qd. for x^4. He used p: for + and m: for −. He would thus write cub. p: 6 pos. equals 20 for what we would express as $x^3 + 6x = 20$. Our equal sign has replaced a word in Cardano. With this notation, Cardano, like Diophantus, gave recipes for changing the form of equations and solving equations by presenting specific cases. For example, he showed that the first step in solving a cubic equation such as

cub. p: 9 pos equals 3 quad. p: 27

(which to us is $x^3 + 9x = 3x^2 + 27$) was to transform it to an equation with no quad (x^2) term. Then he presented a recipe for solving cubic equations that had no square term.

Vieta used letters for both the unknowns and the fixed but undetermined coefficients. He used vowels like A for the unknowns as we would use the later letters of the alphabet like x and consonants like B and Z for fixed but unspecified constants and coefficients as we would use early letters of the alphabet like a and b in

Section 3.10   From the past: Vieta and Symbolic Algebra   181

$x^2 + bx = c$. Vieta did not use exponents. His notation for powers was similar to that of Cardano. However, we will use exponents for clarity here. For example, he gives a quadratic equation as $A^2 + BA = Z^p$. The symbol A is the unknown to be solved for. Retaining a geometric flavor, the constant Z has the label p, which stands for "planar". A^2 and AB, being the products of lengths, represented areas. Thus Z was to be thought of as a two dimensional quantity. The three terms of the equation thus had the same dimension. An example of a cubic equation is given in Problem 1 of Chapter VII of his book *Understanding and Amendment of Equations*.

$$A^3 + 3B^p A = 2Z^s$$

Here the superscript s stands for solid as Z represents a volume, a three-dimensional quantity. Notice that his coefficients are $2Z^s$ and $3B^p$, not just 2 and 3. Using 2 and 3 as multipliers, he simplified his formulas; he found them as integral to the form of the solution to cubic equations. His new notation had already given him insight. Vieta's notation is a great advance on what came before. Not only are all equations of a given form solved at once, but the notation allowed him to see the form hiding under the specific coefficients.

Now let's see how Vieta solved $A^3 + 3B^p A = 2Z^s$. First he applied the substitution $E^2 + EA = B^p$. (Note that E is a vowel and hence a variable.) If we solve for A in the substitution, we get $A = \frac{B^p - E^2}{E} = \frac{B^p}{E} - E$. We will continue the solution in modern notation.

Suppose the original equation is $x^3 + 3px = 2q$. We let $x = \frac{p}{y} - y$. Then we have $x^3 + 3px = \left(\frac{p}{y} - y\right)^3 + 3p\left(\frac{p}{y} - y\right) = \frac{p^3}{y^3} - 3\frac{p^2}{y^2}y + 3\frac{p}{y}y^2 - y^3 + 3p\frac{p}{y} - 3py = \frac{p^3}{y^3} - y^3 = 2q$.

Multiplying through by y^3 gives us the equation $p^3 - y^6 = 2qy^3$, which is a quadratic in y^3 and hence can be solved for y. This then allows us to find x using the original substitution. Here is a numerical example: $x^3 + 6x = 20$. Note that $p = 2$ in this case. Vieta's substitution is $x = \frac{2}{y} - y$, which transforms the original equation into $8 - y^6 = 20y^3$ or $(y^3)^2 + 20y^3 - 8 = 0$. Using the quadratic equation we find $y^3 = -10 \pm 6\sqrt{3}$. Choosing $y^3 = -10 + 6\sqrt{3}$, we find that $y = -1 + \sqrt{3}$ and hence that $x = \frac{2}{\sqrt[3]{-10+6\sqrt{3}}} - \sqrt[3]{-10 + 6\sqrt{3}} = \frac{2}{-1+\sqrt{3}} - (-1 + \sqrt{3}) = \frac{2-4+2\sqrt{3}}{-1+\sqrt{3}} = 2$.

3.10 Exercises

1. How would Vieta express the general form of the equation $x^3 + 8x = 2x^2 + 5$? How would Diophantus have expressed it?

2. In *Understanding and Amendment of Equations*, Vieta gives the following example:

$$\text{Let } A^3 + 3BA^2 + D^p A = Z^s.$$

We might represent it as $x^3 + 3px^2 + qx = r$. His goal was to reduce this to a cubic with no A^2 term.

> To overcome the modification, therefore purge by one third [the coefficient].
> Let $A + B = E$, hence $E - B$ will be A.
> The result is:
> $$E^3 + E(D^p - 3B^2) = Z^s + D^p B - 2B^3.$$

We would make the substitution $y = x + p$ or $x = y - p$. Substitution yields $y^3 + (q - 3p^2)y = r + qp - 2p^3$, which translates exactly to what Vieta produced.

   i. Perform the substitution $A = E - B$ in the equation $A^3 + 3BA^2 + D^p A = Z^s$. Show that it removes the F^2 term.
   ii. Apply this substitution to reduce $x^3 + 12x^2 + 6x = 10$.
3. Solve the cubic equation $x^3 + 9x = 3x^2 + 27$ by first making the substitution that removes the square term and then solve the resulting equation using Vieta's method. Notice that in the present form if one divides both sides of the equation by $x^2 + 9$ the solution falls out.
4. Using Vieta's method, solve the cubic equation $x^3 + 4x = 240$. Leave your solution in terms of cube roots.
5. In an early work, *Five Books of Zetetics*, Vieta proves that given two cubes one can find two other cubes whose sum equals the difference of the two given cubes. He was assuming that these were cubes of rational numbers in all cases.

    Let B^3 and D^3 be the given cubes. Let $B - A$ be the root of the first cube to be found. Let $\frac{B^2 A}{D^2} - D$ be the root of the second. Then A is equal to $\frac{3D^3 B}{B^3 + D^3}$. This allows us to find both cubes. Using Vieta's method find two cubes of rational numbers whose sum equals the difference of the two cubes 8 and 1.
6. In Theorem 1 of Chapter IV of *Understanding and Amendment of Equations*, Vieta states,

    > If $A^3 + B^2 A = B^2 Z$, then there are four continued proportionals, the first of which, whether the greater or the smaller of the extremes, is B and the sum of the second and the fourth is Z, making A the second.

   i. Consider four numbers in continued proportion $a : b : c : d$ such that $a = B$ and $b + d = Z$. Using the fact that $a{:}b = b{:}c = c{:}d$ show that b is a solution to the given equation.
   ii. Solve $x^3 + 4x = 240$ given that the extreme terms in the continued proportion are 2 and 54, that is $2{:}b : c{:}54$.

Bibliography

Bashmakova, I. G. and Mirnova, G. S. " The Literal Calculus of Viete and Descartes," *American Mathematical Monthly*, Vol. 106, No. 3 (March 1999), p. 260–263.

Vieta, F. *The Analytic Art*, translated and edited by T. Richard Witmer. Kent, Ohio: Kent State University Press, 1983.

Chapter 3 Highlights

We opened this chapter with an investigation of how $F[x]$, **the ring of polynomials with coefficients in a field** F, resembles the ring of integers. We formulated the **Division Algorithm for Polynomials** by which we can determine when a polynomial $f(x)$ is a factor of another polynomial $g(x)$. Its corollaries, the **Remainder** and the **Root Theorems**, are particularly useful when $f(x)$ has degree 1. **Euclid's Algorithm for Polynomials** uses the Division Algorithm to determine $\gcd(f(x), g(x))$, a **greatest common divisor of polynomials** $f(x)$ and $g(x)$. In the ring $F[x]$, the **irreducible polynomial** is analogous to the prime in the ring Z. We found the polynomials in $F[x]$ can be factored into irreducible polynomials in a unique way (except for constant multiples and the order of the factors). That is to say, $F[x]$ had **unique factorization**.

Next we investigated factorization in special rings of polynomials. **Gauss's Lemma** allowed us to conclude that if a polynomial with integer coefficients can be factored in $Q[x]$, then it can be factored in $Z[x]$. Tools like **Eisenstien's Criterion** can help determine when a polynomial in $Z[x]$ is irreducible. We can find irreducible polynomials of any degree in $Z[x]$ and in $Q[x]$. But, as the worksheet of Section 3.8 demonstrates, if a polynomial with integer coefficients is reducible, we can determine its factorization in a finite number of steps. Computer algebra systems can exploit this fact very well.

The situation is quite different for the rings $C[x]$ and $Q[x]$. The **Fundamental Theorem of Algebra** asserts that *every* polynomial in $C[x]$ can be factored in linear terms. As a corollary, we find that no polynomial with degree higher than 2 is irreducible in $R[x]$. The proof of the Fundamental Theorem is based on facts from analysis, the study of the real and complex valued functions, because the coefficients are from these fields that are built from the continuum of real numbers. The discrete counting techniques that allow us to factor polynomials in $Z[x]$ and $Q[x]$ are not available for $C[x]$ and $R[x]$.

Lastly, we used the equivalence relation of **polynomial congruence** to construct fields in the same ways we constructed the fields Z_p in Chapter 2. Given an irreducible polynomial $p(x)$ in $F[x]$, we constructed the field $F[x]/\langle p(x) \rangle$ that contains F and contains a root of $p(x)$. Given any polynomial $f(x)$ in $F[x]$, we can iterate our procedure to construct a field in which $f(x)$ factors into linear terms. Ultimately, we leave no polynomial rootless. When $p(x)$ is an irreducible polynomial in $Q[x]$, the question remains, how do the roots of $p(x)$ found in $Q[x]/\langle p(x) \rangle$ relate to the roots of $p(x)$ that we know can be found in C? We will continue our investigation of these issues in Chapter 6. The tools to do so will be honed in Chapters 4 and 5.

Chapter Questions

1. State the definition of the following terms.

    i. degree of a polynomial in $R[x]$
    ii. monic polynomial
    iii. root of a polynomial
    iv. $\gcd(f(x), g(x))$

v. irreducible polynomial
vi. content of a polynomial in $\mathbf{Z}[x]$
vii. primitive polynomial in $\mathbf{Z}[x]$
viii. $f(x) \equiv g(x) \bmod p(x)$

2. Give the statement of each of the following theorems. For each, indicate the strategy of its proof.

   i. The Division Theorem for Polynomials
   ii. The Root Theorem for Polynomial
   iii. The Remainder Theorem for Polynomials
   iv. Unique Factorization Theorem for Polynomials
   v. Gauss's Lemma
   vi. Eisenstein's Criterion

3. State the Fundamental Theorem of Algebra and its corollary for polynomials with real coefficients.

4. Add and multiply the given polynomials:

   a. $p(x) = 4x^2 + 3x + 1$ and $q(x) = 6x + 2$ in $\mathbf{Z}_7[x]$
   b. $p(x) = 2x^2 + 3x + 1$ and $q(x) = 3x + 2$ in $\mathbf{Z}_6[x]$
   c. $p(x) = 2x^2 + 1$ and $q(x) = 2x + 2$ in $\mathbf{Z}_3[x]$
   d. $p(x) = (2 - 3i)x^2 + 3x + i$ and $q(x) = 3ix + 2$ in $\mathbf{C}[x]$

5. Find two polynomials, each of degree 2, in $\mathbf{Z}_6[x]$, whose product has degree 0. Can you repeat the exercise in $\mathbf{Z}_7[x]$? Explain.

6. Suppose that m is a composite natural number and let $n > 0$ be any integer. Find a polynomial of degree n in $\mathbf{Z}_m[x]$ that is a zero divisor.

7. For each of the following, find $q(x)$ and $r(x)$ as in the Division Algorithm for Polynomials for the given $g(x)$ and $p(x)$.

   i. $g(x) = 3x^4 + 4x^3 + x + 1$, $p(x) = 2x^2 + 5x + 1$ in $\mathbf{Z}_7[x]$
   ii. $g(x) = x^5 + x^2 + x + 1$, $p(x) = x + 1$ in $\mathbf{Z}_2[x]$
   iii. $g(x) = x^3 + (3 + i)x + 1$, $p(x) = (x - 2i)$ in $\mathbf{C}[x]$.

8. Use the Remainder Theorem to find the remainder of $x^{10} + x^7 - 3x^2 + 1$ after division by $x + 3$ in $\mathbf{Q}[x]$.

9. Find the remainder of $x^4 + 3x + 2$ after division by $x - 4$ in $\mathbf{Z}_7[x]$.

10. Let $f(x) = x^4 - 9x^2 + 4x + 12$ and let $g(x) = x^3 - 7x + 6$ be polynomials in $\mathbf{Q}[x]$.

    i. Use Euclid's Algorithm to find $\gcd(f, g)$.
    ii. Express $\gcd(f, g)$ as $s(x)f(x) + t(x)g(x)$.

11. Let $f(x) = x^4 + 5x^2 + 4x + 5$ and let $g(x) = x^3 + 6$ be polynomials in $Z_7[x]$.
    i. Use Euclid's Algorithm to find $\gcd(f, g)$
    ii. Express $\gcd(f, g)$ as $s(x)f(x) + t(x)g(x)$.

12. Factor each of the following polynomials into irreducible polynomials
    i. $x^3 - 7x + 6$ in $Q[x]$
    ii. $6x^2 + 6x - 36$ in $Z[x]$
    iii. $x^3 + x^2 + 5x + 5$ in $Z_7[x]$

13. Find the content of the following polynomials
    i. $x^5 + 3x^4 + 5x + 11$
    ii. $6x^3 + 15x + 9$
    iii. $77421x^3 + 335491x^2 + 180649x + 283877$

14. Determine which of the following polynomials are irreducible in $Q[x]$. Factor those that are not irreducible.
    i. $x^5 + 6x^3 + 9x^2 + 21$
    ii. $x^3 + 333x + 222$
    iii. $x^4 + x^3 - 7x + 5$
    iv. $\frac{2}{3}x^3 + \frac{14}{3}x^2 + 4x + 28$

15. Suppose that $p(x) \in C[x]$ and that $p(x) = a_7x^7 + a_6x^6 + \ldots + a_0$. Suppose that $p(x)$ has two roots, each of multiplicity 2, and no other repeated roots. What is the format of the factorization of $p(x)$ into irreducible polynomials in $C[x]$?

16. Suppose that $p(x) \in R[x]$ and that $p(x) = a_7x^7 + a_6x^6 + \ldots + a_0$ has two non-conjugate complex roots that are not real, and no repeated roots. What are the possible formats of the factorization of $p(x)$ into irreducible polynomials in $R[x]$?

17. Use the fact that i is a root of $p(x) = x^3 + (1 + i)x^2 + 3ix + (4 + 2i)$ to factor $p(x)$ into linear factors in $C[x]$.

18. Write out the multiplication table for the field $Z_3[x]/\langle x^2 + x + 2 \rangle$.

19. Let α denote the $[x]$, the equivalence class of the polynomial $f(x) = x$, in $F = Z_3[x]/\langle x^2 + x + 2 \rangle$. Factor $x^2 + x + 2$ in $F[x]$.

20. Find a zero divisor in $Z_3[x]/\langle x^2 + x + 1 \rangle$.

21. Find a field with 25 elements.

22. Find a field that contains Z_5 in which the polynomial $x^3 + x + 1$ factors completely.

4

A FIRST LOOK AT GROUP THEORY

The area of mathematics that investigates algebraic structure is called **abstract algebra**. It distills what is common to familiar algebraic structures. For instance, it determines how and why matrix addition is similar to integer addition. On the other hand, abstract algebra delineates why and how structures differ. For instance, it addresses the question of how and why multiplication in Z_m differs from matrix multiplication. In this chapter, we give an overview of the algebraic structure known as a **group**. It is one of the three principal structures that are the focus of study in abstract algebra. The other two, namely rings and fields, are already familiar to us. In some sense, groups seem simpler than rings and fields because groups have only one operation, and the properties that define a group are very simple. But the subject of groups is incredibly rich, with applications as diverse as arithmetic, geometry, and chemistry.

4.1 INTRODUCTION TO GROUPS

The word "group" is used to describe a set that is equipped with one binary operation.[1] The most familiar example is the set of integers Z under addition. Notice how the definition below captures the most basic algebraic facts about addition.

> **DEFINITION 1.** Let $*$ be a binary operation on a nonempty set G. We say that *G* **is a group under** $*$ if the following properties hold.
>
> 1. The operation $*$ is associative, that is, for all a, b, and c in G, $(a * b) * c = a * (b * c)$.
> 2. There is an element $e \in G$ such that $e * a = a * e = a$ for all $a \in G$.
> 3. For each $a \in G$, there is an element $b \in G$ such that $a * b = b * a = e$.

---
[1] The appendix contains a discussion of binary operations.

A group, namely the set G together with its operation $*$, is denoted as the pair $\{G, *\}$. If there is no confusion about the operation in use, we refer to the group as simply G. We can think of the symbol $*$ as a generic operation. Later, when we deal with specific examples, we will use specific operations symbols such as the addition sign, the multiplication sign, and composition symbol in place of $*$. The element e required in Property 2 of Definition 1 is called the **identity element of G**. It too will be replaced in specific examples. For instance, 0 plays the role of e when our group is the set Z of integers under addition as in Example 1 below. The symbol 1 plays the role of e when our group is the set of nonzero elements of Z_5 under multiplication. (See Example 8.)

Property 3 requires that for each element $a \in G$, there is an element $b \in G$ such that $a * b = b * a = e$. The element b is called the **inverse** of a with respect to $*$. Notice that if b is the inverse of a, then a is the inverse of b. Inverses will be written differently in different contexts. For instance, when our group is the set of integers under addition, $\{Z, +\}$, then the inverse of an integer a is the integer $-a$ because $a + (-a) = 0$. The existence of inverses in a group $\{G, *\}$ means that we can always solve an equation of the form $a * x = c$ for x when a and c are members of the set G. To do this, suppose that the inverse of a is denoted by the symbol b. Then $b * (a * x) = b * c$. We can rewrite the latter equation as $(b * a) * x = b * c$ because of associativity. Since $b * a = e$ and $e * x = x$, we have $x = b * c$. So we have solved for x. Notice how all the group properties come into play in this simple process.

EXAMPLE 1

Consider the set Z under the operation $+$. The operation $+$ is associative (Property 1 of Definition 1). The identity element is the integer 0 since $0 + a = a + 0 = a$ for each element a in Z (Property 2). For each a in Z, there is a corresponding element $b = -a$ such that $a + b = b + a = 0$ (Property 3). Thus Z is a group under addition. ∎

The operation of addition in the group $\{Z, +\}$ is commutative since $a + b = b + a$ for every pair of integers a and b. But, as the next example shows, group operations are not necessarily commutative.

EXAMPLE 2

Let $GL(\mathbf{R}, 2)$ be the set of 2×2 real valued, nonsingular matrices (i.e., matrices with nonzero determinant). Let $\times$ denote the operation of matrix multiplication. Recall that if A and B are nonsingular, then $A \times B$ is also nonsingular because $\det(A \times B) = \det(A)\det(B)$. Thus the operation $\times$ is closed on $GL(\mathbf{R}, 2)$. Since matrix multiplication is associative, Property 1 is satisfied. The identity element required of Property 2 is the matrix $I = \begin{bmatrix} 1 & 0 \\ 0 & 1 \end{bmatrix}$ since $I \times A = A \times I = A$ for any matrix A in

$GL(R, 2)$. For any matrix $A = \begin{bmatrix} a & b \\ c & d \end{bmatrix}$ in $GL(R, 2)$, the matrix $B = \dfrac{1}{\det(A)} \begin{bmatrix} d & -b \\ -c & a \end{bmatrix}$ is also in $GL(R, 2)$ because $\det(B) = \dfrac{1}{\det(A)} \neq 0$. It is easy to check that $A \times B = B \times A = I$ so that Property 3 is satisfied. The set $GL(R, 2)$ is thus a group under matrix multiplication. Since $\begin{bmatrix} 1 & 2 \\ 3 & 4 \end{bmatrix} \times \begin{bmatrix} 5 & 6 \\ 7 & 8 \end{bmatrix} = \begin{bmatrix} 19 & 22 \\ 43 & 50 \end{bmatrix}$ but $\begin{bmatrix} 5 & 6 \\ 7 & 8 \end{bmatrix} \times \begin{bmatrix} 1 & 2 \\ 3 & 4 \end{bmatrix} = \begin{bmatrix} 23 & 34 \\ 31 & 46 \end{bmatrix}$, the operation $\times$ is *not* commutative. The group $GL(R, 2)$ is called the **general linear group of degree 2**. More generally, $GL(R, n)$ denotes the group of nonsingular $n \times n$ matrices with entries from the set of real numbers. The name reflects the fact that such matrices can be thought of as linear transformations from R^n to R^n.

Now let $A = \begin{bmatrix} 3 & 2 \\ 1 & 1 \end{bmatrix}$ and $B = \begin{bmatrix} 1 & -1 \\ 5 & 1 \end{bmatrix}$. We solve for X in the equation $A \times X = B$. To do so, note that $A^{-1} = \begin{bmatrix} 1 & -2 \\ -1 & 3 \end{bmatrix}$ since $\begin{bmatrix} 1 & -2 \\ -1 & 3 \end{bmatrix} \times \begin{bmatrix} 3 & 2 \\ 1 & 1 \end{bmatrix} = \begin{bmatrix} 1 & 0 \\ 0 & 1 \end{bmatrix}$. Now we multiply both sides of the equation $A \times X = B$ on the *left* by A^{-1} to obtain $(A^{-1} \times A) \times X = X = A^{-1} \times B$. Thus $X = A^{-1} \times B = \begin{bmatrix} 1 & -2 \\ -1 & 3 \end{bmatrix} \times \begin{bmatrix} 1 & -1 \\ 5 & 1 \end{bmatrix} = \begin{bmatrix} -9 & -3 \\ 14 & 4 \end{bmatrix}$. Remember that when an operation is not commutative, we must take special care to determine when to multiply a given expression on the left and when to multiply on the right. ∎

We call $\{G, *\}$ an **abelian** group when the operation $*$ is a commutative operation. Thus $\{Z, +\}$ is abelian while $\{GL(R, 2), \times\}$ is a **nonabelian** group. A group $\{G, *\}$ is called finite or infinite depending on whether the set G is finite or infinite. The groups $\{Z, +\}$ and $\{GL(R, 2), \times\}$ are infinite groups. Our next three examples investigate finite groups.

EXAMPLE 3

For each integer $n > 0$, $\{Z_n, +\}$ is a finite abelian group. The operation of addition mod n is associative and commutative. The element 0 is the identity element. The (additive) inverse of any element $a \in Z_n$ that is not 0 is the element $(n - a)$ since $((n - a) + a) \bmod n = 0 \bmod n$. ∎

The Cayley table for $\{Z_4, +\}$ is as shown in Table 1. In a **Cayley table** (or **operation table**) for a finite group $\{G, *\}$, we label each row and column of the table with an element of the group, and enter $x * y$ in the position labeled row x, column y.

TABLE 1: The Cayley Table for $\{\mathbf{Z_4}, +\}$

+	0	1	2	3
0	0	1	2	3
1	1	2	3	0
2	2	3	0	1
3	3	0	1	2

EXAMPLE 4 *(The Group of Parade Commands)*

Groups arise in many situations, not just in the familiar territory of arithmetic. And groups are fun! In this example, our set G is the following set of parade commands: {Attention (A), Right Face (R), Left Face (L), About Face (B)}. By "attention" we mean, "keep facing in your current direction." Our operation $*$ is the operation "followed by." Thus $R * L = A$ since a right face followed by a left face leaves Private Smith facing forward (i.e., back "at attention" and facing her original direction). The operation is associative since we do not need parentheses to interpret "command x followed by command y followed by command z." Table 2 shows that A is the identity element for $*$ and that every element has an inverse.

TABLE 2: The Cayley Table for the Group of Parade Commands

*	A	L	B	R
A	A	L	B	R
L	L	B	R	A
B	B	R	A	L
R	R	A	L	B

From Table 2, we see that the operation is commutative since the operation table is a symmetric matrix, that is, the entry in row i, column j is the same as the entry in row j, column i. Let's solve the equation $L * x = R$. First we find the inverse of L which is R. Then we multiply both sides by R to obtain $(R * L) * x = R * R$ or $x = R * R = B$. Let's check! The command $L * B$ is a Left Face followed by an About Face, which is indeed a Right Face. ∎

Table 2 looks suspiciously like that of $\mathbf{Z_4}$ under addition. If 0 were replaced by A, and 1 replaced by L, and 2 replaced by B, and 3 by R at *every* position in Table 1, then Tables 1 and 2 would be identical. The Group of Parade Commands is just $\{\mathbf{Z_4}, +\}$ in disguise. We take up the task of how to tell when two groups are essentially the same except for notation when we study the concept of isomorphism in Section 4.5.

EXAMPLE 5

Table 3 shows another group of four elements. The set is $K = \{(0,0), (0,1), (1,0), (1,1)\}$. The operation $\oplus$ is performed by adding components mod 2. The operation table shows that each element is its own inverse. So this group is quite different from $\{\mathbf{Z_4}, +\}$. The group $\{K, \oplus\}$ is often called the **Klein 4-Group**.

TABLE 3: The Klein 4-Group

$\oplus$	(0, 0)	(0, 1)	(1, 0)	(1, 1)
(0, 0)	(0, 0)	(0, 1)	(1, 0)	(1, 1)
(0, 1)	(0, 1)	(0, 0)	(1, 1)	(1, 0)
(1, 0)	(1, 0)	(1, 1)	(0, 0)	(0, 1)
(1, 1)	(1, 1)	(1, 0)	(0, 1)	(0, 0)

The number of elements of a group G is called the **order of G**. It is denoted by $|G|$. The group $\{Z, +\}$ has infinite order. The order of $\{K, \oplus\}$ is 4 (i.e., $|K| = 4$).

Generally, the binary operation on a group G is either modeled on multiplication or it is modeled on addition. Some conventions apply in each case. When we are using **multiplication**, usually denoted by the symbols $*$ or $\cdot$, we often write ab for $a * b$. We denote the product $a * a * \ldots * a$ (n copies of a) by a^n and we denote the inverse of the element a with respect to $*$ by a^{-1}. The product of a^{-1} with itself n times is denoted by a^{-n}. We define $a^0 = e$. Then, just as in ordinary arithmetic, $a^n * a^{-n} = a^0 = e$. The ordinary rules of exponents apply. The multiplicative notation is the most common notation for abstract groups.

Caution: The operation $*$ is **not** generally assumed to be commutative. In general, we cannot assume that $ab = ba$ or that $abab = a^2b^2$.

A group G is called an **additive group** if its binary operation is denoted by $+$. **The notation $+$ is only used for binary operations that are commutative.** Instead of exponents, coefficients are used. The sum $x + x$ is denoted by $2x$ even though 2 itself may not be in the group. Similarly, nx denotes $x + x + \ldots + x$, the sum of x with itself n times. Because $+$ is always commutative, $a + b + a + b = 2a + 2b$. The symbol 0 denotes the identity element. The inverse of a with respect to $+$ is denoted by $-a$. Thus $a + (-a) = 0$. The expression $a - b$ means $a + (-b)$.

EXAMPLE 6

Matrix addition makes $\mathbf{M}_{2,2}$, the set of all 2×2 matrices with real entries, into an additive group. Addition is associative and commutative. The identity element is $\mathbf{0} = \begin{bmatrix} 0 & 0 \\ 0 & 0 \end{bmatrix}$. The (additive) inverse of $\begin{bmatrix} a & b \\ c & d \end{bmatrix}$ in $\mathbf{M}_{2,2}$ is $\begin{bmatrix} -a & -b \\ -c & -d \end{bmatrix}$.

EXAMPLE 7

The set $\mathbf{Z}_n$ is *not* a group under multiplication mod n. It has a multiplicative identity, namely 1, but the element 0 has no multiplicative inverse because there is no element x such that $x \cdot 0 = 1$. The sets $\mathbf{Z}$, $\mathbf{Q}$ and $\mathbf{R}$ are all groups under **addition**. None is a group under multiplication because, in each case, the element 0 has no multiplicative inverse.

EXAMPLE 8

Recall from Section 2.4 that an element $x \in Z_n$ is called a "unit" if there is an element $y \in Z_n$ such that $xy = 1$ mod n. Let U_n denote the set of units of the ring Z_n. From Proposition 5 of Section 2.1, we know that an element $x \in U_n$ if and only if x is relatively prime to n. We will now show that the set U_n is a group under the operation of multiplication mod n. First, we must check that the product of two elements x and y in U_n is also in U_n. That is to say, we must check that U_n is **closed** under multiplication. (In general, if $*$ is a binary operation on G, and $S \subseteq G$, we say that S is **closed** under the operation $*$ if, for all x and y in S, $x * y \in S$.) To see that U_n is closed under multiplication mod n, note that if two numbers x and y are relatively prime to n, then xy is also relatively prime to n. (See Exercise 5 of Section 1.3.) Now we check that the three defining properties of a group hold. Since multiplication in Z_n is associative, multiplication in U_n is also associative. The multiplicative identity 1 is in U_n. Lastly, any element $x \in U_n$ has a multiplicative inverse y in U_n, where y is the solution to $xy \equiv 1$ mod n. Thus U_n is a group under multiplication.

For $n = 8$, U_n is the set $\{1, 3, 5, 7\}$. The multiplication table for U_8 is shown in Table 4. Like the Klein 4-Group, U_8 has the property that each element is its own inverse. ∎

TABLE 4: $\{U_8, \cdot\}$

·	1	3	5	7
1	1	3	5	7
3	3	1	7	5
5	5	7	1	3
7	7	5	3	1

The following proposition summarizes some arithmetic properties common to all groups.

PROPOSITION 1 Let $\{G, *\}$ be a group. Then the following properties are true.

  i. G has exactly one identity element.
 ii. If x, y and z are elements of G such that $xy = xz$, then $y = z$.
iii. Each element has exactly one inverse.
 iv. For each $y \in G$, $(y^{-1})^{-1} = y$.
  v. For any x and y in G, $(xy)^{-1} = y^{-1}x^{-1}$. (Note the positions of x and y.)

Note: The analogous statement of (ii) for additive groups would be, "If $x + y = x + z$, then $y = z$." Parts iv and v require similar interpretation.

Proof.
  i. Suppose that e_1 and e_2 are elements in G such that, for all $y \in G$, $e_1 y = y e_1 = y$ and $e_2 y = y e_2 = y$. (So both e_1 and e_2 satisfy the definition of an identity element.) Then $e_1 e_2 = e_2$ since e_1 is an identity element. Also $e_1 e_2 = e_1$ since e_2 is an identity element. Thus $e_1 = e_2$.

ii. Suppose that $xy = xz$. Multiplying on the left by x^{-1}, we have $(x^{-1}x)y = (x^{-1}x)z$, or $ey = ez$. Thus $y = z$.
iii. It follows from (ii) that if $xy = xz = e$, then $y = z$. We denote the unique inverse of x by x^{-1}.
iv. Exercise 8
v. Exercise 9 ▲

Suppose we want to construct a group with three elements $\{e, a, b\}$ and use the operation. What are the possibilities for a group table if we agree that e will be the identity element? The properties presented in Proposition 1 are the key to solving this problem.

·	e	a	b
e	?	?	?
a	?	?	?
b	?	?	?

The row and column labeled by e are easy to fill in since $ex = xe = x$. We also know that $ab \neq a$ because otherwise, we could multiply by a^{-1} on the left of both sides to conclude that $b = e$. But b is not the identity element. Similarly, $ab \neq b$. Thus $ab = e$. The same argument says that $ba = e$. So far we have

·	e	a	b
e	e	a	b
a	a	?	e
b	b	e	?

There is only one way to fill in the remaining entries: $aa \neq a$ since $a \neq e$. Also $aa \neq ab$ since $a \neq b$. Since $ab = e$, $aa \neq e$. Thus $aa = b$. Similarly, $bb = a$.

·	e	a	b
e	e	a	b
a	a	b	e
b	b	e	a

Notice that each row and column has each group element appearing exactly once. If we replace the symbol e by 0, the symbol a by 1, the symbol b by 2, and the symbol · by +, we have the group table for Z_3 under addition. This suggests that there is essentially only one group that has three elements. We will follow up this idea in Section 4.5.

SUMMARY

Groups come equipped with one binary operation, generally denoted by $*$. All groups have three simple properties: the operation is associative, one element serves as an identity element, and each element has an inverse. This section presented many examples of groups. The term "group" covers a rich and diverse class of mathematical contexts. Some groups have finite order, while others are infinite. Some are abelian, and some nonabelian. Some use a notation that is like multiplication, while others use additive notation. But in any group G, we can solve the simple equation $a * x = b$ for x. Proposition 1 presented several other properties that are common to all groups. In the next few sections, we will investigate attributes of groups more closely, and we shall look at special groups such as cyclic groups and permutation groups. We shall apply our knowledge of groups to several interesting contexts that range from counting to geometry.

4.1 Exercises

1. Determine if $\{G, *\}$ is a group. Justify your response fully.
   i. $G = \mathbf{Z}$ and $*$ is subtraction so that $\{G, *\}$ is $\{\mathbf{Z}, -\}$.
   ii. $G = \{2, 4, 1\}$ and $*$ is multiplication mod 7.
   iii. $G = \{2, 4, 1\}$ and $*$ is multiplication mod 5.
2. Is the set of even integers a group under addition? How about the odd integers? Explain.
3. Suppose that G is the set $\{a, b, c\}$ and that $*$ is the associative operation defined by the following table. Show that $\{G, *\}$ is a group.

*	a	b	c
a	b	c	a
b	c	a	b
c	a	b	c

4. Solve for x in the following equations.
   i. $3x = 5$ in U_{11}
   ii. $(0, 1) \oplus x = (1, 0)$ in the Klein 4-Group
   iii. $\begin{bmatrix} 1 & 3 \\ 2 & 4 \end{bmatrix} \times x = \begin{bmatrix} 2 & 1 \\ 0 & 1 \end{bmatrix}$ in $GL(\mathbf{R}, 2)$, the set of nonsingular 2 by 2 matrices with real entries
5. Suppose that $\{G, *\}$ is a group.
   i. For $n > 0$ and $a \in G$, show that $(a^n)^{-1} = (a^{-1})^n$. (What this exercise asks you to do is show that the inverse of a^n is the product of the inverse of a with itself n times. There really is something to prove. Don't take the notation for granted!)
   ii. For any integers i and j, prove that $a^i * a^j = a^{i+j}$.
6. Let G be an additive group and suppose that $a \in G$.
   i. Show that $-(na) = n(-a)$ for any natural number n.
   ii. Show that for any integers m and n, $ma + na = (m + n)a$.
7. Translate $(a^{-3}b^2)$ into additive notation and translate $(5a - 2b)$ into multiplicative notation.
8. Prove Part iv of Proposition 1.
9. Prove Part v of Proposition 1.

10. What is the order of the group U_3? U_8? What is the order of U_n?
11. Solve $3x = 7$ in U_8.
12. Show that in U_8, $a^n = a$ if n is odd and $a^n = 1$ if n is even.
13. Write out the operation tables for U_{12} and for U_{10}. How do they differ?
14. Let G be a group and let $g \in G$. Let $T_g : G \to G$ be the function defined by $T_g(x) = gx$.
    i. Give all the values of $T_g(x)$ when $G = U_8$ and $g = 3$.
    ii. Prove that the map T_g is a bijection for any group G and element $g \in G$.
    iii. Explain how part ii of this exercise guarantees that each element of G appears exactly once in each row of the operation table for G.
    iv. Modify the function T_g to show that each element of G appears exactly once in each column of the operation table for G. (This exercise shows that any finite group table must be a **Latin square**, which is to say that each element occurs in each row and column exactly once.)
15. Assuming that e is the identity element, how many group tables can you construct using the symbols $\{e, a, b, c\}$? Among the tables you produce, indicate essential differences.
16. Assume that x, y and z are elements of a group $\{G, *\}$.
    i. Give an example that shows that, in general, $(xy)^{-1} \neq x^{-1}y^{-1}$. (Investigate a familiar nonabelian group.)
    ii. Prove that $(xy)^{-1} = x^{-1}y^{-1}$ for all x and y in G if and only if G is abelian.
17. Prove that in an abelian group, $(ab)^n = a^n b^n$ for each integer n. Give an example that shows that this is not generally true in nonabelian groups.
18. Prove that if G is a finite group and $x \in G$, then there is a positive integer n such that $x^n = e$.
19. Prove that if G is a group such that $a^2 b^2 = (ab)^2$ for each a and b in G, then G is abelian.

To the Teacher

What's algebra? Ask any student, past or present. You are most likely to hear, "Solving for x." The definition of a group crystallizes what we do to solve for x in the most basic equation, $a * x = b$. Let's think of $a * x = b$ as the equation $5x = 2$ in the group of nonzero rational numbers under multiplication, and let's solve for x, a very typical elementary algebra task.

> To find x, we first find c, the inverse of a, so that $c * a = e$, where e is an identity element. In our example, $e = 1$ and $c = \frac{1}{5}$. *In any group, there is an identity element and every element has an inverse.* Then we multiply both sides of the equation by c to obtain: $c * (a * x) = c * b$. In our example we then have, $\frac{1}{5} \cdot (5 \cdot x) = \frac{1}{5} \cdot 2$. Then we reparenthesize to obtain $c * (a * x) = (c * a) * x = x = c * b$. In our example, we obtain $(\frac{1}{5} \cdot 5) \cdot x = 1 \cdot x = x = \frac{1}{5} \cdot 2 = \frac{2}{5}$. To do this, we need the operation $*$ to be associative. *Every group operation is associative.*

The definition of a group simply distills what we need to do the most elementary process of algebra, solving for x. With that observation, we leap to the conclusion that the process of "solving for x" remains the same even though the contexts for a, b, x, and $*$ can vary widely. For instance, a and b and $*$ can be

- integers and addition
- nonzero real numbers and multiplication
- invertible 2 by 2 matrices and matrix multiplication
- mod 7 integers with mod 7 addition
- units mod 6 and mod 6 multiplication
- permutations and composition

A teacher's job is big: to practice the details of solving for x in equations like $\frac{3x}{5} = 53$ frequently enough so that it does indeed become automatic, and yet to keep the structural aspects firmly in view. Students must know both how and when to use the multiplicative inverse. An experience with Cayley tables, as presented in the context of Example 3 for the younger student, or modular arithmetic for the more advanced, can serve both to solidify familiar algebraic processes and to foster a new level of mathematical sophistication. A look at the multiplication table for $\mathbf{Z_6}$ (not a group) gives many surprises like the fact that $2 \cdot 3 = 0$, and the fact that some, but not all, nonzero elements have multiplicative inverses that don't look like fractions. Solving $5x = 2$ takes some thinking in $\mathbf{Z_6}$. A guess-and-check might yield $x = 4$, but so does multiplying both sides by 5. The lesson is in the role of the inverse—that which multiplies 5 to give 1—rather than in the reflex "multiply both sides by one over five."

Task

Design an activity appropriate for a high school class that focuses on the Klein 4-Group. (A peek at Sections 4.2 and 4.3 might provide an inspiration.)

4.2 GROUPS AND GEOMETRY

In this section we investigate some connections between group theory and geometry. These connections have a long and rich history. We shall focus on groups of symmetries of the regular figures in 3-space. A **symmetry** of a regular figure (polygon or solid) is a rotation of the figure around an axis of symmetry that takes the figure congruently onto itself.

The first figure we investigate is the equilateral triangle. In what follows, we shall always take the triangle to be oriented as in Figure 1a with its base on a horizontal line in the plane of the paper. The triangle has four axes of symmetry. The first three are easy to find. At each vertex there is an axis that passes through it and bisects the opposite side. These are labeled a, b, and c in Figure 1b. When we rotate the triangle 180° around any of these three axes, it comes back onto itself but with its flip side forward. (This motion can also be described as a reflection across the axis line.) The fourth axis is harder to find. It passes through the center of the triangle (where its angle bisectors meet) and it is perpendicular to the plane of the triangle. This axis of symmetry is represented by the star in the center of the triangle drawn in Figure 1b. It is drawn with perspective in Figure 1c where it is labeled P. We can rotate the triangle clockwise around P onto itself through angles of degrees 0, 120, and 240. The front side stays frontward. (These rotations can also be thought of as rotations around the center of the triangle.)

For this part of the discussion, cut out an equilateral triangle and label its vertices with 1, 2, and 3. You might want to distinguish front and back. Trace its outline on a paper and label the axes of symmetry that pass through the vertices and its center. Then you can perform the symmetries (motions) as we discuss them. **Note:** The axes stay put. If, after a series of moves, vertex 2 is at the top, rotation around axis a still rotates the triangle through the fixed vertical axis that now passes through vertex 2.

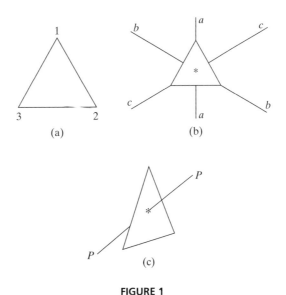

FIGURE 1

We denote the rotations around axes P through 0, 120, and 240 degrees (clockwise) by R_0, R_1, and R_2, respectively. We denote the flips around axes a, b, and c by F_a, F_b, and F_c. There are a total of six symmetries of the equilateral triangle.

We can summarize the action of each symmetry by indicating the new position of each vertex under its action. For instance, the motion R_1 takes $1 \to 2$, $2 \to 3$, and $3 \to 1$. The actions are summarized in the following table. The vertex listed in the top row goes to the position listed in the bottom row.

$$R_0 = \begin{pmatrix} 1 & 2 & 3 \\ 1 & 2 & 3 \end{pmatrix} \quad R_1 = \begin{pmatrix} 1 & 2 & 3 \\ 2 & 3 & 1 \end{pmatrix} \quad R_2 = \begin{pmatrix} 1 & 2 & 3 \\ 3 & 1 & 2 \end{pmatrix}$$

$$F_a = \begin{pmatrix} 1 & 2 & 3 \\ 1 & 3 & 2 \end{pmatrix} \quad F_b = \begin{pmatrix} 1 & 2 & 3 \\ 3 & 2 & 1 \end{pmatrix} \quad F_c = \begin{pmatrix} 1 & 2 & 3 \\ 2 & 1 & 3 \end{pmatrix}$$

A look at the bottom row of each table reveals that each of the symmetries corresponds to exactly one permutation of the symbols 1, 2, and 3. Since there are exactly six permutations of three symbols, we know that our list of symmetries is complete.

If X and Y are symmetries, we define $X * Y$ to be the result of the motion Y followed by the motion X. For example, to perform motion $F_a * R_2$, we first rotate the triangle clockwise 240 degrees and then flip around axis a. The result of performing these two actions in sequence is the same as performing the one motion F_c so that $F_a * R_2 = F_c$. The convention that $X * Y$ requires Y to be executed first resembles function composition and for good reason—it is function composition! Each motion defines a function on the set $\{1, 2, 3\}$ that sends the symbol listed on the top to the symbol listed below it. The effect of the motion Y followed by the motion X on vertex i is the same as the function Y applied to the symbol i followed by (composed with) the function X applied to symbol $Y(i)$.

The table for the operation $*$ on the set of symmetries of the equilateral triangle is given in Table 1.

TABLE 1

$*$	R_0	R_1	R_2	F_a	F_b	F_c
R_0	R_0	R_1	R_2	F_a	F_b	F_c
R_1	R_1	R_2	R_0	F_c	F_a	F_b
R_2	R_2	R_0	R_1	F_b	F_c	F_a
F_a	F_a	F_b	F_c	R_0	R_1	R_2
F_b	F_b	F_c	F_a	R_2	R_0	R_1
F_c	F_c	F_a	F_b	R_1	R_2	R_0

Note: For some authors, rotation is counterclockwise. Also, for some authors, X acts before Y in $X * Y$. For these authors, $X * Y$ means "do X first and then do Y." The resulting tables look different.

The set of symmetries of the triangle is a group under $*$ because of the following reasons.

i. The action is associative because function composition is associative.
ii. R_0 is the identity element for the operation $*$.
iii. Each action has an inverse, an action that undoes it.

We denote the group of symmetries of the equilateral triangle by D_3 and call it the **dihedral group** of the triangle. The word "dihedral" means "two sided." D_3 is a nonabelian group of order six. For $n > 3$, similar analysis produces the dihedral group of symmetries D_n for each regular polygon with n sides. In each case, there are n rotations (including the identity symmetry), and n flips, so that the order of D_n is $2n$. In each case, the group is nonabelian.

Let's look at D_4, the symmetry group of the square. It has eight elements. To see this, pick a vertex—the upper left corner, for instance. It can be placed at any of the four corner positions. Once placed at its new position, we can orient the square in two ways, face up or face down, for a total of eight. Now label the four corners of the square with 1, 2, 3, and 4 as in Figure 2. As with the equilateral triangle, there is an axis of symmetry P through the center of the square, perpendicular to the plane of the square. There are four rotations, R_0, R_1, R_2, and R_3, that rotate the square through 0, 90, 180 and 270 degrees (clockwise), respectively, about P. There are four axes of symmetry in the plane of the square: H (horizontal), V (vertical), D_1 (diagonal going northwest) and D_2 (diagonal going northeast) with 4 corresponding flips with the same name. It is not difficult to see that $H * V = V * H = R_2$. But $H * R_1 = D_2$ while $R_1 * H = D_1$.

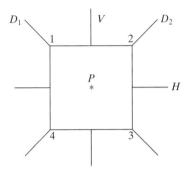

FIGURE 2

The symmetries of the equilateral triangle exhausted the possible permutations of its 3 vertices. The symmetries of the square do not. Many permutations do not correspond to rigid motions of the square onto itself. For instance, the permutation $\begin{pmatrix} 1 & 2 & 3 & 4 \\ 1 & 2 & 4 & 3 \end{pmatrix}$ is not possible since it would require us to keep vertices 1 and 2 fixed while interchanging vertices 3 and 4. (Ouch!)

The group T of symmetries of the tetrahedron has 12 motions. For a quick count, label the vertices of a tetrahedron 1, 2, 3 and 4 as in Figure 3.

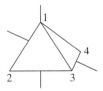

FIGURE 3

We can place vertex 1 in any of four positions. Then, fixing its position, we can rotate 0, 120, or 240 degrees about the axis through the vertex 1 and the center of the triangle formed by vertices 2, 3, and 4. Thus we have a total of $4 \cdot 3 = 12$ motions or symmetries in T. Again, this does not exhaust all 24 permutations on the set of its four vertices. The axes of symmetry of T are as follows. There is an axis from each vertex to the center of the triangle opposite that vertex, for a total of four such axes. In addition to leaving the tetrahedron fixed (the identity motion), we can rotate the tetrahedron through 120 and 240 degrees on each of these axes. This accounts for 9 of the 12 symmetries. Additionally, there is an axis through the midpoints of opposite edges (i.e., edges that do not share a vertex). We can rotate 180 degrees on each of these three axes, thus accounting for all 12 motions. It is interesting that the composition of two motions, each with a different axis of symmetry, still results in a motion that can be accomplished as a rotation about a single axis of symmetry.

SUMMARY

When we hold a model of an equilateral triangle or a tetrahedron, we naturally flip it and rotate it. In this section we discovered that those natural motions, here called **symmetries**, form a group. The group of symmetries of a regular polygon with n vertices is the **dihedral** group D_n. It is a nonabelian group because if you flip then rotate, the vertices end up in a different position from when you rotate then flip. The group D_n has order $2n$. The tetrahedron has four vertices as does square. But its group of symmetries has order 12. As we continue our study of groups, we shall discover that the structure of the group of symmetries of an object can shed light about the structure of the object itself. In the worksheet presented in Section 4.8, we use groups of symmetries to help us solve counting problems. The connection between geometry and algebra is very important.

4.2 Exercises

1. Solve for x in the equation $R_1 * x = F_b$ in D_3, the group of symmetries of the triangle.
2. Solve for x in the equation $x * R_1 = F_b$ in D_3, the group of symmetries of the triangle.
3. If no flips are allowed when rotating the equilateral triangle, what would the operation table be? Can you write out a table that uses only R_0 and flips? What if you restrict your operations further, and use only one flip?
4. Write out each of the symmetries of the square in tabular form, e.g., $R_1 = \begin{pmatrix} 1 & 2 & 3 & 4 \\ 2 & 3 & 4 & 1 \end{pmatrix}$.
5. Finish the Cayley table for D_4.

*	R_0	R_1	R_2	R_3	H	V	D_1	D_2
R_0	R_0							
R_1				R_0				
R_2			R_0					
R_3		R_0						
H					R_0			
V						R_0		
D_1							R_0	
D_2								R_0

6. Solve for x in D_4:
   i. $x * H = V$
   ii. $x * R_3 = D_1$
   iii. $H * x = V$
   iv. $R_3 * x = D_1$
7. If you begin to fill the group table for D_4 with only H and R_2, what other symmetries are you forced to use? What if you started with only H and R_1?
8. Describe the ten symmetries of the regular pentagon.
9. Describe the twelve symmetries of the regular hexagon.
10. How many symmetries does a cube have? Determine the axes of symmetry of a cube.
11. How many symmetries does the regular icosahedron have? How many does the regular dodecahedron have? (See Figure 4.) Why are they so similar? (*Hint.* Put a dot in the middle of each face of an icosahedron. Connect the dots of faces that share a common edge. What do you get?)

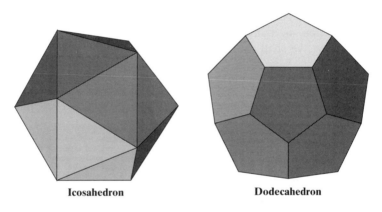

Icosahedron   **Dodecahedron**

FIGURE 4

To the Teacher

This section samples the extremely rich and varied interplay between algebra and geometry. The essay *In the Classroom*, Section 4.9, investigates how algebra can be explored through geometry in the high school classroom.

Task

Design a board game based on the symmetries of the rectangle.

4.3 WORKSHEET 6: FRIEZE GROUPS

Often, in architecture, art and home decoration, we encounter a strip design called a **frieze**—a linear, repeating pattern like those presented in Figure 1. The eye is captured and propelled along the pattern. Artists have given us what seems like an infinite variety of such patterns. But from a mathematical point of view, the variety is limited. In this worksheet, you will explore the types of possible patterns and symmetry groups of these patterns.

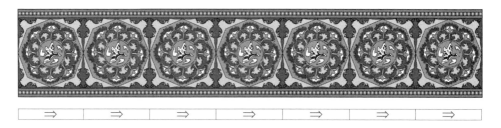

FIGURE 1

The basic unit of design in a frieze is a rectangle that contains exactly one copy of the repeated pattern. That pattern repeats to the right and left along the entire length of the frieze. No subdivision of the basic pattern propagates the design. We can find the basic rectangle by marking the beginning and end of the design. This can be done in many ways, but the length of the basic rectangle will be the same in all cases. We shall call that length d.

Task 1

Mark off a basic rectangle for the following design in several different ways.

In Figure 2, we marked off adjacent congruent rectangles, each containing a copy of the basic pattern. In this not-so-artistic frieze, the design element is the set containment symbol, $\subseteq$. We shall think of a frieze as extending infinitely far to the right and the left. (Don't worry about falling off the wall!) By translating the rectangles congruently onto themselves, we move the entire pattern onto itself. We shall call any rigid movement of a frieze that maps the pattern of the frieze congruently onto itself a **symmetry of the frieze**. The most basic symmetry of a frieze is translation right or left. When we translate, we move the frieze rigidly to the right or left a distance that is a multiple of the minimal distance d.

$\subseteq$	$\subseteq$	$\subseteq$	$\subseteq$
−1	0	+1	+2

$\underbrace{\qquad\qquad\qquad}_{d}$

FIGURE 2

If we number the rectangles as in Figure 2, we can designate translations as T_0, T_1, T_{-1}, etc., where T_i moves rectangle 0 onto rectangle i and all other rectangles are shifted with it. So T_2 shifts rectangle 3 onto rectangle 5 and shifts rectangle 2 onto rectangle 0. Similarly, T_{-3} shifts the rectangle labeled 0 to the left, onto the rectangle labeled -3. The pattern is of course mapped onto a copy itself. The set $T = \{T_i : i \in \mathbf{Z}\}$ is an infinite group under composition. The products are easy to compute: $T_i * T_j = T_{i+j}$. The group T is a subset of the group of symmetries of every frieze pattern.

Now let's take a closer look at the rectangles that contain the basic design.

Task 2

The group of symmetries of a rectangle is the set $\{I, R_{180}, V, H\}$ under composition, where I is the identity map that keeps all points fixed, R_{180} rotates the rectangle 180 degrees around its center, V reflects the rectangle around its vertical axis, and H reflects the rectangle around its horizontal axis.

Apply I, R_{180}, V, and H to each of the following rectangles. In each case, determine whether or not the simple design is mapped onto itself or not. For example, if the rectangle

is subjected to H, the answer is yes because the pattern is mapped onto itself. But if the rectangle is subjected to R_{180}, the answer is no because the orientation of "B" is reversed and appears backward. Similarly, V reverses the orientation of the pattern. Which symmetries maintain the following patterns and which do not?

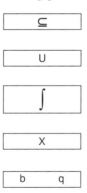

One would think that there would be eight kinds of patterns, namely patterns maintained by

    i. only I
   ii. only I and R_{180}
  iii. only I and V
  iv. only I and H
   v. only I, H and V
  vi. only I, H and R_{180}

vii. only I, V and R_{180}
viii. all of I, V, H, and R_{180}

Task 3

Try drawing a pattern that is maintained by only I, H, and V but **not** by R_{180}. Similarly, try to draw a pattern that is maintained by only I, H, and R_{180} and not by V, and a pattern maintained by only I, V, and R_{180} and not by H. What is the mathematical reason behind your outcome? **Hint.** Think about the Cayley table of the group of symmetries of the rectangle. ∎

The symmetries of the rectangle can be extended to a minimalist frieze that consists of a strip of blank congruent rectangles as in Figure 3. (We left in their position identifiers.) We can consider the center of the frieze to be the center of a rectangle numbered 0 and we can place vertical and horizontal axes through this center. This is indicated by the plus sign + in rectangle 0 in Figure 3. So, along with the identity I (no move) and translations T_i to the right and left, we have H, R_{180} and V. The horizontal flip H places each numbered rectangle onto itself, flipped over and upside-down. But V and R_{180} take the rectangle in position i to position $-i$.

	+		
−1	0	1	2

FIGURE 3

Task 4

Apply T, R_{180}, V, and H to each of the following friezes. In each case, notice whether or not the frieze pattern is mapped onto a copy of itself.

⊆	⊆	⊆	⊆
−1	0	1	2

U	U	U	U
−1	0	1	2

D	D	D	D
−1	0	1	2

H	H	H	H
−1	0	1	2

b q	b q	b q	b q
−1	0	1	2

∎

As with the basic rectangles, some symmetries of the blank frieze leave a design unchanged (the design is said to be "fixed") and others do not.

Task 5

Design a frieze for each of the following conditions. Try to make your frieze both clear and pretty!

1. The design is only fixed by translations.
2. The design is fixed by only reflection around a horizontal axis and translations.
3. The design is fixed by only reflection around a vertical axis and translations.
4. The design is fixed only by rotations of 180 degrees and translations.
5. The design is fixed by reflections across the vertical axis and the horizontal axis and translations. (Why must it also be fixed by rotations of 180 degrees?) ∎

There is one more type of symmetry that maps the design of a frieze onto a copy of the design but *not* the basic rectangle onto a copy of itself. It is called a **glide symmetry** or **glide reflection**. In a glide symmetry, the pattern is translated a distance $d/2$ to the right or left and then reflected across the horizontal axis.

Original:

p		p		p		p	
	b		b		b		b

Translate $d/2$:

p		p		p		p	
	b		b		b		b

Apply H:

	b		b		b		b
p		p		p		p	

Notice that the b's and p's line up in the first and third rows. The design is mapped onto itself even though the rectangles are not.

Task 5 continued

Design a frieze such that

6. The design is fixed only by a glide reflection and translations.
7. The design is fixed only by rotation of 180 degrees, glide reflection, and translation. ∎

Task 6

In tasks 4 and 5, you developed seven distinct types of friezes. Show that there are no others. To do this show that each of the following possibilities is already covered in the list of seven above.

i. A pattern is preserved under glide reflection and H.
ii. A pattern is preserved under glide reflection and any *two* of H, V, and R_{180}.
iii. A pattern is preserved under glide reflection and V.

Part iii is actually more general. Show that a pattern is preserved under glide reflection and V if and only if it is preserved under glide reflection and R_{180}. (**Hint.** Look for a point $1/4$ of the way over from the left edge of the basic design rectangle.) ∎

Conclusion: There are only seven distinct types of frieze patterns.

Five of the seven symmetry groups associated with each of the seven patterns developed in Task 5 are given below, in the corresponding order. (You will find the other two.) The group operation will be denoted by $*$, which can be translated as "followed by." Thus $T_2 * V$ means vertical reflection followed by translation to the right by 2 lengths.

1. $G_1 = \{\ldots, T_{-1}, T_0, T_1, \ldots\}$. In this group, $T_i * T_j = T_{i+j}$.
2. $G_2 = \{\ldots, T_{-1}, T_0, T_1, \ldots\} \cup \{H\}$. In this group, $H * T = T * H$. So this group is abelian.
3. $G_3 = \{\ldots, T_{-1}, T_0, T_1, \ldots\} \cup \{V\}$. It is not abelian since $T_1 * V \neq V * T_1$. Instead, you can check that $T_1 * V = V * T_{-1}$.
4. $G_4 = \{\ldots, T_{-1}, T_0, T_1, \ldots\} \cup \{R_{180}\}$. It is not abelian since $T_1 * R_{180} \neq R_{180} * T_1$. Similar to G_3, $T_1 * R_{180} = R_{180} * T_{-1}$. The groups G_3 and G_4 are similar. But the motions are different. In G_4 the same side of the strip is always outward.
5. $G_5 = \{\ldots, T_{-1}, T_0, T_1, \ldots\} \cup \{H, V, R_{180}\}$. In this group, H, V, and R_{180} commute with each other, $H * T = T * H$, $T_1 * V = V * T_{-1}$ and $T_1 * R_{180} = R_{180} * T_{-1}$.

Task 7

Characterize the elements in the groups G_6 and G_7. Which is abelian? Why?

Task 8

Analyze the pattern symmetries of the following friezes. Pair each with its group of symmetries. You will now see frieze patterns everywhere. Enjoy the art and enjoy your new insight!

4.4 SUBGROUPS

Sometimes we can find groups within groups. These are called subgroups. For example, let's look at the Cayley table for the set $S = \{0, 2, 4\} \subseteq \mathbf{Z_6}$ under addition mod 6 (Table 1).

TABLE 1

+	0	2	4
0	0	2	4
2	2	4	0
4	4	0	2

We already know that addition mod 6 is associative. Since $0 \in S$ and $0 + x = x + 0 = x$ for all x in S, we see that S has an identity element. We can see from Table 1 that every element has an inverse. So $S \subseteq \mathbf{Z_6}$ is itself a group under the operation of addition mod 6, which it inherits from $\mathbf{Z_6}$. We call S a **subgroup** of $\{\mathbf{Z_6}, +\}$. Table 1 does another job. It shows us that S is closed under addition mod 6. Recall that if $\{G, *\}$ is a group and $S \subseteq G$, we say that S is closed under the operation $*$ if $x * y \in S$ for all x and y in S. For instance, the set of even integers is closed under addition because the sum of any two even integers is another even integer. But the set of odd numbers is not closed under addition because the sum of two odd numbers is not odd.

As you have seen, there can be several distinct groups of a given order. Often we can distinguish between two groups of the same order by what kind of subgroups they have. In this section, we investigate how to test whether a subset of a given group is actually a subgroup and we will construct subgroups that contain designated elements.

> **DEFINITION 1.** Let $\{G, *\}$ be a group and let H be a nonempty subset of G. We say that $\{H, *\}$ is **subgroup** of G if H is a group under the operation of G.

In particular, for $H \subseteq G$ to be a subgroup of G,

i. H must be **closed** under the operation $*$,
ii. If $x \in H$, then $x^{-1} \in H$.

The criteria of Definition 1 look a bit sparse. But, as we noted above, since the operation $*$ is associative on G, it remains associative when restricted to H. Since H is not empty, there is an element $x \in H$. So $x^{-1} \in H$ by ii. Since H is closed under $*$ and $e = x * x^{-1}$, the identity element e must also be in H. So criteria i and ii do indeed ensure us that H is a group. We say that H is a **proper subgroup** of G if $H \neq G$.

EXAMPLES

1. For any group $\{G, *\}$, the set $\{e\}$ is a subgroup. It is called the **trivial subgroup**.
2. The set E of even integers is a subgroup of $\{\mathbf{Z}, +\}$. The set E is certainly not empty. Addition on E is closed because the sum of two even integers is even. The negative of any even integer is even and so E contains the (additive) inverse of any element in E.
3. The set of rotations $\{R_0, R_1, R_2\}$ is a subgroup of the group of D_3, the group of symmetries of the equilateral triangle. (No flipping!)

$*$	R_0	R_1	R_2
R_0	R_0	R_1	R_2
R_1	R_1	R_2	R_0
R_2	R_2	R_0	R_1

4. The set $\{1, 2, 4\}$ is a subgroup of the multiplicative group U_7.

$*$	1	2	4
1	1	2	4
2	2	4	1
4	4	1	2

5. For each $n \in \mathbf{Z}$, the set $n\mathbf{Z} = \{ni : i \in \mathbf{Z}\}$ is a subgroup of $\mathbf{Z}$ under addition. For example, if $n = 5$, then $5\mathbf{Z} = \{\ldots, -10, -5, 0, 5, 10, 15, \ldots\}$.
6. The set $K = \{R_0, R_2, H, V\}$ is a subgroup of D_4, the group of symmetries of the square. The set $C = \{R_0, R_1, R_2, R_3\}$ is a different subgroup of D_4, also of order 4.

7. The set $S = \{2^n : n \in \mathbf{Z}\} = \{\ldots, 2^{-3}, 2^{-2}, 2^{-1}, 2^0, 2^1, 2^2, 2^3, \ldots\}$ is closed under multiplication, $1 = 2^0$ is in S, and the inverse of each power of 2 is again a power of 2. Thus S is a nontrivial, proper subgroup of $\{\mathbf{Q}\setminus\{0\}, \cdot\}$, the group of nonzero rational numbers under multiplication.

8. Let S be the subset of matrices in $GL(\mathbf{R}, 2)$ of the form $\begin{bmatrix} a & 0 \\ 0 & d \end{bmatrix}$. (See Example 2 of Section 4.1.) The identity matrix $I = \begin{bmatrix} 1 & 0 \\ 0 & 1 \end{bmatrix}$ is in S. Let $A = \begin{bmatrix} a & 0 \\ 0 & d \end{bmatrix}$ be in S. Since the $\det(A) = ad$, neither a nor d is equal to 0. The inverse of A is $A^{-1} = \begin{bmatrix} a^{-1} & 0 \\ 0 & d^{-1} \end{bmatrix}$, which is also in S. Let $B \in S$ be the matrix $\begin{bmatrix} x & 0 \\ 0 & y \end{bmatrix}$. Then $AB = BA = \begin{bmatrix} ax & 0 \\ 0 & dy \end{bmatrix}$ and we see that S is closed under the operation $\times$. So S is a subgroup of $GL(\mathbf{R}, 2)$. Notice that while $GL(\mathbf{R}, 2)$ is not abelian, the subgroup S is abelian. ∎

If G is a finite group, the task of determining which subsets are actually subgroups is even easier, as the next proposition shows.

PROPOSITION 1 Let $\{G, *\}$ be a *finite group* and let H be a nonempty subset of G. If H is closed under the operation of G, then H is a subgroup of G.

Proof. We need to show that the inverse of each element $a \in H$ is also in H. Let $a \in H$. If $a = e$, then $a^{-1} = a$ and so $a^{-1} \in H$. Suppose $a \neq e$. The set of powers of a, namely the set $\{a, a^2, a^3, \ldots, a^n, \ldots\}$ is contained in H because H is closed under the group operation. Since G is finite, the powers of a are not all distinct. So we can find $i > j > 0$ such that $a^i = a^j$ or, equivalently, $a^{i-j} = e$. Since $a \neq e$, $i - j > 1$ and $i - j - 1 > 0$. Since $a^{i-j-1}a^1 = e$, we see that $a^{i-j-1} = a^{-1}$ and thus $a^{-1} \in H$. Also, since a and a^{-1} are in H, $e \in H$. ▲

EXAMPLE 9

Let $S = \{R_0, R_1, \ldots, R_{n-1}\} \subseteq D_n$, where R_i is rotation in the plane through an angle of $\frac{2\pi i}{n}$. The set S is a subgroup of D_n because the composition of two such rotations is another rotation in S. ∎

Next we show a way of creating subgroups within a given group.

DEFINITION 2. Suppose that a is an element of a group G. The **cyclic subgroup generated by** a is the set $\{a^n : n \in \mathbf{Z}\}$. It is denoted by $\langle a \rangle$.

Using the arithmetic of exponents, it is easy to check that $\langle a \rangle$, the set of *all* integer powers of a, is a subgroup of G. When we can find $a \in G$ such that $\langle a \rangle = G$, then G is called a **cyclic group** and the element a is called a **generator** of G.

EXAMPLES

9. Let $G = \{Q\setminus\{0\}, \cdot\}$ so that G is the multiplicative group of nonzero rational numbers. Then $\langle 2 \rangle = \{2^i : i \in Z\} = \{\ldots, 2^{-2}, 2^{-1}, 2^0, 2^1, 2^2, \ldots\}$. Notice that $\langle 2 \rangle = \langle 2^{-1} \rangle$. A cyclic subgroup may have more than one element that serves as a generator.

10. For a group $\{G, +\}$, the cyclic subgroup $\langle a \rangle = \{na : n \in Z\}$. For example, in $\{Z, +\}$, $\langle 3 \rangle = \{\ldots, -3, 0, 3, 6, 9, \ldots\}$. As a subgroup of Z, $\langle m \rangle = mZ$, the set of all multiples of m.

11. Let $G = \{Z_6, +\}$. Then $\langle 2 \rangle = \{0, 2, 4\}$ and $\langle 3 \rangle = \{0, 3\}$ and $\langle 5 \rangle = \{5, 2 \cdot 5, 3 \cdot 5, 4 \cdot 5, 5 \cdot 5\} = \{5, 4, 3, 2, 1\} = Z_6$. Thus 5 is a generator of Z_6 and so Z_6 is a cyclic group.

12. Let $G = Z$. Then $Z = \langle 1 \rangle = \langle -1 \rangle$. Thus Z is cyclic.

13. All the additive groups $\{Z_m, +\}$ are cyclic. In each case, $Z_m = \langle 1 \rangle$. There are other generators as well. For instance, $\{Z_5, +\} = \langle 2 \rangle = \langle 3 \rangle = \langle 4 \rangle$. However, $\{Z_6, +\} \neq \langle 2 \rangle$ since $\langle 2 \rangle = \{0, 2, 4\}$ in Z_6. ∎

Since every element of every group is a member of the cyclic subgroup that it generates, cyclic subgroups play an important role when we try to determine the structure of a group. We shall pursue that investigation in later sections. Next we expand on the notion of how to generate a subgroup. Instead of using one element $a \in G$ to generate a subgroup, we shall use a set of elements $A \subseteq G$. Before doing so, we need the following proposition, which we ask the reader to prove in the exercises.

PROPOSITION 2 If H_1 and H_2 are subgroups of G, then $H_1 \cap H_2$ is also a subgroup of G. More generally, the intersection of any collection of subgroups of G is a subgroup of G.

> **DEFINITION 3.** Let $\{G, *\}$ be a group and let A be a nonempty set of elements of G. The **subgroup generated by** A is the intersection of all subgroups of G that contain the set A. It is denoted by $\langle A \rangle$.

EXAMPLES

14. Let $A = \{R_1, F_A\}$ be a subset of D_3. The subgroup generated by A, namely $\langle R_1, F_A \rangle$, must contain the identity element R_0. Since it must be closed, it must contain $R_1 * R_1 = R_2$ and it must contain $R_1 * F_A = F_C$ and $R_1 * F_C = F_B$. The group D_3 is not cyclic, but the two-element set $\{R_1, F_A\}$ generates all of D_3. Thus $D_3 = \langle A \rangle$.

15. Let $G = \{Z, +\}$ and let $A = \{4, 6\}$. Since the sums and differences formed from the numbers 4 and 6 are even, it is clear that $\langle 4, 6 \rangle$ is a subset of the even numbers (i.e., $\langle 4, 6 \rangle \subseteq \langle 2 \rangle$). Since $4 \in \langle 4, 6 \rangle$ we must have $-4 \in \langle 4, 6 \rangle$ as well. Similarly, since $6 + (-4) = 2$, we have $2 \in \langle 4, 6 \rangle$. Thus $\langle 2 \rangle \subseteq \langle 4, 6 \rangle$.

The subgroup generated by the two-element set $\{4,6\}$ is in fact cyclic: $\langle 4, 6 \rangle = \langle 2 \rangle$. ∎

DEFINITION 4. Let $a \in G$. The **order** of a, denoted $|a|$, is the order of the subgroup generated by a. That is, $|a| = |\langle a \rangle|$.

Note that $|a|$ is a strictly positive integer or infinity.

EXAMPLES

16. In any group, the order of the identity element e is 1 because $\langle e \rangle = \{e\}$. Thus $|e| = 1$.
17. In $\{\mathbf{Z}, +\}$, the order of any nonzero element x is infinite because if $x \in \mathbf{Z}$ had finite order $m > 0$, then adding x to itself m times we get $(x + x + \ldots + x) = mx = 0$, which cannot be since $m \neq 0$.
18. The order of R_1 in D_3 is 3 since $\langle R_1 \rangle = \{R_0, R_1, R_2\}$. ∎

PROPOSITION 3 Suppose $a \in G$ and that $|a| = m$. Then m is the smallest strictly positive integer such that $a^m = e$.

Proof. First we show that there is some integer $t > 0$ such that $a^t = e$. Since $\langle a \rangle$ is a finite group, the powers of a cannot all be distinct. So we can find some integers i and j such that $i > j$ and $a^i = a^j$, or equivalently, $a^{i-j} = e$. Letting $t = i - j$, we have $a^t = e$. Let r be the smallest strictly positive integer for which $a^r = e$. Then the r members of the set $\{a^0, a^1, \ldots, a^{r-1}\}$ are distinct. Futhermore, if t is any integer, then $t = qr + i$ for some i such that $0 \leq i < r$. So $a^t = (a^r)^t a^i = a^i$, which shows that there are exactly r distinct members of $\langle a \rangle$. Therefore, $r = m$. ▲

COROLLARY 4 Suppose $a \in G$ and that $|a| = m$. Let n be any integer. Then $a^n = e$ if and only if n is a multiple of m.

Proof. We can find integers q and r such that $n = qm + r$ and $0 \leq r < m$. Then $a^n = a^r$. Since m is the smallest strictly positive integer such that $a^m = e$, $a^r = e$ if and only if $r = 0$. ▲

EXAMPLE 20

The order of 9 in $\{\mathbf{Z}_{12}, +\}$ is 4 since $\langle 9 \rangle = \{9, 2 \cdot 9, 3 \cdot 9, 4 \cdot 9\} = \{9, 6, 3, 0\}$. Similarly, the order of 8 in $\{\mathbf{Z}_{12}, +\}$ is 3. ∎

Product Groups

So far we have started with a group G and looked for its subgroups. Now we reverse our point of view. From groups G_1 and G_2, we will construct a bigger group that will, in some sense, contain G_1 and G_2 as subgroups. Suppose two groups G_1 and G_2 have operations $*_1$ and $*_2$ respectively. We can turn the Cartesian product $G_1 \times G_2$

into a group by multiplying componentwise, in much the same way as we add vectors in the plane. In other words, to multiply (x, y) and (s, t), we multiply x and s in G_1 and y and t in G_2:

$$(x, y) * (s, t) = (x *_1 s, y *_2 t).$$

The operation $*$ is associative because $*_1$ and $*_2$ are both associative. The identity element is (e_1, e_2), where e_1 and e_2 are the identity elements of G_1 and G_2, respectively. The inverse of any element (x, y) is (x^{-1}, y^{-1}) where x^{-1} and y^{-1} are the inverses of x and y in G_1 and G_2, respectively. Thus $\{G_1 \times G_2, *\}$ is a group, which leads us to the following definition.

> **DEFINITION 5.** Let $G_1 \times G_2$ be the Cartesian product of the groups G_1 and G_2 with operations $*_1$ and $*_2$ respectively. Let $*$ be the operation on $G_1 \times G_2$ defined by $(x, y) * (s, t) = (x *_1 s, y *_2 t)$. The resulting group $\{G_1 \times G_2, *\}$ is called the **direct product** of G_1 and G_2.

EXAMPLE 21

Let $G_1 = \{\mathbf{Z_2}, +\}$ and $G_2 = \{\mathbf{Z_4}, +\}$. The set of elements of $\mathbf{Z_2} \times \mathbf{Z_4}$ is $\{(0, 0), (1, 0), (0, 1), (1, 1), (0, 2), (1, 2), (0, 3), (1, 3)\}$. To add $(1, 3)$ and $(1, 2)$, we add $1 + 1$ in $\mathbf{Z_2}$ and $3 + 2$ in $\mathbf{Z_4}$. Thus $(1, 3) + (1, 2) = (0, 1)$. The resulting group is clearly abelian since it does not matter in which order we add the components. Although the group has order 8, it has no elements of order 8. The elements of highest order are all of order 4, e.g. $(1, 3) + (1, 3) + (1, 3) + (1, 3) = (0, 0)$. ∎

EXAMPLE 22

Let $G_1 = \{\mathbf{Z_4}, +\}$ and $G_2 = \{\mathbf{U_5}, *\}$. The resulting direct product $\mathbf{Z_4} \times \mathbf{U_5}$ is an abelian group of order 16. Here $(3, 3) * (3, 3) = (2, 4)$ because we add the first components in $\mathbf{Z_4}$ but we multiply the second components in $\mathbf{U_5}$. ∎

We can extend our notion of direct product to the Cartesian product of any finite set of groups, $G_1 \times G_2 \times \ldots \times G_n$. Products are performed componentwise.

EXAMPLE 23

The group $\mathbf{Z_2} \times \mathbf{Z_2} \times \mathbf{Z_2}$ is an abelian group with 8 elements, all of which have order 2. The elements of $\mathbf{Z_2} \times \mathbf{Z_2} \times \mathbf{Z_2}$ are $\{(0, 0, 0), (1, 0, 0), (0, 1, 0), (1, 1, 0), (0, 0, 1), (1, 0, 1), (0, 1, 1), (1, 1, 1)\}$. Any element added to itself is $(0, 0, 0)$. ∎

When each of the groups $G_1, G_2, \ldots, G_n$ uses additive notation, we sometimes denote $G_1 \times G_2 \times \ldots \times G_n$ by $G_1 \oplus G_2 \oplus \ldots \oplus G_n$ and call it the **direct sum** of $G_1, G_2, \ldots, G_n$. So the group given in Example 23 might also be expressed as $\mathbf{Z_2} \oplus \mathbf{Z_2} \oplus \mathbf{Z_2}$.

In any direct product, $G_1 \times G_2 \times \ldots \times G_n$ we can think of the set of elements of the form $\{(x, e_2, e_3, \ldots, e_n) : x \in G_1\}$ as a copy of G_1. In this way, the direct product

$G_1 \times G_2 \times \ldots \times G_n$ contains G_1, or at least a copy of it, as a subgroup. The other groups G_i can similarly be regarded as subgroups of $G_1 \times G_2 \times \ldots \times G_n$.

From Corollary 4, we know that if a is an element of order n in a group G, then $a^m = e$ if and only if m is a multiple of n. Let $x \in G_1$ be an element of order n_1 and $y \in G_2$ be an element of order n_2. In $G_1 \times G_2$, $(x, y)^m = (x^m, y^m)$ and so $(x, y)^m = (e_1, e_2)$ if and only if m is a multiple of both n_1 and n_2. The smallest such number is lcm(n_1, n_2). Thus the order of (x, y) in $G_1 \times G_2$ is lcm(n_1, n_2). We can easily extend these observations to elements of $G_1 \times G_2 \times \ldots \times G_n$ to obtain the following proposition.

PROPOSITION 5 Let $(x_1, x_2, \ldots, x_n)$ be an element of $G_1 \times G_2 \times \ldots \times G_n$ such that the order of x_i in G_i is n_i for each $i = 1, 2, \ldots, n$. Then the order of $(x_1, x_2, \ldots, x_n)$ in $G_1 \times G_2 \times \ldots \times G_n$ is lcm$(n_1, n_2, \ldots, n_n)$.

EXAMPLE 24

In $\mathbf{Z}_8 \times \mathbf{Z}_5$, the element $x = (2, 2)$ has order 20 because the order of 2 in $\mathbf{Z}_8$ is 4 and the order of 2 in $\mathbf{Z}_5$ is 5. However, if we regard $(2, 2)$ as an element of $\mathbf{Z}_8 \times \mathbf{Z}_{12}$, then it has order lcm$(4, 6) = 12$. ∎

SUMMARY

A **subgroup** is a group within a group. In this section, we looked at some very natural examples such as the subgroup of rotations in a dihedral group, the **trivial subgroup** $\{e\}$ in any group $\{G, *\}$, and $n\mathbf{Z} \subseteq \mathbf{Z}$, the subgroup of multiples of a fixed value n in $\mathbf{Z}$. We presented tests to check if a subset H of a group G is actually a subgroup. For instance, in the case that G is finite, all we need to do is check that H is closed under the group operation of G. Then we showed how to construct subgroups. If a is an element of a group G, then $\langle a \rangle$ is **the cyclic subgroup generated by a**. The **order of the element a** is the order of the subgroup $\langle a \rangle$, which is the smallest integer $n > 0$ such that $a^n = e$. More generally, if S is a nonempty subset of G, then $\langle S \rangle$ is the subgroup of G generated by S. It is the intersection of all subgroups of G that contain the elements of S. The Cartesian product of a set of n groups, $G_1 \times G_2 \times \ldots \times G_n$, is a group when we multiply elements component wise. We call it the **direct product** of $G_1, G_2, \ldots, G_n$.

As we have seen, there may be many groups of a given order. For instance, $\{\mathbf{Z}_6, +\}$ and $\{D_3, *\}$ are certainly different groups of order six. To sort out the differences, we often look to see what kind of subgroups we can find. The group $\{D_3, *\}$ has three distinct subgroups of order 2, but $\{\mathbf{Z}_6, +\}$ has only one subgroup of order 2. In the next section, we shall look more closely at the problem of how to decide whether two groups of the same order are intrinsically different or really the same group, except for notation.

4.4 Exercises

1. Determine several proper subgroups of $\{\mathbf{Q} \setminus \{0\}, \cdot\}$.
2. Find two distinct proper subgroups of the additive group $\{\mathbf{Z}_{12}, +\}$.

3. Find four distinct, proper, nontrivial subgroups of the group of symmetries of the equilateral triangle.
4. Show that the Klein 4-Group K has three proper subgroups of order 2, while the additive group $\{Z_4, +\}$ has only one.
5. Show that if H_1 and H_2 are subgroups of G, then $H_1 \cap H_2$ is also a subgroup of G. More generally, prove the intersection of any collection of subgroups of G is a subgroup of G. (This is Proposition 2 of this section.)
6. Consider the group $\{Z_n, +\}$ and let m be a positive integer. Let $H = \{mx : x \in Z_n\}$. Show that H is a subgroup of Z_n. Construct H for
   i. $n = 8, m = 4$
   ii. $n = 8, m = 3$
   iii. $n = 12, m = 3$
   iv. $n = 12, m = 5$
   What conditions on n and m ensure that H is a proper subgroup?
7. Show that the set $\{A \in GL(\mathbf{R}, 2) : \det(A) = 1\}$ is a subgroup of $GL(\mathbf{R}, 2)$.
8. Determine whether or not the set of matrices in $GL(\mathbf{R}, 2)$ of the form $\begin{bmatrix} a & 0 \\ c & d \end{bmatrix}$ is a subgroup of $GL(\mathbf{R}, 2)$.
9. Let G be a group and let H be a nonempty subset of G. Suppose that whenever x and y are elements of H, then $x^{-1}y \in H$. Prove that H is a subgroup of G.
10. Let $\{G, *\}$ be a group. Let $Z(G) = \{x \in G : xy = yx \text{ for each } y \in G\}$. Prove that $Z(G)$ is a subgroup of G. The group $Z(G)$ is called the **center** of G. It is the set of all elements that commute with all elements in G.

    a. Find $Z(G)$ for $G = D_3$ and for $G = D_4$.
    b. What is the center of an abelian group?

11. Find the elements of the cyclic subgroup $\langle 3 \rangle$ in $\{Z_{12}, +\}$.
12. Find $\langle 5 \rangle$ in $\{Z_{12}, +\}$.
13. Find $\langle 5 \rangle$ in the multiplicative group U_{12}.
14. Find $\langle 2, 5 \rangle$ in $\{Z, +\}$.
15. Let $A = \begin{bmatrix} 2 & 0 \\ 0 & 3 \end{bmatrix}$. Characterize the elements in $\langle A \rangle$, the subgroup generated by A in $GL(\mathbf{R}, 2)$.
16. Suppose that $a \neq 0$ is a real number and let $A = \begin{bmatrix} a & 0 \\ a & a \end{bmatrix}$. Characterize the elements in $\langle A \rangle$, the subgroup generated by A in $GL(\mathbf{R}, 2)$.
17. Find $\langle H, V \rangle$ in D_4.
18. Determine when $\langle m, n \rangle = \{Z, +\}$.
19. Find c such that $\langle m, n \rangle = \langle c \rangle$ in $\{Z, +\}$.
20. Determine the order of 3 in Z_{12}. Determine the order of 9 in Z_{12}.
21. Find the order of 3 in Z_7. Find the order of 7 in Z_8. Find the order of 6 in Z_8. In general, what do you think the order of m is in Z_n?
22. Find the order of 5 in U_7 and of 5 in U_{12}.

23. Suppose that a is an element of a group G and that $|a| = m$. Let i be an integer. What is the order of a^i? Express your answer in terms of $\gcd(i, m)$.
24. Is U_{12} cyclic?
25. Is U_{20} cyclic?
26. Construct a group table for $Z_2 \times Z_3$. Is $Z_2 \times Z_3$ cyclic?
27. Find the order of each element in $U_8 \times Z_2$. Is $U_8 \times Z_2$ cyclic?
28. Prove that if G is abelian, then $|ab|$ divides $\text{lcm}(|a|, |b|)$. Give a counterexample to show that this is not the case in a nonabelian group.
29. Find all the subgroups of $U_8 \times Z_5$.
30. Show that if H_1 is a subgroup of G_1 and H_2 is a subgroup of G_2, then $H_1 \times H_2$ is a subgroup of $G_1 \times G_2$. Let H be a subgroup of $G_1 \times G_2$. Is H necessarily of the form $H_1 \times H_2$?
31. Prove that in any group, an element and its inverse have the same order.
32. Find all the cyclic subgroups of D_4 and of D_3.
33. Suppose that x and y are elements of a group G and that $xy = yx$ (x and y commute). Suppose that $|x| = m$ and $|y| = n$. Prove that the order of xy is $\text{lcm}(m, n)$.
34. Let G be a group and H a subgroup of G. Define a relation $\sim$ on G as follows: $x \sim y$ if and only if $y^{-1}x \in H$. Prove that $\sim$ is an equivalence relation.

To the Teacher

Students encounter subgroups every time they use a multiplication table. The "7-times-table" is just part of $7\mathbf{Z}$, a subgroup of $\mathbf{Z}$. They know that 0 is a multiple of 7, that if x is a multiple of 7, then so is $-x$, and that if x and y are both multiples of 7, so $x + y$. All the criteria for a subgroup come naturally. Here are some questions that might make multiplication tables more interesting.

1. If $x = 8$ and $y = 12$ are in a certain multiplication table, what other numbers must be in the table? What are the multiplication tables that this could be?
2. Repeat question 1 for $x = 3$ and $y = 5$.

In this section, we would rephrase the questions 1 and 2 as, "What subgroup of $\mathbf{Z}$ is generated by x and y?" Group theory and its definitions capture familiar scenarios. The answers to the questions rest in number theory. In each case, the key is $\gcd(x, y)$. In the first case, the $\gcd(8, 12)$ is 4 so the table could be the 4-times table or the 2-times table. In the second, it is the 1-times table since $\gcd(3, 5) = 1$. In the latter case, we might say that x and y do not generate a proper subgroup. The process of discovering the answers to questions such as 1 and 2 deliver some good lessons.

- Students will notice that if x and y are in a table, then (extrapolating to the infinitely long times tables) so are their negatives, sums and differences and most surprisingly to them, so is their gcd. Most students do not think of $\gcd(x, y)$ as a number to be expressed as a sum.

- The activity of generating a multiplication table from just two given numbers is an inductive process where sums and differences must be taken of sums and differences previously obtained. This is probably a new skill for the high school student. It is a crucial lesson for future computer scientists. It should prompt some serious questions, most especially, "When do you know you are done?"

Variations can be carried out in $\{Z_n, +\}$ under the guise of clock-arithmetic if necessary.

3. What is the 2-times table mod 6?
4. What is the 2-times table mod 5?

In the language of this section, these questions ask for the subgroup generated by 2 in $\{Z_6, +\}$ and $\{Z_5, +\}$, respectively. Answering questions like 3 and 4 enhances a student's number sense. Why are the answers so different? We know, and the student will discover, that the answer lies in whether 2 shares a factor with the modulus or not.

Task

Develop a manipulative through which students can discover subgroups of $\{Z_n, +\}$. (Beaded necklaces might work.)

4.5 ISOMORPHISM AND HOMOMORPHISM

In the previous sections, we encountered several groups of order 4, for instance $\{Z_4, +\}$, $\{U_8, *\}$, the subgroup $A = \{R_0, R_1, R_2, R_3\}$ of D_4, and the subgroup $B = \{R_0, R_2, H, V\}$ of D_4. The groups Z_4 and A are similar in that both are cyclic: $Z_4 = \langle 1 \rangle$ and $A = \langle R_1 \rangle$. The groups U_8 and B are similar because in each, the product of any element x with itself is the identity. The similarity of A to Z_4 and the similarity of U_8 to B is even deeper. If we rename the elements of A by $R_0 = 0, R_1 = 1, R_2 = 2, R_3 = 3$, and write out the Cayley table for A, it is indistinguishable from the Cayley table for Z_4. (See Figure 1.)

$A, *$	0	1	2	3
0	0	1	2	3
1	1	2	3	0
2	2	3	0	1
3	3	0	1	2

$Z_4, +$	0	1	2	3
0	0	1	2	3
1	1	2	3	0
2	2	3	0	1
3	3	0	1	2

FIGURE 1

The groups $\{Z_4, +\}$ and $\{A, *\}$ differ in notation and interpretation but not in size or structure. They are essentially the same group. A similar correspondence exists between the groups U_8 and B, but **not** between A and B. We now develop tools with which we can determine when two seemingly different groups have the same group structure.

> **DEFINITION 1.** Let $\cdot_1$ be a binary operation on a set S and let $\cdot_2$ be a binary operation on a set T. A function $f : S \to T$ is called **operation preserving** if $f(a \cdot_1 b) = f(a) \cdot_2 f(b)$ for any two elements a and b in S.

EXAMPLES

1. In this example, we let S be the set of real numbers under addition and T be the set of real numbers under multiplication. Thus $S = T = \mathbf{R}$ but $\cdot_1$ is addition and $\cdot_2$ is multiplication. Let $f : S \to T$ be defined by $f(x) = 2^x$. To check that f is operation preserving, note that $f(x + y) = 2^{x+y} = 2^x 2^y = f(x)f(y)$. (We take advantage of operation preserving maps in our everyday arithmetic.)

2. Consider $\mathbf{Z}$ under the operation of addition. The map $f : \mathbf{Z} \to \mathbf{Z}$ defined by $f(x) = 2x$ is operation preserving since $f(x + y) = 2(x + y) = 2x + 2y = f(x) + f(y)$. However, if we consider the same map f but switch to the operation of multiplication, it is **not** operation preserving since $f(xy) = 2xy$ but $f(x)f(y) = 2x2y = 4xy$.

3. The map $f : \mathbf{Z} \to \mathbf{Z}_m$ defined by $f(x) = [x]_m$ preserves both the operations of addition and multiplication because $[x + y]_m = [x]_m + [y]_m$ and $[xy]_m = [x]_m \cdot [y]_m$. ∎

Our goal—to detect when two groups differ only in notation rather than in size or structure—is accomplished when the criteria for the next definition are met.

> **DEFINITION 2.** Groups $\{G_1, \cdot_1\}$ and $\{G_2, \cdot_2\}$ are said to be **isomorphic** if there exists a function $f : G_1 \to G_2$ such that
>
> i. f is a bijection,
> ii. f is operation preserving.
>
> The function f is called an **isomorphism**.

Sometimes we write $G_1 \sim G_2$ when G_1 and G_2 are isomorphic. If $\{G_1, \cdot_1\}$ and $\{G_2, \cdot_2\}$ are isomorphic, the sets G_1 and G_2 have the same size and same algebraic structure. They differ only in notation.

EXAMPLES

4. The tables in Figure 1 show that the groups $\{Z_4, +\}$ and $\{A, *\}$ are isomorphic because the bijection $f : Z_4 \to A$ given by $f(i) = R_i$ is operation preserving.

5. Let $\{E, +\}$ denote the group of even integers under addition. The groups $\{E, +\}$ and $\{Z, +\}$ are isomorphic. The function $f : Z \to E$ defined by $f(x) = 2x$ is a bijection and it is operation preserving as demonstrated in Example 1.
6. Let $B = \{R_0, R_2, H, V\}$, a subgroup of D_4. The groups B and $\{U_8, *\}$ are isomorphic. The bijection f is defined by the correspondence

$$R_0 \to 1$$
$$R_2 \to 3$$
$$H \to 5$$
$$V \to 7.$$

A comparison of the tables in Figure 2 shows that f is operation preserving because the correspondence established by the bijection is maintained in each of the corresponding positions in the tables. ■

$B, *$	R_0	R_2	H	V
R_0	R_0	R_2	H	V
R_2	R_2	R_0	V	H
H	H	V	R_0	R_2
V	V	H	R_2	R_0

$U_8, \cdot$	1	3	5	7
1	1	3	5	7
3	3	1	7	5
5	5	7	1	3
7	7	5	3	1

FIGURE 2

Not every bijection between two groups of order n is an isomorphism. Since there are $n!$ different bijections between any two sets of n elements, a search for which, if any, are actually isomorphisms could be very time consuming. Any information that narrows down the possibilities is useful. The next two propositions give us help in that direction.

PROPOSITION 1 Suppose that $\{G, \cdot_1\}$ and $\{H, \cdot_2\}$ are groups with identity elements e_G and e_H respectively. Suppose that $f : G \to H$ is an isomorphism. Then the following statements hold.

i. $f(e_G) = e_H$.
ii. For any element x in G, $f(x^{-1}) = (f(x))^{-1}$.

Proof of i. Let x be any element of G. Then $e_H \cdot_2 f(x) = f(x) = f(e_G \cdot_1 x) = f(e_G) \cdot_2 f(x)$. Since $e_H \cdot_2 f(x) = f(e_G) \cdot_2 f(x)$, we can multiply by $f(x)^{-1}$ on the right of each side to obtain $e_H = f(e_G)$.

ii. Exercise 7. ▲

Now we know that isomorphisms must map identities to identities and inverses to inverses. But there is much more to say. Isomorphic groups $\{G, \bullet_1\}$ and $\{H, \bullet_2\}$ have identical structures. If G has an element of order m, then so does H. If G is abelian, then so is H, and so forth. These facts (some of which are formalized in the next proposition) help us determine when two groups are **not** isomorphic. For instance, $\{Z_6, +\}$ and D_3, both of order 6, cannot be isomorphic since one is abelian and the other is not. Similarly, $\{Z_4, +\}$ and the Klein 4-Group cannot be isomorphic since Z_4 has an element of order 4 and the Klein 4-Group does not.

PROPOSITION 2 Suppose that $\{G, \bullet_1\}$ and $\{H, \bullet_2\}$ are groups and that $f: G \to H$ is an isomorphism.

i. The order of x in G is the same as the order of $f(x)$ in H.
ii. If K is a subgroup of G, then $f(K) = \{f(x): x \in K\}$ is a subgroup of H.
iii. If G is abelian, then so is H.
iv. If G is cyclic, then so is H.

Proof of i. Suppose that $x \in G$ and that the order of x is n. Since $x^n = e_G$, we have $f(e_G) = e_H = f(x^n) = (f(x))^n$ by Proposition 1. If $0 < m < n$, then $x^m \neq e_G$. Thus $f(x)^m = f(x^m)$, and $f(x^m) \neq e_H$ because f is a one-to-one function. Thus n is the smallest strictly positive integer such that $(f(x))^n = e_H$.

The proofs of parts ii through iv are left as exercises. ▲

EXAMPLE 7

The subgroups $A = \{R_0, R_1, R_2, R_3\}$ and $B = \{R_0, R_2, H, V\}$ of D_4 are **not** isomorphic. In B, the squares of all elements are equal to the identity element, but in A they are not. ■

We think of two isomorphic groups as being really the same group. The mathematical construction that helps us to classify objects that are somehow the same is the equivalence relation.

PROPOSITION 3 Isomorphism on the class of all groups is an equivalence relation.

Proof. A group $\{G, \bullet_1\}$ is related to a group $\{H, \bullet_2\}$ if there is an isomorphism from G to H. We show that the relation is reflexive, transitive, and symmetric.

Reflexivity: For any group $\{G, *\}$, the identity map $id_G: G \to G$, where $id_G(x) = x$, is an operation preserving bijection. So G is related to itself.

Symmetry: Suppose that $\{G, \bullet_1\}$ and $\{H, \bullet_2\}$ are groups and that G is related to H. Let $f: G \to H$ be an isomorphism. Since f is a bijection, it has an inverse such that $f^{-1}: H \to G$, which is also a bijection. We will show that f^{-1} is operation preserving and hence an isomorphism from H to G. In so doing we show that H is related to G, from which we can conclude that the relation is symmetric. Let u and v be elements of H. Since f is a bijection, we can find elements x and y in

G, such that $f(x) = u$ and $f(y) = v$. Let $f^{-1}: H \to G$ be the inverse of f so that $f^{-1}(f(x)) = x$. The map f^{-1} is a bijection. It is operation preserving since,

$$f^{-1}(u \bullet_2 v) = f^{-1}(f(x) \bullet_2 f(y)) = f^{-1}(f(x \bullet_1 y)) = x \bullet_1 y =$$
$$f^{-1}(f(x)) \bullet_1 f^{-1}(f(y)) = f^{-1}(u) \bullet_1 f^{-1}(v).$$

Transitivity: Let $\{G, \bullet_1\}$, $\{H, \bullet_2\}$ and $\{K, \bullet_3\}$ be groups. Suppose that $f: G \to H$ and $g: H \to K$ are isomorphisms so that G is related to H and H is related to K. We will show that the composition $g \circ f: G \to K$ is an isomorphism, thus showing that G is related to K. The composition $g \circ f: G \to K$ is a bijection. It is operation preserving because, for each pair of elements x and y in G,

$$g \circ f(x \bullet_1 y) = g(f(x \bullet_1 y)) = g(f(x) \bullet_2 f(y)) = g(f(x)) \bullet_3 g(f(y)) =$$
$$g \circ f(x) \bullet_3 g \circ f(y). \quad \blacktriangle$$

We can think of all groups within one equivalence class as the same group. "There is only one group of order three," means that all groups of order three are isomorphic to each other. For any prime p, there is only one group of order p, the cyclic group. There are two groups of order four because any group of order four is isomorphic to $\mathbf{Z_4}$ or to the Klein 4-Group. One of the great challenges of mathematics has been to classify all finite groups. Given an integer n, how many different finite groups are there of that order?

Homomorphism

The map $f: \mathbf{Z} \to \mathbf{Z}$ defined by $f(x) = 3x$ preserves the operation of addition but it is not an isomorphism since it is not surjective. Similarly, the map $f: \mathbf{Z} \to \mathbf{Z_3}$ defined by $f(x) = [x]_3$ preserves addition but it is not injective. Such maps, called **homomorphisms**, are tools to investigate the structural relationships between groups that are not necessarily identical. We can also use homomorphisms to construct new groups from known groups, as we shall see in Section 5.2.

> **DEFINITION 3.** Let G_1 and G_2 be groups with operations $\bullet_1$ and $\bullet_2$, with operations respectively. An operation preserving map $f: G_1 \to G_2$ is called a **homomorphism**.

Every isomorphism is a homomorphism but not conversely, as the next examples show.

EXAMPLES

8. There is a homomorphism $f: G_1 \to G_2$ between any pair of groups $\{G_1, \bullet_1\}$ and $\{G_2, \bullet_2\}$ defined by $f(x) = e_{G_2}$ for all $x \in G_1$. It is called the **trivial homomorphism**. (It is not very informative.)

9. Consider the groups $\{Z_6, +\}$ and $\{Z_3, +\}$. In this example, we show that the map $f : Z_6 \to Z_3$ defined by $f([x]_6) = [x]_3$ is a homomorphism. First we make sure that f is well-defined. If $[x]_6 = [y]_6$, then $x = y + 6m$ for some integer m so that $x = y + 3(2m)$. Thus $[x]_3 = [y]_3$ and $f([x]_6) = f([y]_6)$. Now we show that f is operation preserving. Let $[x]_6$ and $[y]_6$ be any two elements in Z_6. Then $f([x]_6 + [y]_6) = f([x+y]_6) = [x+y]_3 = [x]_3 + [y]_3 = f([x]_6) + f([y]_6)$. Thus f is a homomorphism.

10. Let M be a $n \times m$ matrix and let $f : R^m \to R^n$ be the linear transformation defined by $f(\mathbf{x}) = M\mathbf{x}$ for each vector $\mathbf{x}$ in R^m. (The operation is matrix-vector multiplication.) Since $M(\mathbf{x} + \mathbf{y}) = M\mathbf{x} + M\mathbf{y}$, the map is a homomorphism. It is an isomorphism if and only if M is a nonsingular square matrix, i.e., a matrix with nonzero determinant. When M is nonsingular, the equation $M\mathbf{x} = \mathbf{y}$ has a unique solution for each $\mathbf{y}$ in R^n, thus ensuring that f is a bijection and hence an isomorphism. ∎

The results of Proposition 1 hold for homomorphisms as well as for isomorphisms with no change of proof because the proofs rely only on the assumption that the map is operation preserving. Thus for any homomorphism, identities map to identities and inverses map to inverses. But not all of the results of Proposition 2 hold. For instance, a homomorphism does not necessarily preserve the order of an element. The homomorphism $f : Z \to Z_m$ defined by $f(x) = x$ mod m maps any nonzero number (of infinite order) in Z to a number of order m (or less) in Z_m. Some of the properties of homomorphisms are summarized in the next proposition.

PROPOSITION 4 Suppose that $f : G_1 \to G_2$ is a homomorphism between the groups $\{G_1, \cdot_1\}$ and $\{G_2, \cdot_2\}$. Let K be a subgroup of G_1, and H be a subgroup of G_2. Then

  i. If e is the identity element of G_1, then $f(e)$ is the identity element of G_2.
  ii. $f(x^{-1}) = (f(x))^{-1}$.
  iii. $f(K) = \{f(x) : x \in K\}$ is a subgroup of G_2.
  iv. $f^{-1}(H) = \{x \in G_1 : f(x) \in H\}$ is a subgroup of G_1.

The proofs of parts i and ii are identical to those given in Proposition 3. The proof of part iii is left as an exercise. We prove part iv.

Proof of iv. Suppose that x and y are elements of the set $f^{-1}(H)$. We can find elements s and t in H such that $f(x) = s$ and $f(y) = t$. Since f is operation preserving, $f(x \cdot_1 y) = s \cdot_2 t$. Since H is a subgroup of G_2, $s \cdot_2 t \in H$ and so $x \cdot_1 y \in f^{-1}(H)$. Thus $f^{-1}(H)$ is closed under $\cdot_1$. Since $e_2 \in H$, and since $f(e_1) = e_2$, we have $e_1 \in f^{-1}(H)$. If $f(x) = y$ and $y \in H$, then $f(x^{-1}) = y^{-1}$ so that $x^{-1} \in f^{-1}(H)$. Thus $f^{-1}(H)$ is a subgroup of G_1. ▲

EXAMPLE 11

Let $f : Z \to Z_6$ be defined by $f(x) = x$ mod 6. The set $H = \{0, 3\}$ is a subgroup of Z_6. The pre-image $f^{-1}(H) = \{3x : x \in Z\}$ is a subgroup of Z, namely $3Z$. ∎

By Proposition 4, it is always the case that $f^{-1}(\{e_2\})$ is a subgroup of G_1 since $\{e_2\}$ is a subgroup of G_2.

> **DEFINITION.** Let $f : G_1 \to G_2$ be a homomorphism. The **kernel of f** is the set $\{x \in G_1 : f(x) = e_2\}$. It is denoted by **ker(f)**.

EXAMPLE 12

The kernel of the map given in Example 11 is $6\mathbf{Z}$.

Example 10 is drawn from linear algebra, where the kernel of a linear transformation associated with a matrix M is defined as the set of vectors $\{\mathbf{x} : M\mathbf{x} = \mathbf{0}\}$. Our use of the term kernel agrees with this usage. You might recall that the kernel of a linear transformation from $\mathbf{R}^m$ to $\mathbf{R}^n$ is a subspace of $\mathbf{R}^m$ and that a linear transformation is injective if and only if its kernel is the single vector $\mathbf{0}$. Our next theorem shows how these facts about kernels extend to group homomorphisms in general. In Chapter 5 we shall use this special subgroup to construct new groups called quotient groups in much the same way as we constructed the group $\{\mathbf{Z}_m, +\}$ from $\mathbf{Z}$.

THEOREM 5 Let G and H be groups and suppose that $f : G \to H$ is a group homomorphism.

  i. Ker(f) is a subgroup of G.
  ii. The map f is injective if and only if ker(f) = $\{e_G\}$.

Proof of Part i. (The proof of part ii is left as an exercise.) By Definition 1 of Section 4.4, we need to check that ker(f) is closed under the operation of G and that if $x \in \ker(f)$, then $x^{-1} \in \ker(f)$. Let x and y be elements of ker(f). Since $f(xy) = f(x)f(y) = e_H e_H = e_H$, the product xy is also in ker(f). Thus ker(f) is closed under the operation of G. Similarly, for each x in ker(f), we have $f(x^{-1}) = (f(x))^{-1} = (e_H)^{-1} = e_H$. So $x^{-1} \in \ker(f)$. Thus ker(f) is a subgroup of G. ▲

SUMMARY

An **isomorphism** between two groups is an **operation preserving** bijection. When two groups are isomorphic, they are essentially identical, differing only in notation but not in structure. Not surprisingly, we saw that all group properties that we could describe are identical in two isomorphic groups. For instance, corresponding elements have the same order and subgroups correspond. We can sort groups into **isomorphism classes**. A **homomorphism** is also an operation preserving map. We use homomorphism to relate the structures of two given groups. The **kernel of f**, denoted ker(f), is a special subgroup of G. Its structure tells us when a homomorphism is injective.

4.5 Exercises

1. Show that the set of rotations of the equilateral triangle, $\{R_0, R_1, R_2\}$ is a group under $*$. Show that this group is isomorphic to $\{\mathbf{Z_3}, +\}$.

2. Show that the group $\{U_7, *\}$ is isomorphic to $\{Z_6, +\}$.
3. Show that there is an isomorphism between the Klein 4-Group and the subgroup $B = \{R_0, R_2, H, V\}$ of D_3.
4. In Exercise 15 of Section 4.1, you obtained different group tables for the four elements $\{e, a, b, c\}$ and hence different groups. Are any pairs of these isomorphic?
5. How many different isomorphism equivalence classes are there for groups with three elements? four elements? five elements?
6. The group of symmetries of the regular hexagon, D_6, and the group of symmetries of the tetrahedron T are both nonabelian groups of order 12. Are they isomorphic? If so, find an explicit isomorphism. If not, indicate why not.
7. Prove part ii of Proposition 1.
8. Prove parts ii through iv of Proposition 2.
9. Prove that any two cyclic groups of the same order are isomorphic.
10. Find a homomorphism between the additive groups Z_3 and Z_6.
11. Find a homomorphism between the additive groups Z_9 and Z_3.
12. Can you find a nontrivial homomorphism from Z_6 to Z_4? From Z_6 to Z_5?
13. Show that all homomorphisms from Z_m to Z_n are of the form $f(x) = qx \bmod n$ for some $q \in Z_n$. Show that not every map from Z_m to Z_n of the form $f(x) = qx \bmod n$ is a homomorphism.
14. Show that the map $f : Z_m \to Z_n$ is a homomorphism if and only if n divides qm, where $f(x) = qx \bmod n$. (See Exercise 13.)
15. Let $f : G_1 \to G_2$ be a homomorphism and let $x \in G_1$. Suppose that that the order of x is n so that $x^n = e_1$. Prove that the order of $f(x)$ divides n. (You may want to use the Division Algorithm.)
16. Prove part iii of Proposition 4.
17. Prove part ii of Theorem 5.

To the Teacher

Isomorphism is an important concept. Through it, we see when algebraic structures are alike, despite notations that may differ. An informal sense of isomorphism comes naturally to students. They will grasp readily that the set of rotational symmetries of a triangle is really mod 3 addition (perhaps not by name) or that the group given in Example 4 of Section 4.1 is no different from the rotations of a square. (The soldier, viewed from above, might be modeled by a square.) We have already seen that utilizing the isomorphism between, for instance, the set of rotations of a tangible cut-out hexagon is an excellent way to explore mod 6 arithmetic, with all its similarities and differences to integer arithmetic. (See the *In the Classroom* essay, Section 4.9, for a more thorough discussion.)

We confirm isomorphic structure through operation-preserving maps. This idea is not so remote from the high school classroom as it might first seem. Students utilize an operation-preserving map every time they use the rules of exponents (e.g., $a^i a^j = a^{i+j}$). The isomorphism is the map $i \to a^i$ from $\mathbf{Z}$ to the set of powers of a. Once it is established that a^3 means $a \cdot a \cdot a$, the fact that $a^{i+j} = a^i a^j$ follows naturally for positive exponents. (Addition is preserved.) But why use a^{-1} to denote the multiplicative inverse of a? You know that isomorphisms take inverses to inverses. So to extend the isomorphism to all integers, -1 must map to the inverse of a since 1 maps to a. Why is $a^0 = 1$? Since $1 = a^{-1}a^1 = a^0$. Other rules of exponents follow readily from the underlying isomorphism, e.g., $(a^{-1})^n = a^{-n}$ and $\frac{a^i}{a^j} = a^{i-j}$. As teachers we can appreciate just how clever exponential notation is. It transforms multiplication and division into easy addition and subtraction. A look at mathematics without such notation can be found at the beginning of Section 3.10, a historical essay about notation. The older expressions are extremely difficult to decipher because so often we almost completely identify the concepts of our mathematics with the notation for those concepts. Teachers must strike a balance as they ask students to internalize, exploit and respond automatically (but correctly) to notational prompts while always reinforcing the conceptual rigor that supports computation. (Rote and reason go hand and hand.)

Task

Young students often have manipulatives with which they can model the arithmetic of positive and negative numbers. Modify your favorite set to model the arithmetic of exponents.

4.6 CYCLIC GROUPS

In this section we take a closer look at cyclic groups. Recall that $\{G, *\}$ is a **cyclic group** if it contains an element a such that $G = \langle a \rangle = \{a^n : n \in \mathbf{Z}\}$. A multiplicative cyclic group is simply the set of all powers of one of its elements. If $\{G, +\}$ is an additive cyclic group, then $G = \{na : n \in \mathbf{Z}\}$. The element a is called a **generator** of G. We already know most of the essential facts about cyclic groups because, as we shall see, each cyclic group of order n is isomorphic to $\{\mathbf{Z}_n, +\}$ and each infinite cyclic group is isomorphic to $\{\mathbf{Z}, +\}$. Our task is to translate what we know about $\{\mathbf{Z}_n, +\}$ and $\{\mathbf{Z}, +\}$ from number theory into the language of groups.

EXAMPLES

1. $\{\mathbf{Z}, +\}$ has two generators, 1 and -1. Thus $\mathbf{Z} = \langle 1 \rangle = \langle -1 \rangle$. There are no other elements that generate $\mathbf{Z}$ because unless $x = 1$ or -1, there is no way to express 1 as nx for $n \in \mathbf{Z}$.
2. $\{\mathbf{Z}_n, +\}$ is generated by $x = 1$. What its other generators are depends on n, as we shall soon investigate.
3. The additive group $\{\mathbf{Z}_8, +\}$ has four different generators:
$\mathbf{Z}_8 = \langle 1 \rangle = \langle 3 \rangle = \langle 5 \rangle = \langle 7 \rangle$.
Let's check that 3 is indeed a generator.
$3 + 3 = 2 \cdot 3 = \mathbf{6}$ mod 8,
$3 + 3 + 3 = 3 \cdot 3 = 9 = \mathbf{1}$ mod 8,
$4 \cdot 3 = 12 = \mathbf{4}$ mod 8,
$5 \cdot 3 = 15 = \mathbf{7}$ mod 8,

$6 \cdot 3 = 18 = 2 \mod 8$,
$7 \cdot 3 = 21 = 5 \mod 8$,
$8 \cdot 3 = 24 = 0 \mod 8$.

So the set of multiples of 3, taken mod 8, is the set $\{3, 6, 1, 4, 7, 2, 5, 0\} = \mathbf{Z_8}$. ∎

First, we investigate finite cyclic groups. In Chapter 1, we learned that $\operatorname{lcm}(n, x) = \frac{nx}{\gcd(n, x)}$. So the smallest positive integer q for which qx is divisible by n is $q = \left(\frac{n}{\gcd(n, x)}\right)$. This fact is translated into the language of groups in the next proposition. It tells us about the order of elements in $\{\mathbf{Z_n}, +\}$.

PROPOSITION 1 Suppose $n > 0$ and that x is an integer such that $\gcd(x, n) = d$. The order of the element x in $\{\mathbf{Z_n}, +\}$ is $\frac{n}{d}$.

Proof. The smallest integer q such that qx is divisible by n is $q = \frac{n}{d}$. Thus the order of x is $\frac{n}{d}$. ▲

COROLLARY 2 An element $x \in \mathbf{Z_n}$ generates $\{\mathbf{Z_n}, +\}$ if and only if $\gcd(x, n) = 1$.

Proof. An element $x \in \mathbf{Z_n}$ generates the group $\{\mathbf{Z_n}, +\}$ if and only if it has order n. An element $x \in \mathbf{Z_n}$ has order n if and only if $\gcd(x, n) = 1$. ▲

Now let's look at finite cyclic groups that use multiplicative notation. The analysis is similar, but this time it is carried out on the exponent rather than the coefficient of the generator. The next proposition gives us information about when two elements a^i and a^j in a (multiplicative) cyclic group are equal.

PROPOSITION 3 Let $G = \langle a \rangle$ and suppose that $|G| = m$. Then $a^i = a^j$ if and only if $j - i$ is a multiple of m.

Proof. Since $|G| = m$, the order of its generator a is also m. Suppose that $j \geq i$ and that $j - i = pm$. Then $a^{j-i} = (a^m)^p = e$. Multiplying a^{j-i} by a^i, we have $a^j = a^{j-i}a^i = ea^i$ and so $a^j = a^i$. Conversely, suppose that $a^j = a^i$ so that $a^{j-i} = e$. From the Division Algorithm, $(j - i) = pm + k$ where $0 \leq k < m$. Thus $e = a^{j-i} = a^{pm}a^k = (a^m)^p a^k = a^k$. Since $0 \leq k < m$ and since the order of a is m, we conclude that $k = 0$ and $j - i = pm$. ▲

Now we are ready to prove that all finite cyclic groups of order m are isomorphic to $\mathbf{Z_m}$. The isomorphism will allow us to transfer what we know about $\mathbf{Z_m}$ to all cyclic groups.

PROPOSITION 4 Any finite cyclic group of order m is isomorphic to $\{\mathbf{Z_m}, +\}$.

Proof. Assume that $G = \langle a \rangle$ is a finite cyclic group of order m. Let $f: \mathbf{Z_m} \to G$ be defined by $f([i]_m) = a^i$. First we show that the map is well-defined. Suppose

$[i]_m = [j]_m$. Then $i = j + mq$ for some integer q. So $a^i = a^{j+mq} = a^j a^{mq} = a^j$ because a has order m. Thus $f([i]_m) = f([j]_m)$. The map f is surjective because $G = \{a^0, a^1, \ldots, a^{m-1}\} = \{f([0]_m), f([1]_m), \ldots, f([m-1]_m)\}$. It is injective because $a^i = a^j$ if and only if $j - i$ is divisible by m (i.e., $[i]_m = [j]_m$). To see that f is operation preserving, observe that

$$f([i]_m) f([j]_m) = a^i a^j = a^{i+j} = f([i+j]_m) = f([i]_m + [j]_m).$$

Thus f is an isomorphism. ▲

Here are two facts that we can deduce from the fact that a finite cyclic group $\langle a \rangle$ of order m is isomorphic to $\{\mathbf{Z}_m, +\}$.

i. If $G = \langle a \rangle$ and $|G| = m$, then a^i generates G if and only if i and m are relatively prime.
ii. If $G = \langle a \rangle$ and $|G| = m$, then the order of a^i is $\frac{m}{\gcd(m,i)}$.

Part ii of Proposition 4 tells us that the order of any element in a cyclic group divides the order of the group. This is a very natural consequence of number theory. But we shall see in Chapter 5, the conclusion—that the order of an element of a group divides the order of a group—holds for all groups.

Now let's look at infinite cyclic groups.

PROPOSITION 5 Let $G = \langle a \rangle$ and suppose that G is an infinite group. Then all the powers of a are distinct.

Proof. Suppose that $i < j$ and that $a^i = a^j$. Then $e = a^{j-i}$. By the Division Algorithm, any integer n can be expressed as $n = q(j-i) + r$, where $0 \leq r < j - i$. So $a^n = a^{(j-i)q+r} = e^q a^r = a^r$. Thus if $a^i = a^j$, there would be at most $j - i$ distinct elements in G and G could not be infinite. ▲

PROPOSITION 6 Any infinite cyclic group is isomorphic to $\{\mathbf{Z}, +\}$.

Proof. Suppose that $G = \langle a \rangle$ is an infinite cyclic group. Let $f : \mathbf{Z} \to G$ be defined by $f(i) = a^i$. Clearly f is surjective. It is injective because $a^i = a^j$ if and only if $i = j$. To see that f is operation preserving, note that

$$f(i+j) = a^{i+j} = a^i a^j = f(i) f(j). \quad ▲$$

In the next propositions we investigate the structure of cyclic groups. Just knowing the order of the cyclic group lets us know everything about the group: the order and the structure of each of its subgroups, the orders of each of its elements, and the number of elements of any given order.

PROPOSITION 7 Any subgroup of a cyclic group $G = \langle a \rangle$ is cyclic.

Proof. Let H be a subgroup of G. If $H = \{e\} = \langle a^0 \rangle$, we are done. Assume that H is not the trivial subgroup. Then all the elements of H are of the form a^i for some integer i. Let k be the smallest positive integer such that $a^k \in H$. Certainly $\langle a^k \rangle \subseteq H$.

We shall show that $H = \langle a^k \rangle$. Let $y = a^q \in H$. By the Division Algorithm, $q = pk + i$ where $0 \leq i < k$. Since H is a group and $a^{pk} \in H$, we know that a^{-pk} and $a^{-pk}y$ are both elements of H. But $a^{-pk}y = a^{-pk}a^{pk+i} = a^i$. Since $0 \leq i < k$ and k is the smallest strictly positive power of a appearing in H, we must have $i = 0$. Thus $y = a^{pk}$ and $H = \langle a^k \rangle$. ▲

COROLLARY 8 Let $G = \langle a \rangle$ and suppose that $|G| = m$. The order of every subgroup of G divides the order of G. Furthermore, if q divides m, there is a unique subgroup of G of order q, namely $\langle a^{m/q} \rangle$.

Proof. We know that if H is a subgroup of G, then H is cyclic and $H = \langle a^i \rangle$ for some i. The order of H is the order of the element a^i, namely $\frac{m}{\gcd(m,i)}$. Thus the order of H divides m. Now let q be any divisor of m and let $i = \frac{m}{q}$. The order of a^i is q and hence $\langle a^i \rangle$ is a subgroup of order q. We must show that it is the only such subgroup. Suppose that $|\langle a^k \rangle| = q$. Then $kq = mp$ for some integer p. Since $m = iq$ we have $kq = iqp$. Canceling q, we have $k = ip$. Thus $a^k = (a^i)^p$ and so $a^k \in \langle a^i \rangle$. Since a^k generates a cyclic subgroup of $\langle a^i \rangle$ that also has order q, the subgroups $\langle a^k \rangle$ and $\langle a^i \rangle$ are equal. ▲

As we shall see in Section 5.1, the first assertion of Corollary 8 holds for all groups: The order of a subgroup of G must divide the order of G. But the second assertion does not hold generally. For example, there is no element of order 6 in D_3, a group of order 6.

EXAMPLE 4

Let $|\langle a \rangle| = 12$. The divisors of 12 are 1, 2, 3, 4, 6, and 12. The corresponding subgroups of $\langle a \rangle$ are

$$\{a^0\}, \{a^0, a^6\}, \{a^0, a^4, a^8\}, \{a^0, a^3, a^6, a^9\}, \{a^0, a^2, a^4, a^6, a^8, a^{10}\}, \langle a \rangle.$$ ■

SUMMARY

In this section, we capitalized on our notion of isomorphism. We determined that there is basically only one infinite cyclic group because all infinite cyclic groups are isomorphic to $\{\mathbf{Z}, +\}$. Similarly, any cyclic group of order m is isomorphic to $\{\mathbf{Z}_m, +\}$. Thus we can transfer the number theory information from $\{\mathbf{Z}_m, +\}$ to any cyclic group of order m. For instance, a cyclic group $\langle a \rangle$ of order m has $\varphi(m)$ generators, and $\langle a \rangle$ has a unique subgroup of order q for each divisor q of m. Because any element x of any group G generates a cyclic subgroup of G, cyclic groups form the basis of the structure of any group.

4.6 Exercises

1. Let $G = \langle a \rangle$ be a **finite** cyclic group. Prove that $b \in G$ is a generator of G if and only if $|b| = |G|$. (This is NOT true if G is an infinite group. Why not?)

2. Find all the generators of Z_8, of Z_{12}, and of Z_7.
3. Find all the generators of U_7, and of U_{13}. Does U_{20} have a generator?
4. Suppose that $|\langle a \rangle| = n$. Show that for any integer $i \in Z$, $a^i = a^{i \bmod n}$.
5. Find all the generators of each of the following cyclic groups.

   a. The even numbers under addition
   b. $\{Z_{20}, +\}$
   c. $\langle a \rangle$, where $|a| = 15$
   d. $\langle a^3 \rangle$, where $|a| = 15$

6. Find all the subgroups of Z_{18}.
7. Suppose that $|\langle a \rangle| = 13$. Find all the subgroups of $\langle a \rangle$.
8. Find the order of 8 in Z_{12}.
9. Find the order of a^{20} in $\langle a \rangle$ if $|a| = 28$.
10. Find all the subgroups of $\langle a \rangle$ where $|a| = 20$.
11. Let G be a cyclic group of order m and let d be a divisor of m. Prove that the number of elements in G of order d is $\varphi(d)$, where φ is the Euler phi-function. Give an example that verifies this fact.
12. Find all the subgroups of U_{13}.
13. Suppose that $\langle a \rangle$ is an infinite cyclic group. Let m and n be integers. Find a generator for $\langle a^m \rangle \cap \langle a^n \rangle$. Is $\langle a^m \rangle \cup \langle a^n \rangle$ a subgroup of $\langle a \rangle$? If so, what is its generator? Prove your results. Give examples that illustrate these results.
14. How many elements in Z_{60} have order 5? Find them.
15. In the multiplicative group $U_8 = \{1, 3, 5, 7\}$, each element has order 2. It is not a cyclic group. However, $U_7 = \{1, 2, 3, 4, 5, 6\} = \langle 3 \rangle$ is cyclic. Prove that if p is prime, then U_p is a cyclic group. The following steps will help to guide your proof.

Guide to Proof. Let p be a prime. Fermat's Little Theorem states that if a is not a multiple of p, then $a^{p-1} \equiv 1 \bmod p$. Thus the polynomial $x^{p-1} - 1 \in Z_p[x]$ has $p - 1$ distinct roots in the field Z_p. Its roots are all the members of the set $U_p = \{1, 2, \ldots, p - 1\}$.

Step 1. Show that $x^d - 1$ is a factor of $x^{p-1} - 1$ if and only if d is a factor of $p - 1$.

Step 2. Show that if d is a factor of $p - 1$, then $x^d - 1$ has d distinct roots in U_p.

Now suppose that q is prime and that q^m divides $p - 1$. There are $q^m - q^{m-1}$ numbers $x \in U_p$ that are relatively prime to q^m since $\varphi(q^m) = q^m - q^{m-1}$ by Proposition 1 of Section 1.7.

Step 3. Suppose that $x \in U_p$ and $x^{q^m} = 1$. Suppose that d is the smallest positive integer such that $x^d = 1$. Show that $d | q^m$. (Use the Division Algorithm.)

Step 4. Show that there are $q^m - q^{m-1}$ members of U_p that are roots of $x^{q^m} - 1$ but that are not roots of $x^d - 1$ for any $d < q^m$.

In the language of groups, Step 4 means that if x is one of the $q^m - q^{m-1}$ members of U_p described in Step 4, then the order of x is q^m. Now suppose that $p - 1 = q_1^{m_1} q_2^{m_2} \cdots q_n^{m_n}$, where each q_i is a distinct prime. For each i, let x_i be an element of order $q_i^{m_i}$. Let $x = x_1 x_2 \cdots x_n$.

Step 5. Show that the order of $x = x_1 x_2 \cdots x_n$ is $p - 1$.

Thus x is a generator of U_p, which proves the proposition.

16. Find a generator for U_{11}.
17. Show that the converse to the proposition given in Exercise 15 is not true by showing that we can find a composite number m such that U_m is cyclic.

To the Teacher

In some sense, this section is a summary: all finite cyclic groups of the same order are isomorphic and all infinite cyclic groups are isomorphic to $\{Z, +\}$. It captures a progression of understanding and abstraction. We have traveled from the study of remainders via the Division Algorithm, to modular arithmetic, to the set of equivalence classes mod n, to the groups $\{Z_n, +\}$, to the identification of all finite cyclic groups via isomorphism. Quite a trip!

Your overview can help you shape your students' understanding at every level. You will recognize that, whether you are teaching integer division, clock or modular arithmetic, the composition of rotational symmetries, or figuring how to duck out of serving in volleyball rotation, you are dealing with cyclic groups. Here's how you might investigate groups of prime order through arithmetic.

Children can chart the sequence of remainders of $7 \overline{)x}$ as x varies between say 0 and 50 in increments of 1s, 2s, and 3s, and 4s. In every case, the remainders cycle through all the numbers between 0 and 6 periodically. For example, the remainders of $2, 4, 6, 8, \ldots$, after division by 7 are $2, 4, 6, 1, 3, 5, 0, 2, 4, \ldots$. Why? You might answer that it is because 2 (or 3 or 4) is relatively prime to 7 and so Z_7 is generated by 2 (and similarly by 3 and by 4). But for students, we can get more basic. Let's take a necklace with 7 beads as in Figure 1. Starting anywhere, go around the necklace in jumps of 2.

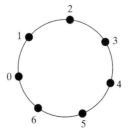

FIGURE 1

You and your students can observe the following.

- It will take 7 jumps to get back to the starting place.

  Reason: Because 2 times the number of jumps must be a multiple of 7 and because 2 and 7 are relatively prime, the number of jumps must be divisible by 7.

- We must visit every bead before getting back to the beginning.

  Reason: If not, we would have visited at least one bead twice in fewer than 7 jumps.

We can track the remainders of 0, 2, 4, 6, ... after division by 7 by going around the necklace in jumps of 2, starting at 0. Just as all the beads are visited, all possible remainders appear. Of course, 7 can be replaced with any prime p and the increment can be any number not divisible by p.

Task

Develop a similar activity for a cyclic group with nonprime order n. Make sure your activity prompts students to notice how the sequence of remainders depends on the greatest common divisor of the number n and the sequence increment.

4.7 PERMUTATION GROUPS

A **permutation** of the n objects in the set $A = \{a_1, \ldots, a_n\}$ is a bijection or one-to-one correspondence from A to itself. For ease of notation, we will usually take A to be the set $\{1, 2, \ldots, n\}$ but A can be any set of n objects. We denote the set of all permutations (bijections) of A to itself by S_n. The set S_n becomes a group under the operation of function composition. In this section we take a close look at these important groups. In some sense, the study of permutation groups is really the study of all finite groups because every finite group is a subgroup of some permutation group as we shall show in Proposition 7 at the end of this section.

A typical element $\sigma \in S_5$ is defined by the following map (read top to bottom):

$$\begin{pmatrix} 1 & 2 & 3 & 4 & 5 \\ 2 & 4 & 1 & 5 & 3 \end{pmatrix}.$$

Here $\sigma(1) = 2$, $\sigma(2) = 4$, and so on. We could denote the map by simply denoting the order of the elements in the bottom row, namely 2, 4, 1, 5, 3, and assume that the top row is in the usual order. Each ordering of the bottom row defines a unique map and each map defines a unique ordering. Thus the conventional use of the word permutation to mean an arrangement of n objects is preserved. There are $n!$ ways to order the elements in the list $\{1, 2, 3, \ldots, n\}$ and so the set S_n contains $n!$ elements.

PROPOSITION 1 For each integer $n > 0$, $\{S_n, \circ\}$ is a group of order $n!$.

Proof. Let $n > 0$. Since the composition of two bijections is a bijection, S_n is closed under the operation of composition. The operation of composition is associative since $\alpha \circ (\beta \circ \gamma)(x) = \alpha((\beta(\gamma(x)))) = (\alpha \circ \beta) \circ \gamma(x)$ for any permutations α, β, and γ in S_n. The identity element for the operation of composition is the identity map id where

$id(x) = x$ for each x in $\{1, 2, \ldots, n\}$. Since every permutation σ is a bijection, its inverse σ^{-1} is also a permutation and hence in S_n. ▲

The $3! = 6$ elements in S_3 are

$$\sigma_1 = \begin{pmatrix} 1 & 2 & 3 \\ 1 & 2 & 3 \end{pmatrix} \quad \sigma_2 = \begin{pmatrix} 1 & 2 & 3 \\ 2 & 3 & 1 \end{pmatrix} \quad \sigma_3 = \begin{pmatrix} 1 & 2 & 3 \\ 3 & 1 & 2 \end{pmatrix}$$

$$\sigma_4 = \begin{pmatrix} 1 & 2 & 3 \\ 1 & 3 & 2 \end{pmatrix} \quad \sigma_5 = \begin{pmatrix} 1 & 2 & 3 \\ 3 & 2 & 1 \end{pmatrix} \quad \sigma_6 = \begin{pmatrix} 1 & 2 & 3 \\ 2 & 1 & 3 \end{pmatrix}.$$

Since the group operation is composition, $\sigma_i \sigma_j$ is computed as σ_j followed by σ_i. Thus $\sigma_2 \sigma_4 = \sigma_6$ and $\sigma_4 \sigma_2 = \sigma_5$. Thus S_3 is not abelian. Setting $\sigma_1 = e$, $\sigma_2 = \sigma$, and $\sigma_4 = \alpha$, we can check that the group elements are e, σ, σ^2, α, $\alpha\sigma$, and $\alpha\sigma^2$. The group S_3 is generated by σ and α. The proper subgroups of S_3 are $\{e\}$, $\{e, \sigma, \sigma^2\}$, $\{e, \alpha\}$, $\{e, \alpha\sigma\}$, and $\{e, \alpha\sigma^2\}$.

In the first exercise of this section, you are asked to write out the group table for S_3. It should look very familiar! The group S_3 is isomorphic to the group of symmetries of the equilateral triangle. The group S_n is called the **nth symmetric group**. For $n \geq 3$, the group of symmetries of the regular n-gon D_n is very naturally a subgroup of S_n because we can regard each of the $2n$ symmetries of the n-gon as a permutation of its n vertices. For $n > 3$, note that $2n < n!$ and so D_n is a proper subgroup of S_n. For $n \geq 3$, D_n is not abelian and hence S_n is not abelian.

Sometimes, improved notation for mathematical objects not only improves writing efficiency, but improves our understanding of the objects. This is the case with the **cycle notation** for permutations that we now present. Let $\sigma \in S_n$ and let $j \to i$ mean that $\sigma(j) = i$. By knowing $j \to i$ for each $j = 1, 2, \ldots, n$, we know σ completely. Take $\sigma = \begin{pmatrix} 1 & 2 & 3 & 4 & 5 \\ 2 & 4 & 1 & 5 & 3 \end{pmatrix}$, for example. We have $1 \to 2 \to 4 \to 5 \to 3 \to 1$. Reading from left to right, we know that $\sigma(2) = 4$ and that $\sigma(3) = 1$ without referring to the larger chart. Efficiency comes from piggybacking the arrows. Now note that we come full cycle with $3 \to 1$. We can further shorten the notation by simply writing $(1, 2, 4, 5, 3)$ with the convention that $\sigma(j)$ is the number immediately to the right of j unless j is the last entry, in which case $\sigma(j)$ is the very first entry. We call $(1, 2, 4, 5, 3)$ a **cycle**. The length of a cycle is the number of entries in it. A cycle of length q is called a **q-cycle**. So the cycle $(1, 2, 4, 5, 3)$ is a 5-cycle that can also be written $(2, 4, 5, 3, 1)$ or $(4, 5, 3, 1, 2)$, etc.

EXAMPLE 1

Suppose that the 6-cycle $(2, 3, 6, 4, 1, 5)$ denotes the permutation π in S_6. Then $\pi(2) = 3$, $\pi(3) = 6$, $\pi(6) = 4$, $\pi(4) = 1$, $\pi(1) = 5$ and $\pi(5) = 2$. (We "cycled" around.) The cycle for π can be denoted in six ways, starting with 1, 2, 3, 4, 5, or 6. For instance, $(6, 4, 1, 5, 2, 3)$ also denotes π. ■

Suppose that we have two cycles $(a_1, a_2, \ldots)$ and $(b_1, b_2, \ldots)$ that denote two permutations α and β in S_n. Then the **product of the two cycles**, namely

$(a_1, a_2, \ldots)(b_1, b_2, \ldots)$, denotes the composition $\alpha \circ \beta$. For instance, we suppose $\alpha = (1, 2, 3, 4, 5)$ and $\beta = (1, 3, 5, 2, 4)$ in S_5. To compute $\alpha \circ \beta(1)$ using the cycle product $(1, 2, 3, 4, 5)(1, 3, 5, 2, 4)$, we locate 1 in the cycle for β on the right, and we see that $1 \to 3$. Now we locate 3 in cycle on the left for α and find that $3 \to 4$. Thus the composition takes $1 \to 4$. Similarly, to compute $\alpha \circ \beta(2)$, we see that $2 \to 4$ in the cycle for β and then $4 \to 5$ in the cycle for α. Thus that $\alpha \circ \beta(2) = 5$. Continuing, we find that product $(1, 2, 3, 4, 5)(1, 3, 5, 2, 4) = (1, 4, 2, 5, 3)$.

Consider the permutation $\gamma = \begin{pmatrix} 1 & 2 & 3 & 4 & 5 \\ 3 & 2 & 1 & 5 & 4 \end{pmatrix}$. Here we have $1 \to 3 \to 1$ and $2 \to 2$ and $4 \to 5 \to 4$. We can denote γ as the product $(2)(3, 1)(4, 5)$ or as $(4, 5)(1, 3)(2)$. The cycles in each expression are **disjoint**; no number appears in more than one set of parentheses. Often, to further shorten notation, we omit cycles of length one. If an element j does not appear in a cycle decomposition of a permutation γ, then we assume that $\gamma(j) = j$. For example, $(2, 3, 1, 5, 6)(4)(7, 8)$ can be rewritten as $(2, 3, 1, 5, 6)(7, 8)$. The identity permutation $e = (1)(2) \cdots (n)$ is usually denoted by (1). (If we left out everything that e fixed, its cycle decomposition would be invisible!)

EXAMPLE 2

Now consider $\alpha = (1, 2, 3)$ and $\beta = (2, 3, 4)$ in S_4. (Remember we often omit cycles of length 1 so that α fixes 4 and β fixes 1.) The composition $\alpha \circ \beta$ or more briefly $\alpha\beta$ is the product of **nondisjoint** cycles: $\alpha\beta = (1, 2, 3)(2, 3, 4)$. To evaluate $\alpha\beta(2)$ for instance, we start with the cycle on the right, to see that $2 \to 3$. We move to the next cycle on the left to find that $3 \to 1$ so that $\alpha\beta(2) = 1$. Similarly $\alpha\beta(1) = 2$, $\alpha\beta(3) = 4$ and $\alpha\beta(4) = 3$. The cycle decomposition of $\alpha\beta$ (i.e., the expression for $\alpha\beta$ as the product of *disjoint* cycles is $(1, 2)(3, 4)$). Now let's compute $\beta\alpha = (2, 3, 4)(1, 2, 3)$. Here $\beta\alpha(1) = 3$, $\beta\alpha(3) = 1$, $\beta\alpha(2) = 4$, and $\beta\alpha(4) = 2$. The cycle decomposition of $\beta\alpha$ is $(1, 3)(2, 4)$. **Clearly, nondisjoint cycles do not necessarily commute.** ∎

> **DEFINITION 1.** A **cycle decomposition** of $\gamma \in S_n$ is an expression for γ as the product of disjoint cycles.

Any permutation can be written as the product of disjoint cycles. Here is an algorithm that produces a cycle decomposition for any $\gamma \in S_n$.

Algorithm

Let L be the list of the n objects permuted by the members of S_n and let $\gamma \in S_n$.

1. If there are no elements left in the list L, **stop**. Else, pick any element $a \in L$. Start the construction of the next (or first) cycle by entering a in a new set of parentheses: (a). Remove a from the list L.
2. Let x be the last element entered into the cycle under construction. Let $b = \gamma(x)$. If $b \neq a$, add b to the cycle immediately after x, remove b from the list L, and return to the beginning of step 2. If $\gamma(x) = a$, end the cycle and return to step 1.

The set of cycles constructed via steps 1 and 2 are a cycle decomposition for γ.

EXAMPLE 3

Consider $\gamma = \begin{pmatrix} 1 & 2 & 3 & 4 & 5 & 6 & 7 & 8 \\ 2 & 5 & 3 & 8 & 7 & 4 & 1 & 6 \end{pmatrix}$ in S_8. Starting with 1, we have $1 \to 2 \to 5 \to 7 \to 1$. Since 3 does not appear yet, we begin a new cycle with 3 and find that $3 \to 3$. The number 3 is fixed, giving us a very short cycle. Continuing onto 4, we have $4 \to 8 \to 6 \to 4$. All the symbols from 1 to 8 are accounted for. A cycle decomposition for γ is (1, 2, 5, 7)(3)(4, 8, 6). The cycle decomposition for γ^{-1} is (7, 5, 2, 1)(3)(6, 8, 4). Just go backward. ∎

It should be clear now that a cycle of length n can be written in n different ways, starting at each of its n entries. For example, $(1, 3, 2) = (3, 2, 1) = (2, 1, 3)$. Notice also that $(2, 3, 1, 5, 6)(7, 8) = (7, 8)(2, 3, 1, 5, 6)$. The disjoint cycles of a given permutation γ can be presented in any order because **disjoint cycles commute**, as the next proposition shows.

PROPOSITION 2 Let β and α be *disjoint* cycles in S_n. Then $\beta\alpha = \alpha\beta$. (Disjoint cycles commute.)

Proof. Suppose that $\alpha = (a_1, a_2, \ldots, a_q)$ and $\beta = (b_1, b_2, \ldots, b_m)$ have no elements in common. We must show that $\beta\alpha(x) = \alpha\beta(x)$ for each $x \in \{1, 2, \ldots, n\}$. If x appears in neither cycle, then it is fixed by both cycles (i.e., $\alpha(x) = x = \beta(x)$). Hence $\beta\alpha(x) = \beta(\alpha(x)) = \beta(x) = x$ and similarly $\alpha\beta(x) = x$. Now let $x = a_i$. Then $\alpha(x) = a_j$, where $j = i + 1$ unless $i = q$, in which case $j = 1$. Both a_i and a_j are fixed by β. Thus $\beta\alpha(a_i) = \beta(a_j) = a_j$ and $\alpha\beta(a_i) = \alpha(a_i) = a_j$. Similarly for $x = b_i$. ▲

When we write a permutation as the product of disjoint cycles, we can vary the starting element of each cycle, the order in which we write the cycles, and we can choose to omit cycles of length one. But there are no other variations.

THEOREM 3 Let $\gamma \in S_n$. Any two expressions of γ as the product of disjoint cycles are identical except perhaps for the order in which the cycles appear.

Proof. Suppose that m is the smallest number of disjoint cycles used in any cycle decomposition of γ. (All cycles of length 1 are included in the count.) We proceed by induction on m. If $m = 1$, then γ can be expressed as a single cycle of length n as follows: $(1, \gamma(1), \gamma(\gamma(1)), \ldots, \gamma^{n-1}(1))$, where γ^i denotes the composition of the map γ with itself i times. Suppose that γ had an alternative expression as $\gamma = \alpha_1\alpha_2 \cdots \alpha_k$, where the α_j are disjoint cycles. Let α_i be the cycle that contains 1. Then the sequence $1 \to \alpha_i(1) \to \alpha_i(\alpha_i(1)) \to \ldots \to \alpha_i^{n-1}(1)$ must be a sequence of n distinct symbols and it must be identical to the sequence $1 \to \gamma(1) \to \gamma(\gamma(1)) \to \ldots \to \gamma^{n-1}(1)$. Thus α_i is a cycle of length n. So $k = 1$ and $\alpha_i = \gamma$.

Now suppose that the assertion is true for permutations that can be expressed in fewer than m cycles. Suppose that $\gamma = \alpha_1\alpha_2 \cdots \alpha_m = \beta_1\beta_2 \cdots \beta_q$. Because disjoint cycles commute, we can renumber the cycles and assume that the symbol 1 appears in α_1 and in β_1. Repeating the argument above shows that $\alpha_1 = \beta_1$. Since the cycles are disjoint, $\alpha_2\alpha_3 \cdots \alpha_m = \beta_2\beta_3 \cdots \beta_q$ must be identical permutations of $n - k$ elements where k is the length of β_1. Now we can apply induction to conclude that, except perhaps for the order in which they occur, the cycles in $\alpha_2\alpha_3 \cdots \alpha_m$ are the same as those which occur in $\beta_2\beta_3 \cdots \beta_q$. ▲

Intuitively, we know that we can produce any rearrangement of n symbols by interchanging symbols two at a time. Try starting with (1, 2, 3, 4, 5, 6, 7, 8, 9). Think of the numbers as written on tiles. Interchange the positions of the tiles two at a time until you obtain the arrangement (3, 2, 5, 1, 6, 7, 4, 9, 8). Suppose that $n > 1$ so that we have more than one symbol. We now show how to express any cycle of length n (and hence any permutation) as the product of 2-cycles. (2-cycles are called **transpositions**.) Let $\beta = (a_1, a_2, a_3, \ldots, a_n)$, where a_i is the symbol in the ith position of the cycle β. In terms of transpositions, $\beta = (a_1, a_n)(a_1, a_{n-1}) \cdots (a_1, a_2)$. Let's check. In the right most parenthesis, $a_1 \to a_2$ and a_2 is fixed by all the other cycles, the correct result. Similarly, a_i is fixed by all until we reach the cycle (a_1, a_i), which sends a_i to a_1. The next cycle sends a_1 to a_{i+1}, which is fixed by the remaining cycles to the left, again the correct result. So we have the following proposition.

PROPOSITION 4 Let $n > 1$. Every permutation in S_n can be expressed as the product of transpositions.

EXAMPLES

4. $(2, 1, 3, 5, 4, 6) = (2, 6)(2, 4)(2, 5)(2, 3)(2, 1)$.
5. The identity e can be written $(1, 2)(1, 2)$.
6. $(1, 2, 3, 6)(7, 5, 4) = (1, 6)(1, 3)(1, 2)(7, 4)(7, 5)$. ■

The expression of a given permutation in terms of transpositions is not unique. For example, $(1, 2, 3, 4) = (1, 4)(1, 3)(1, 2) = (3, 2)(3, 1)(3, 4) = (4, 3)(3, 2)(1, 2)(3, 1)(4, 2)$. What does not change is whether we need an odd or an even number of transpositions to express a given permutation. To show this, we need to introduce some new algebraic equipment, namely, the polynomial:

$$P = (x_1 - x_2)(x_1 - x_3) \cdots (x_1 - x_n)(x_2 - x_3) \cdots (x_{n-1} - x_n) = \prod_{0 \leq i < j \leq n} (x_i - x_j).$$

The polynomial P has $\binom{n}{2} = \dfrac{n(n-1)}{2}$ factors, each of the form $(x_i - x_j)$ where $i < j$. Let $\alpha \in S_n$. We define αP to be the polynomial P with the permutation α applied to the indices of the variables as follows.

$$\alpha P = (x_{\alpha(1)} - x_{\alpha(2)})(x_{\alpha(1)} - x_{\alpha(3)}) \cdots (x_{\alpha(1)} - x_{\alpha(n)})(x_{\alpha(2)} - x_{\alpha(3)}) \cdots (x_{\alpha(n-1)} - x_{\alpha(n)})$$

For example, suppose that $n = 4$ and that $\alpha = (1, 3)(2, 4)$. Then

$$\alpha P = (x_3 - x_4)(x_3 - x_1)(x_3 - x_2)(x_4 - x_1)(x_4 - x_2)(x_1 - x_2).$$

Notice that αP has the same factors as P except perhaps for sign. This reflects the general case. For any α, either $\alpha P = P$ or $\alpha P = -P$ because any permutation is a bijection. If β is another permutation in S_n, then

$$(\beta\alpha)P = \beta(\alpha P) = (x_{\beta\alpha(1)} - x_{\beta\alpha(2)})(x_{\beta\alpha(1)} - x_{\beta\alpha(3)})$$
$$\cdots (x_{\beta\alpha(1)} - x_{\beta\alpha(n)})(x_{\beta\alpha(2)} - x_{\beta\alpha(3)}) \cdots (x_{\beta\alpha(n-1)} - x_{\beta\alpha(n)}).$$

Now suppose that β is a transposition, say (i,j). Let's see what effect β has on αP. It does not affect any factor in which neither i nor j appears as an index. It negates the factor in which both i and j appear because, if $(x_i - x_j)$ occurs in αP, then $(x_j - x_i)$ appears in $(\beta\alpha)P$, or vice versa, if it is $(x_j - x_i)$ that appears in αP. Now suppose that $k \neq i$ and $k \neq j$. If the product of factors $(x_k - x_i)(x_k - x_j)$ occurs in αP then the same pair of factors occurs in $(\beta\alpha)P$ since β simply swaps i and j. There is no change of sign. If the pair of factors $(x_i - x_k)(x_k - x_j)$ occurs in αP, then the pair $(x_j - x_k)(x_k - x_i)$ occurs in $(\beta\alpha)P$. Here **both** of the factors are negated, and again there is no sign change. The net effect is that, when β is a transposition, we have $(\beta\alpha)P = -\alpha P$.

Let γ be any permutation in S_n. Then either $\gamma P = P$ or $\gamma P = -P$. From what we said above, if γ can be expressed as the composition of an even number of transpositions, then $\gamma P = P$. Therefore, **every** expression of γ as the composition of transpositions must be the composition of an even number of transpositions. Similarly, if γ can be expressed as the composition of an odd number of transpositions, then $\gamma P = -P$. In this case, every expression of γ as the composition of transpositions must be the composition of an odd number of transpositions. We summarize our results in the following proposition.

PROPOSITION 5 Suppose that π is a permutation in S_n. Suppose that $\pi = \alpha_1\alpha_2\cdots\alpha_m = \beta_1\beta_2\cdots\beta_q$, where each α_i and each β_j is a transposition. Then either q and m are both even or they are both odd.

We call a permutation **even** if we can express it using an even number of transpositions. Otherwise, it is an **odd** permutation. Thus $(1, 2, 3, 4)$ is an odd permutation since $(1, 2, 3, 4) = (1, 4)(1, 3)(1, 2)$. When we express a cycle of length n as the product of transpositions as demonstrated above, we use $n - 1$ transpositions. Thus a cycle of even length is an odd permutation and a cycle of odd length is an even permutation. The identity permutation is the product of 0 transpositions, and 0 is even. Thus the identity is an even permutation. (For $n > 1$, we can also think of e as $e = (1, 2)(1, 2)$.)

EXAMPLE 7

Let $\gamma = (1, 2, 3)(4, 5)(6, 7, 8, 9)$. Since $(1, 2, 3)$ is even, $(4, 5)$ is odd and $(6, 7, 8, 9)$, is odd, and since the sum of two odd numbers and an even number is even, γ is even. (Remember that we are adding the number of cycles needed to express the product of permutations.) Let's check by explicitly expressing γ as the product of an even number of transpositions:

$$\gamma = (1, 3)(1, 2)(4, 5)(6, 9)(6, 8)(6, 7).$$

So we see that γ is the product of six transpositions and thus γ is even as asserted. ∎

We noted that the identity element in S_n is even. Since the sum of even numbers is even, the product of even permutations is also even. A cycle and its inverse have the same length, they are either both odd or they are both even. So we see that the subset $A_n = \{\alpha \in S_n : \alpha \text{ is even}\}$ is a subgroup of S_n. It is often called the **nth alternating group**.

EXAMPLES

8. $A_3 = \{e, (1,2,3), (1,3,2)\}$
9. The possible forms of cycle decompositions in S_4 are as follows.

    i. $e = (1) = (1,2)(1,2)$ (even)
    ii. $(a)(b)(c,d)$ (odd)
    iii. $(d)(a,b,c)$ (even)
    iv. $(a,b)(c,d)$ (even)
    v. (a,b,c,d) (odd) ∎

Let's count the number of permutations of each form. There is only one identity element. There are $\binom{4}{2} = 6$ permutations of the form $(a)(b)(c,d)$. To see this, notice that we need only choose the two elements for the transposition, in any order. The remaining two elements are fixed. There are 8 permutations of the form $(d)(a,b,c)$ because we can choose the one fixed element in 4 ways. The remaining three elements can be arranged into two distinct cycles. There are 3 permutations of the form $(a,b)(c,d)$ because the permutation is determined by choosing one pair from 4 elements for the first transposition. This can be done in 6 ways. The remaining two must form the other transposition. However, since disjoint transpositions commute, we have counted each distinct permutation twice. There are 6 permutations of the form (a,b,c,d) because there are 4! ways to arrange the symbols in the cycle. But a given permutation can be expressed starting with any of the four elements. Thus there are $4!/4 = 6$ permutations that are 4-cycles. Summing, we have a total of $1 + 6 + 8 + 3 + 6 = 24$ permutations. Of these, $1 + 8 + 3 = 12$ are even.

THEOREM 6   For $n > 1$, the number of elements in A_n is $n!/2$.

Proof. For any odd permutation σ, the composition $(1,2)\sigma$ is even. So there are at least as many even permutations as there are odd ones. Similarly, if σ is an even permutation, $(1,2)\sigma$ is odd and so there are at least as many odd permutations as evens. So the number of even permutations is the same as the number of odd permutations. ▲

The alternating groups are particularly important in many applications. For instance, A_4 is isomorphic to the group of symmetries of the tetrahedron. In some sense, the entire study of finite groups can be carried out on permutation groups because, as the next proposition shows, every group is isomorphic to a subgroup of a permutation group S_n.

PROPOSITION 7 **Cayley's Theorem.** Any group of order n is isomorphic to a subgroup of S_n.

Proof. Let G be a group of order n and list the elements of G as $\{g_1, g_2, \ldots, g_n\}$. We shall show that G is isomorphic to a subgroup of S_n, where S_n is considered to be the group of permutations on the symbols $\{g_1, g_2, \ldots, g_n\}$. For each g_i, let $T_{g_i} : G \to G$ be defined by $T_{g_i}(x) = g_i x$. Clearly, if $g_i x = g_i y$, then $x = y$. Also, $T_{g_i}(g_i^{-1} x) = x$. So T_{g_i} is injective and surjective and hence a permutation of the symbols $\{g_1, g_2, \ldots, g_n\}$. What we must show is that the map $g_i \to T_{g_i}$ from G to S_n is injective and operation preserving. For injectivity, let $x \in G$. Suppose that $T_{g_i} = T_{g_j}$. Then $T_{g_i}(x) = T_{g_j}(x)$ or $g_i x = g_j x$. Thus $g_i = g_j$ and so the map $g_i \to T_{g_i}$ is injective. To see that it is operation preserving, we note that $g_i g_j \to T_{g_i g_j}$ and that $T_{g_i g_j} = T_{g_i} \circ T_{g_j}$. ▲

SUMMARY

The origins of group theory began with the group S_n, the set of all permutations of n symbols under composition. (See Section 4.10 for more about the origins of groups.) The group S_n has $n!$ members and it is nonabelian for $n > 2$. It has a rich and complicated structure (as it must) because every finite group of order n can be regarded as a subgroup of S_n. Any permutation $\gamma \in S_n$ has a **cycle decomposition** whereby it can be expressed as the product of disjoint **cycles**. This simple notation gives us insight into the structure of S_n. We also found that we could express any permutation as the product of 2-cycles or **transpositions**. Furthermore, all permutations can be classified as **odd** or **even permutations**. The subset $A_n \subseteq S_n$ of all even permutations is a group called the *n*th alternating group. Historically, the structure of A_5 proved to be the key to proving that not all polynomials are "solvable."

4.7 Exercises

1. Write out the group table for S_3.

2. Find a subgroup of S_4 of order 6. Find a subgroup of S_4 of order 8 and another of order 12. (The latter two groups are already familiar as groups of symmetries of geometric objects.)

3. With π as in Example 1, find $\pi(6)$. Show that the cycle $(3, 6, 4, 1, 5, 2)$ also denotes π. Find a cycle that denotes π and starts with 6.

4. Show that if β is the cycle $(a_1, a_2, \ldots, a_n)$, then $\beta^{-1} = (a_n, a_{n-1}, \ldots, a_1)$. Give examples. What does this say for 2-cycles?

5. Suppose that $(2, 3, 1, 5, 6)(4)(7, 8)$ is the cycle decomposition of a permutation α in S_8. Find $\alpha(i)$ for $i = 1, 2, \ldots, 8$.

6. Write out the following permutation as a product of disjoint cycles.
$$\alpha = \begin{pmatrix} 1 & 2 & 3 & 4 & 5 & 6 & 7 & 8 \\ 1 & 5 & 3 & 4 & 7 & 8 & 2 & 6 \end{pmatrix}$$
7. Let $\beta = (1, 2, 3)(4, 5)$ and $\alpha = (1, 2)(3, 4)(5)$. Compute $\beta\alpha$ and $\alpha\beta$ and write out the cycle decomposition of each.
8. (Important!) Let $\beta = (a_1, a_2, \ldots, a_n)$ be a cycle. Prove that the order of β is n. Let β and α be *disjoint* cycles of lengths n and m respectively. Prove that the order of $\beta\alpha$ is lcm(n, m).
9. Determine the forms of each of the cycle decompositions of permutations in S_5. Classify each as even or odd. Determine the number of permutations of each form. Determine the number of even permutations.
10. The 12 elements of A_4 are
    $(1) = e$,
    $(1, 2)(3, 4)$
    $(1, 3)(2, 4)$
    $(1, 4)(2, 3)$
    $(1, 2, 3)$
    $(2, 3, 4)$
    $(1, 4, 2)$
    $(1, 3, 4)$
    $(1, 3, 2)$
    $(1, 4, 3)$
    $(2, 4, 3)$
    $(1, 2, 4)$
    Labeling these α_1 through α_{12}, or perhaps more simply, 1 through 12, form the Cayley table for A_4.
11. Find a subgroup of order 8 of S_4. Are the permutations all odd? Can they ever be all odd in a subgroup?
12. Find the order of each element in A_4.
13. Find the order of each element in S_5.
14. Show that every 3-cycle is a square (i.e., every 3-cycle can be written as some permutation composed with itself).
15. For $n > 2$, show that A_n is generated by 3-cycles.
16. Suppose that $m \geq 3$. Show that any 3-cycle in S_m can be expressed as the product of two m-cycles.

To the Teacher

The word "permutation" carries at least two distinct but related meanings in mathematics. Most commonly, the word means an arrangement of a set of distinct symbols. We ask, "How many permutations (arrangements) are there of the letters in the word MATH?" But the word "permutation" also means bijection (i.e., a permutation is a one-to-one correspondence from a set A onto itself). The bracket notation that we used in this section to denote a bijection, e.g., $\begin{pmatrix} a & b & c & d & e \\ b & e & d & c & a \end{pmatrix}$, reconciles the two meanings.

Every arrangement of the symbols a, b, c, d, and e defines a bijection when it appears on the bottom line of the bracket and conversely, every bottom line is an arrangement.

One-to-one and onto mappings are a standard precalculus topic. Coupling the study of bijections (permutations) of finite sets with the study of symmetries of regular figures provides a dynamic to the otherwise somewhat dry topic. And gone are the horizontal and vertical line tests! Let's see what a student can learn from a cube. We will start by considering a symmetry of a cube as a permutation of its eight corners. (You will think of it as an element of S_8.)

1. One-to-one means no two corners can go to the same position and onto means that each position must be occupied by some corner. The concepts of one-to-one and onto are hurdles for most students. Repackaging the ideas many ways helps solidify their meanings.

2. Counting: There are clearly 24 ways we can situate a cube because a given face can be in any of six orientations (up, down, right, left, front or back) and once oriented, rotated into one of four positions. Students can search for the $(8! - 24)$ permutations that do not correspond to symmetries. There are only 40,296 of them!

3. Function composition has a manipulative! The correspondence of the composition of the symmetries of the cube with the composition of the permutations that model them allows the teacher to investigate a wealth of concepts. Among these might be

   i. Noncommutative operations,
   ii. The notion of the inverse of a function,
   iii. The notion of repeated composition and order.

4. The same symmetries can be modeled by permutations of different sets. We can, for instance, treat each symmetry of a cube as a permutation of six faces (a subgroup of S_6) or of its 12 edges (a very proper subgroup of S_{12}). The fact that there are 24 symmetries and $24 = 4!$ suggests that the symmetries might be modeled by permuting just four objects. This is indeed true. The four objects are the four diagonals that attach opposite corners of the cube. (See Figure 1.)

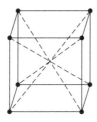

FIGURE 1

Task

Show exactly how each symmetry of the cube can be modeled by a distinct permutation of the diagonals of a cube.

4.8 WORKSHEET 7: GROUPS AND COUNTING

Suppose we had an equilateral triangle with vertices numbered 1 (left), 2 (top), and 3 (right) drawn on a paper. We could color each of its vertices with one of two colors, say black and white, in $2^3 = 8$ different ways. However, if we cut the triangle out of the paper, we could rotate and flip it, subjecting it to all the symmetries in the group of symmetries of the triangle. Once out of the paper, the coloring $c_1 = $ **WBB** of the left, top and right vertices respectively, would be considered the same as the coloring $c_2 = $ **BWB** because the motion of rotating the triangle 120 degrees clockwise moves the first coloring **WBB** into the second coloring **BWB**. Similarly, if we flip the triangle colored $c_3 = $ **BWW** around its vertical axis, **BWW** becomes $c_4 = $ **WWB**. Figure 1 groups all the colorings into sets of equivalent colorings (i.e., colorings that can be moved into one another via the group of symmetries of the triangle). Notice that the set of colorings is partitioned into nonoverlapping subsets. By defining two colorings c_i and c_j to be equivalent if and only if there is a symmetry that moves one to the other, we establish an equivalence relation on the set of all colorings.

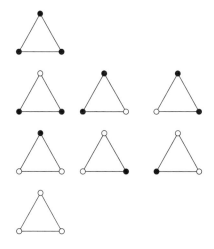

FIGURE 1

The number of nonequivalent black and white colorings (2-colorings) of the vertices of the triangle reduces to four, one chosen from each of the four equivalence classes. (See Figure 2.) Each one of the original eight colorings can be flipped or rotated into one of these four; no two of the four can be moved into each other.

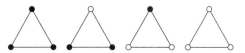

FIGURE 2 Four nonequivalent 2-colorings of the triangle

Task 1

For each of the eight triangles in Figure 1, list all the symmetries of the triangle that do not change the positions of its colors. Such symmetries are said to "fix" its coloring.

Task 2

The vertices of a square can be colored black or white in 16 different ways. Figure 3 shows three of them. Sort the 16 colorings into sets of equivalent colorings. As with the triangle, two colorings are equivalent when we can find a symmetry of the square that moves one into the other. Use all eight symmetries of the square.

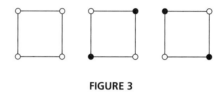

FIGURE 3

Task 3

For each of the sixteen 2-colorings of the square, list the symmetries that do not change the position of its colors (i.e., list the symmetries that fix it). Show that each list is a subgroup of D_4.

Task 4

List all 27 ways that the vertices of the equilateral triangle can be colored using three colors, say red, white, and blue.

Task 5

Sort the 3-colorings of the equilateral triangle into sets of equivalent colorings using all six symmetries of the equilateral triangle.

Task 6

Suppose that we lived in Flatland with no third dimension. In Flatland we could not flip the triangle. A Flatland dweller would think that the equilateral triangle had only three symmetries, $\{R_0, R_{120}, R_{240}\}$. List the sets of equivalent 3-colorings of the equilateral triangle when only rotations are available. Which colorings were equivalent in Task 4 but are no longer equivalent?

The square has eighty-one 3-colorings. The task of sorting them begins to look like a challenge! Let's attack it with basic counting techniques. We will build on what we know about 2-colorings.

Task 7

Sort the 3-colorings of the square using the entire group of eight symmetries (i.e., with rotations and flips). Here's some help with the counting strategies.

1. Count all the colorings that use one color.
2. Count all the colorings that use two colors. There are $_3C_2 = 3$ ways to choose 2 colors and we know that there are six ways to two-color the square. But wait! Remember the cardinal rule of counting: **Don't count things twice.**
3. Every coloring that uses exactly three colors must color two vertices with the same color. (This is the **pigeonhole principle**: If there are more pigeons than pigeonholes, some pigeonhole must be occupied by more than one pigeon.)

    i. The two same-colored vertices can be adjacent or on a diagonal (two situations).
    ii. Pick the color for the same-colored vertices in three ways.
    iii. Determine if orientation front to back makes a difference. (Are you using only rotations or rotations and flips?)

Task 8

Sort the 3-colorings of the square only using the subgroup of rotations (i.e., with no flips).

The number of sets of equivalent colorings is 24 with no flips or 21 with flips. Clearly, the counting challenge will grow greatly as we increase the complexity of the figure or increase the number of colors. But we can utilize the work we did in Tasks 1 and 3. For any coloring c, let $f(c)$ denote the number of symmetries that fix c. Notice that the same number of symmetries fixes each of the colorings in a set of equivalent colorings. So $f(c)$ is the same for each c in a set of equivalent c's. Notice also that if c is in a set of n equivalent colorings, then $nf(c)$ is equal to the size of the group of symmetries in use.

Task 9

Verify that $nf(c) = |G|$ for the following:

i. c is BBW and G is the set of all six symmetries of the equilateral triangle.
ii. c is BBWW and G is the set of all eight symmetries of the square.
iii. c is RWBB (a 3-coloring of the square) and G is all eight symmetries of the square.

Using the ideas of Task 9, we can develop a strategy for counting. Suppose that $\{c_1, c_2, \ldots, c_n\}$ is an entire equivalence class of colorings using symmetry group G. Then $nf(c_1) + nf(c_2) + \ldots + nf(c_n) = n|G|$. Canceling n, we have $\sum_{i=1}^{n} f(c_i) = |G|$

and $\dfrac{1}{|G|}\sum_{i=1}^{n} f(c_i) \cdot 1$. We can think of this sum of 1 as representing 1 set of equivalent colorings. If we let C denote the set of all colorings, then the sum $\dfrac{1}{|G|}\sum_{c \in C} f(c)$ contributes a count of 1 for each distinct set of equivalent colorings. Thus $\dfrac{1}{|G|}\sum_{c \in C} f(c)$ counts the number of nonequivalent colorings.

(1) $$\dfrac{1}{|G|}\sum_{c \in C} f(c) = \text{number of nonequivalent colorings.}$$

Let's count the number of distinct black and white colorings of the equilateral triangle using formula (1). The group G is the group of all symmetries of the triangle so that $|G| = 6$. Table 1 matches each coloring with the symmetries that fix it as indicated by the ∗. (The symmetries are given as permutations of the vertices of the triangle labeled 1, 2, and 3.)

TABLE 1

	(1)(2)(3)	(1, 2, 3)	(1, 3, 2)	(1)(2, 3)	(2)(1, 3)	(3)(1, 2)	$f(c)$
BBB	∗	∗	∗	∗	∗	∗	6
BBW	∗				∗		2
BWB	∗			∗			2
WBB	∗					∗	2
WWB	∗				∗		2
WBW	∗			∗			2
BWW	∗					∗	2
WWW	∗	∗	∗	∗	∗	∗	6
							sum = 24

Task 10

Verify that formula (1) gives the number of distinct 2-colorings of the vertices of the square when G is the group of all eight symmetries of the square.

In formula (1), we count the number of stars in Table 1. As given, the formula asks us to count up the stars row by row. But we can just as well count column by column. The stars in each column are the colorings that are fixed by the symmetry named at the top of the column. For instance, (1)(2, 3) flips the triangle, keeping vertex 1 fixed. It fixes any coloring in which the vertices 2 and 3 are colored with the same color, namely WWW, BBB, BWB and WBW. Let $n(\gamma)$ denote the number of colorings fixed by the symmetry γ. Counting column by column, formula (1) becomes

(2) $$\dfrac{1}{|G|}\sum_{\gamma \in G} n(\gamma) = \text{number of nonequivalent colorings.}$$

Task 11

Use formula (2) to compute the number of distinct black and white colorings of the square.

We'll get you started. The symmetries of the square are written as permutations of its vertices labeled 1, 2, 3, and 4. Thus (1, 2, 3, 4) denotes rotation by 90 degrees.

$n((1)(2)(3)(4)) = 16$ since the identity fixes all colorings.

$n((1,2,3,4)) = 2$ and $n((1,4,3,2)) = 2$ since the rotations through 90 and 270 degrees fix only BBBB and WWWW.

$n((1,3)(2,4)) = 4$ since a rotation through 180 degrees fixes BBBB, WWWW, BWBW, and WBWB.

$n((1)(3)(2,4)) = ?$

$n((1,3)(2)(4)) = ?$

$n((1,4)(2,3)) = ?$

$n((1,2)(3,4)) = ?$

The sum in formula (2) = ?

Counting with formula (2) may not look easier to start. But if we look at each permutation γ as being given by its cycle decomposition, we can compute $n(\gamma)$ without even referring to the figure! The vertices within each cycle of γ must be colored identically and each cycle can be colored independently. **Thus if the cycle decomposition of γ consists of i disjoint cycles, and we are using m colors, then γ fixes exactly m^i colorings.** The following example shows just how easy it is to apply formula 2. It is followed by a set of new counting problems for you to do.

EXAMPLE 1

Let us compute the number of ways we can color the six beads on a necklace with 3 colors, if we only allow the necklace to rotate. Numbering the beads 1 through 6, the cycle decomposition of the rotations and the number of colorings each fixes is as follows:

γ	$n(\gamma)$
(1)(2)(3)(4)(5)(6)	3^6
(1, 2, 3, 4, 5, 6)	3^1
(1, 3, 5)(2, 4, 6)	3^2
(1, 4)(2, 5)(3, 6)	3^3
(1, 5, 3)(2, 6, 4)	3^2
(1, 6, 5, 4, 3, 2)	3^1

The total number of distinct colorings is $(3^6 + 3 + 9 + 27 + 9 + 3)/6 = 130$. ■

Task 12

Solve the following problems.

1. Find the number of ways to 4-color the corners of a square (a) allowing only rotation and (b) allowing all eight rigid motions.
2. How many distinct 3-colorings are there of the necklace in Example 1 if we allow flips as well as rotations?
3. How many ways can you 2-color the squares of a 3 by 3 checkerboard

    (a) allowing only rotations?
    (b) allowing rotations and flips?

4. How many ways can you 3-color n bands on a baton with each of the bands of equal length?
5. How many ways are there to r-color a necklace with n beads, allowing only rotations?
6. George Pólya[1] was prompted to develop the ideas of this section by problems in chemistry that are similar to the following problem.

    Two molecules are called isomers if they have the same chemical formula but differ in their configuration. A benzene ring is a set of six carbon atoms bonded so as to form a hexagon. A hydrogen atom is bonded to each carbon atom.

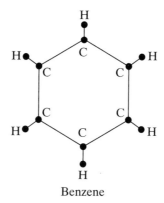

Benzene

A variety of chemically different molecules can be obtained from a benzene ring by replacing all or some of its hydrogen atoms with either fluorine or chlorine. How many distinct isomers can be so formed?

Task 13

To find out why formula 1 works, two questions need to be answered. They are given below. Your job is to carry out the details of answering these questions, which follow. Extensive guidance is provided.

---

[1] G. Pólya and R. C. Read, *Combinatorial Enumeration of Groups, Graphs, and Chemical Compounds*, Springer-Verlag, New York, 1987.

Question 1. If c_1 and c_2 are equivalent colorings of a figure with symmetry group G, why is it that $f(c_1) = f(c_2)$?

The colorings c_1 and c_2 are equivalent because there is a symmetry γ that moves c_1 to c_2. Note that γ^{-1} moves c_2 to c_1. Suppose that σ fixes c_1. Then $\gamma \sigma \gamma^{-1}$ fixes c_2. Show that the set of symmetries that fix c_1 are in one-to-one correspondence with the set of symmetries that fix c_2.

Question 2. Suppose that c_1 is in a set of n equivalent colorings. Why is it that $nf(c_1) = |G|$?

Suppose that $\{c_1, c_2, \ldots, c_n\}$ is an equivalence class of colorings. For each i, let $S_i = \{\gamma \in G : \gamma(c_1) = c_i\}$. So S_i is the set of symmetries that take c_1 to c_i. Each member of G occurs in exactly one S_i. (Why?) Notice that S_1 is the set of all symmetries that fix c_1. Thus the number of elements in S_1 is $f(c_1)$. We can show that $nf(c_1) = |G|$ by showing that each S_i contains the same number of elements as S_1. Suppose that γ_i is a symmetry that moves c_1 to c_i and suppose that σ fixes c_1. Then $\gamma_i \circ \sigma$ also moves c_1 to c_i. Show that if δ_i also moves c_1 to c_i, then $\delta_i = \gamma_i \circ \tau$ where τ fixes c_1. (**Hint.** You can solve for τ because G is a group.) Use these observations to show that the elements of S_i are in one-to-one correspondence with the elements of S_1.

4.9 IN THE CLASSROOM: GROUPS OF SYMMETRIES OF REGULAR POLYGONS

The study of finite groups, especially cyclic and permutation groups, has been a principal focus of Chapter 4. At first glance, this seems to be remote from the high school algebra curriculum. But it is not remote from the geometry curriculum. One prominent area of overlap in geometry is the study of symmetry.

Typically, students in geometry learn the following aspects of symmetry:
- Transformations—reflections, rotations, translations, glide reflections, dilations—under different representations (in coordinate geometry, with vectors, as functions on the vertices, with matrices);
- Properties of transformations;
- Composition of transformations;
- Reflectional and rotational symmetry—lines of symmetry and centers and angles of rotational symmetry;
- Applications to congruence and similarity.

Students may get the opportunity to explore these ideas using mirrors or MIRAs (a see-through reflecting device), paper folding, and with a dynamical geometry software program such as *The Geometer's Sketchpad* (Jackiw, 1995) and *Cabri Geometry II* (Texas Instruments, 1994).

In its *Principles and Standards for School Mathematics* (NCTM 2000), the National Council of Teachers of Mathematics frames this study in terms of its overall standards. The Geometry Standard states that "*instructional programs from*

Section 4.9   In the Classroom: Groups of Symmetries of Regular Polygons   247

prekindergarten through grade 12 should enable all students to—apply transformations and use symmetry to analyze mathematical situations" (NCTM, p. 41). More detailed expectations by grade band are that students should

(Grades 6–8) (NCTM, p. 232)
- *Describe sizes, positions, and orientations of shapes under informal transformations such as flips, turns, slides, and scaling;*
- *Examine the congruence, similarity, and line or rotational symmetry of objects using transformations;*

(Grades 9–12) (NCTM, p. 308)
- *Understand and represent translations, reflections, rotations, and dilations of objects in the plane by using sketches, coordinates, vectors, function notation, and matrices;*
- *Use various representations to help understand the effects of simple transformations and their compositions.*

There are many sophisticated algebraic ideas from Chapter 4 that can be brought to the high school classroom. Through symmetry explorations, students can explore concepts such as order, noncommutativity, groups, subgroups, cyclic groups, and generators in a very hands-on geometrical way, without necessarily attaching specific mathematical names to them. What follows is a discussion of such topics and how they can be approached with high school students.

Counting Symmetries

All regular polygons possess reflectional and rotational symmetry. Students can be asked to count how many symmetries a regular n-gon has. When identifying lines of symmetry for reflection, students are likely to be able to tell you (correctly) that when n is odd, the n lines of symmetry are the lines that bisect a vertex angle and its opposite side. But when n is even, a student might think that there are $2n$ lines of symmetry: n angle bisectors, and n side bisectors (see Figure 1). However, each of these lines has been counted twice. It is always a good lesson in combinatorics to avoid counting twice.

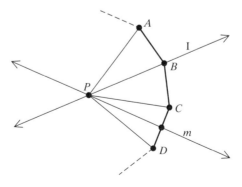

FIGURE 1

Regular n-gons also have rotational symmetry. We rotate *clockwise* (this choice is arbitrary) around P, the center of the polygon. Each rotation of the regular n-gon onto itself is completely determined by looking at one vertex, call it A, and rotating the polygon so that the vertex maps onto another vertex. Since all sides are congruent, when we draw line segments from P to each vertex, we form n congruent isosceles triangles. The measures of the central angles formed by any vertex, P, and the next vertex as you traverse the polygon clockwise are therefore the same and equal to $(360/n)°$. This is the minimal angle of rotation that maps the n-gon onto itself. Any other rotational symmetry is a multiple of $(360/n)°$. Since vertex A can be rotated to n different vertices, there are n rotational symmetries. Students should note that it takes n rotations of $(360/n)°$ to get back to the original labeling of the n-gon, which is the same as having rotated the figure by $0°$. Continuing to rotate beyond $360°$ will result in labelings that have already been encountered. In fact, this is a good application of modular or clock arithmetic. For a regular n-gon, there will be n rotational symmetries $R_0, R_{\frac{360}{n}}, R_{\frac{360(2)}{n}}, \ldots, R_{\frac{360(n-1)}{n}}$.

Have we accounted for all possible symmetries of the regular n-gon? Recognize that the position of the regular n-gon under any symmetry is determined by the location of any particular vertex and whether the n-gon is facing up or facing down. Considering any vertex, there are n positions (vertices) on which it may land and then the n-gon can be either face up or face down. This gives us a total of $2n$ possibilities for the given vertex that determines the rest of the vertices. Since we have found $2n$ motions that result in $2n$ different positions and orientations (face up or down), there are no more.

Therefore, there are $2n$ symmetries of a regular n-gon, and we have counted them: n reflectional symmetries and n rotational symmetries. From Section 4.2, you know that this set of symmetries forms a group called the dihedral group of order $2n$, D_n.

We can also explore symmetries of regular solids as well. Consider the tetrahedron with a labeling of the vertices by 1, 2, 3, and 4. We can quickly count that there are 12 symmetries by noting that any one of the four vertices can be at the apex of the tetrahedron and once one is fixed at the apex, the opposite face can be in any of three positions. This gives a total of $4 \cdot 3 = 12$ symmetries of the tetrahedron. These twelve symmetries are rotations. There is the identity rotation, or a rotation by $0°$. There are two rotations of $120°$ and $240°$ around each of the four axes that pass through each vertex and the center of the opposite face, which gives us eight more symmetries. There are also three rotations of $180°$ around the three axes through the midpoints of edges with no common vertex. For example, there is an axis of symmetry that goes through the midpoint of the edge connecting vertex 1 and vertex 2 and the midpoint of the edge connecting vertex 3 and vertex 4 as in Figure 2. Since there are only 12 symmetries and we have described 12, we have found them all.

Students often regard symmetry as a property of a figure. We regard it as a movement which preserves the figure and as such, we can do one movement followed by another movement, that is, composition of movements. The composition of two symmetries is another symmetry. What is remarkable is that the composition of any

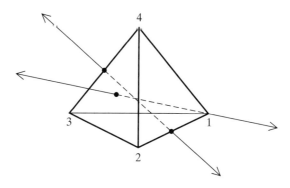

FIGURE 2

two of these rotations of the tetrahedron is once again one of these 12 rotations since there are only these 12 symmetries.

To explore the compositions in greater detail, we can manipulate a tetrahedron or we can think of each of these rotational symmetries as a permutation of the vertices. We note that these symmetries represent only 12 of the 24 total permutations of the four vertices. A rotation of 120 or 240 degrees about an axis through a vertex and the center of the triangle opposite that vertex is a permutation of the other three vertices. For example, rotating about the axis through vertex 1 and the center of the opposite triangle by 120 and 240 degrees permutes the vertices 2, 3, and 4. We get the permutations, written in cycle notation, $(2, 3, 4)$ and $(2, 4, 3)$, respectively. Likewise, we can get the permutations, $(1, 3, 4)$ and $(1, 4, 3)$ using the axis through vertex 2; $(1, 2, 4)$ and $(1, 4, 2)$ using the axis through vertex 3; and $(1, 2, 3)$ and $(1, 3, 2)$ using the axis through vertex 4. A rotation of $180°$ about an axis through the midpoints of edges without common vertices exchanges the vertices of each edge. For example, rotating 180 degrees about the axis connecting the midpoint of the edge containing vertex 1 and vertex 2 and the midpoint of the edge containing vertex 3 and vertex 4 exchanges vertex 1 and vertex 2 and exchanges vertex 3 and vertex 4, giving the permutation $(1, 2)(3, 4)$. Likewise, rotating about the other two such axes produces the permutations $(1, 4)(2, 3)$, and $(1, 3)(2, 4)$. Composition of these symmetries is then accomplished by multiplying these permutations. For example, a rotation of 120 degrees about the axis through vertex 1 followed by a rotation of 180 degrees about the axis through the midpoints of the edge containing vertices 1 and 2 and the edge containing vertices 3 and 4 is the product $(1, 2)(3, 4)(2, 3, 4) = (1, 2, 4)$, the rotation of 120 degrees about the axis through vertex 3. One can check the other compositions as well. We note that this group of symmetries is the nth alternating group, A_4, which was discussed in Section 4.7.

Cayley Tables and the Algebra Associated with Them

Students have had experience forming addition and multiplication tables. They can build on that experience to form the Cayley tables for sets of symmetries under the operation of "followed by", which for us is composition. Starting with small n and working with cutouts of the regular n-gon, students can compose symmetries and fill in Cayley tables of these compositions, as is done in Section 4.2 for the equilateral triangle and the square. Students can be asked to complete the entire Cayley table, fill in a few blanks, or verify several compositions in a complete Cayley table. Then students can be asked to investigate any patterns or phenomena that they can see in the table. Make sure to specify that you want the entry involving the symmetry in the ith row, S_i, and the symmetry in the jth column, S_j, of the Cayley table to be $S_i S_j$, where S_j is done first and S_i is done second, that is, S_j followed by S_i. We will see further connections to function composition in the discussions that follow.

We want students to compare and contrast patterns that they find in these symmetry composition Cayley tables with patterns found in familiar addition and multiplication tables. Ask students to look at addition and multiplication tables and make note of the patterns they see there and the connection to properties of real numbers (closure, commutativity, associativity, identity, inverses) that they know about. Then ask students whether similar patterns appear in these new tables of symmetry composition, which would mean that this set under composition has those same properties. There will be interesting similarities and very apparent differences. What might students observe or be asked to investigate?

- **Closure.** When we compose any two symmetries, we get another one of our symmetries. This means the set of symmetries is closed under composition. You might compare this with addition of whole numbers, which is closed, but subtraction of whole numbers, which is not.
- **Existence of an identity.** Students can note that if you compose the rotation by 0 degrees, R_0, with any symmetry, that you get the other symmetry back again. Students have experience with additive and multiplicative identities for real numbers and can relate that to R_0 being the identity for composition of symmetries of a regular n-gon.
- **Existence of inverses.** Students have experience with additive and multiplicative inverses of real numbers. We extend this experience by asking them to determine if each of our symmetries has an inverse. Given any symmetry, can we find a symmetry with which we can compose to get the identity symmetry, R_0? Students can observe from the Cayley table that every row and every column contains R_0 and hence every symmetry does in fact have an inverse. Students may also observe that any reflection is its own inverse and that any rotation's inverse is another rotation. In fact, $R_\alpha^{-1} = R_{360-\alpha}$ for $\alpha = 0, 360/n, 360 \cdot 2/n, \ldots, 360(n-1)/n$.
- **Associativity.** Students can be asked to verify $S_i(S_j S_k) = (S_i S_j)S_k$ for several symmetries S_i, S_j, and S_k.

We know these properties mean that the set of symmetries of a regular n-gon form a group under composition, which is the dihedral group, D_n. Sometime in this investigation, the issue of whether the composition of these symmetries is commutative or not will come up. In fact, composition is not commutative.

- **Noncommutativity.** If students had to complete the Cayley table on their own, there is a strong possibility that they tried to get away with less work and only did half of the compositions and then assumed (incorrectly) that the compositions in the other order produced the same results. Have them recheck the compositions. If you gave them a completed Cayley table, they can verify that composition is indeed not commutative. In fact, comparing the addition and multiplication tables of whole numbers to this Cayley table, they can see that their table is not symmetric along the diagonal from the upper left corner to the bottom right corner. Order does matter!

In fact, we say the set of symmetries of a regular n-gon forms a noncommutative group under composition. Other observations students might make from the Cayley tables are as follows:

- Every symmetry appears in each row and in each column and appears only once in each.
- Any rotation composed with another rotation yields a rotation. Any reflection composed with a rotation yields a reflection. Investigate why this is true. Students have investigated the fact that the composition of symmetries is closed and must therefore produce another symmetry. They can note that rotations are orientation preserving—after a rotation, the order of labels in relation to one another when you read the labels on the vertices clockwise, is the same—and that reflections do not preserve orientation. Therefore, if we compose two rotations, orientation is not changed so the composition is a rotation. If we do a reflection and orientation changes and then do a rotation, this changed orientation stays. If we do a rotation and orientation is not changed, it is once we follow that by a reflection. Therefore the composition of a rotation and a reflection in either order must be a reflection.
- **Order of elements.** You might ask students to investigate the minimum number of times you would have to compose a symmetry with itself until you get back the identity symmetry, R_0. It would also be easy to explain that this is called the **order** of the symmetry. Students can compute the orders of each of the symmetries and look for patterns. For example, the identity has order 1, every reflection has order 2, and depending on the choice of n used, students may make interesting observations about the orders of the rotations. You might discuss the connection between order 2 elements and elements that are their own inverses. As they find orders of various rotations, the discussion might naturally lead to one regarding generators of subgroups of rotations. We discuss this further below.
- **Solving equations involving symmetries.** Consider the simple equation $3x = 12$. This is automatic for a high school student. But how can we relate such an equation with how we might solve similar equations such as $Sym_1 \cdot X = Sym_2$, where Sym_1 and Sym_2 are two symmetries of a regular n-gon, $\cdot$ means composition, and we are looking for what symmetry X will make the equation true? How do we know that we can always solve such equations and that there will be a unique solution? (Every row of our Cayley table contains each of the $2n$ symmetries and contains each only once.) You can give students specific equations based on a particular n and ask them to solve them and discuss their process. Some students may simply use the Cayley table and find which first move they

can compose with Sym_1 to get Sym_2. Make sure that they are composing in the proper order! Others can make use of multiplying both sides by the inverse of Sym_1 under composition to get $X = Sym_1^{-1} Sym_2$. Again, order matters.

Representations of Symmetries as Permutations of Vertices

While the concept of function can be found throughout the mathematics curriculum, research shows that this topic is one that is very difficult for students to understand (Tall, 1996; Sierpinska, 1992; Markovits, Eylon, and Bruckheimer, 1988; Dreyfus and Eisenberg, 1982). Students are often locked into ideas of functions strictly in terms of specific formulas, graphs, and making use of the vertical line test. We can use symmetries to help extend students' notion of function. Any symmetry of a regular n-gon can be expressed as a permutation of the vertices. Since a permutation is just a function from the set of vertices to itself that is both one-to-one and onto, the study of symmetries provides an avenue for reinforcing and expanding student understanding of these topics. Many current popular high school geometry books do denote transformations in function notation. For example, given a figure and a line of reflection, m, if vertex A is mapped to vertex B by the reflection over line m, then we write $R_m(A) = B$.

Let us expand beyond just using function notation. Consider a regular n-gon with vertices $1, 2, 3, \ldots, n$ labeled in a clockwise fashion around the figure. Have students consider any symmetry of a regular n-gon. As we perform the symmetry on the n-gon, every vertex either stays where it is, in which case we have the identity function from the set $\{1, 2, 3, \ldots, n\}$ to itself, or it must be mapped to one and only one other vertex. No vertex can physically be mapped to two different vertices. Here is the definition of "function" in use. Now, since the n-gon must land back on itself, every vertex must have some vertex that maps *onto* it. Given any vertex, it is impossible to map more than one vertex to it, and here is the definition of "one-to-one" in use. Students can make arrow maps indicating where each vertex gets mapped by the symmetry or they can use function notation. They can also quickly pick up the array notation of Chapter 4. Composition of symmetries is then just a reinforcement of composition of functions.

It is worth exploring with students whether *any* permutation of the set $\{1, 2, 3, \ldots, n\}$ is indeed a symmetry of a regular n-gon. We have seen in Chapter 4 that this is true when $n = 3$, but not when $n = 4$. Ask students to make up one-to-one correspondences of the set $\{1, 2, 3, 4\}$ onto itself and determine whether they can use a square to find a corresponding symmetry. Students can do some counting arguments to see that the number of symmetries of a square and the number of permutations of the set $\{1, 2, 3, 4\}$ are not even the same. We know there are four reflections and four rotations, giving us $2 \cdot 4 = 8$ symmetries of a square. When we permute the set $\{1, 2, 3, 4\}$, there are four choices for where we can send 1. Once that is determined, in order to make a one-to-one mapping, there are only three choices left for where to map 2. Once that is determined, there are only two choices of where to map 3, and finally only one choice left for where to map 4. Thus, there are $4! = 4 \cdot 3 \cdot 2 \cdot 1 = 24$ different permutations of the set $\{1, 2, 3, 4\}$. Note that $2 \cdot 3$, the number of symmetries of an equilateral triangle is equal to $3!$, the number of permutations of 3 things. Students can see that $n! > 2n$ for all $n > 3$, since $n! = n(n-1) \cdot \ldots \cdot 2 \cdot 1$ has more factors than $2n$ when $n > 3$. So there are more permutations of n objects than there are symmetries of a regular n-gon when $n > 3$. In Section 4.7, the group of permutations of n objects is denoted S_n and

Section 4.9   In the Classroom: Groups of Symmetries of Regular Polygons   253

has order $n!$. Since D_n is a subset of S_n and is a group itself, D_n is a subgroup of S_n. In fact, we now know that D_n is a proper subgroup of S_n, for $n > 3$.

Subgroup Structures

By looking at Cayley tables, students have determined that the set of all symmetries of the regular n-gon satisfy the properties which make it a group. We can now investigate whether it is possible to take subsets of the symmetries and still have closure under composition, associativity, identity, and inverses. Clearly, any such subset we create will have to contain R_0, so that is a good starting point. It should be noted that associativity of composition on this subset will be inherited by the associativity of the composition on the entire set. So we really need students to concentrate on making sure that the composition of any two symmetries in this subset must lie in this subset and that for any symmetry that we put into this subset, we must also have its inverse. Since the composition of two rotations is a rotation and the inverse of a rotation is a rotation, the subset of the n rotational symmetries of the regular n-gon, $\{R_0, R_{\frac{360}{n}}, R_{\frac{360(2)}{n}}, \ldots, R_{\frac{360(n-1)}{n}}\}$, is a good example of a subgroup of D_n. In fact, it is not hard for students to take a regular n-gon and determine that we can use the symmetry $R_{\frac{360}{n}}$, which rotates the n-gon by $360/n°$, the minimal angle of symmetry, and compose it with itself repeatedly and can "generate" all of the n rotational symmetries. Let $R = R_{\frac{360}{n}}$ and let R^i denote the composition of R with itself i times. Then the subgroup of rotations is really $\{R, R^2, R^3, \ldots, R^{n-1}, R^n = R_0\}$. From Section 4.6, we know this means that the subgroup of all rotations is in fact a cyclic subgroup of D_n and that $R_{\frac{360}{n}}$ is a generator of this group.

Any subgroup of this cyclic subgroup of rotations will also be a cyclic subgroup of D_n. We can find these using the results in Section 4.6. How would high school students find these? We look at the example of a regular hexagon. We have six angles of 60 degrees about the center O of the hexagon. This means that we have six rotational symmetries $\{R, R^2, R^3, R^4, R^5, R^6 = R_0\}$ with $R = R_{\frac{360}{6}} = R_{60}$ forming a cyclic subgroup of D_6. Now, students can determine as in Figure 3 that we also have three angles of 120 degrees about the center O of the hexagon which will give us the cyclic subgroup $\{R_{120}, R_{240}, R_0\} = \langle R_{120} \rangle$. We also have two angles of 180 degrees about the center O of the hexagon, which will give us the cyclic subgroup $\{R_0, R_{180}\}$. We also have the trivial subgroup $\{R_0\}$.

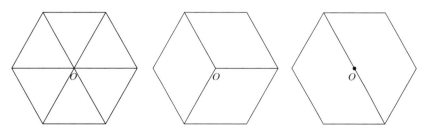

FIGURE 3

Since we could take all the rotations and have closure under composition, associativity, identity, and inverses, students might then think that they can put R_0 into a subset with all the n reflections and still satisfy these properties. Let them try this. What is the problem? Why is this subset not a subgroup? We have already seen that the composition of two reflections is a rotation so this subset fails to be closed under composition. However, if we consider a set that contains the identity, R_0, and any *one* of the reflectional symmetries, call it F, then $\{R_0, F\}$ is a subgroup of D_n. In fact, it is a cyclic subgroup generated by F.

Having seen examples of cyclic groups, students can be asked to explore why the entire set of $2n$ symmetries forming D_n is not a cyclic group. We recall that any reflection has order 2 and will only generate the identity and itself, any rotation composed with itself only generates other rotations, and the identity will only generate itself. Therefore, there is no candidate in D_n for a single generator of this set; we cannot use just one rotation or just one reflection to get all of the rotations and reflections.

Now, we can investigate whether it is possible to put together a subset of the symmetries that contains R_0 and *both* reflections and rotations (but not the entire set of them) and still maintain closure and inverses. Start by putting R_0 in the subset. If you put in any other rotation, you will need to put in all the rotations it generates. Perhaps it might be easier to add in a reflection. Since a reflection has order 2, it will not generate anything that is not already in our subset. What happens if you want to add in another reflection? Now you must concern yourself with the composition of those two reflections, which we know will be a rotation. In fact, depending on the two reflections and the order in which they are composed, we may now have to add in two rotations and any further compositions that would result by their inclusion. In the high school curriculum, students often investigate the fact that the result of reflecting a figure over two intersecting lines is a rotation about the point at which the two lines intersect with angle of rotation $\pm 2x°$, where $x°$ is the measure of the nonobtuse angle between the two lines.

For high school students, perhaps we might start with the square and look for easy subgroup examples that contain both reflections and rotations. Let's start with a subset containing R_0, and H (reflection on the horizontal axis). If we want to add in another reflection, it might be good to start with V (reflection on the vertical axis). This makes things easier since VH and HV are the same and equal to R_{180}, which we determine by physically performing the composition or using the result above since the angle between V and H is $90°$. We can include R_{180} in our subset and since it has order 2, it will only generate itself and R_0 which we already have. When R_{180} is composed with either H or V, it will produce the other one. Therefore, the set $\{R_0, H, V, R_{180}\}$ is closed and since each nonidentity element has order 2, it is also closed under inverses and is therefore a subgroup of D_4. A similar argument shows that the subset containing R_0, R_{180}, and the reflections over the two diagonals of the square is also a subgroup. Students can explore other possible subsets and test whether they are subgroups.

Number Theory: Generators of Cyclic Groups of Rotations and Connections with Finite Fields Z_p

While investigating rotations and the fact that the set of rotations is a cyclic subgroup of the group of symmetries of a regular n-gon, we can also investigate whether rotations other than the rotation by $360/n°$ can be used to generate all the others and if so, which rotations? In general, can cyclic groups have more than one generator? Have students start with a particular n. Ask them to list the n rotational symmetries. Have them choose one of the rotations, R, at a time and compose it with itself repeatedly, making use of the Cayley table to help in recording which other rotations it generates. Which degree rotations are also generators? Since we know that $R_{360/n}$ is a generator, any of these new generators can be written as powers (compositions) of $R_{360/n}$. Look at the powers of $R_{360/n}$ that are generators. Students should see that the powers of $R_{360/n}$ that are also generators of the cyclic group of all rotations are those which are relatively prime to n—that is, those powers for which the greatest common divisor with n is 1. Students who chose a prime n will then have all the rotations being generators.

Finding the Generators of Dihedral Groups

We have noted that the group of symmetries of the regular n-gon is not cyclic. By this we mean that there is no single symmetry that can be used to generate all $2n$ symmetries. Students can investigate whether there is some minimum number of symmetries that can be used to generate all of them. That is, can we take some subset of the symmetries and compose them to produce all of D_n? In fact, D_n is generated by two elements, $R_{360/n}$ and any one of the reflections, call it F. Clearly by including $R_{360/n}$, we can compose it with itself and get all the rotations. So how do we get the n reflections as compositions of $R_{360/n}$ and F? Consider any reflection. Once we know where it sends one vertex, we will know where all the other vertices map to since reflections reverse orientation. Thus, any reflection is determined by where it sends one vertex of the regular n-gon. Suppose we have a reflection F^*, different than F and vertex A in our polygon. Let us figure out how to get F^* by making compositions using only F and $R_{360/n}$. Suppose we know that $F^*(A) = A'$, where A' is some vertex of the polygon other than A. Since F is a symmetry we know it to be a onto function from the set of vertices to the set of vertices, which means that there is some vertex B of the polygon such that $F(B) = A'$. We also know that there exists an i such that $R_{360/n}^i(A) = B$. Therefore, $FR_{360/n}^i(A) = A' = F^*(A)$. In other words, we can rotate vertex A around the polygon using $R_{360/n}$ until it is mapped to vertex B which is opposite A' in relation to the line of reflection for F and then use F to map B to A'. (In Figure 4, the vertices of our n-gon lie around the circle.) The net effect of these compositions is to map A to A', which is what F^* does. We have shown that every symmetry in D_n can be written as $F^i R_{360/n}^j$, for some $0 \leq j < n - 1$ and $i = 0$ or 1.

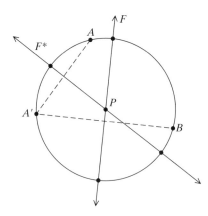

FIGURE 4

While Chapter 4 consisted of many advanced algebraic ideas, we have seen ways that high school students can explore these concepts and work with them in a very hands-on way.

4.9 Exercises

1. Verify that D_4 is indeed generated by R_{90} and any one of the four reflections.
2. In Section 4.2, you were asked to describe the 12 symmetries of the regular hexagon that make up D_6. In this section, we found all of the cyclic subgroups of D_6 that consist only of rotations. In addition, we have the six cyclic subgroups consisting of R_0 and any one of the six reflectional symmetries.

   Let us now find other subgroups of D_6 that contain both rotations and reflections. We will do this geometrically by investigating polygons inscribed in the regular hexagon. Consider regular hexagon $ABCDEF$ and equilateral triangle ACE. Some but not all of the symmetries of the hexagon are also symmetries of this triangle. For example, R_{120} is a symmetry of both the hexagon and triangle ACE but R_{60} is a symmetry only of the hexagon. Those symmetries of the hexagon that are also symmetries of this equilateral triangle form a subgroup of D_6.

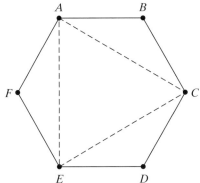

Section 4.9 In the Classroom: Groups of Symmetries of Regular Polygons 257

Let F_a, F_b, F_c, F_d, F_e, and F_f be the reflectional symmetries over the indicated axes.

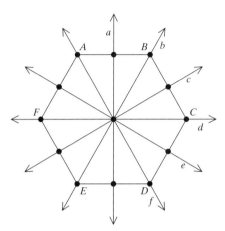

i. What subgroup of D_6 do we obtain by looking at the symmetry group of triangle ACE?
ii. What subgroup of D_6 do we get from the symmetry group of triangle BDF?
iii. Now consider rectangle $ABDE$. What subgroup of D_6 do we obtain by looking at the symmetry group of this rectangle?

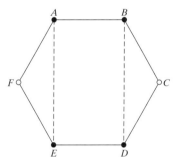

iv. Determine the subgroups that come from the symmetry groups of the other rectangles in this hexagon.
v. For each reflectional symmetry, find a polygon whose symmetry group is the subgroup of D_6 consisting of R_0 and that reflection.
vi. Verify that $\{R_0, R_{120}, R_{240}, F_a, F_c, F_e\}$ is a subgroup of D_6 and try to find a polygon in the hexagon whose symmetry group is this group.

3. Find real-life objects whose symmetry group is D_6.
4. Verify that D_4 is a subgroup of D_8 by finding a polygon inscribed inside a regular octagon whose symmetry group is D_4.
5. Verify that D_4 and D_6 are subgroups of D_{12} by finding polygons inscribed inside a regular dodecagon whose symmetry groups are D_4 and D_6.

6. Give an example of a polygon that has the symmetry group D_5 but is not a regular pentagon.
7. Develop an activity for high school students to explore and discover some of the ideas suggested in this section.

Bibliography

Budden, F. J. *The Fascination of Groups*. London: Cambridge University Press, 1972.

Dreyfus, Tommy, and Theodore Eisenberg, "Intuitive Functional Concepts: A Baseline Study on Intuitions." *Journal for Research in Mathematics Education* 13 (November 1982): 3360–80.

Jackiw, N. (1995). *The Geometer's Sketchpad*. [Computer software]. Emeryville, Calif.: Key Curriculum Press.

Markovits, Zvia, Bat Sheva Eylon, and Maxim Bruckheimer. "Difficulties Students Have with the Function Concept." In *The Ideas of Algebra K–12*, 1988 Yearbook of the National Council of Teachers of Mathematics (NCTM), edited by Arthur Coxford, pp. 43–60. Reston, Va.: NCTM, 1988.

National Council of Teachers of Mathematics. *Principles and Standards for School Mathematics*. Reston, Va.: National Council of Teachers of Mathematics, 2000.

Sierpinska, Anna. "On Understanding the Notion of Function." In *The Concept of Function: Aspects of Epistemology and Pedagogy*, edited by Guershon Harel and Ed Dubinsky, pp. 25–58. Washington, D.C.: Mathematics Association of America, 1992.

Tall, David. "Functions and Calculus." In *International Handbook of Mathematics Education, Part 1*, vol. 4, edited by Alan J. Bishop, Ken Clements, Christine Keitel, Jeremy Kilpatrick, and Colette Laborde, pp. 289–325. Dordrecht, Netherlands: Kluwer Academic Publishers, 1996.

Texas Instruments (1994). *Cabri Geometry II*. [Computer software]. Dallas, Texas.

4.10 *FROM THE PAST: ÉVARISTE GALOIS AND THE THEORY OF GROUPS—PART 1*

Today, Évariste Galois (1811–1832) is remembered for his deep mathematical discoveries and for his very short dramatic life. He lived in Paris in the period following the French Revolution and the reign of Napoleon. He took an active part in the radical politics of the day, to the point of spending time in prison for his political activities. He died in a duel that has alternately been described as suicide due to romantic difficulties, as suicide for the sake of political action, or as a plot by the forces of the French government. The sources for his life are plentiful enough to back each of these interpretations but not plentiful enough to decide between them. Galois was also unlucky in his mathematical career. His work was not published in his lifetime because of its depth and difficulty, and because of the unwillingness of the mathematical establishment to consider his advanced views. Luckily, he was able to put his discoveries into articles that eventually reached the eyes of sympathetic mathematicians. Though it took many years, he was eventually credited with the development of group theory and the branch of algebra now called Galois Theory.

One of the mathematical problems that interested Galois was the solution of polynomial equations. In particular, he continued the 300-year-old quest to determine when a given polynomial's roots could be expressed in terms of its coefficients, rational numbers, the four arithmetic operations, and the extraction of roots. Galois gave a complete though difficult answer to this question. On the way he initiated the theory of groups. His group theory was not a theory of abstract algebraic objects. It was a concrete study of the problem of the roots of polynomials. It is not exactly the theory of groups as we know it, but the family resemblance is clear.

Building on the work of Lagrange and Newton, Galois considered functions of the roots of a polynomial. For example, let $f(x) = (x - x_1)(x - x_2)(x - x_3) = x^3 + ax^2 + bx + c$. Then $a = -(x_1 + x_2 + x_3)$. Thus a and the rest of the coefficients of $f(x)$ are functions of the roots x_1, x_2, and x_3. By considering permutations of the roots and their effect on values of such functions of the roots, it was sometimes possible to express the roots themselves in terms of the coefficients of the polynomial. For example, consider the quadratic formula. It gives the roots of $ax^2 + bx + c = 0$ as a function of the coefficients. The roots are $x = \frac{-b \pm \sqrt{b^2 - 4ac}}{2a}$. This was Galois' starting point.

Galois took a fixed number of letters, and let them stand for the roots of a polynomial. For a quartic polynomial, we will use the letters a, b, c, and d to denote its roots. He called any arrangement of these letters a **permutation**. Thus $c\ a\ d\ b$ was a **permutation**. We can think of these arrangements as 4-tuples in which each letter appears once. A **substitution** was the result of substituting one letter for another. For example starting from the **permutation** $a\ b\ c\ d$ and applying the **substitution** that carries $a \to b, b \to d, c \to a, d \to c$ yields the **permutation** (new arrangement) $b\ d\ a\ c$. We can represent this **substitution** by $\begin{pmatrix} a & b & c & d \\ b & d & a & c \end{pmatrix}$. (Today we would use the term "permutation" for Galois' **substitution** and the term "arrangement" for his **permutation**.) We will use Galois' terminology in this article and keep it in boldface. So be aware!

Galois was interested in arrays of **permutations**. The following is an example of an array with four letters.

$$\begin{array}{cccc} a & b & c & d \\ b & c & d & a \\ c & d & a & b \\ d & a & b & c \end{array}$$

Array 1

He considered this array to be four **substitutions** with the top row of the array as the beginning. It was a short way of writing the following set of **substitutions**:

$$\begin{pmatrix} a & b & c & d \\ a & b & c & d \end{pmatrix}, \begin{pmatrix} a & b & c & d \\ b & c & d & a \end{pmatrix}, \begin{pmatrix} a & b & c & d \\ c & d & a & b \end{pmatrix}, \begin{pmatrix} a & b & c & d \\ d & a & b & c \end{pmatrix}.$$

The first **substitution** is the identity **substitution** since it carries each letter to itself. Note that each row of the array appears as the bottom row in one of the **substitutions** while the first row of the array is the top row of each **substitution**. When an array was closed under composition, Galois called it a "group." For example, consider just rows 2 and 3:

$$\begin{matrix} b & c & d & a \\ c & d & a & b \end{matrix}.$$

These two rows represent the **substitutions** $\begin{pmatrix} a & b & c & d \\ b & c & d & a \end{pmatrix}$ and $\begin{pmatrix} a & b & c & d \\ c & d & a & b \end{pmatrix}$. When composed, they yield the substitution $\begin{pmatrix} a & b & c & d \\ d & a & b & c \end{pmatrix}$, which is the substitution defined by the first and the fourth row. You can check that Array 1 defines a set of four **substitutions** that is closed under composition. Hence we have a group in the sense of Galois. We also have a group in the modern sense also because any nonempty subset of S_n that is closed under composition is a subgroup of S_n. (See Proposition 1 of Section 4.4.) Now label the **substitutions** e, α, β, and χ as follows.

$$e = \begin{pmatrix} a & b & c & d \\ a & b & c & d \end{pmatrix}$$

$$\alpha = \begin{pmatrix} a & b & c & d \\ b & c & d & a \end{pmatrix}$$

$$\beta = \begin{pmatrix} a & b & c & d \\ c & d & a & b \end{pmatrix}$$

$$\chi = \begin{pmatrix} a & b & c & d \\ d & a & b & c \end{pmatrix}$$

Their compositions form the Cayley table given in Table 1.

TABLE 1

$\circ$	e	α	β	χ
e	e	α	β	χ
α	α	β	χ	e
β	β	χ	e	α
χ	χ	e	α	β

You can see from Table 1 that the subgroup of S_4 defined by Array 1 is isomorphic to the group $\{\mathbf{Z_4}, +\}$. All of Galois' groups were subgroups of S_n for some n.

We will call the array that defines a group a **presentation** of that group. If we rearrange the columns of Array 1, we have exactly the same **substitutions**, and hence the same group. It is a different "presentation" of that group. For example, if we simply switch columns 1 and 2 in Array 1, we obtain the Array 2.

$$\begin{array}{cccc} b & a & c & d \\ c & b & d & a \\ d & c & a & b \\ a & d & b & c \end{array}$$

Array 2

These **permutations** form exactly the same set of **substitutions** as the original array. In the original case the top two rows give the **substitution** $\begin{pmatrix} a & b & c & d \\ b & c & d & a \end{pmatrix}$. In the second form of the array the top two rows give the **substitution** $\begin{pmatrix} b & a & c & d \\ c & b & d & a \end{pmatrix}$. These **substitutions** are equal because b goes to c in both cases, and a goes to b, etc. So Array 2 is simply a different presentation of the group Z_4 as first presented in Array 1.

Galois found that he could list the six different **presentations** of the group presented in Array 1 in such a way that every **permutation** of the symbols a, b, c, and d appeared exactly once. (See Figure 1.)

$$\begin{array}{cccc cccc cccc} a & b & c & d & \quad b & a & c & d & \quad c & b & a & d \\ b & c & d & a & \quad c & b & d & a & \quad d & c & b & a \\ c & d & a & b & \quad d & c & a & b & \quad a & d & c & b \\ d & a & b & c & \quad a & d & b & c & \quad b & a & d & c \\ \\ d & b & c & a & \quad a & c & b & d & \quad a & b & d & c \\ a & c & d & b & \quad b & d & c & a & \quad b & c & a & d \\ b & d & a & c & \quad c & a & d & b & \quad c & d & b & a \\ c & a & b & d & \quad d & b & a & c & \quad d & a & c & b \end{array}$$

FIGURE 1

To form a second array, we can use any **permutation** not found in the first array as the first row of a new presentation. For instance, $b\ a\ c\ d$ is not found in the first array; it became the top row of the second array in Table 1. To form a third array, simply pick as a top permutation any permutation not found in the first two arrays. Continuing this process will lead to a set of six different presentations of the original group. All 24 permutations will be found in the set of six arrays exactly once. This set of six arrays is certainly not unique. We could have also used $d\ c\ b\ a$, which is not in the first array, as the top row of the second presentation. From this permutation we would obtain Array 3, which is not listed in the six arrays given in Figure 1.

$$\begin{array}{cccc} d & c & b & a \\ a & d & c & b \\ b & a & d & c \\ c & b & a & d \end{array}$$

Array 3

262   Chapter 4   A First Look at Group Theory

Let's generalize a bit. Suppose that G is a subgroup of S_n that is presented in two ways, in Array A and in Array B. Then the following statements hold.

Statement i. If the first row of Array B does not appear anywhere in Array A, then none of the rows of Array A appear in Array B.

Statement ii. If the first row of Array A does appear somewhere in Array B, then all the rows of Array A appear in Array B. In fact, we can always rearrange the rows of a presentation in any order.

We will show that the contrapositive of the first statement holds and you will tackle Statement ii in the third task of three tasks given below.

Proof of Statement i.   Suppose that a row i from Array A is identical to row j in Array B. Let α be the **substitution** that row j defines in Array B and let β be the substitution that row i defines in Array A. These must both be in G and so $\alpha^{-1} \circ \beta$ must also be in G. The **substitution** $\alpha^{-1} \circ \beta$ goes from row 1 of Array A to row 1 of Array B. So row 1 of Array B must appear in Array A.   ▲

Task 1

Find a **presentation** of the cyclic subgroup of S_5 generated by the **substitution**

$$\begin{pmatrix} a & b & c & d & e \\ b & c & d & a & e \end{pmatrix}.$$

Task 2

Find two other presentations of the subgroup of S_3 presented in the following array. Do this in such a way that no permutation appears more than once.

$$\begin{array}{ccc} a & b & c \\ a & c & b \end{array}$$

Task 3

Prove Statement ii.

We will continue to explore the relationships that exist between the different presentations of a group in Section 5.7. Now let's see how Galois used groups of permutations to analyze polynomials. The roots of a polynomial $p(x)$ are not necessarily found in the field from which its coefficients are found. For instance, the polynomial $p(x) = x^4 - x^2 - 2$ has its coefficients in Q. However, its roots, $\pm\sqrt{2}, \pm i$, are in the larger field $Q(\sqrt{2}, i) = \{a + bi + c\sqrt{2} + d\sqrt{2}i : a, b, c, d \in Q\}$, which is itself a subfield of the complex numbers. (We leave it as an exercise to show that the set $Q(\sqrt{2}, i)$ is indeed a field.) More generally, suppose that a polynomial $f(x) \in Q[x]$ has roots in a field F and that $Q \subseteq F \subseteq C$. Galois considered bijections from F to F that preserved both the operations of addition and multiplication. Such maps are called **automorphisms**. We shall study automorphisms more closely in Chapter 6. There we shall prove that if $\varphi : F \to F$ is an automorphism and if a is a root of

Section 4.10 From the Past: Évariste Galois and the Theory of Groups—Part 1

$f(x)$, then $\varphi(a)$ must also be a root of $f(x)$. In other words, φ is a permutation of the roots of $f(x)$. The set of all automorphisms is a group under composition, but not all possible permutations of roots of $f(x)$ can occur as a result of an automorphism. Let's look again at the example of $p(x) = x^4 - x^2 - 2$ and its roots $\pm\sqrt{2}, \pm i$. Let σ be the function from $Q(\sqrt{2}, i)$ to itself that satisfies $\sigma(\sqrt{2}) = -\sqrt{2}$, $\sigma(i) = i$, and $\sigma(r) = r$ for every rational number r. The map σ is an automorphism. (We leave the details as an exercise.) Similarly, let γ be the function satisfying $\gamma(\sqrt{2}) = \sqrt{2}$, $\gamma(i) = -i$, and $\gamma(r) = r$ for every rational number r. The map γ is also an automorphism. The set $\{e, \sigma, \gamma, \sigma\gamma\}$ forms a group under the operation of composition. As permutations of the roots, this group is presented as follows.

$$\begin{array}{cccc} \sqrt{2} & -\sqrt{2} & i & -i \\ -\sqrt{2} & \sqrt{2} & i & -i \\ \sqrt{2} & -\sqrt{2} & -i & i \\ -\sqrt{2} & \sqrt{2} & -i & i \end{array}$$

It is a **presentation** of the Klein 4-Group. What is important to note that there is no operation preserving function on $Q(\sqrt{2}, i)$ that is the identity on Q and sends $\sqrt{2}$ to i or to $-i$. If there were such a function σ, then we would have $2 = \sigma(2) = \sigma(\sqrt{2}\sqrt{2}) = \sigma(\sqrt{2})\sigma(\sqrt{2}) = i \cdot i = -1$, which is not true. The group structure reflects the fact that the polynomial $p(x) = x^4 - x^2 - 2$ can be factored as $p(x) = (x^2 - 2)(x^2 + 1)$.

A second example that gives rise to a different group of order 4 comes from the polynomial $q(x) = x^4 + x^3 + x^2 + x + 1$. The polynomial $q(x)$ is a factor of the polynomial $x^5 - 1$. Thus the roots of $q(x)$ are the fifth roots of unity different from 1. One of these roots is given by

$$\gamma = \left(\frac{\sqrt{5} - 1}{4}\right) + i\left(\frac{\sqrt{2}(\sqrt{5 + \sqrt{5}})}{4}\right).$$

(You can find this by applying the Law of Cosines to the regular pentagon.) The other roots of $q(x)$ are powers of γ, namely γ^2, γ^3, and γ^4. Let σ be the operation preserving function on the field of numbers $Q(\gamma) = \{a + b\gamma + c\gamma^2 + d\gamma^3 + e\gamma^4 : a, b, c, d, e \in Q\}$ such that $\sigma(x) = x$ for any rational number x and such that $\sigma(\gamma) = \gamma^3$. Then σ sends the set $(\gamma, \gamma^2, \gamma^3, \gamma^4)$ to the set $(\gamma^3, \gamma, \gamma^4, \gamma^2)$ in order. Similarly, σ^2 sends $(\gamma, \gamma^2, \gamma^3, \gamma^4)$ to $(\gamma^4, \gamma^3, \gamma^2, \gamma)$, while σ^3 carries $(\gamma, \gamma^2, \gamma^3, \gamma^4)$ to $(\gamma^2, \gamma^4, \gamma, \gamma^3)$. Finally, σ^4 is the identity. Thus the group $\{\sigma, \sigma^2, \sigma^3, \sigma^4\}$ is a group under composition that is isomorphic to $(\mathbb{Z}_4, +)$. Array 4 is a presentation of this group.

$$\begin{array}{cccc} \gamma & \gamma^2 & \gamma^3 & \gamma^4 \\ \gamma^3 & \gamma & \gamma^4 & \gamma^2 \\ \gamma^4 & \gamma^3 & \gamma^2 & \gamma \\ \gamma^2 & \gamma^4 & \gamma & \gamma^3 \end{array}$$

Array 4

The last example was of particular importance to Galois since it presented a cyclic group. Galois was able to determine whether a polynomial was solvable based on the structure of the group of permutations of its roots that arise from automorphisms. We shall return to just how this is done in Chapter 6.

Postscript on the Definition of a Group

We have seen how Galois defined a group based on what he needed to solve the problem of the roots of a polynomial. He was not particularly interested in groups in and of themselves, but in certain relations between the permutations of roots of a given polynomial. He needed the properties of subgroups to accomplish his task. In the years following Galois' death, other mathematicians, studying problems in geometry, found that certain sets of linear transformations had properties similar to those of Galois' groups. They concentrated on their geometric problems rather than the similarities to Galois' work. In 1854 the English mathematician Arthur Cayley published a paper titled, *On the Theory of Groups, as Depending on the Symbolic Equation* $\theta^n = 1$. He looked at groups as structures independent of any application of the elements of the group. He defined a group as a set of symbols

$$1, \alpha, \beta, \ldots$$

each of which could have any meaning whatever. The essential point was that there was some associative operation that combined two of the elements of the group to form a third. In his words,

> [T]he symbols $\theta, \phi, \ldots$ are in general such that $\theta \cdot \phi\chi = \theta\phi \cdot \chi$, etc., so that $\theta\phi\chi$, $\theta\phi\chi\omega$ etc. have a definite signification independent of the particular mode of compounding the symbols.

Cayley assumed that each of his groups had a finite number of elements, and that he had a closed and associative operation with an identity element. By what we proved in Proposition 1 of Section 4.4, these assumptions imply that Cayley's definition of a group is equivalent to the modern definition of a group. Cayley did not stop with the definition but showed groups with the same number of elements could be different. He began with the statement,

> Suppose that the group $1, \alpha, \beta, \ldots$ contains n symbols, it may be shown that each of these symbols satisfies the equation $\theta^n = 1$; so that a group may be considered as representing a system of roots of this symbolic binomial equation.

Letting $n = 4$, he showed by what are now called *Cayley tables* that there were two different groups with four elements. (He did the same with $n = 6$, where he found a nonabelian group.) Using the symbols $1, \alpha, \beta, \gamma$ for his groups of order 4, he gave the following two tables:

TABLE 1

$\cdot$	1	α	β	γ
1	1	α	β	γ
α	α	β	γ	1
β	β	γ	1	α
γ	γ	1	α	β

TABLE 2

·	1	α	β	γ
1	1	α	β	γ
α	α	1	γ	β
β	β	γ	1	α
γ	γ	β	α	1

These tables define groups that are isomorphic to $(\mathbf{Z}_4, +)$ and the Klein 4-Group respectively.

As students of mathematics, we sometimes work as if definitions come first, out of the blue, before theory that utilizes them. "Let [...] be defined as .... Prove that [...] is ...." But in practice, definitions often emerge as distillations of what is either needed by or common to evolving theories. But once a definition carves out a good and rich idea, new roads into mathematics are opened. Such was the case with the word "group," as defined by Cayley.

BIBLIOGRAPHIC NOTE

Two very good references for Galois' work that cover far more than is in this essay are the following:

1. Edwards, Harold. *Galois Theory*. New York: Springer-Verlag, 1984.
2. Radloff, Ivo. "Évariste Galois: Principles and Applications," *Historia Mathematica*, 29 (2002),114–137.

Both works present Galois' accomplishments from a historical perspective with a translation to modern language.

4.10 Exercises

1. Find a different subgroup of S_4 isomorphic to $(\mathbf{Z}_2, +)$.
2. Given the 2 by 3 array below, find two other 2 by 3 arrays by switching columns so that when placed together into a 6 by 3 array they form the entire group S_3.

$$\begin{array}{ccc} a & b & c \\ a & c & b \end{array}$$

3. Find the three **substitutions** represented by the array below and show that it forms a group:

$$\begin{array}{ccc} a & b & c \\ c & a & b \\ b & c & a \end{array}$$

4. Find **substitutions** of orders 2, 3, 4, and 5 in S_5.

5. Find the roots of the polynomial $p(x) = x^4 - 12x^2 + 35$. What group corresponds to the operation preserving functions of the field over the rationals that contains these roots?
6. Find the roots of the polynomial $p(x) = x^4 + 1$. What group corresponds to the operation preserving functions of the field over the rationals that contains these roots?

Chapter 4 Highlights

Although the definition of the mathematical term **group** is simple, the variety, complexity, and utility of the structures covered by the term are enormous. In this chapter, we first covered the vocabulary and general properties of abstract groups. We sorted groups according to the their **orders**, as finite or infinite, and as **abelian** or **nonabelian**. We investigated **symmetry groups** of regular geometric figures. We found groups within groups—**subgroups**—and we developed criteria for detecting and generating subgroups. In particular, if G is a group and $a \in G$, then $\langle a \rangle$ is called the **cyclic** subgroup generated by a. The **order of the element** a is equal to the order of the subgroup $\langle a \rangle$, namely, $|a| = |\langle a \rangle|$. Two groups G and H differ only in notation when there is an **isomorphism** between them, namely, an **operation preserving** bijection from G to H. More generally, **homomorphisms** between groups detect structural similarities. We ended the chapter with a closer investigation of two important types of groups, cyclic groups and permutation groups. We discovered that all infinite cyclic groups are isomorphic to $\mathbf{Z}$ under addition and all cyclic groups of order n are isomorphic to $\mathbf{Z}_n$. The nonabelian **permutation groups** S_n are more challenging. We developed the **cycle decomposition** notation whereby we could express permutations as the products of disjoint, commuting cycles. We saw that any permutation α could be classified as an **even** or **odd permutation**, according to the number of **transpositions** needed to express it. The nth **alternating subgroup** A_n of S_n is the subgroup of all even permutations.

Chapter 4 Questions

1. Give the definition of the each following terms.

    i. Group
    ii. Inverse
    iii. Identity
    iv. Abelian group
    v. Order of a group
    vi. Order of an element in a group
    vii. Subgroup
    viii. Subgroup generated by the set S
    ix. Cyclic group
    x. Isomorphism

xi. Homomorphism

xii. Kernel

xiii. Cycle decomposition

xiv. Even and odd permutation

xv. Transposition

xvi. Cayley table

2. Suppose that $\{G, *\}$ is a group with identity element e. Finish its Cayley table that begins as follows.

*	e	a	b	c	d
e					
a		c			
b			a		
c		b		d	
d					b

3. Is the set $\{0, 2, 4, 6\}$ a group under addition mod 8? Justify your answer.

4. Suppose that $\{G, *\}$ is a group with identity element e. Suppose that $x^2 = e$ for every x in G. Prove that G is an abelian group.

5. Suppose that $\{G, *\}$ and $\{H, \bullet\}$ are both groups.

   i. Show that $\{G \times H, \otimes\}$ is a group where $G \times H$ is the Cartesian product of the sets G and H and the operation $\otimes$ is defined as $(x, y) \otimes (s, t) = (x * s, y \bullet t)$.

   ii. Write out the Cayley table for $G \times H$ when $\{G, *\} = \{\mathbf{Z}_3, +\}$ and $\{H, \bullet\} = \{U_8, \bullet\}$.

6. Suppose that a regular n-gon is drawn so that it has a vertical axis of symmetry. Suppose that the vertices of a regular n-gon are labeled 1 through n going clockwise. Suppose that R denotes clockwise rotation through $360/n$ degrees. Suppose that F denotes rotation through 180 degrees around the vertical axis of symmetry. If vertex 1 starts out on the vertical axis, where does vertex 1 end up under $R * F$ versus $F * R$? How does your observation show that D_n is nonabelian for $n \geq 3$?

7. Solve for x in the group $D_4 : x * V = R_3$.

8. Find the symmetry group of a nonsquare rectangle. (It has order 4.) Write down its group table. Is it abelian?

9. Find all the symmetries of the following figure. It is a regular hexagon with an equilateral triangle inscribed.

10. List all the subgroups of D_6.
11. List all the subgroups of $\{Z_{12}, +\}$.
12. Suppose that the vertices of a regular hexagon are labeled 1 through 6. List all the elements in the subgroup of D_6 generated by the set $S = \{F_1, R_{120}\}$, where R_{120} is rotation through 120 degrees and F_1 is a rotation of 180 degrees about the axis through vertex 1 and 4.
13. Decide if U_{13} is cyclic. What about U_{15}?
14. Let $\{G, *\}$ be a group and $a \in G$ be an element of order n. Prove that if m and n are relatively prime, then a^m is a generator of $\langle a \rangle$.
15. Find the subgroup generated by the matrix $A = \begin{bmatrix} 0 & -1 \\ -1 & 0 \end{bmatrix}$ in $GL(\mathbf{R}, 2)$ (i.e., find $\langle A \rangle$).
16. Let $A = \begin{bmatrix} 1 & 0 \\ 1 & 1 \end{bmatrix}$. Characterize the elements in $\langle A \rangle$, the subgroup generated by A in $GL(\mathbf{R}, 2)$.
17. Either find a generator $U_8 \times Z_5$ or show that it is not cyclic.
18. Show that if H_1 is a subgroup of G_1 and H_2 is a subgroup of G_2, then $H_1 \times H_2$ is a subgroup of $G_1 \times G_2$. Let H be a subgroup of. $G_1 \times G_2$. Is H necessarily of the form $H_1 \times H_2$?
19. Let G be the group $\{\frac{1}{2^n} : n \in \mathbf{Z}\}$ under multiplication. Find an isomorphism between $\{\mathbf{Z}, +\}$ and G.
20. Is the group of symmetries of the rectangle isomorphic to $\{Z_4, +\}$ or the Klein 4-Group?
21. Let $f : G_1 \to G_2$ be a surjective homomorphism. Prove that f is an isomorphism if and only if $\ker(f) = \{e_{G_1}\}$.
22. Let G be a group and let $g \in G$. Define $T_g : G \to G$ to be the map defined by $T_g(x) = g^{-1}xg$.

    i. Prove that T_g is an isomorphism.
    ii. Let $G = D_4$ and $g = R_1$. Find $T_g(x)$ for each x in D_4.
    iii. Repeat Part ii with $g = R_2$. What accounts for the difference?

23. Suppose that $G = \langle a \rangle$ is a cyclic group of order 20. Find all the generators of G. Find all the proper subgroups of G.
24. Suppose that G is a cyclic group of order n and that i is a positive integer relatively prime to n. Prove that a^i is a generator of G.

25. Suppose that $G = \langle a \rangle$ is a cyclic group of order n. How many distinct isomorphisms are there from G to G?
26. Suppose that $G = \langle a \rangle$ is a cyclic group of order n. What are the generators of the subgroup $\langle a^i \rangle$? Give an example for a specific value of n.
27. Find the cycle decomposition for the following map: $\begin{pmatrix} 1 & 2 & 3 & 4 & 5 & 6 & 7 & 8 \\ 2 & 3 & 8 & 1 & 5 & 7 & 6 & 4 \end{pmatrix}$.
28. Find the cycle decomposition of $\alpha \circ \alpha$, where $\alpha = (1, 3, 5)(2, 4, 6, 7, 8)$.
29. How many distinct permutations are there of the form $(a, b, c)(e, f, g)(h)$ in S_7?
30. Let $1 \leq i \leq n$. Show that there is an element of order i in S_n.
31. The cycle decomposition of a permutation β in S_7 is $(1, 4, 5)(3, 7, 6, 2)$. Express β as the product of transpositions.

5
NEW STRUCTURES FROM OLD

In Chapter 2, we constructed $\mathbf{Z}_m$ from $\mathbf{Z}$. In this chapter, we generalize that construction to other groups and rings. First, let's review the process of constructing $\mathbf{Z}_m$ using the vocabulary of Chapter 4. For each positive integer m, the set $m\mathbf{Z} = \{\ldots, -2m, -m, 0, m, 2m, 3m, \ldots\}$ is a subgroup of $\mathbf{Z}$. The objects in the group $\{\mathbf{Z}_m, +\}$ are the equivalence classes of the following relation: Two elements x and y are related (congruent) if and only if their difference $x - y$ is in the subgroup $m\mathbf{Z}$. In this chapter, we start with an arbitrary group $\{G, *\}$ and a special kind of subgroup H called a "normal" subgroup, and we construct a new group G/H called a "quotient group." The objects in G/H are the equivalence classes of the relation in which x is related to y if and only if $xy^{-1} \in H$. (Note the change to multiplicative notation.) Similarly, starting with a ring R and a special subring I called an "ideal," we will construct a new ring R/I, called a "quotient ring." Both of these constructions are important tools for investigating the various structures that groups and rings possess. And both are crucial to understanding the link between abstract algebra and classical algebra—solving polynomial equations.

5.1 COSETS

Let G be the group of integers and $H = m\mathbf{Z}$ for some $m \in \mathbf{N}$. Let $a \in \mathbf{Z}$. We can think of a member of the group $\{\mathbf{Z}_m, +\}$ as a set of the form $[a]_m = \{a + x : x \in H\}$. Now let G be an arbitrary group that uses the conventions of multiplication for its operation. Let $H \subseteq G$ be a subgroup, and $a \in G$. The analogous set is $[a] = \{ax : x \in H\}$. Such a set is called a **coset**. It will turn out to be the building block of our new structure, the quotient group. But we must be a bit careful. Since, in general, a group $\{G, *\}$ is not necessarily abelian, we must distinguish between ax and xa. That nuance is reflected in the next definition.

> **DEFINITION 1.** Let $\{G, *\}$ be a group and H a subgroup of G. Let $a \in G$. The set $aH = \{ax : x \in H\}$ is called the **left coset** of H containing a. Similarly, the set $Ha = \{xa : x \in H\}$ is called the **right coset** of H containing a.

In aH, all the elements of the subgroup H are operated on the left by a. In Ha, all the elements of H are operated on the right by a. The element a is in both aH and Ha because $e \in H$ and $a = ae = ea$.

EXAMPLE 1

Let $G = \{S_3, \circ\}$. Let $H = \{e, (1,2)\}$.

$a \in G$	left coset aH	right coset Ha
e	$\{ee, e(1,2)\} = \{e, (1,2)\} = H$	$\{ee, (1,2)e\} = \{e, (1,2)\} = H$
$(1,2,3)$	$\{(1,2,3)e, (1,2,3)(1,2)\} = \{(1,2,3), (1,3)\}$	$\{e(1,2,3), (1,2)(1,2,3)\} = \{(1,2,3), (2,3)\}$
$(1,3,2)$	$\{(1,3,2)e, (1,3,2)(1,2)\} = \{(1,3,2), (2,3)\}$	$\{e(1,3,2), (1,2)(1,3,2)\} = \{(1,3,2), (1,3)\}$
$(1,2)$	$\{((1,2)e, (1,2)(1,2)\} = \{(1,2), e)\} = H$	$\{e(1,2), (1,2)(1,2)\} = \{(1,2), e\} = H$
$(1,3)$	$\{(1,3)e, (1,3)(1,2)\} = \{(1,3), (1,2,3)\}$	$\{e(1,3), (1,2)(1,3)\} = \{(1,3), (1,3,2)\}$
$(2,3)$	$\{(2,3)e, (2,3)(1,2)\} = \{(2,3), (1,3,2)\}$	$\{e(2,3), (1,2)(2,3)\} = \{(2,3), (1,2,3)\}$

Notice that, in general, $aH \neq Ha$. For instance, $(1,2,3)H = \{(1,2,3), (1,3)\}$ but $H(1,2,3) = \{(1,2,3), (2,3)\}$. The right coset containing a is not necessarily equal to the left coset containing a because, in general, $ax \neq xa$. However, even though $a \neq b$, the coset aH can equal the coset bH. For instance, $(1,2,3)H = (1,3)H$. ∎

When H is a subgroup of a group $\{G, +\}$ that uses addition, we often write $a + H$ and $H + a$ for the left and right cosets of a. Since $+$ is always commutative, $a + H$ always equals $H + a$.

EXAMPLE 2

The set $H = \{0, 3\}$ is a subgroup of $\{\mathbf{Z_6}, +\}$. Here are the left cosets of H in $\mathbf{Z_6}$.

$$0 + H = \{0 + 0, 0 + 3\} = \{0, 3\}$$
$$1 + H = \{1 + 0, 1 + 3\} = \{1, 4\}$$
$$2 + H = \{2 + 0, 2 + 3\} = \{2, 5\}$$
$$3 + H = \{3 + 0, 3 + 3\} = \{3, 0\}$$
$$4 + H = \{4 + 0, 4 + 3\} = \{4, 1\}$$
$$5 + H = \{5 + 0, 5 + 3\} = \{5, 2\}$$

We see that left cosets are either disjoint like $1 + H$ and $2 + H$ or identical like $1 + H$ and $4 + H$. A typical right coset would be $H + 1 = \{0 + 1, 3 + 1\} = \{1, 4\}$. In this example $a + H = H + a$ for each $a \in H$ because addition is commutative. ∎

Two important things to notice in both Examples 1 and 2 are:

i. The number of elements in any coset (right or left) is the same as the number of elements in H;
ii. The left cosets partition G. That is to say, two left cosets are either disjoint or identical. Furthermore, every element of G is in some left coset. Similarly, the right cosets partition G.

In Example 2, the fact that G is partitioned by the left cosets of H points to an underlying equivalence relation. In fact, G is partitioned into disjoint sets, each the *same size* as H. The following proposition generalizes our observations. It is stated for left cosets but it holds true if all left cosets are replaced by right cosets.

PROPOSITION 1 Let $H \subseteq G$ be a subgroup of G and let $a \in G$.

i. There is a one-to-one correspondence between the members of H and the members of the left coset aH.
ii. Every element x of G is in some left coset of H.
iii. If a and b are elements of G, then either $aH = bH$ or $aH \cap bH = \emptyset$.

Proof.

i. Let $a \in G$ and define $\varphi \colon H \to aH$ by $\varphi(h) = ah$ for each $h \in H$. The map φ is injective because when $ah_1 = ah_2$, multiplying on both sides by a^{-1} shows that $h_1 = h_2$. The map is surjective because if $x \in aH$, then $x = ah$ for some $h \in H$ and $\varphi(h) = ah = x$. Since φ is a bijection, H and aH have the same cardinality.
ii. The element x is in the coset xH because $e \in H$.
iii. Suppose that $aH \cap bH \neq \emptyset$. We shall show that $aH \subseteq bH$. Let $z \in aH \cap bH$. We can find h_1 and h_2 in H such that $z = ah_1$ and $z = bh_2$. Since $ah_1 = bh_2$, we solve for a to find that $a = bh_2h_1^{-1}$. Let y be any element in aH. Then $y = ah$ for some $h \in H$. We can substitute for a and write $y = bh_2h_1^{-1}h$. Since $h_2h_1^{-1}h \in H$, $y \in bH$ and we have shown that $aH \subseteq bH$. A similar argument shows that $bH \subseteq aH$. Thus $bH = aH$. ▲

Parts 2 and 3 of Proposition 1 show us that the set of left cosets of H partition G. Thus the relation R on G defined by "aRb if and only if a and b are in the same coset" is an equivalence relation on G. When we studied the mod m equivalence relation on $\mathbf{Z}$, we saw that two integers x and y are in the same equivalence class mod m if and only if $x - y \in m\mathbf{Z}$. The criterion for determining when two elements a and b are in the same coset is similar once we translate into multiplicative notation and take care of a possible lack of commutativity.

PROPOSITION 2 Let H be a subgroup of G and let a and b be elements of G. Then $aH = bH$ if and only if $b^{-1}a \in H$.

Proof. First suppose that $b^{-1}a \in H$. Then $b^{-1}a = h$ for some $h \in H$ and $bh = bb^{-1}a = a$. Thus $a \in bH$. Since a is in both aH and bH, $aH = bH$. Conversely, assume that $aH = bH$. Since $a \in aH$ it must be that $a \in bH$ and $a = bh$ for some $h \in H$. Solving for h, we see that $b^{-1}a = h$ and so $b^{-1}a \in H$. ▲

Notice that if G were an additive group, then the term "$b^{-1}a$" would be written as "$a - b$." The criterion for a and b to be in the same equivalence class would then be that $a - b \in H$, which is just how we defined the mod m relation on $\mathbf{Z}$ for $H = m\mathbf{Z}$.

COROLLARY 3 The coset aH is equal to H if and only if $a \in H$.

Proof. We may take $b = e$ in Proposition 2. ▲

EXAMPLE 3

Let's check Proposition 2 with the cosets of $H = \{e, (1, 2)\}$ given in Example 1. You can look on the chart to see that the cosets of $a = (1, 3, 2)$ and of $b = (3, 2)$ are identical (i.e., $aH = bH$). To verify Proposition 2, we compute $b^{-1}a$ and check that it is in H.

$$b^{-1}a = (3, 2)(1, 3, 2) = (1, 2)(3) = (1, 2)$$

Indeed $(1, 2)$ is an element of H. ■

Finally, we give a criterion to determine when $aH = Ha$.

PROPOSITION 3 Let H be a subgroup of G and let $a \in G$. Let $aHa^{-1} = \{aha^{-1} : h \in H\}$. Then $aH = Ha$ if and only if $aHa^{-1} = H$.

Proof. Exercise 15.

In Section 5.2 we shall use cosets to construct new groups. We will use the remainder of this section to investigate important numerical implications of the above propositions. The first, known as **Lagrange's Theorem**, gives us very useful numerical information about finite groups.

THEOREM 4 (*Lagrange's Theorem*) Let G be a finite group and H be a subgroup of G. Suppose that $|H| = m$ and $|G| = n$. Then m divides n. Furthermore, the number of distinct left (similarly right) cosets of H in G is n/m.

Proof. The theorem follows immediately from our observation that the left (right) cosets of H partition G into disjoint subsets, each the size of H. Thus the order of G is a multiple of the order of H, namely, $n = pm$ where p is the number of distinct cosets of H. ▲

A group of order 24 *may* have proper subgroups of orders 1, 2, 3, 4, 6, 8, or 12 but not one of order 7. The converse need not hold. A number n may divide $|G|$, but G *need not* have a subgroup of that order. No counterexample is found among cyclic groups! (Why?) But the group A_4 has order 12 but no subgroup of order 6. There is a guided proof of this fact in the exercises.

DEFINITION 2. Let H be a subgroup of G. The number of distinct left (similarly right) cosets of H is called the **index** of H in G. It is denoted by $|G:H|$.

EXAMPLE 4

The distinct (left) cosets of the subgroup $H = 5\mathbf{Z}$ in $\mathbf{Z}$ are $\{0+5\mathbf{Z}, 1+5\mathbf{Z}, 2+5\mathbf{Z}, 3+5\mathbf{Z}, 4+5\mathbf{Z}\}$. Therefore, $|\mathbf{Z}:5\mathbf{Z}| = 5$. Similarly, if $H = \{e, (1,2)\}$, then $|S_3:H| = \frac{6}{2} = 3$. ■

COROLLARY 4 Let G be a finite group and let $a \in G$. Then the order of a divides the order of G.

Proof. The order of an element a is equal to the order of the subgroup generated by a, which is to say that $|a| = |\langle a \rangle|$. Since $\langle a \rangle$ is a subgroup of G, the corollary follows from Lagrange's Theorem. ▲

EXAMPLE 5

The forms of the cycle decompositions of permutations in S_4 are $(1), (a,b)(c)(d), (a,b)(c,d), (a,b,c)(d), (a,b,c,d)$. The orders of permutations of these forms are, respectively, 1, 2, 2, 3, and 4. Each order divides 24. The numbers 6, 8, and 12 are also divisors of 24 but there are no elements of order 6, 8, or 12 in S_4. ■

COROLLARY 5 Let G be a finite group and let $a \in G$. Then $a^{|G|} = e$.

Proof. By Lagrange's Theorem, $|G| = k|a|$ for some k. Thus $a^{|G|} = (a^{|a|})^k = e^k = e$. ▲

EXAMPLE 6

Let p be a prime. The group $\{U_p, \cdot\}$ is a group of order $p-1$. Thus for any $a \in U_p$, we have $a^{p-1} = e$. Restated in terms of congruences, this says that $a^{p-1} \equiv 1 \bmod p$ for any integer a that is not divisible by p. We have just proved Fermat's Little Theorem (Corollary 4 of Section 1.7) via Lagrange's Theorem! ■

COROLLARY 6 Every group of prime order is cyclic.

Proof. Suppose G is a group and $|G| = p$ where p is prime. Let $a \in G$. Then the only possible orders for a are 1 and p. In the first case, $a = e$ and in the second, $G = \langle a \rangle$. ▲

SUMMARY

The purpose of this section is to explore the properties of the **right and left cosets** of a subgroup $H \subseteq G$. They are analogous to the equivalence classes $[a]_m$ that form the elements of $\mathbf{Z}_m$. We found that the left cosets of H partition G into subsets that all have the same cardinality as H. The number of left cosets is called the **index** of H in G. (All our observations are of course true for right cosets as well.) In the case that G is a finite group, we used the coset partition to deduce **Lagrange's Theorem**, which states that the order of any subgroup $H \subseteq G$ must divide the order of G. As a corollary, we found that the order of any element of G must divide the order of G. The concept of coset gave us important structural information about finite groups. In the next section, we will use cosets to construct new groups called quotient groups.

5.1 Exercises

1. Let H be a subset of a group G and $a \in G$. Prove that if G is abelian, then $aH = Ha$ for all $a \in G$.
2. Let $H = \{e, (1, 3)\}$. Find the left and right cosets of H in S_3.
3. Let $H = \{e, (1, 2, 3), (1, 3, 2)\}$. Find the left and right cosets of H in S_3.
4. Let $S = \{R_0, V\}$. Find the left and right cosets of S in D_4.
5. Let $S = \{R_0, R_1, R_2, R_3\}$. Find the left and right cosets of S in D_4.
6. Find the right and left cosets of $4\mathbf{Z}_8 = \{0, 4\}$ in $\mathbf{Z}_8$. (Here we use additive notation so that a coset has the form $a + H = \{a + h : h \in H\}$.)
7. Reformulate Propositions 1 and 2 and Corollary 3 for right cosets. Give examples.
8. Show that the left cosets of H are the equivalence classes of the equivalence relation on G defined by bRa if and only if $a^{-1}b \in H$.
9. Reprove Euler's Theorem (Theorem 3 of Section 1.7) using Corollary 5.
10. Find the right and left cosets of A_4 in S_4.
11. Let $H = \{R_0, R_2\}$ in D_4. Show that for each $x \in D_4, xH = Hx$.
12. Find all the cosets of $\{0, 4\}$ in $\{\mathbf{Z}_8, +\}$.
13. Find all the cosets of $\{0, 2, 4\}$ in $\{\mathbf{Z}_6, +\}$.
14. Find all the cosets of $\{1, 10\}$ in U_{11}.
15. Prove Proposition 3.
16. Suppose that $G = \langle x \rangle$ and that $|\langle x \rangle| = 12$. Find the cosets of $\langle x^4 \rangle$.
17. Suppose that G is a group that has more than one element but no proper, nontrivial subgroups. Prove that $|G|$ is prime.
18. Let p be a prime number and let G be a group of order p^2. Prove the following statement. Either G is cyclic or each element in G has order 1 or order p.
19. Prove that A_4 has no subgroup of order 6 by carrying out the following steps.
    a. Show that A_4 has eight elements of order 3, and 3 of order 2.
    b. Let a be any element in A_4 of order 3. Suppose that H is a subgroup of A_4 of order 6. Show that two of the cosets H, aH and a^2H must be the same set.
    c. Show that having any two of H, aH and a^2H equal implies that $a \in H$.

d. Show that part c implies that H has 8 elements, contradicting the assumption that it has 6.

To the Teacher

"Coset" is not a word likely to enter the vocabulary of a high school student. Yet, like so many advanced concepts in mathematics, it is a notion grounded in very basic mathematics. Students encounter cosets when they study arithmetic sequences, a standard topic in high school algebra. What we have studied as the coset $a + d\mathbf{Z}$, they study as the sequence $\{a, a + d, a + 2d, \ldots\}$. We would extend the pattern to the left as well: $\{\ldots, a - 2d, a - d, a, a + d, a + 2d, \ldots\}$. They could too. Typical high school activities might include the following.

1. Extend the pattern $\{4, 9, 14, 19, \ldots\}$.
   (The coset in question is $4 + 5\mathbf{Z}$.)
2. Determine if the sequence $\{2, 6, 11, 17, \ldots\}$ is arithmetic.
   (No!)
3. Sum the first 1000 terms of the sequence $\{2, 5, 8, 11, \ldots\}$.
   (Answer: $1000 \cdot 2 + 3 \cdot (0 + 1 + \ldots + 999) = 2000 + 3 \cdot \frac{999 \cdot 1000}{2}$
   $= 1,500,500$.)
4. How many matchsticks are needed to build a sequence of n hexagons such as the following?

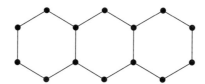

   What is the answer when $n = 20$? Here the sequence is $\{6, 6 + 5, 6 + 2 \cdot 5, \ldots\}$.
   (Answer: $6 + 19 \cdot 5$)

Let's ask the high school student to extend a given arithmetic sequence both forward and backward (subtracting multiples of d) so that an arithmetic sequence always looks like $\{\ldots, a - 2d, a - d, a, a + d, a + 2d, \ldots\}$. Then we can ask questions that anticipate or suggest some new and some familiar facets of algebra. With d fixed at say 5 or 7, we can ask the following.

5. How many different arithmetic sequences are possible?
   (Answer: There are only d distinct sequences. This may surprise students at first. But it points to the underlying partition of the integers into the cosets of the subgroup $d\mathbf{Z}$.)
6. Given a random integer x, which arithmetic sequence does it fall into?
   (Answer: The students would have to develop a way to name the distinct sequences just as we have chosen representatives to denote cosets. They can then use the Division Algorithm to write x as $a + id$ and find its place.)

7. Pick an x from one sequence and a y from another. In what sequence does the sum $x + y$ appear?

(Answer: Find the coset of $x + y$ mod d. This will of course remind the student of clock or modular arithmetic. For you, it looks back to the construction of the group $\{Z_n, +\}$ and forward to the generalization of that construction we will present in Section 5.2.)

Task

Develop a portfolio of word problems that involve arithmetic sequences.

5.2 NORMAL SUBGROUPS AND QUOTIENT GROUPS

Our goal is to construct a new group from a group G and a subgroup H in much the same way as we constructed $\mathbf{Z}_m$ from $\mathbf{Z}$ and its subgroup $m\mathbf{Z}$. The key to defining addition $[a]_m + [b]_m$ in $\mathbf{Z}_m$, is the fact that $[x + y]_m$ is the same coset no matter what choice $x \in [a]_m$ and $y \in [b]_m$ we make. To obtain an operation that is similarly well-defined in the more general case, the subgroup H must be "normal" in G as defined below. Historically, mathematicians like Lagrange and Ruffini knew a great deal about sets of permutations and their relevance to the theory of polynomial equations. Galois took these investigations in the new direction of what we now call abstract algebra by realizing the importance of group and subgroup structures in the set of S_n, and in particular, by realizing the importance of normal subgroups.

> **DEFINITION 1.** Let G be a group and let H be a subgroup of G. We call H a **normal subgroup** if $aH = Ha$ for each $a \in G$. We denote the fact that H is normal in G by the notation $H \triangleleft G$.

The right and left cosets aH and Ha are equal when for each $h_1 \in H$, we can find $h_2 \in H$ such that $ah_1 = h_2 a$ and conversely, for each $h_2 \in H$, we can find $h_1 \in H$ such that $ah_1 = h_2 a$. Note that h_1 and h_2 do not have to be equal.

EXAMPLE 1

The subgroup $H = \{(1), (1, 2, 3), (1, 3, 2)\}$ is normal in S_3. To check, we must verify that $aH = Ha$ for each a in S_3. We will check the case where $a = (1, 2)$ and leave the other cases to the reader.

$$aH = \{(1, 2)(1), (1, 2)(1, 2, 3), (1, 2)(1, 3, 2)\} = \{(1, 2), (2, 3), (1, 3)\}$$

$$Ha = \{(1)(1, 2), (1, 2, 3)(1, 2), (1, 3, 2)(1, 2)\} = \{(1, 2), (1, 3), (2, 3)\}$$ ∎

If G is abelian, then every subgroup H of G is normal. That is why the concept of normality did not arise explicitly when we constructed $\mathbf{Z}_m$ from $\mathbf{Z}$. The following proposition gives an alternative formulation of normality.

PROPOSITION 1 A subgroup H of G is normal if and only if $x^{-1}hx \in H$ for each $x \in G$ and $h \in H$.

Proof. Exercise 4.

Putting Proposition 1 to work, we obtain the following result that says every group homomorphism produces a normal subgroup.

PROPOSITION 2 Let $f: G \to H$ be a group homomorphism and let K be the kernel of f. Then K is a normal subgroup of G.

Proof. Suppose that $x \in K$ and that $y \in G$. Since f is operation preserving, it maps inverses to inverses. Since f maps x to e_H, we have

$$f(y^{-1}xy) = f(y^{-1})f(x)f(y) = f(y^{-1})f(y) = f(y^{-1}y) = e_H.$$

Thus $y^{-1}xy \in K$. By Proposition 1, K is normal in G. ▲

EXAMPLE 2

Let $f: S_n \to \mathbf{Z}_2$ be defined by $f(\alpha) = 0$ if α is even and $f(\alpha) = 1$ if α is odd. It is easy to check that f is a group homomorphism and that the kernel of f is A_n, the subgroup of even permutations. From Proposition 2, we conclude that $A_n \triangleleft S_n$. ■

Now we show how to construct the **quotient group** G/H from G and a **normal subgroup** H. (Sometimes G/H is called a **factor group**. However, it is neither a group of quotients nor a group of factors!) The set G/H is the set of cosets of H in G:

$$G/H = \{aH : a \in G\}.$$

Let aH and bH be members of G/H. The product $(aH) * (bH)$ is defined as follows:

$$(aH) * (bH) = abH.$$

In words, the product of the coset that contains a with the coset that contains b is the coset that contains ab. The next proposition shows that the operation $*$ is **well-defined**, which is to say that the outcome abH depends only on the cosets aH and bH, and not on the particular representatives a and b.

PROPOSITION 3 Suppose that $H \triangleleft G$ and that aH and bH are cosets in G/H. Suppose that $aH = cH$ and that $bH = dH$. Then $abH = cdH$.

Proof. Since $aH = cH$, $a = ch_1$ for some $h_1 \in H$. Similarly, $b = dh_2$ for some $h_2 \in H$. Since H is normal, we can find $h_3 \in H$ such that $h_1 d = dh_3$. Substituting, we have

$$ab = (ch_1)(dh_2) = c(h_1d)h_2 = c(dh_3)h_2 = cd(h_3h_2).$$

Thus $ab \in cdH$. Since the cosets abH and cdH are not disjoint, they are equal. Thus $abH = cdH$. ▲

The arguments above may seem fussy. But the fact that H is normal in G is crucial to the construction. The following example shows that if H is **not** normal, the operation $aH * bH = abH$ is **not** well-defined.

EXAMPLE 3

Let $H = \{e, (1, 2)\}$. To see that H is **not** normal in S_3, note that $(1, 3)H = \{(1, 3), (1, 2, 3)\}$ but $H(1, 3) = \{(1, 3), (1, 3, 2)\}$. Now let $a = (1, 3), b = (2, 3), c = (1, 2, 3)$ and $d = (1, 3, 2)$. Then

$$aH = cH = \{(1, 3), (1, 2, 3)\}$$
$$bH = dH = \{(2, 3), (1, 3, 2)\}.$$

For the product $aH * bH$ to be well-defined, we must have $abH = cdH$:

$$ab = (1, 3, 2) \text{ and } abH = \{(1, 3, 2), (2, 3)\}$$
$$cd = (1, 2, 3)(1, 3, 2) = e \text{ and } cdH = H.$$

Clearly, $abH \neq cdH$. So the operation is **not** well-defined. ∎

PROPOSITION 4 $\{G/H, *\}$ is a group when H is a normal subgroup of G.

Proof. The operation is associative since $(aH * bH) * (cH) = (abH) * (cH)$ $= abcH = (aH) * (bcH) = (aH) * (bH * cH)$. The coset $H = eH$ is the identity element in G/H since $eHaH = eaH = aH$ for each coset aH in G/H. The inverse of aH in G/H is $a^{-1}H$ since $aH * a^{-1}H = aa^{-1}H = eH = H$. ▲

EXAMPLE 4

Now let's actually construct a quotient group. Let's take $G = D_4$, the group of the eight symmetries of the square. Let $K = \{e, R_2\}$, where R_2 denotes rotation through 180 degrees. Recall that R_2 commutes with all other elements in D_4. Thus $K \triangleleft D_4$. From Lagrange's Theorem, we know that the number of distinct cosets in D_4/K is $8/2 = 4$. Thus the order of the group D_4/K is 4. The cosets are as follows:

$$K = \{e, R_2\}$$
$$R_1 K = \{R_1, R_3\}$$
$$HK = \{H, V\}$$
$$D_1 K = \{D_1, D_2\}.$$

The Cayley table for D_4/K follows.

Section 5.2 Normal Subgroups and Quotient Groups

$*$	K	R_1K	HK	D_1K
K	K	R_1K	HK	D_1K
R_1K	R_1K	K	D_1K	HK
HK	HK	D_1K	K	R_1K
D_1K	D_1K	HK	R_1K	K

Since each element in D_4/K has order 2, it is isomorphic to the Klein 4-Group. ∎

EXAMPLE 5

Let $G = \mathbf{Z}$ and let $H = 3\mathbf{Z}$. Since $\mathbf{Z}$ is abelian, H is necessarily normal in G. The cosets of H are $0 + 3\mathbf{Z} = 3\mathbf{Z}$, $1 + 3\mathbf{Z}$, $2 + 3\mathbf{Z}$. The group table is

$+$	$0 + 3\mathbf{Z}$	$1 + 3\mathbf{Z}$	$2 + 3\mathbf{Z}$
$0 + 3\mathbf{Z}$	$0 + 3\mathbf{Z}$	$1 + 3\mathbf{Z}$	$2 + 3\mathbf{Z}$
$1 + 3\mathbf{Z}$	$1 + 3\mathbf{Z}$	$2 + 3\mathbf{Z}$	$0 + 3\mathbf{Z}$
$2 + 3\mathbf{Z}$	$2 + 3\mathbf{Z}$	$0 + 3\mathbf{Z}$	$1 + 3\mathbf{Z}$

The group is clearly isomorphic to $\mathbf{Z}_3$. In fact, each of the groups $\mathbf{Z}_n$ can be constructed as the quotient group $\mathbf{Z}/n\mathbf{Z}$. ∎

We now end this section with two very different uses of the material developed in this section. First we relate quotient groups to homomorphisms and establish what is known as the First Isomorphism Theorem. Then we derive the existence of elements of prime order p in any group G for which p divides $|G|$.

The First Isomorphism Theorem

If we can find an isomorphism between two groups G and H, we know that their structures are identical. If there is only a homomorphism from G to H, the relation between the structures of the two groups is less obvious. The First Isomorphism Theorem helps clarify the issue. Since the kernel K of a homomorphism $f: G \to H$ is normal in G, we can form the quotient group G/K. This is the key to relating the structures of G and H.

THEOREM 5 (*The First Isomorphism Theorem*) Let $f: G \to H$ be a group homomorphism with kernel K. Let $F: G/K \to H$ be defined by $F(xK) = f(x)$. Then F is an isomorphism from G/K onto $f(G) = \{f(x) \in H : x \in G\}$.

Proof. First we must show that F is well-defined. Suppose that $yK = xK$. Then $x^{-1}y \in K$ and $e_H = f(x^{-1}y) = f(x)^{-1}f(y)$. Thus $f(x) = f(y)$ and $F(xK) = F(yK)$. Clearly, F maps G/K onto the group $f(G)$. Since $F(xKyK) = F(xyK) = f(xy) = f(x)f(y) = F(xK)F(yK)$, the map F is operation preserving. What remains to be shown is that F is injective. Suppose that $F(xK) = F(yK)$. Then $f(x) = f(y)$ and $e_H = f(x^{-1}y)$ so that $x^{-1}y \in K$. Thus $xK = yK$, showing that F is injective. ▲

EXAMPLE 6

Let $f: \mathbf{Z} \to \mathbf{Z_6}$ be the group homomorphism defined by $f(x) = 2x \mod 6$. The image $f(\mathbf{Z}) = \{0, 2, 4\} \subseteq \mathbf{Z_6}$. This subgroup of $\mathbf{Z_6}$ is isomorphic to $\mathbf{Z_3}$. The kernel of f is the set $3\mathbf{Z}$ and the quotient group $\mathbf{Z}/3\mathbf{Z}$ is also isomorphic to $\mathbf{Z_3}$. ∎

Elements of Prime Order p

We know that if d divides the order of a cyclic group G, then G contains an element of order d. If G is not cyclic, the result is not always true. For instance, although the order of S_5 is 120, a look at the possible cycle decompositions tells us that no element has order greater than 6. However, if p is *prime* and p divides the order of G, then G must have an element of order p. The proof of the next theorem establishes the assertion for abelian groups. It utilizes the results of this section. But the result holds in all finite groups. We take this opportunity to present the more general proof even though it requires quite a different approach.

THEOREM 6 Let G be a finite abelian group of order n and let p be a prime that divides n. Then G contains an element of order p.

Proof. The proof is by induction on n, the order of G. The assertion is clearly true if $|G| = 2$ since 2 is prime and G must be cyclic. Now assume the assertion is true for all groups with order less than $|G|$. Let $x \neq e$ be an element of G. Suppose that $|x| = m$ and let q be a prime that divides m. Let $y = x^{m/q}$. The order of y is the prime q. If $p = q$ we are done. If not, let $H = \langle y \rangle$. Since G is abelian, $H \triangleleft G$. The order of $G/H = n/q$ and p divides n/q. By our induction hypothesis, we can find an element $zH \in G/H$ of order p, which means $(zH)^p = z^p \langle y \rangle = \{z^p y^0, z^p y, \ldots, z^p y^{q-1}\} = H$. Thus for some $i, 0 \leq i \leq q-1, z^p = y^i$. If $i = 0$, we are done because $z^p = e$. If not, z^q has order p since $(z^q)^p = (z^p)^q = (y^i)^q = (y^q)^i = e$. ▲

The following more general result is known as **Cauchy's Theorem.**

THEOREM 7 (*Cauchy's Theorem*) Let G be any finite group of order n and let p be a prime that divides n. Then G contains an element of order p.

Proof. Let S be the set of all p-tuples $(x_1, x_2, \ldots, x_p)$ of elements of G with the property that the product $x_1 x_2 \cdots x_p = e$. Our goal will be to find an element $x \neq e$ such that the p-tuple $(x, x, \ldots, x) \in S$. First, let's count the elements in S. There are exactly n^{p-1} ways to choose an ordered set $\{x_1, \ldots, x_{p-1}\}$ of $p-1$ elements of G and exactly one way to choose x_p such that $x_1 x_2 \cdots x_{p-1} x_p = e$. Thus $|S| = n^{p-1}$ and p divides n^{p-1} since p divides n. Let $(x_1, x_2, \ldots, x_p)$ be in S. Then $(x_2, x_3, \ldots, x_p, x_1)$ is also in S because $x_1^{-1} x_1 (x_2 x_3 \cdots x_p x_1) = x_1^{-1}(x_1 x_2 x_3 \cdots x_p) x_1 = e$. Iterating, we also have $(x_{i+1}, x_{i+2}, \ldots, x_p, x_1, \ldots, x_i) \in S$ for each $i = 1, \ldots, p-1$. Now suppose that $0 < i < p-1$ and that when we add i to each index, identical p-tuples ensue: $(x_1, x_2, \ldots, x_p) = (x_{i+1}, x_{i+2}, \ldots, x_p, x_1, \ldots, x_i)$. By repeatedly shifting the indices up by i, we have $(x_1, x_2, \ldots, x_p) = (x_{i+1}, x_{i+2}, \ldots, x_p, x_1, \ldots, x_i) = (x_{2i+1}, x_{2i+2}, \ldots, x_p, x_1, \ldots, x_{2i}) = \ldots = (x_{ji+1}, x_{ji+2}, \ldots, x_p, x_1, \ldots, x_{ji})$, for

$j = 0, \ldots, p-1$ with all the indices reduced modulo p. Thus $x_1 = x_{i+1} = x_{2i+1} = \ldots = x_{ji+1}$ for $j = 0, \ldots, p-1$. Since $0 < i < p-1$, the indices $(j_1 i + 1) \bmod p = (j_2 i + 1) \bmod p$ if and only if p divides $(j_1 - j_2)$. Thus the p indices $1, 1+i, 1+2i, \ldots, 1+(p-1)i$ are all distinct modulo p. So if $(x_1, x_2, \ldots, x_p) = (x_{i+1}, x_{i+2}, \ldots, x_p, x_1, \ldots, x_i)$ for $0 < i < p-1$, then $x_1 = x_2 = \ldots = x_p$. Next we deduce that such a p-tuple must exist.

The relation whereby $(x_1, x_2, \ldots, x_p) R (y_1, y_2, \ldots, y_p)$ if and only if $(y_1, y_2, \ldots, y_p) = (x_{i+1}, x_{i+2}, \ldots, x_p, x_1, \ldots, x_i)$ for some $i = 0, \ldots, p-1$ is easily seen to be an equivalence relation on S. The equivalence class of a p-tuple $(x_1, x_2, \ldots, x_p)$ either contains p distinct p-tuples of the form $(x_{i+1}, x_{i+2}, \ldots, x_p, x_1, \ldots, x_i)$ or exactly one p-tuple of the form $(x, x, \ldots, x)$. There is at least one equivalence class that contains exactly one element, namely, $(e, e, \ldots, e)$. A simple counting argument will show that there must be at least one other (actually at least $p-1$ others). We know that $|S|$ is divisible by p and that $|S|$ is the sum of the cardinalities of equivalence classes of the relation R. Each of the equivalence classes has cardinality p or 1. If there was only one equivalence class with cardinality 1, we would have $|S| = qp + 1$ for some integer q. But $qp + 1$ is not divisible by p. So we can conclude that there must be a p-tuple of the form $(x, x, \ldots, x)$ such that $x \neq e$ and for which $x^p = e$. ▲

EXAMPLE 7

Let $G = S_5$, which contains $5! = 120$ elements. Cauchy's Theorem guarantees us that we can find elements of orders 2, 3, and 5, which are the only prime divisors of 120. These are easy to find. The 2, 3, and 5-cycles have orders 2, 3, and 5. ∎

SUMMARY

In this section, we focused on **normal** subgroups. The distinctive property of a normal subgroup $H \subseteq G$, namely that $aH = Ha$ for each $a \in G$, ensures that the operation $aH * bH = abH$ on the cosets of H is well-defined. The resulting structure is the **quotient group** G/H. We saw that the kernel of a group homomorphism is normal and we used that to state and prove the **First Isomorphism Theorem**. We also used the techniques of this section to prove that if p is a prime that divides the order of a finite group G, then G has an element of order p. We did this first for abelian groups, and then more generally for all finite groups with **Cauchy's Theorem**. The concepts and constructions of this section are in some ways pivotal historically. With them, we can investigate the structure of various groups with increased subtlety. Galois utilized the concepts of normality and quotient group to solve the problem of the insolvability of the quintic.

5.2 Exercises

1. Find all the normal subgroups of S_3.
2. Find all the normal subgroups of $\{Z_6, +\}$.
3. Find all the normal subgroups of D_4.
4. Prove Proposition 1.

5. Suppose that H is a subgroup of a group G and that $a \in G$. Prove that aHa^{-1} is a subgroup of G.

6. Let H and K be subgroups of G. We say that K and H are **conjugate** if there exists $x \in G$ such that $K = xHx^{-1} = \{xhx^{-1} \in G : h \in H\}$.
   i. Find all the conjugates of $H = \{e, (1,2)\}$ in S_3.
   ii. Find the conjugates of $K = \{e, V\}$ in D_4.
   iii. Find all the conjugates of $\{e, R_2\}$ in D_4.

7. Prove that H is a normal subgroup of G if and only if $xHx^{-1} = H$ for all $x \in G$.

8. For any group G, the **center of G** is the set $Z(G) = \{x \in G : xy = yx$ for all $y \in G\}$.
   i. Prove that $Z(G)$ is a subgroup of G.
   ii. Prove that $Z(G)$ is normal in G.
   iii. Find the center of D_4.
   iv. Find the center of D_5.

9. Prove that the set of rotations in the dihedral group D_n is a normal subgroup. (This fact can be argued geometrically, with no calculations! Think about it!)

10. Suppose that H is a subgroup of G and that $|G:H| = 2$ (i.e., the index of H in G is 2). Prove that H is a normal subgroup of G.

11. Determine if the subgroup $H = \{e, (1,2)(4,3), (1,3)(2,4), (1,4)(2,3)\}$ is normal in A_4.

12. Let $G = \mathbf{Z}_8$ and let $H = \{0, 4\}$. Write out the group table for G/H.

13. Let $G = S_3$ and $A_3 = \{e, (1,2,3), (1,3,2)\}$. Write out the Cayley table for S_3/A_3.

14. Let $G = A_4$ and let $H = \{e, (1,2)(3,4), (1,3)(2,4), (2,3)(1,4)\}$. Write out the group table for G/H.

15. Let $G = \mathbf{Z}_{18}$ and $H = \{0, 6, 12\}$. Write out the group table for G/H.

16. Prove that a quotient group of a cyclic group is cyclic.

17. Determine the cosets of $(\mathbf{Z} \oplus \mathbf{Z})/\langle(2,2)\rangle$ and determine if the quotient group is cyclic.

18. Find the order of U_{35}. Find the order of the elements of prime order guaranteed by Theorem 6.

19. Is there a divisor d of $|U_{35}|$ such that no element in U_{35} has order d?

20. Prove that the relation established in the proof of Theorem 7 is an equivalence relation.

21. Let $\widetilde{G}_i$ be the set $\{(e_1, e_2, \ldots, e_{i-1}, g, e_{i+1}, \ldots, e_n) : g \in G_i\}$. Is the set $\widetilde{G}_i$ normal in $G_1 \times G_2 \times \ldots \times G_n$?

22. Let $\widetilde{G}_i$ be as in Exercise 21. Use the First Isomorphism Theorem to show that $(G_1 \times G_2 \times \ldots \times G_n)/\widetilde{G}_i$ is isomorphic to $G_1 \times G_2 \times \ldots \times G_{i-1} \times G_{i+1} \times G_n$.

23. Suppose that G and K are finite groups with relatively prime orders. Use the First Isomorphism Theorem to prove that there can be no nontrivial homomorphism from G to K.

To the Teacher

The focus of this section was to construct a quotient group from a group $\{G, *\}$ and a subgroup H in much the same way as we constructed $\mathbf{Z}_n$ and $n\mathbf{Z}$. To do this, we discovered that, in order to transfer the operation $*$ to the quotient group G/H, the subgroup H had to be normal in G. We had to make the operation $*$ well-defined on G/H. The question is subtle. You probably had to step back and ask, okay, just what is at issue here? Why can't we always say $xH * yH = xyH$?

Young math doers face the question of well-defined operations early on, every time they encounter fractions. Pizza cutting makes it easy for young students to see that $\frac{1}{2} = \frac{2}{4}$. The lesson here is that a number thought of as an amount of a unit pizza can have many representations. The problem of many representations is what makes us struggle with the issue of well-defined operations in contexts as varied as fractions and quotient groups. Eventually students should be able to determine if $\frac{16973}{26231}$ equals $\frac{5951}{9197}$. (Yes) The criterion that $\frac{a}{b} = \frac{c}{d}$ if and only if $ad = bc$ is a long way from pizza cutting. But it lies behind all operations involving fractions.

- Why is $\frac{2}{3} + \frac{1}{5} = \frac{4}{6} + \frac{2}{10}$?
- How do you add $\frac{659}{2741}$ and $\frac{253}{1001}$? Why?
- Why is $\frac{23}{41} \cdot \frac{17}{53}$ equal to $\frac{391}{2173}$?
- Why is $\frac{23}{41} \div \frac{17}{53}$ equal to $\frac{1219}{697}$?

Your struggle with transferring the operation $*$ from G to G/H should prompt a sensitivity (if not sympathy) to the hurdles that a young math student faces transferring integer arithmetic to rational numbers. Your task will be to make that transition natural, both as a matter of computation and comprehension.

Challenge: Develop student-friendly but rigorous answers to the questions posed above.

5.3 CONJUGACY IN S_n

Here's another way to think about normal subgroups. Let x and y be elements of a group G. If there is an element $g \in G$ such that $x = g^{-1}yg$, then x is said to be **conjugate** to y. The relation R defined by xRy whenever x is conjugate to y, establishes an equivalence relation on G. The equivalence class or **conjugacy class** of x is the set $\{g^{-1}xg : g \in G\}$. A subgroup $H \subseteq G$ is normal if and only if whenever $x \in H$ then all elements of G that are conjugate to x are also in H. In other words, H is normal in G if and only if the entire conjugacy class of x is contained in H for each $x \in H$. So to check if a subgroup is normal, we can check that it is the union of conjugacy classes.

It is particularly easy to determine when elements of S_n are conjugate. Two permutations α and β in S_n are conjugate when they have the same cycle structure. That is to say, the expressions for α and β in terms of disjoint cycles must have the

EXAMPLE 1

Suppose that the members of S_7 permute the members of the set $\{a, b, c, d, e, f, g\}$ and that $\alpha = (a, b, c)(d, e)(f, g)$ and $\beta = (c, d, g)(e, a)(b, f)$. Note that α and β have the same cycle structure. To show that α and β are conjugate, we must find γ such that $\alpha = \gamma^{-1}\beta\gamma$. In general, to find γ, we simply write the cycles of α and β one above the other so that cycles of the same size align. We drop the parentheses, and let γ be the resulting one-to-one correspondence. For our example, the resulting γ is as follows.

$$\begin{array}{ccccccc} a & b & c & d & e & f & g \\ \downarrow & \downarrow & \downarrow & \downarrow & \downarrow & \downarrow & \downarrow \\ c & d & g & e & a & b & f \end{array}$$

(Note that γ is not unique because the order of elements within a cycle is not unique.) The cycle decomposition of γ is (a, c, g, f, b, d, e). To see that $\alpha = \gamma^{-1}\beta\gamma$, pick any element in the top row, say e, and trace where $\gamma^{-1}\beta\gamma$ maps it to. So $\gamma(e) = a$ and $\beta\gamma(e) = e$ and finally $\gamma^{-1}\beta\gamma(e) = d$, which is indeed equal to $\alpha(e)$. (In the above diagram, start at e in the top row. Go down, left, and up.)

On the other hand, suppose that $\alpha = \gamma^{-1}\beta\gamma$. Let's check that α and β have the same cycle structure. For each integer $j, \alpha^j = \gamma^{-1}\beta^j\gamma$. Suppose that the symbol x is in a cycle of length j in α. Then $x = \alpha^j(x)$ if and only if $x = \gamma^{-1}\beta^j\gamma(x)$. Composing with γ, we have $\gamma(x) = \beta^j\gamma(x)$. Thus x is in a cycle of length j in α if and only if $\gamma(x)$ is in a cycle of length j in β. ∎

EXAMPLE 2

In this example we find all the normal subgroups of S_4 using the fact that a normal subgroup must be the union of conjugacy classes. First, we count the number of elements in each conjugacy class. We shall denote a conjugacy class by the sequence of lengths of its cycles.

Cycle Structure	Example	Number	Reason
1–1–1–1	$(a)(b)(c)(d)$	1	Only the identity element has this structure.
4	(a, b, c, d)	6	Assuming a is listed first, there are 3! ways to complete the cycle.
3–1	$(a, b, c)(d)$	8	There are four ways to place a symbol in the 1-cycle and then 2! ways to form the 3-cycle from the remaining symbols.
2–2	$(a, b)(c, d)$	3	There are 3 ways to choose the entry paired with a.
2–1–1	$(a, b)(c)(d)$	6	There are $\binom{4}{2}$ ways to choose a pair.

The number of elements n in any normal subgroup of S_4 must be $1 + 6x + 8y + 3z + 6w$, where x, y, z, and w can be 0 or 1. (It's all or nothing for a conjugacy class!) Furthermore, n must divide $|S_4| = 24$. The possibilities are $1, 1 + 3 = 4, 1 + 3 + 8 = 12$, and $1 + 6 + 8 + 3 + 6 = 24$. In each case, we can find a corresponding normal subgroup. For $n = 1$, the subgroup is $\{e\}$. For $n = 4$, the subgroup is $K = \{e, (a, b)(c, d), (a, c)(b, d), (a, d)(c, b)\}$. For $n = 12$, the subgroup is A_4. For $n = 24$, the subgroup (not proper) is S_4. ∎

Let $H \subseteq S_4$ be the subgroup $\{e, (a, b)(c, d)\}$. From Example 2, we have the following sequence of subgroups, each normal in the next.

$$\{e\} \triangleleft H \triangleleft K \triangleleft A_4 \triangleleft S_4$$

For each adjacent pair, we can form a quotient group.

$$H/\{e\} = H \sim \mathbf{Z}_2$$
$$K/H \sim \mathbf{Z}_2$$
$$A_4/K \sim \mathbf{Z}_3$$
$$S_4/A_4 \sim \mathbf{Z}_2$$

In each case, the quotient group is cyclic. Groups like S_4 that have such a sequence of subgroups are called **solvable groups**.

> **DEFINITION 1.** A group G is called **solvable** if it has a finite sequence of subgroups $H_0 = \{e\}, H_1, \ldots, H_n = G$, where $H_i \triangleleft H_{i+1}$ and the quotient group H_{i+1}/H_i is cyclic for each $i = 0, \ldots, n-1$.

The terminology might seem strange since we do not seem to be solving for anything. But in Chapter 6, we shall show the relation between solvable groups and the process of solving for the roots of a polynomial. The terminology anticipates that association.

EXAMPLE 3

We show that the only nontrivial normal subgroup of S_5 is A_5. Any subgroup of S_5 of order 60 must be normal. (See Exercise 10 of Section 5.2.) This example also shows that A_5 is the only subgroup of S_5 of order 60. As in Example 2, the argument is primarily numerical. So first let's determine the size of the conjugacy classes in S_5.

Type	Example	Size	Parity
1–1–1–1–1	e	1	even
5	(a, b, c, d, e)	24	even
4–1	$(a, b, c, d)(e)$	30	odd
3–2	$(a, b, c)(d, e)$	20	odd
3–1–1	$(a, b, c)(d)(e)$	20	even
2–2–1	$(a, b)(c, d)(e)$	15	even
2–1–1–1	$(a, b)(c)(d)(e)$	10	odd

The size n of a normal subgroup H must be $n = 1 + 24x + 30y + 20z + 20w + 15s + 10t$ where the variables can assume values of 0 or 1. There are three ways in which $n > 1$ and n divides 120.

1. $n = 1 + 24 + 20 + 15 = 60$ when x, z, and s are equal to 1. However, there is no corresponding subgroup. If a permutation of the form $\alpha = (a, b, c)(d, e)$ is in a subgroup H, then α^2 must also be in H. But α^2 is of the form $(a, c, b)(d)(e)$. So all three cycles must be in H and hence $w = 1$. But then $n = 80$, which is impossible.
2. $n = 1 + 24 + 20 + 15 = 60$ when x, w, and s are equal to 1. In this case, $H = A_5$. There are no other ways that $1 + 24x + 30y + 20z + 20w + 15s + 10t = 60$. Thus A_5 is the **only** subgroup of A_5 of order 60.
3. $n = 1 + 24 + 15 = 40$ when only x and s are equal to 1. But the product $(a, b, c, d, e)(a, b)(c, d)(e) = (a, c, e)(b)(d)$ must also be in H so that $w = 1$ and again $H = A_5$. ∎

A group with no nontrivial proper normal subgroups is called a **simple group**. A nonabelian group G that is a simple group cannot be solvable because its only normal subgroup is $\{e\}$ and the quotient group $G/\{e\}$ is not cyclic.

EXAMPLE 4

The group A_5 is a simple group. Again, the key is a counting argument. But it's a bit more subtle. Two element α and β are conjugate in A_5 if and only if we can find an *even* permutation γ such that $\alpha = \gamma^{-1}\beta\gamma$. If two permutations are conjugate in A_5, they are conjugate in S_5 (and have the same cycle structure) but two even permutations that are conjugate in S_5 are not necessarily conjugate in A_5.

The facts about A_5 are as follows:

i. All three cycles are conjugate and the size of the conjugacy class of any 3-cycle is 20.
ii. All cycles of the form 2–2–1 are conjugate and the size of the conjugacy class of any such cycle is 15.
iii. The cycles of the form $\alpha = (a, b, c, d, e)$ and $\beta = (a, c, e, b, d)$ are not conjugate in A_5 because there is no even permutation γ such that $\alpha = \gamma^{-1}\beta\gamma$. However, each of α and β is in a conjugacy class of size 12.

(We will ask the reader to complete the details in the Exercises.) The number of elements in a normal subgroup H would be $n = 1 + 20x + 15y + 12z + 12w$ where n divides 60. The only possibilities are $n = 1$ and $n = 60$. Thus A_5 has no nontrivial, proper normal subgroups. ∎

EXAMPLE 5

Putting the results of the above examples together, we find that S_5 **is not a solvable group**. Its only normal subgroup is A_5, which is not cyclic, and A_5 has no nontrivial proper normal subgroups. ∎

The fact that S_5 is not solvable is critical to the argument that, for polynomials of degree 5 and greater, there can be no analog to the quadratic formula to solve for its roots. In such arguments, S_5 appears as the group of permutations of the five roots of a polynomial of degree 5.

EXAMPLE 6

Suppose that H is a subgroup of S_5 that contains a 5-cycle and a 2-cycle. We will show that $H = S_5$. Suppose that $\alpha = (x_1, x_2, x_3, x_4, x_5)$ and $\beta = (x_1, x_i)$ are both in H. The powers of α are $\alpha^2 = (x_1, x_3, x_5, x_2, x_4)$, $\alpha^3 = (x_1, x_4, x_2, x_5, x_3)$, and $\alpha^4 = (x_1, x_5, x_4, x_3, x_2)$. So some power of α starts with the symbol x_1 immediately followed by the symbol x_i. For ease of notation, let's assume that $\alpha = (a, b, c, d, e)$ and $\beta = (a, b)$. What we know so far is that H contains an element of order 5 and an element of order 2. Thus the order of H is divisible by 10. The composition $\beta \alpha = (a, b)(a, b, c, d, e) = (a)(b, c, d, e)$, which is a 4-cycle of order 4. The composition $\beta \alpha^2 = (a, b)(a, c, e, b, d) = (a, c, e)(b, d)$ has order 6. Now we know that the order of H is divisible by 4, 6, and 5 and hence the order is at least 60. There is only one subgroup of order 60, namely A_5. But H contains a 2-cycle and A_5 does not. So H must have order greater than 60, namely 120. Thus $H = S_5$. ∎

EXAMPLE 7

The group G of symmetries of the regular dodecahedron has order 60. To count them, notice that any of its 12 faces can be moved to the top position, and any vertex in the top face can be rotated into five different positions. A symmetry that rotates the dodecahedron around the axis through the center of a face has order five. A symmetry that rotates it around an axis through a vertex and its antipodal vertex has order 3 because three faces come together at a vertex. (See Figure 1.)

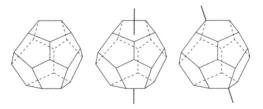

FIGURE 1. Dodecahedron and dodecahedron with axes

Figure 2 is a two-dimensional map of the three-dimensional dodecahedron. (We will refer to the shaded face and labels shortly.) You can count eleven faces. To flatten the figure, it's as if we punctured the twelfth face and expanded it. The whole area outside the boundary of the outer pentagon represents the twelfth face. Let's think of the lightly shaded face as the one that occupies the top position if the dodecahedron were set on a table. Any face can be moved into the top position using a combination of rotations about a vertex located at position A (an order three symmetry) and rotations about the geometric center of the top face (an order five symmetry). For instance,

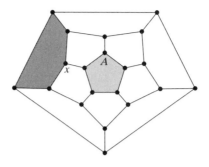

FIGURE 2

suppose we want to move the darkly shaded face in Figure 2 into the top position in such a way that vertex x landed in the position A. We could proceed as follows.

i. First rotate 120 degrees counterclockwise about A.
ii. Rotate 144 degrees counterclockwise about the center of the top face.
iii. Rotate 240 degrees counterclockwise about A.

You can follow the results of the moves in Figure 3. Note that x is a specific vertex on the dodecahedron and the dark shading refers to a specific face whereas the light shading indicates whatever face is on top and A denotes a position on that top face. We moved vertex x into position A and the darkly shaded face into the top position.

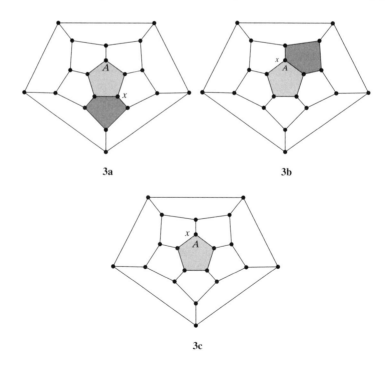

FIGURE 3

Example 7 shows that the group of symmetries of the dodecahedron is generated by an element of order 5 and an element of order 3. In the exercises, you are asked to show that A_5 is generated by a 5-cycle and a 3-cycle. This suggests that these two groups might be isomorphic. They are! It also suggests that there are five objects permuted by the symmetries of the dodecahedron. The five objects are five inscribed cubes. One such inscribed cube is illustrated in Figure 4a. A two-dimensional map is given in Figure 4b. Each of its 12 edges is on a different face of the solid. (The protruding lines represent the one edge on the face represented by the outer region.) There are five different cubes, one for each diagonal of any given face. Exactly two inscribed cubes meet at each vertex. For clarity, we have drawn just one of the five cubes.

FIGURE 4a. Dodecahedron with cube inscribed

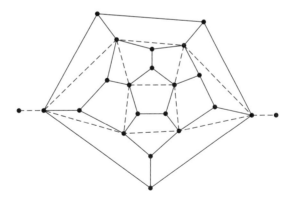

FIGURE 4b. Map of a cube

A rotation of the dodecahedron around an axis through the center of a face causes a cyclic permutation of the diagonals of that face and hence of their attached cubes. Exactly two cubes share any vertex of the dodecahedron. These are indicated by the bold and dashed lines in Figure 5a. The two cubes are fixed by any rotation about an axis through the shared vertex. The other three cubes are cyclically permuted. This is illustrated in Figure 5b, by the three different dashed lines. So the five distinct line types in the diagram represent the five distinct cubes.

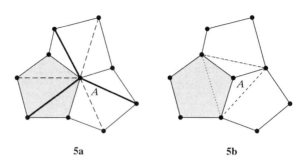

FIGURE 5

SUMMARY

A subgroup $H \subseteq G$ is normal if and only if the **conjugate** of any element $x \in H$ is also in H. Thus a normal subgroup is the union of **conjugacy classes**. In the permutation groups S_n, conjugacy is particularly easy to determine: Two permutations are conjugate if and only if they have the same cycle structure. By counting the number of permutations of a given cycle structure, we determined all the normal subgroups of S_4 and we saw that S_4 was a **solvable** group. By a similar argument, we saw that A_5 is the only normal subgroup of S_5 and that it is a **simple** group. Thus S_5 is *not* a solvable group, a fact that will be the key to proving that, in general, the quintic polynomial is not solvable in terms of radicals. We ended by showing that the group A_5 is isomorphic to the group of symmetries of the dodecahedron, regarded as the set of even permutations of five cubes inscribed in the dodecahedron.

5.3 Exercises

1. For each of the following pairs α and β, find γ such that $\alpha = \gamma^{-1}\beta\gamma$:
   i. $(a, b, c)(d, e, f)$ and $(a, c, d)(b, e, f)$
   ii. (a, b, c, d, e, f) and (a, c, d, b, e, f)
2. For each pair given in Exercise 1, show that γ is not unique.
3. Find a pair of permutations α and β with the same cycle structure for which there exists both even and odd maps γ such that $\alpha = \gamma^{-1}\beta\gamma$.
4. Find the number of elements in the conjugacy class of $(a, b, c)(d, e, f)$ in S_6.
5. Let $K = \{e, (a, b)(c, d), (a, c)(b, d), (a, d)(c, b)\}$ be regarded as a subgroup of A_4. Prove that K is a normal subgroup.
6. Determine if the given group is simple.
   i. Z_5
   ii. D_4, the group of symmetries of the square
   iii. T, the group of symmetries of the tetrahedron

7. Determine which of the following groups are solvable, and give the appropriate sequence of subgroups.
   i. Z_5
   ii. D_4, the group of symmetries of the square
   iii. T, the group of symmetries of the tetrahedron
8. Prove that all 3-cycles are conjugate in A_5.
9. Prove that all permutations with the cycle structure $2-2-1$ are conjugate in A_5.
10. Prove that the cycles (a, b, c, d, e) and (a, c, e, b, d) are **not** conjugate in A_5 but that each is in a conjugacy class with 12 elements.
11. Prove that A_5 is generated by a 5-cycle and a 3-cycle.

To the Teacher

One of the principal goals of the high school algebra agenda is solving polynomial equations. It's nowhere in sight in this section. But it is through the material developed in this section, in particular the fact that S_5 is not a solvable group, that Galois could show that polynomials of degree five or higher could not be solved in terms of roots and rational expressions of their coefficients. There's no "quintic formula" to follow up on the quadratic formula. In a sense, it put limits on the high school curriculum. How the concepts of this chapter merge with the solving of polynomials is the subject of Chapter 6. It's not an easy task. But there is still some interface with the mathematics that high school students experience.

A tangible model for normality and for conjugacy can be found in the dihedral groups, thought of as the groups of symmetries of the regular polygons. The subgroup of rotations is normal in any dihedral group. The only elements outside that subgroup are the flips. To flip, then rotate and then flip again—well, you might as well just rotate! (Here again we see that what seem like mathematically advanced notions are grounded in very basic mathematical experiences.) On the other hand, all the subgroups of the form {id, Flip} are conjugate. Here's how we might "feel" that fact. Suppose you had a pentagon and you needed to flip it around axis 1, the axis through vertex 1. The gears on axis 1 are stuck but axis 2 is working fine. Just rotate until vertex 1 to vertex 2, then flip, and then rotate back. {id, Flip1} = Rotate^{-1} {id, Flip2}Rotate. (See Figure 6.)

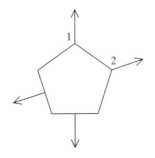

FIGURE 6

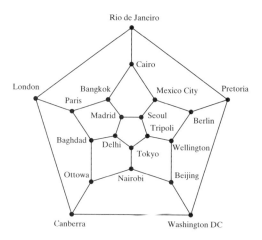

FIGURE 7

This section connects to the high school curriculum in other ways. High school students are often introduced to graph theory—the study of connecting the dots. They might notice that the skeleton of the dodecahedron given in Figure 7 is the graph in Hamilton's famous "Around the World Puzzle."

Puzzle: Start at any dot in Figure 7 and "go around the world," traveling on the edges and visiting each dot exactly once before returning home. (Sir William Rowan Hamilton published it as a puzzle in 1859.)

Students investigating the five regular solids might be pointed to the five distinct cubes that hide in the dodecahedron (see Figure 4). We use them to confirm the fact that the symmetry group of the dodecahedron is A_5. But just noting their presence is literally an eye-opener, intriguing both mathematically and artistically.

Task

Join geography to mathematics by assigning international cities to the dots in Hamilton's puzzle. Make the puzzle a bit tougher by designating the first few cities on a diplomat's route.

5.4 SUBRINGS, IDEALS, AND QUOTIENT RINGS

In this section, we take another look at rings. This time, we look at substructures and we look at homomorphisms between rings. As with groups it is natural to look for subsets of a given ring that are also rings with the operations they inherit. For example, the set of even integers, $2\mathbf{Z}$, is a subset of $\mathbf{Z}$ that is also a ring. Since rings are abelian groups under addition, any subring is a normal subgroup under addition. Using the same notion of cosets as we did for groups, we can build the associated quotient structure that will naturally be a group under addition. In this section, we shall investigate how to extend the operation of multiplication to the quotient structure so that we may construct new rings called quotient rings. To do this, we need special subrings called "ideals." Just as normal subgroups occur naturally as the kernels of group homomorphisms, ideals occur as kernels of ring homomorphisms (i.e., maps between rings that preserve both addition and multiplication).

> **DEFINITION 1.** Let R be a ring and let S be a subset of R that is a subgroup of R under addition. We call S a **subring** of R if S is closed under multiplication. If R is a field, then S is a **subfield** of R if it is a subring of R and the inverse of each nonzero element of S is also in S.

If $S \neq R$, we call S a **proper subring** of R. Every ring has what is called the **trivial subring**, namely the set, $\{0\}$.

EXAMPLES

1. Let n be any integer. The set $n\mathbf{Z} = \{nx : x \in \mathbf{Z}\}$ is a subring of $\mathbf{Z}$ since it is a subgroup of the additive group $\mathbf{Z}$ and it is closed under multiplication. Note that for $n \neq \pm 1$, $n\mathbf{Z}$ is not a ring with unity.

2. Let $\mathbf{R}$ be the set of real numbers and let $\mathbf{Q}(\sqrt{2}) = \{a + b\sqrt{2} : a \in \mathbf{Q} \text{ and } b \in \mathbf{Q}\}$. It is not difficult to verify that $\mathbf{Q}(\sqrt{2})$ is a subring of the ring of real numbers. For instance, to see that $\mathbf{Q}(\sqrt{2})$ is closed under multiplication, note that $(a + b\sqrt{2})(c + d\sqrt{2}) = (ac + 2bd) + (ad + bc)\sqrt{2}$. Actually, $\mathbf{Q}(\sqrt{2})$ is a subfield of $\mathbf{R}$. If a and b are not both zero, then $(a + b\sqrt{2})^{-1} = \left(\dfrac{a}{a^2 - 2b^2}\right) - \left(\dfrac{b}{a^2 - 2b^2}\right)\sqrt{2}$. (Since $\sqrt{2}$ is irrational, denominator $a^2 - 2b^2$ cannot equal 0 when a and b are rational.)

3. The set $\{0, 2, 4\}$ is a subring of $\mathbf{Z}_6$. In fact, the set $q\mathbf{Z}_n = \{qx \bmod n : x \in \mathbf{Z}_n \text{ and } q \in \mathbf{Z}\}$ is a subring of $\mathbf{Z}_n$. However, $q\mathbf{Z}_n$ is not always a proper subring. (Try $3\mathbf{Z}_7$.)

4. The set $\mathbf{Z}[i] = \{a + bi : a, b \in \mathbf{Z}\}$ is a subring of $\mathbf{C}$, the field of complex numbers. In the complex plane, the elements of $\mathbf{Z}[i]$ occur on the lattice of points with integer coordinates. The set is known as the **Gaussian integers**. The subring $\mathbf{Z}[i]$ must be an integral domain because $\mathbf{C}$ is an integral domain. However, $\mathbf{Z}[i]$ is **not** a subfield of the complex numbers. In fact, the only units of $\mathbf{Z}[i]$ are the numbers $1, -1, i$ and $-i$. Clearly, these numbers are units since each has a multiplicative inverse in $\mathbf{Z}[i]$, namely $1, -1, -i$ and i, respectively. To see that they are the only units in $\mathbf{Z}[i]$, recall that in $\mathbf{C}$, the inverse of a number $a + bi$ is $\dfrac{a}{a^2 + b^2} - \dfrac{b}{a^2 + b^2}i$. In order for both $\dfrac{a}{a^2 + b^2}$ and $\dfrac{b}{a^2 + b^2}$ to be integers, one of a or b must be 0 and the other must be 1 or -1. Only the complex numbers $1, -1, i$, and $-i$ have this property. Alternatively, we can take advantage of the absolute value function on $\mathbf{C}$. It has the property

that $|zw|^2 = |z|^2|w|^2$ for z and w in $\mathbf{C}$. When both a and b are integers, $|a+bi|^2 = a^2+b^2$ is an integer. Thus, when both z and w are in $\mathbf{Z}[i]$, the only way that $|zw|^2 = |1|^2 = 1$ is for $|z| = 1$ and $|w| = 1$. This occurs only for z and w equal to $1, -1, i$, or $-i$. ∎

Next, we will discuss a special type of subring called an **ideal**. An ideal in a ring plays the same role as a normal subgroup does in a group. Using ideals, we will then construct quotient rings.

> **DEFINITION 1.** Let S be a subring of a ring R. We say that S is an **ideal** of R if ax and xa are in S for all $x \in S$ and $a \in R$.

An ideal is more than just closed under multiplication. To check that S is an ideal we need to check that the products ax and xa are in S for x in S and any a in the bigger set R. (If S is a commutative ring, we need only check ax.)

EXAMPLES

5. For any n in $\mathbf{Z}$, the set $n\mathbf{Z}$ is an ideal of $\mathbf{Z}$. We already noted that $n\mathbf{Z}$ is a subring of $\mathbf{Z}$. To check that it is an ideal, let $x \in n\mathbf{Z}$, and $a \in \mathbf{Z}$. Then $x = ny$ for some y in $\mathbf{Z}$, and so $ax = nay$. Thus $ax \in n\mathbf{Z}$, which shows that $n\mathbf{Z}$ is an ideal.

6. Let $\widetilde{\mathbf{M}}_{2,2}$ denote the set of 2×2 matrices with integer entries and let I denote the subset of all 2×2 matrices with even entries. The set I is certainly closed under addition and multiplication. Further, if A is any matrix in $\widetilde{\mathbf{M}}_{2,2}$, and B is any matrix in I, both AB and BA have even entries. Thus I is an ideal.

7. The only ideals of a field F are the trivial ideal $S = \{0\}$ and the whole set F itself. To see this, suppose that I is a nontrivial ideal in F. Then I has at least one nonzero element x. Since F is a field, x has an inverse and $x^{-1}x = 1$ is in I. Since $1 \in I$, then $y \cdot 1 = y$ is an element of I for any $y \in F$. Thus $I = F$.

8. Let R be any ring and let $\{x_1, \ldots, x_n\}$ be a subset of R. The **ideal I generated by $\{x_1, \ldots, x_n\}$** is the intersection of all the ideals of R that contain the set $\{x_1, \ldots, x_n\}$. It is denoted by $\langle x_1, \ldots, x_n \rangle$. If R is a commutative ring with unity, then $I = \{a_1x_1 + a_2x_2 + \ldots + a_nx_n : a_1, \ldots, a_n \in R\}$.

9. Let $\mathbf{Z}[x]$ denote the ring of polynomials with integer coefficients. The set $I = \{a_0 + a_1x + \ldots + a_nx^n : a_0 \text{ is even}\}$ is an ideal. Its elements are all polynomials in which the constant term a_0 is even. In the notation given in Example 8, $I = \langle 2, x \rangle$. ∎

Cyclic subgroups are particularly important substructures for group theory. The analogous structure for a ring is an ideal generated by one element. They are equally important in ring theory.

> **DEFINITION 2.** Let R be a commutative ring with unity. An ideal $I \subseteq R$ is called a **principal ideal** if there is an element x in I such that $I = \langle x \rangle$.

Any principal ideal is of the form $\langle x \rangle = \{ax : a \in R\}$. The element x is called the **generator** of $\langle x \rangle$.

EXAMPLES

10. The subrings of $\mathbf{Z}$ are of the form $n\mathbf{Z}$. Hence all subrings of $\mathbf{Z}$ are ideals that are also principal ideals. The ideal $n\mathbf{Z}$ can also be denoted as $\langle n \rangle$.

11. The ideal I in $\mathbf{Z}[x]$ described in Example 9 is not a principal ideal. Note that the polynomials $p(x) = x$ and $q(x) = 2$ are both in I. If I were a principal ideal, we could find an element $g(x)$ in $\mathbf{Z}[x]$ such that $I = \langle g(x) \rangle$. If $\deg(g(x)) > 0$, all multiples of $g(x)$ would have degree greater than 0. Then 2 could not be an element of I. So $g(x)$ must be a constant polynomial and, in fact, it must be that $g(x) = 2$ or $g(x) = -2$. But $\langle -2 \rangle = \langle 2 \rangle = \{p(x) = a_0 + a_1 x + \ldots + a_n x^n : a_i$ is even for all $i\}$. So the polynomial $p(x) = x$ cannot be in $\langle 2 \rangle$. Thus I is **not principal** because it not possible to find one polynomial $g(x)$ such that $I = \langle g(x) \rangle$.

12. Let F be a field. In the ring $F[x]$ of polynomials with coefficients in F, all ideals are principal ideals. To see this, let I be an ideal of $F[x]$ and let $p(x)$ be a polynomial in I of least degree. Then $I = \langle p(x) \rangle$. Any other polynomial $g(x) \in I$ can be expressed as $g(x) = q(x)p(x) + r(x)$, where either $r(x) = 0$ or $\deg(r(x)) < \deg(p(x))$. Since $q(x)p(x)$ and $g(x)$ are both in I, the remainder $r(x)$ is also in I because $r(x) = g(x) - q(x)p(x)$. If $r(x) \neq 0$, the fact that $\deg(r(x)) < \deg(p(x))$ would contradict our choice of $p(x)$. ∎

Ring Homomorphisms

In Section 5.2, we saw that normal subgroups arose most naturally as kernels of group homomorphisms. Ideals are the analogous structures for rings and they too arise as kernels of ring homomorphisms.

> **DEFINITION 4.** Let R_1 and R_2 be rings. A function $f : R_1 \to R_2$ is a **ring homomorphism** if it preserves both multiplication and addition:
>
> $$f(a+b) = f(a) + f(b) \text{ and } f(ab) = f(a)f(b).$$
>
> When f is also a bijection, f is called a **ring isomorphism**.

EXAMPLES

13. The map $f : \mathbf{C} \to \mathbf{C}$ defined by $f(a + bi) = a - bi$ is a ring isomorphism of the set of complex numbers $\mathbf{C}$ to itself. To see that the map preserves addition, note:

$$f((a+bi) + (c+di)) = f((a+c) + (b+d)i) = (a+c) - (b+d)i$$
$$= (a - bi) + (c - di) = f(a+bi) + f(c+di).$$

To see that the map preserves multiplication, note:

$$f((a+bi)(c+di)) = f((ac-bd)+(ad+bc)i) = (ac-bd)-(ad+bc)i$$
$$= (a-bi)(c-di) = f(a+bi)f(c+di).$$

14. The map $f: \mathbf{Z} \to \mathbf{Z}_m$ defined by $f(x) = x \bmod m$ is a ring homomorphism. We leave the details, all familiar facts from number theory, to the reader.

15. The map $f: \mathbf{Z}_3 \to \mathbf{Z}_6$ defined by $f(x) = 2x \bmod 6$ is **not** a ring homomorphism because it fails to preserve multiplication: $f(xy) = 2xy \bmod 6$ but $f(x)f(y) = 4xy \bmod 6$. (However, the map preserves addition.)

16. The map $f: \mathbf{Z}_9 \to \mathbf{Z}_{12}$ defined by $f(x) = 4x \bmod 12$ is a ring homomorphism. The details are left as an exercise. ■

Ring homomorphisms have many of the same basic properties as group homomorphisms. Some of these are summarized in the following proposition.

PROPOSITION 1 Let $f: R_1 \to R_2$ be a **ring** homomorphism.
  i. If 0 is the additive identity element of R_1, then $f(0)$ is the additive identity element of R_2.
  ii. The set $f(R_1)$ is subring of R_2.
  iii. If 1 is the multiplicative identity element for R_1, then $f(1)$ is the multiplicative identity element for $f(R_1)$.

Proof.
  i. Because f preserves addition, f is a group homomorphism between the additive groups R_1 and R_2. By Proposition 4 of Section 4.5, $f(0)$ is the additive identity element of R_2.
  ii. By Proposition 4 of Section 4.5, $f(R_1)$ is a subgroup of R_2 under addition. We need to check that $f(R_1)$ is closed under multiplication. Let x and y be elements of R_1. Then $f(x)f(y) = f(xy)$ is an element of R_2. Thus $f(R_1)$ is closed under multiplication and is hence a subring of R_2.
  iii. Let $y \in f(R_1)$ and suppose that $f(x) = y$. Notice that $f(1 \cdot x) = f(x \cdot 1) = y$, and because f is operation preserving, we have $f(1 \cdot x) = f(1) \cdot f(x) = f(x) \cdot f(1)$. Thus, $y = f(1) \cdot y = y \cdot f(1)$ for all y in $f(R_1)$. ▲

Part iii of Proposition 1 is a bit subtle. Consider $f: \mathbf{Z}_3 \to \mathbf{Z}_6$ defined by $f(0) = 0$, $f(1) = 4$, and $f(2) = 2$. You can check that f preserves addition and multiplication and that $f(\mathbf{Z}_3) = \{0, 2, 4\}$ is a subring of $\mathbf{Z}_6$. It looks like there is no unity element in $\{0, 2, 4\}$. However, Proposition 1 says that $f(1) = 4$ should be a unity element of $f(\mathbf{Z}_3)$, but not necessarily the unit element of $\mathbf{Z}_6$. Let's check.

$$4 \cdot 0 = 0$$
$$4 \cdot 2 \bmod 6 = 8 \bmod 6 = 2$$
$$4 \cdot 4 \bmod 6 = 16 \bmod 6 = 4$$

So $f(1)$ is indeed a unity element for the subring $f(\mathbf{Z}_3)$, but not for the entire ring $\mathbf{Z}_6$.

The concepts of ideals and homomorphisms are linked through the notion of kernel, just as normal subgroups are linked to the notion of group homomorphism. The next definition and Proposition 2 accomplish the connection.

> **DEFINITION 5.** Let $f: R_1 \to R_2$ be a ring homomorphism. The **kernel of f** is the set $\{x \in R_1 : f(x) = 0\}$. It is denoted by $\ker(f)$.

EXAMPLES

17. Let $f: \mathbf{Z} \to \mathbf{Z}_m$ be defined by $f(x) = x \bmod m$. Then $\ker(f) = m\mathbf{Z}$ because $f(x) = 0 \bmod m$ if and only if x is a multiple of m.
18. Let $f: \mathbf{C} \to \mathbf{C}$ defined by $f(a+bi) = a - bi$. Then $\ker(f) = \{0\}$ because f is a ring isomorphism and hence one-to-one.

PROPOSITION 2 Let $f: R_1 \to R_2$ be a ring homomorphism. The kernel of f, namely the set $\ker(f) = \{x \in R_1 : f(x) = 0\}$, is an ideal in R_1.

Proof. To check that K is a subring of R_1, we note that if x and y are in K, then $x+y$ and xy are also in K since $f(x+y) = f(x) + f(y) = 0 + 0 = 0$ and $f(xy) = f(x)f(y) = 0 \cdot 0 = 0$. Now let a be any element of R_1 and x be any element of K. Then $f(ax) = f(a)f(x) = f(a) \cdot 0 = 0$. Similarly, $f(xa) = 0$. Thus $ax \in \ker(f)$ and $xa \in \ker(f)$. So $ker(f)$ is an ideal. ▲

Quotient Rings

Suppose that I is an ideal of R. The ring R is an abelian group under addition and I is a normal subgroup of R. So we can form the additive quotient group R/I. Its elements are the cosets of the form $x + I$. Recall that $a + I = b + I$ if and only if $a - b \in I$. Addition is defined by $(x+I) + (y+I) = (x+y) + I$. Our ultimate goal is to make R/I into a *ring*. So we need to define multiplication also. Let's define multiplication in the following natural way:

$$(x+I)(y+I) = xy + I.$$

For R/I to be a ring, multiplication must be well-defined. So suppose that a and b are elements of R such that $a + I = x + I$ and $b + I = y + I$. Then $x - a$ and $y - b$ are both in I. We must check that $ab + I = xy + I$, or, equivalently, that $xy - ab \in I$. To do this, let $x = a + s$ and $y = b + t$, where both s and t are elements of I. Then the product $xy = ab + at + sb + st$. Thus $(xy - ab) = at + sb + st$. Since I is a subring, the term st is in I. Since I is an *ideal*, we know that at and sb are in I, even though a and b are not necessarily in I. (It is here that we need the special properties of an ideal.) Thus $(xy - ab) \in I$ and $xy + I = ab + I$. Multiplication is thus well-defined. The associative and distributive laws hold in R/I because they hold in R. Thus R/I is a ring. It is called the **quotient ring of R modulo I**. If R is a commutative ring, then so is R/I. If R is a ring with unity 1 and I is proper, then R/I is a ring with $1 + I$ as unity.

EXAMPLES

19. The quotient group $\mathbf{Z}/n\mathbf{Z}$ is isomorphic to the group $\{\mathbf{Z}_n, +\}$. The mapping $x \to x + n\mathbf{Z}$ also preserves multiplication. Thus $\mathbf{Z}/n\mathbf{Z}$ is isomorphic to the ring $\mathbf{Z}_n$. Even though $\mathbf{Z}$ is an integral domain, the quotient ring $\mathbf{Z}/n\mathbf{Z}$ is an integral domain only when n is a prime, in which case it is a field.

20. The set $6\mathbf{Z}$ is an ideal in the ring $2\mathbf{Z}$. The elements of the quotient ring $2\mathbf{Z}/6\mathbf{Z}$ are $0 + 6\mathbf{Z}, 2 + 6\mathbf{Z}$, and $4 + 6\mathbf{Z}$. The operations on the quotient ring are addition and multiplication mod 6. The multiplication table that follows (with its elements listed out of the usual order) shows that $4 + 6\mathbf{Z}$ is the multiplicative identity element. In fact, the ring $2\mathbf{Z}/6\mathbf{Z}$ is isomorphic to the ring $\mathbf{Z}_3$ under the correspondence $0 \to 0 + 6\mathbf{Z}$, $1 \to 4 + 6\mathbf{Z}$, $2 \to 2 + 6\mathbf{Z}$. Even though $2\mathbf{Z}$ is a ring without unity, the quotient ring $2\mathbf{Z}/6\mathbf{Z}$ is a field.

*	0	4	2
0	0	0	0
4	0	4	2
2	0	2	4

21. Let $\mathbf{Z}[i]$ be the set of Gaussian integers described in Example 4 and let $I = \langle 3 - i \rangle$. Since $(3 - i) - 0 \in I$, the element $3 - i + I = 0 + I$ in $\mathbf{Z}[i]/I$. Adding $i + I$ to both sides, we obtain $3 + I = i + I$ in $\mathbf{Z}[i]/I$. So we can substitute 3 for i in every occurrence of i. For example, $2 + 5i + I = 2 + 15 + I = 17 + I$. Thus every coset can be written as $x + I$ for $x \in \mathbf{Z}$. But we can go further. Since $3 + I = i + I$, we can square both sides to obtain $9 + I = -1 + I$ or $10 + I = 0 + I$. So $17 + I = 7 + I$ and more generally, $x + I$ can be replaced by $x \mod 10 + I$ for every integer x. Thus the ring $\mathbf{Z}[i]/I$ has at most 10 elements: $0 + I, 1 + I, \ldots, 9 + I$. To see that there are exactly 10 distinct cosets in $\mathbf{Z}[i]/I$, we must show that none of the cosets $1 + I, \ldots, 9 + I$ is equal to $0 + I$. This is the same as showing that none of the integers $1, 2, \ldots, 9$ is in I. (Details are left as an exercise.) The quotient ring $\mathbf{Z}[i]/I$ is thus isomorphic to the ring $\mathbf{Z}_{10}$, not entirely obvious at the start! This is another example of where the quotient ring formed from an integral domain fails to be an integral domain. ■

We saw that even if a ring R was an integral domain, a quotient ring R/I need not be an integral domain. Suppose that $x + I$ and $y + I$ are nonzero elements of R/I so that x and y are not elements of I. Even though x and y are not in I, their product xy *might* be in I making $xy + I = 0 + I$. Avoiding this situation would guarantee that R/I is an integral domain when R is an integral domain, which prompts the following definition.

DEFINITION 2. Let R be an integral domain and let I be a proper ideal in R. Then I is called a **prime ideal** if, whenever a product $xy \in I$, then either $x \in I$ or $y \in I$.

EXAMPLES

22. The ideal $n\mathbf{Z}$ is a prime ideal in $\mathbf{Z}$ if and only if n is a prime number (hence the terminology). For suppose that n is prime and that xy is in $n\mathbf{Z}$. Since xy is a multiple of the prime n, either x or y must be divisible by n. So either x or y is in $n\mathbf{Z}$. Thus $\langle 3 \rangle$ is a prime ideal in $\mathbf{Z}$ while $\langle 4 \rangle$ is not.

23. In the ring of Gaussian integers $\mathbf{Z}[i]$, the ideal $\langle 2 \rangle$ is not a prime ideal. To see this, notice $|z|^2 \geq 4$ for any nonzero z in $\langle 2 \rangle$. Let $x = (1+i)$ and $y = (1-i)$. Then $|x|^2 = |y|^2 = 2$ so that neither x nor y is an element of $\langle 2 \rangle$. But $xy = 2$ is an element of $\langle 2 \rangle$. ∎

THEOREM 1 Let R be an integral domain and $I \subseteq R$ be an ideal. The quotient ring R/I is an integral domain if and only if I is a prime ideal.

Proof. Suppose that I is a prime ideal. Suppose that $x + I$, and $y + I$ are nonzero elements of R/I so that neither x nor y is an element of I. Since I is a prime ideal, $xy \notin I$ and $xy + I \neq 0 + I$. Hence, R/I is an integral domain. Conversely, suppose that R/I is an integral domain and let x and y be elements of R that are not elements of I. Since $(x+I)(y+I) \neq 0 + I$, the product $xy \notin I$. Thus I is a prime ideal. ▲

Now we look to see what properties an ideal $I \subseteq R$ must have in order to guarantee that its quotient ring R/I is a field. The crucial observation is the relation between ideals in R and ideals in R/I. Suppose that A is an ideal and that $I \subseteq A \subseteq R$. It is easy to see that the set $B = \{x + I : x \in A\}$ is an ideal of R/I. Suppose that A is a proper ideal of R and that $z \in R$ is an element that is not in A. Then $z + I \notin B$ so that B is a proper ideal of R/I. But if R/I is a field, R/I has no proper ideal other than $\langle 0 + I \rangle$. Hence, if R/I is a field and $I \subseteq A \subseteq R$, then either $A = I$ in which case $B = \langle 0 + I \rangle$, or $A = R$, in which case $B = R/I$. The following definition captures the property I must have in order for R/I to be a field.

DEFINITION 4. Let R be an integral domain and let $I \subseteq R$ be a proper ideal of R. We call I a **maximal ideal** if, for any ideal A such that $I \subseteq A \subseteq R$, either $A = I$ or $A = R$.

Thus, if I is a maximal ideal, then I is not properly contained in any other proper ideal.

EXAMPLE 24

In $\mathbf{Z}$, the ideals $n\mathbf{Z}$ are maximal if and only if n is prime. (In $\mathbf{Z}$, the set of prime ideals and the set of maximal ideals are identical.) For suppose that the ideal

$m\mathbf{Z} \subseteq \mathbf{Z}$. If m is composite and properly factors as nq, then $m\mathbf{Z} \subseteq n\mathbf{Z} \subseteq \mathbf{Z}$ with each containment proper. So $m\mathbf{Z}$ is not maximal. On the other hand, suppose m is prime and $m\mathbf{Z} \subseteq n\mathbf{Z} \subseteq \mathbf{Z}$. Then $m = nq$ for some q in $\mathbf{Z}$. Since m is prime, either $n = \pm 1$, in which case $n\mathbf{Z} = \mathbf{Z}$, or $n = \pm m$, in which case $n\mathbf{Z} = m\mathbf{Z}$. ∎

From our discussion, we see that in order for R/I to be a field, it is necessary that I be a maximal ideal. The next theorem shows that the condition is also sufficient.

THEOREM 2 Let R be an integral domain and suppose that $I \subseteq R$ is a maximal ideal. Then the quotient ring R/I is a field.

Proof. The multiplicative identity in R/I is $1 + I$. Let $a + I$ be an element in R/I with $a \notin I$. We need to find the multiplicative inverse of $a + I$. That is, we need to find $x \in R$ such that $ax + I = 1 + I$ or, equivalently, such that $ax - 1 \in I$. We will exploit the fact that I is a maximal ideal. So let $A = \{ax + y : x \in R \text{ and } y \in I\}$. It is easy to see that A is an ideal. Letting $x = 0$, we see that $I \subseteq A$. Letting $x = 1$ and $y = 0$, we see that $a \in A$. Since $a \notin I$, the containment $I \subseteq A$ is proper. Since I is maximal, $A = R$. We can thus find x and y such that $1 = ax + y$, or regrouping, $ax - 1 = -y$. Since $ax - 1 \in I$, we see that the inverse of $a + I$ in R/I is $x + I$. ▲

COROLLARY 3 Every maximal ideal is a prime ideal.

Proof. Since every field is an integral domain, every maximal ideal is a prime ideal. ▲

EXAMPLE 25

Let F be a field and let $R = F[x]$, the ring of polynomials with coefficients in F. Because the ring $F[x]$ has a Division Algorithm similar to that for the ring $\mathbf{Z}$, many of its algebra properties are similar to those of $\mathbf{Z}$. As we have seen, in both rings, every ideal is a principal ideal. Similarly, in $F[x]$ an ideal $\langle p(x) \rangle$ is a prime if and only if $p(x)$ is an irreducible polynomial. Furthermore, an ideal is prime if and only if it is a maximal ideal. (We ask the reader to prove these facts in the exercises. The arguments are very similar to those in Example 24.) So suppose that $p(x)$ is an irreducible polynomial in $F[x]$. By Theorem 2, $F[x]/\langle p(x) \rangle$ is a field. But we knew this already from Theorem 3 of Section 3.7. The cosets of the quotient ring $F[x]/\langle p(x) \rangle$ are the congruence classes of the mod $p(x)$ relation and the operations on the quotient ring $F[x]/\langle p(x) \rangle$ are defined as in Section 3.7. This section generalizes the familiar modular relations on the rings $\mathbf{Z}$ and $F[x]$ to arbitrary rings. ∎

SUMMARY

A subset S of a ring R is a **subring** of R if it is a subgroup for the operation of addition and if it is also closed under multiplication. If, in addition, ax and xa are elements of S for every $x \in S$ and $a \in R$, then S is an **ideal** of R. Ideals are to rings as normal subgroups are to groups. They arise naturally as the kernels of **ring homomorphisms**. A **principal ideal** $\langle x \rangle$ is generated by one element. If $I \subseteq R$ is an ideal, then the operations of addition and multiplication defined on its cosets are well-defined. The resulting structure R/I is a **quotient ring**. When R is an integral domain, R/I is an integral domain if and only if I is a **prime ideal** and R/I is a field if and only if I is a **maximal ideal**.

5.4 Exercises

1. Find all the subrings of Z_6.
2. Find all the subrings of Z_{12}.
3. For what values of q is qZ_n a proper subring of Z_n? Prove your answer.
4. The set $Q(\sqrt{3}) = \{a + b\sqrt{3} : a, b \in Q\}$ is a subring of the field of complex numbers C.
   i. Show that the element $2 + \sqrt{3}$ has a multiplicative inverse in $Q(\sqrt{3})$.
   ii. Show that $Q(\sqrt{3})$ is a subfield of C by showing that every nonzero element of $Q(\sqrt{3})$ has a multiplicative inverse that is also an element of $Q(\sqrt{3})$.
5. Is the set $S = \{a + bi : a \text{ is an even integer}\}$ a subring of $Z[i]$?
6. Let F be a field. Show that the intersection of any collection of subfields of F is a field.
7. Show that the set $S = \{a + bi : a \text{ and } b \text{ are even}\}$ is an ideal of $Z[i]$.
8. The set $T = \{a + bi : b = 0\}$ is a subring of $Z[i]$. Is it an ideal?
9. Let I_1 and I_2 be ideals in a ring R. Prove that the intersection $I = I_1 \cap I_2$ is an ideal of R.
10. Show that the map $f : Z_9 \to Z_{12}$ defined by $f(x) = 4x \mod 12$ is a ring homomorphism. (Make sure you check that it is well-defined.)
11. In Exercise 14 of Section 4.5, you were asked to show that the map $f : Z_m \to Z_n$, where $f(x) = qx \mod n$ is a *group* homomorphism if and only if n divides qm. What additional property must q satisfy so that f is a *ring* homomorphism?
12. Find a nontrivial ring homomorphism from Z_6 to Z_{10}. (*Hint.* Such a map is completely determined by its value at $1 \in Z_6$.)
13. Show that map $f : Q[x] \to Q(\sqrt{3})$ defined by $f(p(x)) = p(\sqrt{3})$ is a ring homomorphism. Find its kernel.
14. Challenge: Show that the field F_3 obtained in Exercise 11 of Section 3.7 is isomorphic to the field F_1 of Chart 1 of Section 3.7.
15. Finish the details of Example 21 by showing that if $a = 1, 2, \ldots, 9$, then $a \notin I$. (You must show that a cannot be expressed as $(x + iy)(3 - i)$ for any integers x and y.)
16. List all distinct cosets in the quotient ring $Z[i]/\langle 2+i \rangle$. Give the addition and multiplication table for the ring. Is it a field?
17. Let $I \subseteq Z \times Z$ be the ideal $\{(x, y): x \text{ and } y \text{ are even}\}$. Find all the distinct cosets of $(Z \times Z)/I$ and give the multiplication and addition tables for the quotient rings.

18. Determine which of the following are prime ideals in $Z[i]$ and which are maximal ideals.
    i. $\langle 3 \rangle$
    ii. $\langle 3 + i \rangle$
    iii. $\langle 1 + i \rangle$
19. Show that the ideal $\langle 3x, 2 \rangle$ is not a principal ideal of $Z[x]$.
20. Give an example of a quotient ring $Z[i]/I$ that is not a field.
21. Let F be a field. Prove that an ideal $\langle p(x) \rangle$ in $F[x]$ is a prime ideal if and only if $p(x)$ is an irreducible polynomial.
22. Let F be a field. Prove that an ideal $\langle p(x) \rangle$ in $F[x]$ is a prime ideal if and only if $\langle p(x) \rangle$ is a maximal ideal.

To the Teacher

Ideals are to quotient rings what normal subgroups are to quotient groups. An ideal I of a ring R is a subring of R with some extra properties needed to make multiplication in the R/I well-defined. An ideal I of the ring Z is just an extended "times table"—all multiples of a fixed number n. Multiplication in Z/I is just multiplication mod n. Again we find an advanced algebraic concept grounded in school mathematics.

Students use concepts like subring, subfield and ideal, without their technical vocabulary. Most students are aware of the subring sequence $Z \subseteq Q \subseteq R \subseteq C$ but they are not usually aware that the sequence can be expanded. As we saw in Example 2, $Q(\sqrt{2})$ is a field that is a subfield of R so that we have $Q \subseteq Q(\sqrt{2}) \subseteq R$ and hence $Z \subseteq Q \subseteq Q(\sqrt{2}) \subseteq R \subseteq C$. Where would a student encounter $Q(\sqrt{2})$? When they "rationalize a denominator" of, say, $\frac{5}{3+\sqrt{2}}$. What they are really doing is utilizing the fact that $3 + \sqrt{2}$ has an inverse of the form $a + b\sqrt{2}$ in the field $Q(\sqrt{2})$. Of course there are many such intermediate fields such as $Q(\sqrt{p})$ for any prime p.

Task

Develop a student-friendly way to "rationalize" expressions such as $\frac{1}{a+b\sqrt[3]{p}}$, where p is a prime. (**Hint.** Suppose $a \neq 0$. You need to multiply numerator and denominator by an expression of the form $c + d\sqrt[3]{p} + e(\sqrt[3]{p})^2$. You can choose $c = 1$. The field is $Q(\sqrt[3]{p}) = \{a + b\sqrt[3]{p} + c(\sqrt[3]{p})^2 : a, b, c \in Q\}$.

5.5 WORKSHEET 8: ON THE RING OF 2×2 MATRICES OVER THE INTEGERS

Note: This worksheet assumes that the reader has studied linear algebra.

The set $M = \left\{ \begin{bmatrix} a & b \\ c & d \end{bmatrix} : a, b, c, d \in Z \right\}$ is a ring under the operations of matrix addition and multiplication.

Task 1

Show that M is a noncommutative ring. (Find two matrices A and B such that $AB \neq BA$.)

The ring M has units, zero divisors, and elements that are neither. The goal of this worksheet is to find examples of each kind of element. In the ring M, the multiplicative identity is $I = \begin{bmatrix} 1 & 0 \\ 0 & 1 \end{bmatrix}$ and the additive identity is $\mathbf{0} = \begin{bmatrix} 0 & 0 \\ 0 & 0 \end{bmatrix}$. Thus a unit will be a matrix $A = \begin{bmatrix} a & b \\ c & d \end{bmatrix}$ (with integer entries) for which there is another matrix $B = \begin{bmatrix} u & v \\ w & z \end{bmatrix}$ (again with integer entries) such that $\begin{bmatrix} a & b \\ c & d \end{bmatrix} \begin{bmatrix} u & v \\ w & z \end{bmatrix} = I$. There is a simple test to determine if a matrix is a unit that uses the determinant of the matrix. Recall that $\det \begin{bmatrix} a & b \\ c & d \end{bmatrix} = ad - bc$.

Task 2

i. Suppose that $ad - bc = 1$ for matrix $A = \begin{bmatrix} a & b \\ c & d \end{bmatrix}$ in M. Find the matrix $\begin{bmatrix} u & v \\ w & z \end{bmatrix}$ such that $\begin{bmatrix} a & b \\ c & d \end{bmatrix} \begin{bmatrix} u & v \\ w & z \end{bmatrix} = I$, and express $u, v, w,$ and z in terms of $a, b, c,$ and d. Remember the matrix $\begin{bmatrix} u & v \\ w & z \end{bmatrix}$ **must** have integer coefficients.

ii. Modify your derivation for $ad - bc = -1$.

iii. Show that if $ad - bc \neq \pm 1$, the matrix A is not a unit. (Notice that A may be invertible as a matrix with rational entries. But in our ring, all entries are integers.)

If we look at the other elements of M, they fall into two categories: those with $ad - bc = 0$ and those with $ad - bc$ equal to an integer other than 1, 0, or -1.

Task 3

Suppose that $ad - bc = 0$ for matrix $A = \begin{bmatrix} a & b \\ c & d \end{bmatrix}$ in M. Show the rows (and it turns out, the columns) of the matrix are each integer multiples of the same pair of numbers. So if $ad - bc = 0$, then $\begin{bmatrix} a & b \\ c & d \end{bmatrix} = \begin{bmatrix} me & mp \\ ne & np \end{bmatrix}$ for some integers m, n, e, p. To begin, try an example like $\begin{bmatrix} 4 & 10 \\ 6 & 15 \end{bmatrix}$.

Task 4

The matrix $\begin{bmatrix} 2 & 3 \\ 4 & 6 \end{bmatrix}$ has determinant 0. Find a matrix $\begin{bmatrix} u & v \\ w & z \end{bmatrix}$ that is different from the zero matrix such that $\begin{bmatrix} 2 & 3 \\ 4 & 6 \end{bmatrix} \begin{bmatrix} u & v \\ w & z \end{bmatrix} = \begin{bmatrix} 0 & 0 \\ 0 & 0 \end{bmatrix}$. This shows that $\begin{bmatrix} 2 & 3 \\ 4 & 6 \end{bmatrix}$ is a zero divisor of M. Suppose that $A = \begin{bmatrix} a & b \\ c & d \end{bmatrix}$ is **not** the zero matrix $\begin{bmatrix} 0 & 0 \\ 0 & 0 \end{bmatrix}$. Adapt your solution to show that if $ad - bc = 0$, then A is a zero divisor.

What matrices are unaccounted for? Those for which the determinant is either greater than 1 or less than -1. (The determinant must be an integer.) These matrices are units in the ring of 2×2 matrices over the rational numbers but not in M.

Task 5

Caution! A matrix A may be a zero divisor with $AB = 0$ but BA is different from 0. Find such a pair A and B.

Task 6

The matrix $A = \begin{bmatrix} 2 & 1 \\ 1 & 2 \end{bmatrix}$ has determinant is $2 \cdot 2 - 1 \cdot 1 = 3$. Find a matrix $B = \begin{bmatrix} u & v \\ w & z \end{bmatrix}$ with coefficients that may be rational numbers that satisfies $\begin{bmatrix} 2 & 1 \\ 1 & 2 \end{bmatrix} \begin{bmatrix} u & v \\ w & z \end{bmatrix} = \begin{bmatrix} 1 & 0 \\ 0 & 1 \end{bmatrix}$. Use the existence of B to show that A cannot be a zero divisor. More generally, show that matrices with determinants that are different from 0 are never zero divisors.

5.6 PRIMES, IRREDUCIBLES, AND THE GAUSSIAN INTEGERS

The arithmetic of the natural numbers determines its algebraic structure. The Division Algorithm forces additive subgroups to be cyclic and hence all ideals to be principal. Euclid's Lemma forces all ideals generated by a prime to be maximal. In this section we investigate the interplay of arithmetic and algebraic structure in the less familiar context of $\mathbf{Z}[i]$, the ring of Gaussian integers.

The number 2 is both an integer and a Gaussian integer. But as a Gaussian integer, it can be factored into the product of two factors, neither of which is a unit: $2 = (1+i)(1-i)$. The Gaussian integer $x = 2+i$ has no such nontrivial factorization. Note that $|x|^2 = 4 + 1 = 5$. If $x = zw$, where both z and w were nonunits, then $|z|^2$ and $|w|^2$ would both be integers greater than 1, with product 5, which is impossible. Thus 2, regarded as a Gaussian integer, can be "reduced" into two factors that are not units whereas $2 + i$ cannot. This leads us to the following definition.

> **DEFINITION 1.** Let R be an integral domain and $x \in R$. Assume that $x \neq 0$ and $x \neq 1$. We say that x is **irreducible** in R if, whenever $x = wz$ in R, then either w or z is a unit. If x is irreducible and $x = wz$, where w is a unit, we call x and z **associates**.

The discussion from the previous paragraph can be summarized as follows: the number $2 + i$ is irreducible in $Z[i]$ but the number 2 is not. The definition of irreducible resembles the definition of what it means for a number to be a prime integer in Z. In fact, a number in Z is irreducible if and only if it is a prime or the negative of a prime. For more general integral domains, a different property of the prime integers lays claim to the adjective "prime." The general definition of what it means to be "prime" reflects how the prime integers behave with respect to division. As with integers, we say that x divides y in an integral domain R if $y = sx$ for some s in R. The following definition will evoke Euclid's Lemma.

> **DEFINITION 2.** Let x be an element of an integral domain R. We say that x is **prime** if, whenever x divides ab, then x divides a or x divides b.

In the ring Z, a number is prime in the sense of Definition 2 if and only if it is irreducible in the sense of Definition 1. But that does not happen in general. The notions of "prime" and "irreducible" are not equivalent.

EXAMPLE 1

In this example, we show that the notions of prime element and irreducible element do not always coincide. The ring we explore is $Z[\sqrt{-5}]$, which is a subring of C. Its elements form the set $\{a + b\sqrt{-5} : a, b \in Z\}$. In this ring, the only units are 1 and -1 because all other nonzero members of the ring have absolute value greater than 1. Let $z = 2 + \sqrt{-5}$. We will show that z is irreducible but not prime. Note that $|z|^2 = 2^2 + 5 = 9$. If $z = st$, where neither of s nor t were a unit, then we would have $|s|^2 = |t|^2 = 3$. However, there are no integer solutions to the equation $a^2 + 5b^2 = 3$. Thus z is irreducible. However, $9 = (2 + \sqrt{-5})(2 - \sqrt{-5})$; therefore, z divides 9, or z divides $3 \cdot 3$. However, it is not the case that z divides 3 because $|z| > |3|$. So in the ring $Z[\sqrt{-5}]$, the number $z = 2 + \sqrt{-5}$ is irreducible but not prime. ∎

In the above example, we found an irreducible number that was not prime. However, the next proposition tells us that, in an integral domain, we cannot find a prime that is not irreducible.

PROPOSITION 1 In an integral domain R, every prime element is irreducible.

Proof. Let z be an element in R that is prime. Suppose that $z = xy$. Since z divides xy, it must be the case that z divides x or z divides y. Let us assume that z divides x so that $x = sz$. Thus $z \cdot 1 = szy$. We can cancel z (R is an integral domain) to obtain $1 = sy$. Thus y is a unit and we have shown that z is irreducible. ▲

The Division Algorithm is the central tool with which we investigate the integers. The ring of Gaussian Integers also has a Division Algorithm that will prove to be a useful tool. Before we investigate it, we look a bit further into the Division Algorithm for integers. Suppose we divide two integers, say 23 by 4. As a rational number the quotient is 23/4 which is between 5 and 6, but closer to 6. Letting $a = 23$ and $b = 4$, and using the Division Algorithm, we obtain $a = qb + r$, with $q = 5$ and $r = 3$, so that $23 = 5 \cdot 4 + 3$. But if we ask that the quotient q be the integer that is nearest to the rational number 23/4, we would set $q = 6$ and write $23 = 6 \cdot 4 - 1$. It is not difficult to prove the following variation of the Division Algorithm.

THEOREM 2 Let a and b be integers, $b \neq 0$. Let q be the integer nearest to the rational number a/b. Then $a = qb + r$, where $|r| < b$.

The expression is not unique. For instance, letting $a = 22$ and $b = 4$, we can set $q = 5$ or 6, since 5 and 6 are each a distance 1/2 from 22/4. In one case, we have $22 = 5 \cdot 4 + 2$ with $r = 2$ and in the other, $22 = 6 \cdot 4 - 2$ with $r = -2$. In both cases, we have $|r| < 4$.

We next consider a Division Algorithm for the Gaussian integers. The function $N(a + bi) = |a + bi|^2 = a^2 + b^2$ plays the role that the absolute value plays in Theorem 2. We leave it as an exercise that, for Gaussian integers x and y, the following properties hold.

1. $N(x) = 0$ iff $x = 0$
2. $N(x) \geq 0$
3. $N(xy) = N(x)N(y)$

THEOREM 3 Let $x = a + bi$ and $y = s + ti$ be Gaussian integers, with $y \neq 0$. We can find Gaussian integers $q = q_1 + q_2 i$ and $r = r_1 + r_2 i$ such that $x = qy + r$ and $N(r) < N(y)$.

Proof. Regard x and y as complex numbers and divide x by y in $\mathbf{C}$. Then $z = x/y = z_1 + z_2 i$ where z_1 and z_2 are rational numbers that are not necessarily integers. Now let q_1 be the integer that is nearest to z_1 and let q_2 be the integer that is nearest to z_2 so that $|z_1 - q_1| \leq 1/2$ and $|z_2 - q_2| \leq 1/2$. Let $q = q_1 + q_2 i$ and $r = x - qy$ so that q and r are Gaussian integers. Then we have $x = qy + r$. We must show that $N(r) < N(y)$, or equivalently, $N(r)/N(y) = N(r/y) < 1$. Now $r/y = (x - qy)/y = z - q$. Therefore $N(r/y) = N((x - qy)/y) = N(z - q) \leq (z_1 - q_1)^2 + (z_2 - q_2)^2 \leq 1/4 + 1/4 = 1/2$. ▲

EXAMPLE 2

Let $x = 2 + 5i$ and $y = 2 - 2i$. In the field of complex numbers, the quotient $x/y = -3/4 + 7i/4$. Then $q_1 = -1$ and $q_2 = 2$ are the closest integers to $-3/4$ and $7/4$, respectively. So we set $q = -1 + 2i$ and $r = x - qy = -i$. The conclusion of the theorem is satisfied because $x = qy + r$ and $N(r) < N(y)$. ∎

We saw that every ideal I in the set of integers is a principal ideal. This fact comes directly from the Division Algorithm. Recall that to find the generator of the ideal I, we find an element that has the smallest nonzero absolute value x. Then every element y in the ideal must be of the form qx. If not, we could find a nonzero remainder r of y after division by x (also in I) of smaller absolute value than x, contradicting our choice of x. We can apply that argument almost verbatim to $\mathbf{Z}[i]$. The function $N(x)$ allows us to utilize the well-ordering of the natural numbers because $N(x)$ is a positive integer for every nonzero Gaussian integer x.

THEOREM 4 Every ideal in $\mathbf{Z}[i]$ is a principal ideal.

Proof. Let $I \subseteq \mathbf{Z}[i]$ be an ideal. Let $x \in I$ be an element in I for which $N(x)$ has the smallest nonzero value. We show that $I = \langle x \rangle$ and is therefore a principal ideal. Suppose that $y \in I$. We need to show that $y = qx$ for some q in $\mathbf{Z}[i]$. By the Theorem 3 we can find q and r such that $y = qx + r$ and $0 \leq N(r) < N(x)$. Since $r = y - qx$, r is an element of I. Since $N(r) < N(x)$, it must be the case that $N(r) = 0$ because of our choice of x. Thus $r = 0$, $y = qx$, and I is principal. ▲

> **DEFINITION 3.** An integral domain in which all ideals are principal ideals is called a **principal ideal domain** (or **P.I.D.** for short).

From the ideal structure of $\mathbf{Z}[i]$, we can deduce an important number theoretical fact: Every irreducible element in $\mathbf{Z}[i]$ is prime. (We already know from Proposition 1 that every prime is irreducible.) That is the conclusion of our next theorem. It is formulated more generally since it holds for all integral domains in which all ideals are principal (i.e., for all P.I.D.s). Notice how the proof resembles the proofs of Chapter 1.

THEOREM 5 Let R be an integral domain in which every ideal is principal. If $x \in R$ is irreducible, then x is prime.

Proof. Let x be an irreducible element in R. Suppose that x divides ab. We must show that x divides a or x divides b. Let $I = \{as + xt : s, t \in R\}$. Clearly I is an ideal and hence principal and so $I = \langle y \rangle$ for some $y \in R$. Letting $s = 0$ and $t = 1$, we see that $x \in I$. Similarly, $a \in I$. Thus we can find $q \in R$ such that $x = qy$. Since x is irreducible, either q or y is a unit. Suppose first that q is a unit in R so that $q^{-1}x = y$. Since $a \in I$ and $a = zy = zq^{-1}x$ for some $z \in R$, it is clear that x divides a. Now suppose that y is a unit. Then $y^{-1}y \in I$ and so $I = \langle 1 \rangle = R$. Thus $1 = as + xt$ for some value of s and t in R. Multiplying the latter expression through by b, we have $b = abs + bxt$. Since x divides ab, it is clear that x divides $abs + axt$. So if y is a unit, then x divides b. ▲

In the foregoing discussion, we have used the algebraic fact that $\mathbf{Z}[i]$ is a P.I.D. to prove the important fact about its arithmetic—every irreducible element is prime. Algebraic structure gives us information about arithmetic structure and conversely. The arithmetic similarity of the set of integers $\mathbf{Z}$ and the Gaussian integers $\mathbf{Z}[i]$ goes further. Every Gaussian integer that is not 0 or a unit can be factored uniquely into a product of irreducible elements (primes) and the factorization is unique up to units. The proof of the analogous fact for $\mathbf{Z}$ implicitly used the well-ordering of the natural numbers. To replace that, we could use the function $N(a+bi) = a^2 + b^2$ which only assumes nonnegative integer values, and the second principle of induction. In very sketchy detail, let x be a Gaussian integer with $N(x) = n$. If x is not irreducible, then $x = yz$ with $1 < N(y) < n$ and $1 < N(z) < n$. Now apply induction to y and z. Uniqueness follows from Definition 2. Alternatively, we present a more general algebraic proof which applies to all P.I.D.s. But first, a helpful and crucial lemma.

LEMMA 6 Let R be a P.I.D. and let $I_1 \subseteq I_2 \subseteq \ldots \subseteq I_i \subseteq \ldots$ be a chain of proper ideals of R, each properly contained in the next. The chain must be finite. That is, there must be some n such that for all $i \geq 0, I_{n+i} = I_n$.

Proof. Let I be the union of all the ideals in the chain. Then I is a proper ideal (exercise) and $I_i \subseteq I$ for each i. Since R is a P.I.D., we can find an element x such that $I = \langle x \rangle$. Since I is the union all the ideals in the chain, each element of I must be an element of some member of the chain. So we can find n such that $x \in I_n$. Since $x \in I_n$, we have $\langle x \rangle \subseteq I_n$. But $I_n \subseteq \langle x \rangle$ since $I = \langle x \rangle$. Thus $I = I_n$ and the chain must terminate at I_n. ▲

Here is an example of a chain of proper ideals $\mathbf{Z}$: $12\mathbf{Z} \subseteq 6\mathbf{Z} \subseteq 3\mathbf{Z}$. We cannot continue the chain because 3 is prime and $3\mathbf{Z}$ is a maximal ideal. So $3\mathbf{Z}$ is not properly contained in any proper ideal of $\mathbf{Z}$. If we didn't already know that every natural number $n > 1$ had a prime factor, we could prove it just using the fact that $\mathbf{Z}$ is a P.I.D. To do so, we will suppose the contrary and obtain a contradiction to Lemma 5. Suppose that $n > 1$ and that n had no prime factor. Then $n = n_1 m_1$ where neither n_1 nor m_1 is prime and neither, a unit. Continuing, we could find $n_1 = n_2 m_2$, and $n_2 = n_3 m_3$, etc. So for each natural number i, we would have $n_i = n_{i+1} m_{i+1}$ with neither n_{i+1} nor m_{i+1} equal to ± 1. Then, for each i, the set $n_i \mathbf{Z}$ would be a proper ideal of $\mathbf{Z}$ and we would have a chain of ideals $n_1 \mathbf{Z} \subseteq n_2 \mathbf{Z} \subseteq \ldots \subseteq n_i \mathbf{Z} \subseteq \ldots$, each properly contained in the next. But the chain would not be finite, contradicting Lemma 5. We will use a very similar argument in the next theorem.

THEOREM 7 Let R be a principal ideal domain. Any element x in R that is not zero or a unit can be factored into a product of irreducible elements (primes). Furthermore, that factorization is unique up to associates.

Proof. Let x be a element of R that is not a unit and not 0. We begin by showing that x has at least one irreducible factor. If x is irreducible, then we are done. If not, then $x = x_1 y_1$. If x_1 is irreducible, we are done. If not we continue, and factor x_1 as $x_1 = x_2 y_2$, etc. where in each case, x_i and y_i are not units. Let $I_i = \langle x_i \rangle$. Then we have a sequence of ideals $I_1 \subseteq I_2 \subseteq \ldots \subseteq I_i \subseteq \ldots$. We note that $x \in I_i$ for each i. It must be the case that for some n, I_n is a maximal ideal. If not, we would be able to construct an infinitely long chain of ideals, each properly contained in the next, contradicting Lemma 6. Now if $I_n = \langle r_1 \rangle$ is maximal, then r_1 is irreducible (exercise). Because $x \in I_n, x = q_1 r_1$ and we have shown that r_1 is an irreducible factor of x. Now suppose that q_1 is not a unit. We can repeat the above argument and find an irreducible element r_2 such that $q = r_2 q_2$. Continuing, we can find a sequence of elements q_i such that $\langle x \rangle \subseteq \langle q_1 \rangle \subseteq \langle q_2 \rangle \subseteq \ldots$. Again, Lemma 6 guarantees that the chain must be finite $\langle x \rangle \subseteq \langle q_1 \rangle \subseteq \langle q_2 \rangle \subseteq \ldots \subseteq \langle q_n \rangle$. So $\langle q_n \rangle$ is maximal and q_n is irreducible. Thus we can factor x as $x = q_n r_1 r_2 \cdots r_n$, where each factor is irreducible.

We must show that the factorization is unique up to associates. So suppose that $x = p_1 p_2 \cdots p_n = q_1 q_2 \cdots q_m$. We proceed by induction on n. Suppose $n = 1$. Then $m = 1$ since p_1 is irreducible. So $p_1 = q_1$. Now suppose the factorization is unique for any element that can be factored as the product of $n - 1$ or fewer irreducible factors elements. Let $x = p_1 p_2 \cdots p_n = q_1 q_2 \cdots q_m$. Since p_1 is *prime*, p_1 must divide one of the q_i's, say q_1 (we can renumber). Since q_1 is irreducible, $q_1 = u p_1$ where u is a unit. So canceling p_1 from both sides, we have $y = p_2 p_3 \cdots p_n = u_1 q_2 q_3 \cdots q_m$. We can apply the induction hypothesis to y to conclude that $m = n$ and (perhaps renumbering) that there are units u_i such that $p_i = u_i q_i$ for each i. ▲

EXAMPLE 3

Since $\mathbf{Z}[i]$ is a P.I.D., Theorem 7 guarantees that every element that is not zero or a unit can be factored uniquely up to associates. Let $x = 5 + 5i = (1+i)(1+2i)(1-2i) = (-1+i)(-1+2i)(-2+i)$. Each factor is irreducible and the factorizations may look quite different. But in $\mathbf{Z}[i]$, there are four units, $1, -1, i$, and $-i$. Thus $i(1+i) = (-1+i)$ and $i(1+2i) = (-2+i)$ and $-1(1-2i) = (-1+2i)$. The factors are associates and the product of the units is 1. ■

SUMMARY

In a ring R, an **irreducible** element is not necessarily a **prime** element. The concepts differ. Irreducibility is about how an element can be factored and primality is about how an element divides other elements. In many familiar rings, such as $\mathbf{Z}$ and $\mathbf{Q}[x]$, the concepts coincide. In this section, we explored a ring property that ensures that all irreducible elements are prime. (We proved that it is always the case that primes are irreducible.) In a **Principal Ideal Domain** (P.I.D.), i.e., an integral domain in which every ideal is principal, every irreducible element must be prime. To see that $\mathbf{Z}[i]$ is a P.I.D., we developed a version of the Division Algorithm for $\mathbf{Z}[i]$. As a further consequence of the fact that every irreducible element of a P.I.D. is also a prime element, we showed that, in a P.I.D., every element can be uniquely factored (up to units) as the product of primes. Thus $\mathbf{Z}[i]$, the set of Gaussian integers, shares almost all of the fundamental algebraic features of the integers $\mathbf{Z}$.

5.6 Exercises

1. For each of the following, determine if the given number is irreducible in $\mathbf{Z}[i]$ or, if not, find a proper factorization.
   i. $5 + 5i$
   ii. 13
   iii. $2 - i$
   iv. $2 - 2i$

v. $2 + 3i$
vi. $3 + 4i$
2. Factor the following elements of $\mathbf{Z}[i]$ into irreducible factors.
   i. $-3 + 11i$
   ii. $5 - 5i$
   iii. $-8 + 6i$
3. Determine if the given number is irreducible in the ring $\mathbf{Z}(\sqrt{-5})$. If it is not irreducible, find a proper factorization.
   i. $-14 + 7\sqrt{-5}$
   ii. 26
   iii. 20
   iv. $7 + \sqrt{-5}$
   v. $-4 + 2\sqrt{-5}$
4. Show that the number $1 + 2\sqrt{-5}$ is irreducible in the ring $\mathbf{Z}(\sqrt{-5})$ but not prime.
5. Show that the number $2 + 3\sqrt{-5}$ is irreducible in the ring $\mathbf{Z}(\sqrt{-5})$, but not prime.
6. Show that in $\mathbf{Z}(\sqrt{-5})$, the number 49 does not factor uniquely into irreducible elements.
7. Carry out the Division Algorithm as formulated in Theorem 3 for the following.
   i. $a = 5 + 7i$ and $b = 2 + i$
   ii. $a = 9 + i$ and $b = 3 - 2i$
8. Carry out the Division Algorithm as formulated in Theorem 3 for $a = 3 + 3i$ and $b = 2 + 2i$ in two different ways, thus illustrating that the remainder in not unique.
9. The ring $\mathbf{Z}[i]$ is equipped for the Euclidean Algorithm because it has a Division Algorithm and a way to measure the size of the remainder, i.e. the function $N(x)$. Use it to find the greatest common divisor of the following pairs of numbers.
   i. $x = 1 + 3i$ and $y = 5 + i$
   ii. $x = 3 + 4i$ and $y = 4 + 7i$
10. The set $I = \{x = a(3 + 4i) + b(4 + 7i) : a \text{ and } b \text{ are in } \mathbf{Z}[i]\}$ is an ideal in $\mathbf{Z}[i]$. Find its generator z to show that I is a principal ideal.
11. The set $I = \{x = a(1 + 3i) + b(5 + i) : a \text{ and } b \text{ are in } \mathbf{Z}[i]\}$ is an ideal in $\mathbf{Z}[i]$. Find its generator z to show that I is a principal ideal.
12. Let R be a P.I.D. and $x \in R$. Prove that the ideal $\langle x \rangle$ is maximal if and only if x is irreducible.
13. In $\mathbf{Z}[i]$, let $I = \langle 2 + i \rangle$. List all the distinct cosets of the quotient ring $\mathbf{Z}[i]/I$. Write out the multiplication table for the ring. Is it a field? Why?
14. Let x be an element in $\mathbf{Z}[i]$ that is not a unit and not 0. Let $I = \langle x \rangle$. Prove that the quotient ring $\mathbf{Z}[i]/I$ is a finite ring. (Note the similarity with $\mathbf{Z}/I$.) *Hint.* Use the Division Algorithm given in Theorem 3.
15. Prove that $\mathbf{Z}[i]$ is a P.I.D. directly, using the function $N(x)$, as sketched in the discussion preceding Theorem 4.

16. Let the function $N(a + bi) = |a + bi|^2 = a^2 + b^2$ be defined on $\mathbf{Z}[i]$. Show that N has the following properties:
   i. $N(x) = 0$ iff $x = 0$
   ii. $N(x) \geq 0$
   iii. $N(xy) = N(x)N(y)$

17. Let R be an integral domain and let $I_1 \subseteq I_2 \subseteq \ldots \subseteq I_i \subseteq \ldots$ be a chain of ideals in R. Prove that their union is a proper ideal in R.

To the Teacher

In this section, we tweezed apart two notions—that of a prime element and that of an irreducible element. "Irreducibility" is concerned with whether a number is divisible by smaller numbers (nonunits). "Primeness" is concerned with how a number divides other numbers: A number is prime if whenever it divides a product ab, then it divides one of a or b. The two notions coincide for whole numbers and also for polynomials as we saw in Chapter 3.

Even relatively young math doers easily and thoroughly identify the two notions of irreducibility and primeness. We start students with the notion of an irreducible number, a number that can't be factored nontrivially. The notion is easy and readily checked—just a matter of evidence for small numbers like 7 and 11. Experience quickly confirms that there is only one way to factor a whole number into the product of irreducible numbers. If p is irreducible and divides ab, it's simply a matter of checking the inventory in the unique list of primes that divide a and b to find which of a or b it is that p divides. So if p is irreducible, then p is prime in the sense of this section.

The concept of irreducibility seems to dominate school arithmetic. We use it when we find common denominators, gcd's with factor trees, etc. But here's a problem that relies on the concept of primeness.

> **Problem:** Suppose that the area of a rectangle is 924 but neither side exceeds 40. Find the dimensions of the rectangle. (Assume that the lengths of the sides are whole numbers.)

Tasks

1. Show how primeness solves the given problem.
2. Review the proof of the fundamental theorem of arithmetic. Indicate how primeness is used to prove uniqueness of factorization.
3. Formulate several simple word problems that utilize the concept of primeness.

5.7 WORKSHEET 9: POLYNOMIAL IDEALS

The ring of polynomials in one variable over a field has a well-understood structure. Let's recall the situation for $\mathbf{Q}[x]$. (We will use the field of rational numbers exclusively in this worksheet.) We know that if I is an ideal in $\mathbf{Q}[x]$, then I is a principal ideal. That is, I is generated by a single polynomial $p(x)$ so that $I = \langle p(x) \rangle$ for some polynomial $p(x)$ in $\mathbf{Q}[x]$.

Task 1

Let $I = \langle x^2 + 2 \rangle$. Determine whether or not $x^4 + 2x^3 + x^2 + 4x - 2 \in I$. (**Hint.** Use the Division Algorithm for Polynomials).

Things get much harder when there are two variables. The ideal structure is more complex and the Division Algorithm no longer holds in such a nice form. Consider the polynomial ring $Q[x, y]$. The elements of $Q[x, y]$ are expressions such as $3x^3 + xy^2 + 7y^3 - 5x^2y + y$, namely the sum of monomials of the form $a_{i,j} x^i y^j$, where $a_{i,j} \in Q$. We can regard a polynomial $p(x, y)$ in $Q[x, y]$ as a polynomial in $Q[x][y]$ (i.e., a polynomial in the variable y with coefficients in the ring $Q[x]$). As such, we would write $3x^3 + xy^2 + 7y^3 - 5x^2y + y$ as $a_3 y^3 + a_2 y^2 + a_1 y + a_0$, where $a_3 = 7, a_2 = x, a_1 = (-5x^2 + 1)$, and $a_0 = 3x^3$. Note that the coefficients are in $Q[x]$, which is *not* a field. So the Division Algorithm from Section 3.1 does not apply directly. Similarly, we can regard $3x^3 + xy^2 + 7y^3 - 5x^2y + y$ as an element of $Q[y][x]$.

The ideals of this ring are no longer necessarily principal because there are ideals which can not be generated by a single polynomial. A very simple example is $I = \langle x, y \rangle$. The ideal I consists of all elements of the polynomial ring in which the constant term is 0. To see this, note that if $p(x, y) \in I$, then $p(x, y) = y \cdot f(x, y) + x \cdot g(x, y)$ and so $p(0, 0) = 0$. If $p(x, y) = 0$, then the constant term $a_{0,0}$ must equal 0. Conversely, if $a_{0,0}$ is 0, then every nonzero term of $p(x, y)$ has a factor of x^i or y^j where i or j is not 0. We can factor a y from any term such that $j \neq 0$. All other terms have a factor of x. Thus we can express $p(x, y)$ as $y \cdot q(x, y) + x \cdot t(x)$, which is an element of I.

Task 2

The ideal $\langle x, y \rangle$ in the paragraph above may be described as the set of polynomials $\{p(x, y) \in Q[x, y] | p(0, 0) = 0\}$. Describe the ideal $I = \langle x - 2, y + 3 \rangle$ in a similar fashion.

For simple ideals like the two above the question of whether a given polynomial is in the ideal is quite easy to answer. We simply evaluate the polynomial at a particular pair of values of x and y. Now suppose that $I = \langle xy + 1, x + 1 \rangle$. Let's determine if $xy^2 + y + 2x + 2$ is in I. We will look to find polynomials g and h so that $xy^2 + y + 2x + 2 = g(x, y)(xy + 1) + h(x, y)(x + 1)$. We can try to factor out one of the two generators. Factoring $xy + 1$ from the first two terms of $xy^2 + y + 2x + 2$ yields $xy^2 + y + 2x + 2 = y(xy + 1) + 2x + 2$. Factoring $x + 1$ out of $2x + 2$ yields $xy^2 + y + 2x + 2 = y(xy + 1) + 2(x + 1)$. Thus $xy^2 + y + 2x + 2$ is in the ideal. However, if we start factoring with the other generator, namely $x + 1$, we get $xy^2 + y + 2x + 2 = (y^2 + 2)(x + 1) + (y - y^2)$. It would not be obvious that the polynomial is in the ideal.

Task 3

Show that $x^3 + 2x^2y - y$ is in the ideal $I = \langle x + y, x^2 - 1 \rangle$. (In this case using the generators in either order will work.)

Another important algebraic object is the quotient of a ring of polynomials modulo an ideal. Again, in one variable, this object is very manageable. But what happens with more than one variable? Consider $Q[x, y]$ with the ideal $I = \langle xy - 1 \rangle$. Let's see how we can we represent the elements of $Q[x, y]/I$. In this case, since $xy - 1 = 0$, $xy = 1$ in $Q[x, y]/I$. Thus we can reduce every polynomial in two variables to the sum of a polynomial in x and a polynomial in y. For example, let $p(x, y) = x^3 + y^2 + 2x^2y + 3xy^3 + 3x + 4y + 7xy + 4$. In the quotient $x^2y = x(xy)$, which is the same as x, while xy^3 is the same as y^2. Thus the quotient p may be represented by $p(x, y) = x^3 + y^2 + 2x + 3y^2 + 3x + 4y + 7 + 4 = x^3 + 4y^2 + 5x + 4y + 11$. Furthermore, since $xy = 1$ in $Q[x, y]/I$, we may think of y as the multiplicative inverse of x in the quotient. Thus $Q[x, y]/I$ may be thought of as the set of all expressions of the form

$$\sum_{k=-n}^{m} a_k x^k = \frac{a_{-n}}{x^n} + \frac{a_{-n+1}}{x^{n-1}} + \ldots + a_0 + \ldots + a_m x^m.$$

Task 4

Let $Q[x, y]/I$ be as in Task 3. Consider the following polynomials in $Q[x, y]$. Express their representatives in the quotient $Q[x, y]/I$ in the form at the end of the example.

i. $2x^3 + 3x^2y - 4xy^2 + 6y^3$
ii. $x^5 y^3$
iii. $x^4 y^8$

Task 5

Let $Q[x, y]/I$ be as in Task 3. The task now is to verify that multiplication is a well-defined operation as one passes to the quotient. Suppose that p and q are two polynomials in $Q[x, y]$. Then pq is also a polynomial in $Q[x, y]$. Passing to the quotient, $Q[x, y]/I$, can be done in two ways. We can reduce p and q in the quotient and then multiply in the quotient or multiply the original polynomials and then reduce the product in the quotient. We should get the same expression either way. Perform these two operations with $p(x, y) = x^3 + 2x^2y + y^2$ and $q(x, y) = x^2 - xy + y^2$.

Task 6

The final task with $Q[x, y]/I$ is to find the invertible elements of the ring. As a starting point the expression $x + 1$ is in the quotient. Is it invertible? That is, is there an element z of the quotient that satisfies $z(x + 1) = 1$? (**Hint.** $Q[x, y]/I$ is not a field.)

5.8 IN THE CLASSROOM: WHY RATIONALIZE?

In *Principles and Standards for School Mathematics*, the National Council of Teachers of Mathematics outlines a vision for school mathematics in which mathematical ideas are connected and expanded upon so that, over time, students develop a deeper understanding of material and a greater ability to apply mathematics. Chapter 5 echoes this program by building up new algebraic structures from old ones, such as quotient groups from groups and normal subgroups and quotient rings from rings and ideals. On first consideration, the new structures and concepts introduced in this chapter—cosets, normal subgroups, quotient groups, conjugacy, subrings, ideals, and quotient rings—seem to have evolved well beyond the mathematics found in the secondary school curriculum. However, as we have seen in some of our *To the Teacher* essays in this chapter, these new concepts are firmly grounded in familiar mathematics. For instance, we saw that cosets of the form $a + d\mathbf{Z}$ are really just arithmetic sequences. Similarly, showing that the operation $*$ in the quotient group G/H is well-defined echoes a student's early experience with operations on fractions. The fact that S_5 is not a solvable group will be critical in showing that there is no formula like the quadratic formula, so familiar to high school students, that can be used to solve for the roots of polynomials of degree five or higher.

In this essay, we will explore the connection between the expansion of the subring sequence $\mathbf{Z} \subseteq \mathbf{Q} \subseteq \mathbf{R} \subseteq \mathbf{C}$ to include fields such as $\mathbf{Q}(\sqrt{2})$ and the technique of "rationalizing the denominator" that is taught in high school. We begin by reflecting on the way in which the mathematics curriculum in grades pre-K to 12 expands the set of numbers that students know about and can work with. Students begin their mathematical journey by learning to count small finite sets of natural (counting) numbers. Then they realize that they could theoretically count natural numbers forever. Over the course of their mathematics instruction, their world of numbers grows to include whole numbers, rational numbers, integers, irrational numbers, real numbers, and ultimately complex numbers. Therefore, by the end of high school, students are familiar with the subring sequence $\mathbf{Z} \subseteq \mathbf{Q} \subseteq \mathbf{R} \subseteq \mathbf{C}$. Once complex numbers are studied, high school mathematics seems to leave the impression that this sequence is now complete in the sense that it cannot be expanded further. In this chapter, we expanded on this sequence by looking at fields such as $\mathbf{Q}(\sqrt{2}) = \{a + b\sqrt{2} : a, b \in \mathbf{Q}\}$ and the sequence $\mathbf{Z} \subseteq \mathbf{Q} \subseteq \mathbf{Q}(\sqrt{2}) \subseteq \mathbf{R} \subseteq \mathbf{C}$. While this notation is not used in a high school curriculum, fields such as $\mathbf{Q}(\sqrt{2})$ are encountered in subtle ways.

A common technique taught in high school algebra is called "rationalizing the denominator." This means to write an equivalent form of a fraction with a rational number as its denominator. For example, the fraction $\frac{3}{\sqrt{2}}$ is multiplied by $\frac{\sqrt{2}}{\sqrt{2}}$ to get $\frac{3\sqrt{2}}{2}$. Fractions involving variables are also used. A student might be asked to rationalize $\frac{3}{\sqrt{2x}}$, $x > 0$, to get $\frac{3}{\sqrt{2x}} \cdot \frac{\sqrt{2x}}{\sqrt{2x}} = \frac{3\sqrt{2x}}{2x}$. Then the denominators become more complicated, as in $\frac{5}{3+\sqrt{2}}$. Students learn to use the conjugate, in our case $3 - \sqrt{2}$, to get $\frac{5}{3+\sqrt{2}} \cdot \frac{3-\sqrt{2}}{3-\sqrt{2}} = \frac{15-5\sqrt{2}}{7}$. Typically, square roots dominate the examples given to high school students, but some problems involving other roots may be given as challenges or extension questions.

The technique of rationalizing the denominator has students working fields like $\mathbf{Q}(\sqrt{2})$, even if it is not explicitly made known. An expression such as $\frac{1}{3+\sqrt{2}}$ can be rationalized because $3 + \sqrt{2}$ is a nonzero element of the *field* $Q(\sqrt{2})$. Its multiplicative inverse can be found in $\mathbf{Q}(\sqrt{2})$ and must be of the form $a + b\sqrt{2}$. Conveniently, we can find a and b by using this process of rationalizing the denominator. (Recall, we rewrite $\frac{1}{3+\sqrt{2}}$ as $\frac{1}{3+\sqrt{2}} \cdot \frac{3-\sqrt{2}}{3-\sqrt{2}} = \frac{3-\sqrt{2}}{7} = \frac{3}{7} - \frac{1}{7}\sqrt{2} \in \mathbf{Q}(\sqrt{2})$.)

Students can quickly get proficient at rationalizing the denominators of quotients involving square roots such as the one above. A rote multiplication of the conjugate divided by itself appears to be the trick. A problem arises if students experience this technique devoid of an explanation of the mathematics that makes it work. They will likely try to rationalize the denominator of problems like $\frac{1}{1+\sqrt[3]{2}}$ by incorrectly multiplying by $\frac{1-\sqrt[3]{2}}{1-\sqrt[3]{2}}$. This is akin to the issue of teaching the FOIL (Firsts, Outsides, Insides, Lasts) method for multiplying two binomials and not explaining that it is really a repeated use of the distributive property. While each technique is simple to learn, they have very limited scopes of use that do not allow students to tackle more complicated problems.

If we focus instead on *why* we use the conjugate in rationalizing the denominators of these quotients with square roots, we see that the basic mathematical idea underlying the technique is factorization, which is a significant topic of study in high school mathematics. There is no reason to give students a rote method that does not extend easily when they have the tools to understand the more general idea behind the technique. We use the conjugate for rationalizing examples like $\frac{5}{3+\sqrt{2}}$ because of the factorization of the difference of perfect squares, $a^2 - b^2 = (a+b)(a-b)$. This formula tells us that when a or b are square roots, we can multiply $a+b$ or $a-b$ by the other (which is its conjugate) so that the product will involve only the square of a and b, eliminating the radical. Likewise, we can use the following factorizations to "rationalize" quotients such as $\frac{1}{1+\sqrt[3]{2}}$:

(1) $$a^3 - b^3 = (a-b)(a^2 + ab + b^2);$$
(2) $$a^3 + b^3 = (a+b)(a^2 - ab + b^2).$$

For example, to rationalize the denominator of $\frac{1}{1+\sqrt[3]{2}}$, we use formula (2). Let $a = 1$ and $b = \sqrt[3]{2}$. Note that $a^2 - ab + b^2 = 1 - \sqrt[3]{2} + \sqrt[3]{4}$ and that $(1 - \sqrt[3]{2} + \sqrt[3]{4})(1 + \sqrt[3]{2}) = a^3 + b^3 = 3$. Multiply the numerator and denominator by $a^2 - ab + b^2 = 1 - \sqrt[3]{2} + \sqrt[3]{4}$ to obtain the following calculations:

$$\frac{1}{1+\sqrt[3]{2}} = \left(\frac{1}{1+\sqrt[3]{2}}\right)\left(\frac{1 - \sqrt[3]{2} + \sqrt[3]{4}}{1 - \sqrt[3]{2} + \sqrt[3]{4}}\right) = \frac{1 - \sqrt[3]{2} + \sqrt[3]{4}}{3}.$$

In this case, you are working in the field $\mathbf{Q}(\sqrt[3]{2}) = \{a + b\sqrt[3]{2} + c(\sqrt[3]{2})^2 : a, b, c \in \mathbf{Q}\}$.

For p prime, you can rationalize quotients involving expressions of the form $c + d\sqrt[n]{p}$ for larger n by dividing $a^n - b^n$ by $a - b$ or $a^n + b^n$ by $a + b$ to find the other factor we need. This involves working in the field $\mathbf{Q}(\sqrt[n]{p}) = \{a_0 + a_1\sqrt[n]{p} + a_2(\sqrt[n]{p})^2 + \ldots + a_{n-1}(\sqrt[n]{p})^{n-1} : a_i \in \mathbf{Q}\}$.

Once students learn the technique of rationalizing denominators they are often told not to leave radicals in the denominators of quotients, much like they are expected to put fraction answers in simplest form. Some students may believe that there is some rule that you cannot leave radicals in the denominator of a quotient. However, to give this impression would be misleading. The technique of rationalizing the denominator remains from the days when all arithmetic was done by hand. It is easier to approximate a quotient like $\frac{15-5\sqrt{2}}{7}$, which requires dividing by 7, than the equivalent $\frac{5}{3+\sqrt{2}}$, which requires dividing by an approximation of an infinite decimal. Today, calculators and computers do the approximations. Rationalizing the denominator is not needed to make the expression any simpler before approximating it.

One might then wonder why many current high school algebra textbooks have sections whose sole purpose is to teach the technique of rationalizing the denominator, often with few examples of when students will encounter a need for this technique. Perhaps such sections remain in current textbooks from pretechnology days. However, while calculators can now approximate these quotients without the need for rationalizing their denominators first, we find that the calculators themselves require an ability of students to recognize equivalent forms. For example, if a student enters an expression like $\frac{5}{3+\sqrt{2}}$ on a TI-89 or TI-92, the calculator will rationalize the denominator and output $\frac{-5(\sqrt{2}-3)}{7}$. Likewise, if a student uses one of these graphing calculators to find trigonometric ratios, the outputs will always rationalize the denominators. For example, $\sin(\pi/4) = \sqrt{2}/2$. A student who is computing this ratio using the isosceles triangle in Figure 1 will find that $\sin(\pi/4) = 1/\sqrt{2}$. That student will need to recognize $1/\sqrt{2}$ and $\sqrt{2}/2$ as equivalent forms.

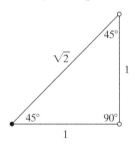

FIGURE 1

Beyond merely rewriting answers to recognize equivalent forms, the technique of multiplying by the conjugate divided by itself is useful in other areas. In calculus, the technique is used when computing the derivative with the limit definition. For example, when finding the slope of the curve $f(x) = \sqrt{x}$ at $x = 4$ using the limit definition, we get $f'(4) = \lim_{h \to 0} \frac{f(4+h)-f(4)}{h} = \lim_{h \to 0} \frac{\sqrt{4+h}-2}{h}$. We can compute this limit by hand and multiply by $\frac{\sqrt{4+h}+2}{\sqrt{4+h}+2}$ to rationalize the numerator in this case. This enables us to simplify the quotient and eliminate the h in the denominator.

In addition, this technique lends itself nicely to other algebraic manipulations as well. Since $\boldsymbol{C} = \boldsymbol{R}(i) = \{a + bi : a, b \in \boldsymbol{R}\}$ we can use the conjugate to show that any nonzero complex number has a multiplicative inverse in $\boldsymbol{C}$. Given any $a + bi, a, b \in \boldsymbol{R}$ with $a \neq 0$ or $b \neq 0$, we can find the multiplicative inverse of $\frac{1}{a+bi}$ by again multiplying by the conjugate $a - bi$ divided by itself to get $\frac{1}{a+bi} = \frac{1}{a+bi} \cdot \frac{a-bi}{a-bi} = \frac{a-bi}{a^2+b^2} = \frac{a}{a^2+b^2} - \frac{b}{a^2+b^2}i$.

We have seen then how the technique of rationalizing the denominator has high school students working in fields like $\mathbf{Q}(\sqrt{2})$. We have seen how to focus on the mathematics underlying the technique so that it can be more easily generalized. We have also looked at several areas in which this technique gets used by high school students and so despite the fact that it is no longer needed to do approximations by hand, it has merit.

5.8 Exercises

1. We gave examples in precalculus and calculus in which students might encounter a need for the technique of rationalizing. Try to find other situations in the secondary mathematics curriculum in which radicals in the denominators of quotients appear.
2. Rationalize the denominator of $\frac{1}{a+b\sqrt{2}}$ where $a \neq 0$ or $b \neq 0$ to verify that a general formula for $(a + b\sqrt{2})^{-1}$ is $\frac{a}{(a^2-2b^2)} + \frac{-b}{(a^2-2b^2)}\sqrt{2}$.
3. In addition to rationalizing the denominator, we can verify the formula for $(a + b\sqrt{2})^{-1}$ in Exercise 2 by setting up the system of equations that arises from $(a + b\sqrt{2})(c + d\sqrt{2}) = 1$ and using matrix techniques to solve the system. Try this.
4. Rationalize the denominator of the following quotients:
   i. $\dfrac{1}{3 - 2\sqrt[4]{5}}$
   ii. $\dfrac{1}{\sqrt{2} + \sqrt{3} + \sqrt{5}}$
   iii. $\dfrac{1}{\sqrt[3]{4} - \sqrt[3]{2} + 1}$
   iv. $\dfrac{1}{a + b\sqrt[3]{p}}$, where p is a prime and $a, b \in \mathbf{Q}$.

Bibliography

National Council of Teachers of Mathematics. *Principles and Standards for School Mathematics*. Reston, Va.: National Council of Teachers of Mathematics, 2000.

Schuette, Paul H. "Rationalizing the Denominator: Why Bother?" *Mathematics and Computer Education* 32 no.1 (Winter 1998): 19–29.

5.9 FROM THE PAST: ÉVARISTE GALOIS AND THE THEORY OF GROUPS—PART 2

Galois needed more than just the idea of groups to succeed in his goal of determining which polynomials had roots that could be expressed in radical form. He needed to work with the relations between a group and its subgroups. In this section we explore how Galois developed the notion of a normal subgroup. In Part 1 (Section 4.10) we saw that the presentation of a group given by an array could be changed by interchanging its columns. The group stayed the same but the individual **permutations** in its presentation changed. Let's look more closely at the presentations given in Table 1 of Part 1. The first two of those presentations are given as Array 1 and Array 2 below. Both present the same subgroup G of S_4. Let us denote the **substitutions** defined by each array as $\{\alpha_1, \alpha_2, \alpha_3, \alpha_4\}$, where in each case α_i maps the top row to the ith row. (Notice that we have arranged the rows in each array so that α_i does indeed denote the same **substitution** in each array.)

$$\begin{array}{cccc} a & b & c & d \\ b & c & d & a \\ c & d & a & b \\ d & a & b & c \end{array} \qquad \begin{array}{cccc} b & a & c & d \\ c & b & d & a \\ d & c & a & b \\ a & d & b & c \end{array}$$

Array 1   **Array 2**

Consider the first **permutation** of Array 1, $a\,b\,c\,d$, and the first **permutation** of Array 2, $b\,a\,c\,d$. The following **substitution** does *not* occur in the subgroup G.

$$\gamma_1 = \begin{pmatrix} a & b & c & d \\ b & a & c & d \end{pmatrix}$$

Similarly, using the remaining rows of the second presentation, none of the following **substitutions** are in G.

$$\gamma_2 = \begin{pmatrix} a & b & c & d \\ c & b & d & a \end{pmatrix}, \quad \gamma_3 = \begin{pmatrix} a & b & c & d \\ d & c & a & b \end{pmatrix}, \quad \gamma_4 = \begin{pmatrix} a & b & c & d \\ a & d & b & c \end{pmatrix}$$

However, in each case, $\gamma_i = \alpha_i \circ \gamma_1$. So the **substitutions** obtained using the top **permutation** of Array 1 and the rows of Array 2 form the right coset $G\gamma_1$.

Task 1

Interpret and explain the following statement.

> Suppose a subgroup G of S_n is presented in two arrays, Array A and Array B. The set of **substitutions** that map the top row of Array A to each of the **permutations** in Array B can be regarded as a right coset of G.

Galois noticed that sometimes the interchange of columns that transforms a given group presentation into another could be accomplished by the action of a **substitution** on the permutations of the original array (with a possible interchange of rows). This turned out to be an important property that held for some, but not all, groups. Let's investigate it. Suppose that G is a group of four elements presented as follows.

$$\begin{array}{cccc} a & b & c & d \\ c & d & a & b \\ b & a & d & c \\ d & c & b & a \end{array}$$

Array 3

Section 5.9   From the Past: Évariste Galois and the Theory of Groups—Part 2

The group G has a two-element subgroup H that is presented by Array 4.

$$\begin{array}{cccc} a & b & c & d \\ c & d & a & b \end{array}$$
Array 4

Interchanging columns 1 and 2 and columns 3 and 4 gives Array 5, which is the same as the bottom two rows of Array 3.

$$\begin{array}{cccc} b & a & d & c \\ d & c & b & a \end{array}$$
Array 5

As we did previously, let γ_1 be the **substitution** $\begin{pmatrix} a & b & c & d \\ b & a & d & c \end{pmatrix}$ from the first row of Array 4 to the first row of Array 5. The **substitution** γ_1 is in G but not H. But the **substitution** γ_1 applied to the two **permutations** of Array 4 gives Array 5. To apply γ_1, we literally substitute the letters in the bottom row for the top. So $a\,b\,c\,d$ becomes $b\,a\,d\,c$ and $c\,d\,a\,b$ becomes $d\,c\,b\,a$. In each case, a is replaced by b, b is replaced by a, c is replaced by d, and d is replaced by c.

Suppose we denote the **substitutions** in the subgroup H by α_1 (the identity map) and α_2. As before, let $\gamma_2 = \begin{pmatrix} a & b & c & d \\ d & c & b & a \end{pmatrix}$ be the **substitution** from the first row of Array 4 to the second row of Array 5. Then $\gamma_1 = \gamma_1 \circ \alpha_1$ and $\gamma_2 = \gamma_1 \circ \alpha_2$. In Task 1, we saw that $\{\gamma_1, \gamma_2\}$ could be regarded as a *right* coset of H, namely $H\gamma_1$. Now we see that $\{\gamma_1, \gamma_2\} = \{\gamma_1 \circ \alpha_1, \gamma_1 \circ \alpha_2\}$, which is a *left* coset, namely $\gamma_1 H$. In our terms, H is a normal subgroup of G.

More generally, suppose that the presentations of a subgroup $H \subseteq G$ partition the presentation of G as in Table 1 of Part 1. (See Section 4.10.) Galois noted that in some cases, each of the presentations of H in the partition of G could be obtained from any other by applying a single **substitution** as was done for Arrays 4 and 5. Galois singled these out as particularly important. They are the special subgroups that we call "normal subgroups" today.

Galois presented the following example in his *Mémoire sur les Conditions de Resolubilité des Equations par Radicaux*. Taken together, the 12 **permutations** of Array 6 present A_4. The first four lines present a subgroup K of A_4. The two other sets of four are presentations of K that partition A_4.

$$\begin{array}{cccc} a & b & c & d \\ b & a & d & c \\ c & d & a & b \\ d & c & b & a \\ \\ a & c & d & b \\ c & a & b & d \\ d & b & a & c \\ b & d & c & a \end{array}$$

$$\begin{matrix} a & d & b & c \\ d & a & c & b \\ b & c & a & d \\ c & b & d & a \end{matrix}$$
Array 6

Applying the **substitution** $\begin{pmatrix} a & b & c & d \\ a & c & d & b \end{pmatrix}$ to each of the **permutations** in the first array (the first four lines) yields the **permutations** that appear in the second array. It will be left as an exercise to find the **substitution** that transforms the first array to the third. These two **substitutions** show that the first array, regarded as a subgroup of the group formed by all three arrays, is a normal subgroup.

Task 2

Show that there are four **substitutions** that take the first array to the second. *Hint.* Apply the **substitution** $\begin{pmatrix} a & b & c & d \\ c & a & b & d \end{pmatrix}$ to the first array. Note that the **permutations** might be presented in a different order but the order in which the **permutations** are presented does not change the group of **substitutions**.

Task 3

Let $\varphi = \begin{pmatrix} a & b & c & d \\ c & a & b & d \end{pmatrix}$. This **substitution**, which comes from the second row of the second array, is an element of A_4. Its inverse $\phi^{-1} = \begin{pmatrix} a & b & c & d \\ b & c & a & d \end{pmatrix}$ is also in A_4 as the **substitution** produced by the third row of the third array. Show that if β is any **substitution** from the first array, then $\phi\beta\phi^{-1}$ is another **substitution** from the first array.

Let K represent the subgroup of A_4 formed by the first array. Task 3 shows that $\phi K \phi^{-1} = K$. This holds for any other **substitution** from A_4. Thus K is a normal subgroup of A_4 in both our modern sense and in the sense of Galois.

To see that not all column interchanges can be accomplished by **substitutions**, consider the group S_3 as given by the array

$$\begin{matrix} a & b & c \\ c & b & a \\ b & a & c \\ b & c & a \\ a & c & b \\ c & a & b \end{matrix}$$

and let H be the subgroup given by the following.

$$\begin{matrix} a & b & c \\ c & b & a \end{matrix}$$

By interchanging columns 1 and 2 and again columns 2 and 3 we have two other arrays for the same group: $\begin{smallmatrix} b & a & c \\ b & c & a \end{smallmatrix}$ and $\begin{smallmatrix} a & c & b \\ c & a & b \end{smallmatrix}$. There is no **substitution** that will accomplish either column interchange. Any **substitution** that changes the value of b will give an array where another letter is above itself. Any **substitution** that leaves b alone will leave it in the middle column. In neither case is one of the other arrays produced. And indeed the subgroup $\begin{smallmatrix} a & b & c \\ c & b & a \end{smallmatrix}$ is not a normal subgroup of S_3.

Task 4

Let H be the subgroup of S_3 given by the array $\begin{smallmatrix} a & b & c \\ c & b & a \end{smallmatrix}$. Show that $\begin{smallmatrix} b & a & c \\ b & c & a \end{smallmatrix}$ and $\begin{smallmatrix} a & c & b \\ c & a & b \end{smallmatrix}$ are right cosets of H by finding **substitutions** α and β such that in S_3 the two arrays are given by $H\alpha$ and $H\beta$. Now show that neither array represents a left coset of H in S_3.

Galois did not develop an abstract theory of groups and normal subgroups. He used what he needed to give a general picture of the roots of a polynomial and the internal relations among the roots. It would be nearly a century before our definitions of group and normal subgroup became standard.

5.9 Exercises

1. Using Galois' definition, show that the subgroup $\begin{smallmatrix} a & b & c \\ b & c & a \\ c & a & b \end{smallmatrix}$ is normal in S_3.

2. Find the **substitution** that transforms $\begin{smallmatrix} a & b & c & d \\ b & a & d & c \\ c & d & a & b \\ d & c & b & a \end{smallmatrix}$ into $\begin{smallmatrix} a & d & b & c \\ d & a & c & b \\ b & c & a & d \\ c & b & d & a \end{smallmatrix}$.

   Remember that you may have to change the order of the rows after making the **substitution**.

3. Suppose that A_1 and A_2 are arrays with A_2 simply A_1 with its columns permuted in some nontrivial way. Further suppose that the array $\begin{smallmatrix} A_1 \\ A_2 \end{smallmatrix}$ is a group. Show that A_1 is normal in $\begin{smallmatrix} A_1 \\ A_2 \end{smallmatrix}$. (We know that A_1 is a subgroup of index 2 in the whole group and thus it must be normal. Restated, the question is, How would Galois have shown normality?)

Chapter 5 Highlights

The principal goal of this chapter is to investigate the construction of quotient groups and quotient rings—new groups and rings from old ones—in much the same way as we constructed the ring $\mathbf{Z}_n$ from the ring $\mathbf{Z}$. For a group G, we start with a **normal** subgroup $H \subseteq G$. Its **cosets** are the elements of the quotient group

G/H. The normality of H guarantees that the product of two cosets, defined as $aH \cdot bH = abH$, is well-defined. We found that a normal subgroup is the union of the **conjugacy classes** of its elements. For a permutation group, a conjugacy class is a set of permutations that all have the same cycle structure. This simple characterization allowed us to deduce that the A_5 is the only normal subgroup of S_5 and that S_5 is not a **solvable group**. These facts are important to the theory of polynomials.

Extending our investigation to rings, we found that any subring I of a ring R is a normal subgroup for the commutative operation of addition on the cosets of the form $a + I$. However, for multiplication, defined by $(a + I) \cdot (b + I) = ab + I$, to be well-defined, the subring I must be an **ideal**. We investigated several special types of ideals, namely **principal** ideals, **prime** ideals and **maximal** ideals. When R is an integral domain, the quotient ring R/I is an integral domain if and only if I is a prime ideal, and R/I is a field if and only if I is a maximal ideal.

Along the way, we learned several interesting and important facts about groups and rings. For instance, Lagrange's Theorem ensures us that the order of a subgroup $H \subseteq G$ must divide the order of G. We investigated new rings such as the **Gaussian integers**, which provided us with an example of a **Principal Ideal Domain**. We discovered that in some rings, the notions of an **irreducible** element and a **prime** element were not identical.

In summary, we looked much more closely at the structure of groups and rings. We continue that investigation in the next chapter where the focus will return to polynomial rings.

Chapter Questions

1. State the definitions of the following terms.

    i. right and left cosets of a subgroup H of a group G
    ii. the index $[G : H]$ of a subgroup H in a group G
    iii. a normal subgroup H of a group G
    iv. the quotient group G/H of a group G by a normal subgroup H
    v. an ideal I in a ring R
    vi. principal ideal
    vii. prime ideal
    viii. maximal ideal
    ix. an irreducible element in a ring R
    x. a prime element in a ring R

2. Let $G = \{\mathbf{Z}_{12}, +\}$ and let $H = \{0, 3, 6, 9\}$. Find all the right cosets and all the left cosets of H.

3. The group D_6 is the group of symmetries of the regular hexagon. Let H be the subgroup of all the rotations of the regular hexagon. Find the left and right cosets of H.

4. The set $H = \{R_0, R_{120}, R_{240}\}$ is a subgroup of D_6, considered as the group of symmetries of the regular hexagon. Find all the right cosets of H and find all the left cosets of H.
5. Let G be a group and let H be a subgroup of index 2. Prove that each right coset of H is also a left coset of H.
6. Find all the proper normal subgroups of D_6.
7. The set $H = \{1, 10\}$ is a subgroup of U_{11}. Show that the quotient group U_{11}/H is cyclic.
8. Show that the subgroup $H = \{e, R_{180}\}$ is a normal subgroup of D_6. Show that the quotient group D_6/H is isomorphic to S_3.
9. Let G be a group and let H be a subgroup of index 2. Prove that H is a normal subgroup of G.
10. Suppose that $G_1, G_2, \ldots, G_n$ are groups and that $G = G_1 \times G_2 \times \ldots \times G_n$, the direct product. Prove that for each $i = 1, \ldots, n$, the group G_i is isomorphic to a subgroup of $G_1 \times G_2 \times \ldots \times G_n$.
11. For each of the following pairs α and β, find γ such that $\alpha = \gamma^{-1} \beta \gamma$:

   i. $\alpha = (a, b, c)(d, e)(f)$ and $\beta = (c, f, b)(a, e)(d)$
   ii. $\alpha = (a, c, b, f, e, d)$ and $\beta = (f, d, e, a, b, c)$

12. Find the number of elements in the conjugacy class of the permutation $(a, b)(c, d)(e, f)$ in S_6.
13. Is A_3 a simple group? Why?
14. Is A_4 a simple group? Why?
15. Determine whether D_6 is a solvable group.
16. Find all the ideals of the ring $\mathbf{Z}_{18}$.
17. Let S be the subset of $\mathbf{Z}[x]$ of all polynomials such that all their odd indexed coefficients are 0. Show that S is a subring of $\mathbf{Z}[x]$ but not an ideal.
18. Find a ring homomorphism from $\mathbf{Z}_8$ to $\mathbf{Z}_{12}$. Prove that your map preserves both the operations of addition and multiplication.
19. Let $I \subseteq \mathbf{Z}[i]$ be the ideal $\langle 1 + 2i \rangle$. Determine how many elements are in the quotient ring $\mathbf{Z}[i]/I$. (*Hint.* Show that occurrence $2i$ can be replaced by -1 and that $(-1)^2 + I = (2i)^2 + I$. Thus $1 + I = -4 + I$ or $5 + I = 0 + I$.) Give several examples of how to compute sums and products in this quotient ring.
20. Determine whether the ideal $\langle 1 + i \rangle$ is a maximal ideal in $\mathbf{Z}[i]$.
21. Is the ideal $\langle 5 \rangle$ a prime ideal in $\mathbf{Z}[i]$? Why?
22. Is the ideal $\langle x^3 + x + 1 \rangle$ a maximal ideal in $\mathbf{Z}_5[x]$? Why?
23. Determine whether the given number is irreducible in $\mathbf{Z}[i]$.

   i. $2 + 3i$
   ii. $3 + 4i$

24. Show that $1 + 2\sqrt{-5}$ is irreducible in $\mathbf{Z}[\sqrt{-5}]$ but that it is not prime.

6

LOOKING FORWARD AND BACK

In previous chapters, we studied integers and polynomials, the most familiar objects of algebra. We do very tangible and active math on these objects: we divide, factor, find roots, etc. Then we studied abstract structures, like groups and rings. There we were presented with a very different kind of mathematics, the study of mathematical properties rather than mathematical computation. In this chapter, we shall see how these two studies link to solve the classical problem, "Is the quintic polynomial solvable by radicals?" To get to the answer, we shall use all our tools, old and new. Number theory, modular arithmetic, groups (especially permutation groups), rings, fields, isomorphisms—all these concepts and techniques will come into play. Along the way, we will take an interesting digression into the distant past to solve some problems left unsolved in antiquity. Can you square a circle? Trisect an angle? Double a cube? Interpretation to follow!

6.1 EXTENSION FIELDS

At the end of Chapter 3, we looked at a way to construct fields. We took an irreducible polynomial $p(x)$ with coefficients in a field F, and constructed $F_1 = F[x]/\langle p(x) \rangle$. By associating each element $a \in F$ with the equivalence class of the constant polynomial $f(x) = a$ in F_1, we can regard F as a subfield of F_1. The bigger field F_1 is called an **extension field** of F. In this section, we continue our investigation of extension fields. Our goal is to formulate the question "Are all polynomials solvable by radicals?" in terms of our algebraic vocabulary. The answer is "No!" but it was not historically and is not currently an easy answer to arrive at. For a glimpse at how to arrive at that answer, steer a course to Section 6.5.

Suppose that K is a field and that F is a subfield of K. Recall that F is a subset of K that is also a field using the same operations as in K. We call the pair $F \subseteq K$ a **field extension**, and we call K an **extension field** of F, or simply an **extension** of F. We sometimes use the following diagram to denote a field extension $F \subseteq K$:

$$
\begin{array}{c}
K \\
| \\
F
\end{array}
$$

EXAMPLE 1

The field of real numbers R is an extension field of the field of rational numbers Q, and the field of complex numbers C is an extension field of R. Since $Q \subseteq R \subseteq C$ we can stack our extension field diagram as follows:

$$\begin{array}{c} C \\ | \\ R \\ | \\ Q \end{array}$$

∎

EXAMPLE 2

Let F be a field and $S = \left\{ \frac{p(x)}{q(x)} : p(x), q(x) \in F[x], q(x) \neq 0 \right\}$. Note that $q(x) \neq 0$ means that $q(x)$ is not the zero polynomial. Typical members of S are $\frac{3}{1}$, $\frac{x^2+1}{x+1}$ and $\frac{x+1}{1}$. Such expressions are called **rational functions** or **algebraic fractions**. We know that two numerical fractions $\frac{p}{q}$ and $\frac{s}{t}$ are equal rational numbers if and only if $pt = qs$. (Thus $\frac{2}{3} = \frac{4}{6}$ since $2 \cdot 6 = 3 \cdot 4$.) Similarly, the equivalence relation $\approx$ on S defined by $\frac{p(x)}{q(x)} \approx \frac{s(x)}{t(x)}$ if and only if $p(x)t(x) = s(x)q(x)$ determines when two algebraic fractions are equal. The set of equivalence classes of this relation is denoted by $F(x)$. So $F[x]$ denotes the set of polynomials with coefficients in the field F and $F(x)$ denotes the set of rational functions with coefficients in F. By identifying each member a of the coefficient field F with the element $\frac{a}{1}$ in $F(x)$, we can consider F to be a subset of $F(x)$. When addition and multiplication of algebraic fractions are carried out as in high school algebra, $F(x)$ becomes a field and hence an extension field of F. ∎

Note: As algebraic fractions in $R(x)$, the expressions $\frac{1}{1}$ and $\frac{x}{x}$ are equal since $x \cdot 1 = x \cdot 1$. However, considered as functions on the real line they are different because they have different domains. The function $f(x) = \frac{1}{1}$ is defined at each point in R but the function $g(x) = \frac{x}{x}$ is not defined at $x = 0$.

EXAMPLE 3

Suppose that F is a field and that $p(x)$ is an *irreducible* polynomial in $F[x]$. For each constant $a \in F$, let $[a]$ denote the equivalence class of the constant polynomial $q(x) = a$ in the quotient field $F[x]/\langle p(x) \rangle$. Then the map $a \to [a]$ is a ring isomorphism of F onto a subfield of $F[x]/\langle p(x) \rangle$. By identifying each $a \in F$ with its image $[a] \in F[x]/\langle p(x) \rangle$, we can regard $F[x]/\langle p(x) \rangle$ as an extension field of F. The diagram is as follows.

$$\begin{array}{c} F[x]/\langle p(x) \rangle \\ | \\ F \end{array}$$

∎

EXAMPLE 4

The polynomial $p(x) = x^2 - 2$ is irreducible in $\boldsymbol{Q}[x]$. The field $\boldsymbol{Q}_1 = \boldsymbol{Q}[x]/\langle x^2 - 2\rangle$ is an extension field of $\boldsymbol{Q}$ as explained in Example 3. Suppose that the equivalence class $[x]$ is denoted by α. The set of elements of $\boldsymbol{Q}_1$ is the set $\{a + b\alpha : a, b \in \boldsymbol{Q}\}$. In this field, $\alpha^2 = 2$. That is because $[x^2 - 2] = [0]$ in $\boldsymbol{Q}[x]/\langle x^2 - 2\rangle$ and therefore $[x]^2 = [2]$. As in Example 3, we identify $[2]$ with 2, and now we replace $[x]$ with the symbol α. So $\alpha^2 = 2$. Here is an example of a computation done with the new symbol α.

$$(1 + 3\alpha)(2 - \alpha) = 2 + 5\alpha - 3\alpha^2 = 2 + 5\alpha - 3 \cdot 2 = -4 + 5\alpha$$

The inverse of the element $1 + 3\alpha$ is $\frac{-1}{17} + \frac{3}{17}\alpha$. ∎

EXAMPLE 5

The set $\boldsymbol{Q}(\sqrt{2}) = \{a + b\sqrt{2} : a \text{ and } b \text{ in } \boldsymbol{Q}\}$ is a subset of the real numbers $\boldsymbol{R}$. It is clearly a subring of $\boldsymbol{R}$. To see that it is a subfield, we note that if $a + b\sqrt{2}$ is not zero, then its multiplicative inverse $\frac{a}{a^2 - 2b^2} - \frac{b\sqrt{2}}{a^2 - 2b^2}$ is also in $\boldsymbol{Q}(\sqrt{2})$. (Note that if a and b are rational numbers that are not both zero, then $a^2 - 2b^2$ cannot equal zero.) Thus $\boldsymbol{Q}(\sqrt{2})$ is an extension field of $\boldsymbol{Q}$, and $\boldsymbol{R}$ is an extension field of $\boldsymbol{Q}(\sqrt{2})$.

∎

EXAMPLE 6

Let $\boldsymbol{Q}_1 = \boldsymbol{Q}[x]/\langle x^2 - 2\rangle$ be as in Example 4 and let $\boldsymbol{Q}(\sqrt{2})$ be as in Example 5. The map $f: \boldsymbol{Q}(\sqrt{2}) \to \boldsymbol{Q}_1$ defined by $f(a + b\sqrt{2}) = a + b\alpha$ is a ring isomorphism. Addition is clearly preserved. To see that multiplication is preserved, first note that $(\sqrt{2})^2 = 2$ in $\boldsymbol{Q}(\sqrt{2})$ and $\alpha^2 = 2$ in $\boldsymbol{Q}_1$. Thus

$$\begin{aligned}f((a + b\sqrt{2})(c + d\sqrt{2})) &= f((ac + 2bd) + (ad + bc)\sqrt{2}) \\ &= (ac + 2bd) + (ad + bc)\alpha \\ &= (a + b\alpha)(c + d\alpha) \\ &= f(a + b\sqrt{2})f(c + d\sqrt{2}).\end{aligned}$$

It is also easy to see that f is one-to-one and onto. The fact that $\boldsymbol{Q}[x]/\langle x^2 - 2\rangle$ and $\boldsymbol{Q}(\sqrt{2})$ are isomorphic reflects a more general situation as we shall soon show. ∎

We have seen that if K is an extension field of F, there may be many intermediate fields, fields that are subfields of K but also extension fields of F. Next we show

how to construct such fields. In Exercise 6 of Section 5.4, you were asked for a proof of the fact that the intersection of subfields is again a subfield. The following definition assumes that basic fact.

> **DEFINITION 1.** Let K be an extension of the field F and let $\alpha_1, \alpha_2, \ldots, \alpha_n$ be elements of K. $F(\alpha_1, \alpha_2, \ldots, \alpha_n)$ is the intersection of all subfields of K that contain both F and the set $\{\alpha_1, \alpha_2, \ldots, \alpha_n\}$.

$F(\alpha_1, \alpha_2, \ldots, \alpha_n)$ is a subfield of K (though not necessarily proper) and it is an extension field of F. Sometimes we say that $F(\alpha_1, \alpha_2, \ldots, \alpha_n)$ is the **smallest subfield of K that contains F and the set of elements $\{\alpha_1, \alpha_2, \ldots, \alpha_n\}$** because any field that contains F and contains each of the elements in the set $\{\alpha_1, \alpha_2, \ldots, \alpha_n\}$ must contain all the elements in $F(\alpha_1, \alpha_2, \ldots, \alpha_n)$. We also say that $F(\alpha_1, \alpha_2, \ldots, \alpha_n)$ is **the field F with the elements $\alpha_1, \alpha_2, \ldots, \alpha_n$ adjoined**.

Suppose that K is an extension field of F and that α is an element of K that is not an element of F. What does $F(\alpha)$ look like? There are two possibilities depending on whether or not α is a root of a polynomial $p(x) \in F[x]$. First assume that α is **not** the root of any nonzero polynomial in $F[x]$. The field $F(\alpha)$ must contain all members of the set $S = \left\{ \frac{p(\alpha)}{q(\alpha)} : p(x), q(x) \in F[x], q(x) \neq 0 \right\}$. (By assumption, $q(\alpha) \neq 0$ for any nonzero polynomial $q(x) \in F[x]$.) Note that $\frac{p(\alpha)}{q(\alpha)} = \frac{s(\alpha)}{t(\alpha)}$ if and only if $p(\alpha)t(\alpha) = s(\alpha)q(\alpha)$ in F. The set S is closed under addition, subtraction, and multiplication; S contains the additive and multiplicative identities $\frac{0}{1}$ and $\frac{1}{1}$, respectively; and S contains the inverses of its nonzero elements. Thus S is a field itself and so S is a subfield of K. Since the elements of S must be contained in any subfield of K that contains both α and F, we have $S = F(\alpha)$. Notice that $F(\alpha)$ is isomorphic to $F(x)$, the field of rational functions over F defined in Example 2.

EXAMPLE 7

The irrational number $\pi \in \mathbf{R}$ is not the root of any polynomial in $\mathbf{Q}[x]$. When we adjoin π to $\mathbf{Q}$, the elements of the field $\mathbf{Q}(\pi)$ are all of the form $\left\{ \frac{p(\pi)}{q(\pi)} : p(x), q(x) \in F[x], q(x) \neq 0 \right\}$. A typical element might be $\frac{\pi^2 + 2\pi + 1}{\pi^3 + 3}$. ∎

Now suppose that α is a root of an irreducible polynomial $p(x) = b_n x^n + b_{n-1} x^{n-1} + \ldots + b_0$ of degree $n > 0$ in $F[x]$. The field $F(\alpha)$ must contain all elements of the form $a_0 + a_1 \alpha + a_2 \alpha^2 + \ldots + a_{n-1} \alpha^{n-1}$. In the next proposition we show that there are no other elements in $F(\alpha)$.

PROPOSITION 1 Let α be a root of the irreducible polynomial $p(x)$ in $F[x]$ of degree $n > 0$ given by $p(x) = b_n x^n + b_{n-1} x^{n-1} + \ldots + b_0$. Then $F(\alpha) = \{a_0 + a_1 \alpha + a_2 \alpha^2 + \ldots + a_{n-1} \alpha^{n-1} : a_0, a_1, \ldots, a_{n-1} \in F\}$.

Proof. Let $S = \{u_0 + u_1\alpha + a_2\alpha^2 + \ldots + a_{n-1}\alpha^{n-1} \cdot a_0, a_1, \ldots, a_{n-1} \in F\}$. Then $S \subseteq F(\alpha)$ and it is closed under addition. Since $p(\alpha) = b_n\alpha^n + b_{n-1}\alpha^{n-1} + \ldots + b_0 = 0$ and $b_n \neq 0$, we can solve for α^n and replace any occurrence of α^n in any product of members of S by

$$\frac{-1}{b_n}(b_{n-1}\alpha^{n-1} + \ldots + b_0).$$

Thus S is closed under multiplication. So far we have shown that S is a subring of K. We now show S is a subfield by showing that the multiplicative inverse of any nonzero element in S is also in S. Let $y = a_0 + a_1\alpha + a_2\alpha^2 + \ldots + a_{n-1}\alpha^{n-1}$ be a nonzero element of S and let $g(x) = a_0 + a_1x + a_2x^2 + \ldots + a_{n-1}x^{n-1}$. So $y = g(\alpha)$. Since $p(x)$ is irreducible in $F[x]$, $g(x)$ and $p(x)$ are relatively prime. We can find polynomials $s(x)$ and $t(x)$ such that $s(x)g(x) + t(x)p(x) = 1$. Let $z = s(\alpha)$. Then $1 = g(\alpha)s(\alpha) + t(\alpha)p(\alpha)$. Since $p(\alpha) = 0$ and $g(\alpha)s(\alpha) = 1$, we see that $yz = g(\alpha)s(\alpha) = 1$ so that z is the inverse of y in S. Since S is a field containing α and F, $F(\alpha) \subseteq S$, which proves that $S = F(\alpha)$. ▲

The next corollary is immediate.

COROLLARY 2 Suppose that K is an extension field of F and that $\beta \in K$ is a root of an irreducible polynomial $p(x) \in F[x]$. Then $F(\beta)$ is isomorphic to $F[x]/\langle p(x) \rangle$.

Whether we adjoin a root β of $p(x)$ from an extension K of F to form $F(\beta)$ or we "invent" a root $\alpha = [x]$ of $p(x)$ by constructing the quotient structure $F[x]/\langle p(x) \rangle$, we end up with isomorphic fields.

COROLLARY 3 Suppose that γ and β are distinct roots of an irreducible polynomial $p(x) \in F[x]$ in some extension K of F. Then $F(\gamma)$ and $F(\beta)$ are isomorphic.

Proof. The fields $F(\alpha)$ and $F(\beta)$ are both isomorphic to $F[x]/\langle p(x) \rangle$. ▲

EXAMPLE 8

The three cube roots of 2 in C are the real number $\alpha_1 = 2^{1/3}$ and the complex numbers $\alpha_2 = 2^{1/3}\left(\frac{-1+\sqrt{-3}}{2}\right)$ and $\alpha_3 = 2^{1/3}\left(\frac{-1-\sqrt{-3}}{2}\right)$. The above corollary assures us that even though these roots seem quite different, the fields $Q(\alpha_1), Q(\alpha_2)$ and $Q(\alpha_3)$ are isomorphic. In that sense, the roots α_1, α_2, and α_3 are interchangeable. (Notice also how each is obtainable from the other through multiplication by a cube root of unity.) However, even though they are isomorphic, the fields $Q(\alpha_1), Q(\alpha_2)$, and $Q(\alpha_3)$ are distinct subfields of C. Their intersection is simply Q. ■

The goal of the rest of this section is to formulate the notion that a polynomial is "solvable by radicals" in terms of the vocabulary field extensions. To keep our discussion focused and concrete, we will assume that all fields are subfields of the complex numbers. So if F is a field under discussion, it will be assumed that

$Q \subseteq F \subseteq C$. Recall that each polynomial $p(x)$ in $C[x]$ has a root α in C. If $p(x)$ is irreducible in $F[x]$, the extension field $F[x]/\langle p(x) \rangle$ is isomorphic to $F(\alpha)$ (i.e., the field F with α adjoined). Thus $Q \subseteq F \subseteq F(\alpha) \subseteq C$.

EXAMPLE 9

Suppose that $q(x) = (x^2 + bx + c) \in Q[x]$ is irreducible and hence has no root in Q. Its roots are $\frac{-b \pm \sqrt{b^2 - 4c}}{2}$ by the quadratic formula. Let D be the rational number $b^2 - 4c$. Both roots of $q(x)$ are in $Q(\sqrt{D})$. Thus we find the roots of $q(x)$ in the extension field of Q obtained by adjoining a root (or radical) of an element of Q to Q. This is the sense in which quadratics are "solvable by radicals." For example, if $p(x) = x^2 - 3x + 5$, then $D = -11$. The roots of $x^2 - 3x + 5$ are not in Q but they are in $Q(\sqrt{-11})$. ∎

Let F be a field and $a \in F$. A field of the form $F(\sqrt[p]{a})$ where $\sqrt[p]{a}$ is a root (or radical) of $x^p - a$ in C is called a **radical extension** of F. If p is prime we call $F(\sqrt[p]{a})$ a **prime radical extension**. It is in such extensions that we find the roots of a polynomial that is "solvable by radicals." For instance, we saw in Example 9 that the roots of an irreducible quadratic are contained in a prime radical extension of Q, namely $Q(\sqrt{D})$, where $\sqrt{D}$ is the root of $x^2 - D$. In the next proposition we show that if $x^p - a$ has no root in F, then it is irreducible in $F[x]$. The field $F(\sqrt[p]{a})$ is thus isomorphic to $F[x]/\langle x^p - a \rangle$. Note: The result does not hold for polynomials in general since, for instance, $x^4 - 5x^2 + 6$ has no root in Q but it is not irreducible in $Q[x]$. It factors as $(x^2 - 2)(x^2 - 3)$. The special form $x^p - a$ is important.

PROPOSITION 4 Let p be a prime and let $a \in F$. Suppose that the polynomial $p(x) = x^p - a$ has no root in F. Then the polynomial $p(x)$ is irreducible in $F[x]$.

Proof. The polynomial $p(x) = x^p - a$ factors in $C[x]$ as $(x - \sqrt[p]{a})(x - \xi\sqrt[p]{a})(x - \xi^2\sqrt[p]{a}) \cdots (x - \xi^{p-1}\sqrt[p]{a})$, where $\sqrt[p]{a} = a^{1/p}$ is any root of $x^p - a$ in C and $\xi \neq 1$ is a pth root of unity ($\xi^p = 1$). If $p(x)$ was not irreducible in $F[x]$, we could factor $p(x)$ as $p(x) = f_1(x)f_2(x) \cdots f_n(x)$, where $1 < \deg(f_i(x)) < p$ for each i. Note that the coefficients of each $f_i(x)$ are all in F. Suppose that the degree of $f_1(x)$ is k. Since $f_1(x)$ is the product of a subset of the factors of $p(x)$, we have $f_1(x) = (x - \xi^{p_1}a^{1/p})(x - \xi^{p_2}a^{1/p}) \cdots (x - \xi^{p_k}a^{1/p}) = x^k + bx^{k-1} + \ldots + \xi^q a^{k/p}$, where q is the remainder of $p_1 + p_2 + \ldots + p_k$ after division by p. This implies that $\xi^q a^{k/p} \in F$, which cannot occur as we now show. Since $1 < k < p$, the numbers k and p are relatively prime. Let s and t be integers such that $1 = sk + tp$, or $1 - tp = sk$. Then $(\xi^q a^{k/p})^s = \xi^{qs} a^{ks/p} = \xi^{qs} a^{(1-tp)/p} = \xi^{qs} a^{1/p} a^{-t}$. If $\xi^{qs} a^{1/p} a^{-t} \in F$ and $a^t \in F$, their product $\xi^{qs} a^{1/p}$ is in F. But this implies $(\xi^{qs} a^{1/p})^p = a$, which is impossible since, by hypothesis, the polynomial $x^p - a$ has no root in F. ▲

Suppose that $F_0 \subseteq F_1 \subseteq \ldots \subseteq F_n$ is a sequence of field extensions such that each F_i is a radical extension of F_{i-1}. We call $F_0 \subseteq F_1 \subseteq \ldots \subseteq F_n$ a **radical tower**. If each extension is a prime radical extension, then $F_0 \subseteq F_1 \subseteq \ldots \subseteq F_n$ is called a **prime**

radical tower. We call F_n the **last field** in the tower and F_0 the **base field** of the tower. The vocabulary reflects the following graphic.

$$F_n$$
$$|$$
$$F_{n-1}$$
$$|$$
$$\cdot$$
$$\cdot$$
$$\cdot$$
$$|$$
$$F_0$$

EXAMPLE 10

The field $Q(\sqrt{2})$ is a prime radical extension of Q since $\sqrt{2}$ is a root of the irreducible polynomial $x^2 - 2 \in Q[x]$. The tower $Q \subseteq Q(\sqrt{2}) \subseteq Q\left(\sqrt{2}, \sqrt{1+\sqrt{2}}\right)$ is a prime radical tower since $Q\left(\sqrt{2}, \sqrt{1+\sqrt{2}}\right) = Q(\sqrt{2})\left(\sqrt{1+\sqrt{2}}\right)$ and $\sqrt{1+\sqrt{2}}$ is the root of the irreducible polynomial $x^2 - (1+\sqrt{2}) \in Q(\sqrt{2})[x]$. Thus $Q\left(\sqrt{2}, \sqrt{1+\sqrt{2}}\right)$ is the last field in a prime radical tower over the base field Q.

$$Q\left(\sqrt{2}, \sqrt{1+\sqrt{2}}\right)$$
$$|$$
$$Q(\sqrt{2})$$
$$|$$
$$Q$$

■

Any radical extension can be decomposed into a prime radical tower. For instance if $n = pq$, and $y = \sqrt[n]{x}$, then $F(\sqrt[pq]{x})$ is the last field in the tower $F \subseteq F(\sqrt[p]{x}) \subseteq F(\sqrt[p]{x})(\sqrt[q]{y})$. So a prime radical tower can replace any radical tower.

We would expect that a polynomial in $Q[x]$ that is "solvable by radicals" to have roots that look something like $r = 3 + \sqrt{5 + \sqrt[3]{2 + 3\sqrt{-7}}}$. Such a root r would be found in an extension field of Q constructed as follows. First we adjoin $\sqrt{-7}$ to Q and let $F_1 = Q(\sqrt{-7})$ so that F_1 is a prime radical extension of Q. Then, letting $a \in F$ be the element $2 + 3\sqrt{-7}$, we adjoin $\sqrt[3]{a}$ to F_1 and let $F_2 = F_1(\sqrt[3]{a})$ so that F_2 is a prime radical extension of F_1. Finally, letting $b = 5 + \sqrt[3]{2 + 3\sqrt{-7}}$, we adjoin $\sqrt{b}$ to F_2 and let $F_3 = F_2(\sqrt{b})$, again a prime radical extension of F_2. In short, we build a radical tower $Q \subseteq F_1 \subseteq F_2 \subseteq F_3$ and find the root r in the last field of a prime radical tower. The definition that follows captures this idea.

> **DEFINITION.** A polynomial in $F[x]$ is said to be **solvable by radicals** if all its roots can be found in the last field of a prime radical tower over the base field F.

334 Chapter 6 Looking Forward and Back

The outstanding question in algebra until 1830 was whether every polynomial had a root in the last field of a radical tower over its field of coefficients. The next example shows that the answer is yes for cubic polynomials in $Q[x]$. The answer is also yes for quartic polynomials. However, for polynomials of degree 5 and greater, the answer is no, as both Galois and Abel showed at approximately the same time (c. 1830) in very different ways. We explore the question further in Section 6.5.

EXAMPLE 11

In Section 3.6, we showed how to find the roots of any cubic polynomial in $Q[x]$. (The reader should review that material for this example.) For simplicity (but without loss of generality) let us consider a cubic equation of the form $x^3 + px + q = 0$. First, we adjoin the complex cube roots of unity. They are in the extension field $Q_1 = Q(\sqrt{-3})$, which is a prime radical extension of Q. Next, let $D = \frac{p^3}{27} + \frac{q^2}{4}$ and let $Q_2 = Q_1(\sqrt{D})$. If $\sqrt{D}$ is an element of Q_1, then $Q_1 = Q_2$. If $\sqrt{D}$ is not in Q_1, then Q_2 is a prime radical extension of Q_1. Finally, let $A = \frac{-q}{2} + \sqrt{\frac{p^3}{27} + \frac{q^2}{4}} = \frac{-q}{2} + \sqrt{D}$ and let $Q_3 = Q_2(\sqrt[3]{A})$. If $\sqrt[3]{A}$ is an element of Q_2, then $Q_2 = Q_3$. If $\sqrt[3]{A}$ is not in Q_2, then $Q_3 = Q_2(\sqrt[3]{A})$ is a prime radical extension of Q_1. Now we have a prime radical tower $Q \subseteq Q_1 \subseteq Q_2 \subseteq Q_3$ and the roots of $x^3 + px + q$ are contained in the last field in the tower. Thus $x^3 + px + q$ (and all cubic equations) are solvable by radicals. Let's check. According to our construction, we should be able to find the roots of the polynomial $x^3 + 3x + 2$ in the field $Q_3 = Q\left(\sqrt{-3}, \sqrt{2}, \sqrt[3]{-1+\sqrt{2}}\right)$ since $D = 2$ and $-q/2 = -1$. By Cardano's method or by an appeal to your favorite CAS, you can determine that the roots of $x^3 + 3x + 2$ are

$$-(1+\sqrt{2})^{1/3} + (-1+\sqrt{2})^{-1/3},$$

$$\frac{1}{2}\left((1+\sqrt{2})^{1/3} - (-1+\sqrt{2})^{1/3} + \sqrt{-3}\left((1+\sqrt{2})^{1/3} + (-1+\sqrt{2})^{1/3}\right)\right), \text{ and}$$

$$\frac{1}{2}\left((1+\sqrt{2})^{1/3} - (-1+\sqrt{2})^{1/3} - \sqrt{-3}\left((1+\sqrt{2})^{1/3} + (-1+\sqrt{2})^{1/3}\right)\right).$$

It looks like we are off by a minus sign in the expression $z = (1+\sqrt{2})^{1/3}$. However, since $(1+\sqrt{2})^{1/3}(-1+\sqrt{2})^{1/3} = ((1+\sqrt{2})(-1+\sqrt{2}))^{1/3} = 1$, we see that z is the inverse of $(-1+\sqrt{2})^{1/3}$ and indeed in Q_3. ∎

SUMMARY

Our goal in this section is to formulate what it means for a polynomial $p(x) \in F[x]$ to be "solvable by radicals" in algebraic terms. To that end, we introduced the notion of an **extension field** of the field F. We saw that if $p(x)$ is irreducible, then we could regard $F[x]/\langle p(x)\rangle$ as an extension field of F. Alternatively, if K is an extension field of F, and α is a root of $p(x)$ in K, then the subfield $F(\alpha) \subseteq K$, i.e. **the field F with α adjoined**, is an extension of F that is isomorphic to $F[x]/\langle p(x)\rangle$. When $p(x) \in F[x]$ is irreducible and of the form $x^n - a$, the extension field $F(\sqrt[n]{a})$ is called a **radical extension of F** or **prime radical extension** in the case that n is prime. A **radical**

tower is a sequence of field extensions $F = F_0 \subseteq F_1 \subseteq \ldots \subseteq F_q$ in which $F_i \subseteq F_{i+1}$ is a radical extension. A polynomial $f(x) \in F[x]$ is **solvable by radicals** if its roots can be found in the last field of a radical tower over the base field F. In our next section we continue our investigation of extension fields.

6.1 Exercises

1. Suppose that p is a prime number. Then $\sqrt{p} \in R$. Let $F = \{a + b\sqrt{p} : a$ and b in $Q\}$. Prove that $F \subseteq R$ is an extension field of Q.
2. Characterize the elements of $Q(i)$, considered as a subfield of the complex numbers.
3. Find five distinct but isomorphic subfields of C that are extension fields of Q.
4. Can you use the two distinct roots in C of $p(x) = x^2 - 2 \in Q[x]$ to find two distinct subfields of C? Generalize to any irreducible quadratic in $Q[x]$.
5. Let α and β be elements in an extension K of F. Prove that $F(\alpha, \beta) = F(\alpha)(\beta)$.
6. Show that the field $Q[x]/\langle x^2 - 2 \rangle$ is **not** isomorphic to $Q(\sqrt{3})$.
7. Characterize the elements in $Q(e)$. (No polynomial in $Q[x]$ has the natural number e as a root.)
8. Exhibit a radical tower over Q, the last field of which contains the roots of a given cubic $x^3 + ax^2 + bx + c$. (It might be easier to transform the given cubic into the form $y^3 + py + q$ first.)
   i. $x^3 + 6x + 4$
   ii. $x^3 + 3x^2 + 3x + 12$
   iii. $x^3 - 9x^2 + 21x - 13$
9. Show that all quartic (polynomials of degree four) equations in $Q[x]$ are solvable by radicals. (See Section 3.6.)

To the Teacher

In this section, we formulated the notion of what it means for a polynomial to be "solvable by radicals" in algebraic terms. We have a way of asking, "Can the staple of high school algebra, the quadratic formula, be extended to polynomials of higher degree?" In Section 6.3, using similar algebraic tools, we will address an ancient question, "What lengths are constructible with a straightedge and compass?" In both cases, we are asking deep questions about the number line. What are its members like? How do we sort them? The historical evolution of our sense of number is mimicked by the student's progression.

1. Natural numbers and ratios of natural numbers come to us, well, naturally! But if we consider all lengths to be numbers, then not all numbers are expressible as ratios of natural numbers. For instance, the length of the diagonal of a unit square is not a rational number.
2. Perhaps all lengths are constructible with ruler and compass. Certainly, the diagonal of a unit square, is such a length. (See how in Section 6.4.) But the length of the edge of a cube, the volume of which is 2, cannot be constructed

from a unit length using ruler and compass. (See Section 6.3.) So not all lengths can be obtained from a unit length and the simple tools that draw lines and circles. The relation between numbers and geometry is more complicated.

3. Perhaps all numbers arise as the roots of polynomials. No. The ratio of the circumference of a circle to its diameter, namely π, is not. It is not constructible, it is not rational, and it is not the root of any polynomial. Such numbers are called "transcendental" numbers. The natural number e is another such number. In fact, most real numbers are transcendental numbers. The set of all possible real roots to polynomials with rational coefficients is called the set of real "algebraic" numbers. It is a countable subset of the real numbers. Its complement, the set of transcendental numbers, is an uncountable set.

4. Perhaps the "algebraic numbers" can all be expressed in terms of the sum, product and roots of the rational numbers, numbers like $\frac{2+\sqrt[5]{7+\sqrt{2}}}{5}$. That is what we mean by "solvable" and that is what we would expect, extrapolating from the quadratic formula. But the answer is no, as Abel and Galois proved in 1830.

Discovering the structure of the continuum—the real line—is an ongoing and difficult task for both the student and the researcher. We hope that we, the authors, have raised the future teacher's awareness of these issues and provided the tools with which you can guide your students.

Let's close this section with a polynomial quiz. Here are the roots. What are the polynomials? This twist on the usual "solve for x" can give a world of insight.

i. Find a quadratic polynomial with rational coefficients that has $2 + 5\sqrt{3}$ as a root.
ii. Find *all* quadratic polynomials with rational coefficients that have $2 + 5\sqrt{3}$ as a root.
iii. Is $1 + 2 \cdot 2^{1/3} - 2^{2/3}$ the root of a quadratic polynomial with rational coefficients?
iv. Find a polynomial with rational coefficients that has $1 + 2 \cdot 2^{1/3} - 2^{2/3}$ as a root.
v. Are all members of $\boldsymbol{Q}(2^{1/3})$ roots of cubic polynomials with rational coefficients?

6.2 THE DEGREE OF AN EXTENSION

In this section we investigate **the degree of a field extension**. It is not really a new idea. We already know, for instance, that each element $a + b2^{1/3} + c2^{2/3}$ in the field $\boldsymbol{Q}(2^{1/3})$ is uniquely defined by the three constants $a, b,$ and c in $\boldsymbol{Q}$. The degree of the extension $\boldsymbol{Q} \subseteq \boldsymbol{Q}(2^{1/3})$ is 3. The degree will help us to determine if it is possible to find the root of a particular polynomial in a given field extension. Our investigation will lead us to answer several intriguing questions left open by ancient Greek mathematics. Can a cube be doubled? Can an angle be trisected? Can a circle be squared? (See Section 6.3 for both an interpretation of these questions and their answers.)

Note: This section assumes that the reader is acquainted with elementary linear algebra. For readers who are not, but who are interested in how to solve the classical problems mentioned, it is enough to consider extensions of the form $F \subseteq F(\alpha)$, where α is the root of an irreducible polynomial $p(x)$ in $F[x]$. The degree of the extension is simply the degree of the polynomial $p(x)$. The conclusion of Theorem 4 (without its proof) is what we need to solve the ancient problems.

In a field extension $F \subseteq F(\alpha)$ where α is a root of an irreducible polynomial $p(x)$ in $F[x]$ of degree n, each element y of $F(\alpha)$ can be expressed as $y = a_0 + a_1\alpha + \ldots + a_{n-1}\alpha^{n-1}$, where $a_i \in F$. To designate an element y of $F(\alpha)$, we need only designate $(a_0, a_1, \ldots, a_{n-1})$, the n-tuple of its coefficients. For instance, $(1, 2, -3)$ stands for $1 + 2\alpha - 3\alpha^2$. The n-tuples or **vectors** are in one-to-one correspondence with members of $F(\alpha)$. If the vector $(a_0, a_1, \ldots, a_{n-1})$ designates $y \in F(\alpha)$ and $(b_0, b_1, \ldots, b_{n-1})$ designates $z \in F(\alpha)$, then the vector $(a_0 + b_0, \ldots, a_{n-1} + b_{n-1})$ designates $y + z$ since $y + z = (a_0 + b_0) + (a_1 + b_1)\alpha + \ldots + (a_{n-1} + b_{n-1})\alpha^{n-1}$. Similarly, if $k \in F$, then $(kb_0, kb_1, \ldots, kb_{n-1})$ is the vector of coefficients of kz. Adding in $F(\alpha)$ or multiplying by a constant k is the same as performing the analogous operations on the associated vectors. So it is natural to think of $F(\alpha)$ as an n dimensional **vector space** over F. Here is a brief reminder of what it means to be a vector space over a field F.

DEFINITION. Let F be a field. A set V is a **vector space over F** if

(1) V is a group under addition.
(2) For each element $a \in F$ and $v \in V$, there is an element $av \in V$.
(3) For every s and t in F and v and w in V, the following hold.
 i. $s(v + w) = sv + sw$
 ii. $(s + t)w = sw + tw$
 iii. $s(tw) = (st)w$
 iv. $1v = v$

EXAMPLES

1. The most familiar examples of vector spaces are the vector spaces $\mathbf{R}^n$ of elementary linear algebra. Here the field is $\mathbf{R}$.
2. The set of all polynomials with complex coefficients is a vector space over $\mathbf{C}$.
3. The set of all polynomials with coefficients in the field $\mathbf{Z}_p$ for any prime p is a vector space over $\mathbf{Z}_p$.

PROPOSITION 1 If $F \subseteq K$ is any field extension, then K is a vector space over F.

Proof. Certainly any field K is a group under addition so that (1) holds. Let $a \in F$ and $v \in K$. Since $F \subseteq K$, we have $av \in K$ so that (2) holds. Since K is a field, the distributive and associative properties described in (3) also hold. ▲

The basic concepts of spanning set, basis and dimension apply to extension fields $F \subseteq K$.

> **DEFINITION 1.** The **degree of a field extension** $F \subseteq K$ is the dimension of K, regarded as a vector space over F. It is denoted $[K : F]$. If $[K : F]$ is a finite number, then K is called a **finite extension** of F.

An extension $F \subseteq K$ may be finite or infinite. For example, consider $Q(\pi)$, the field rational numbers Q with π adjoined. If $[Q(\pi) : Q]$ were finite, then for some $n \in N$, the numbers $1, \pi, \pi^2, \ldots, \pi^n$ would be linearly dependent. So we could find rational numbers $a_0, a_1, \ldots, a_n$, not all zero, such that $a_0 + a_1\pi + \ldots + a_n\pi^n = 0$. That would imply that π was the root of the polynomial $p(x) = a_0 + a_1x + \ldots + a_nx^n$. However in 1882, Ferdinand Lindemann proved that π is not the root of any polynomial with rational coefficients. (It was a difficult fact to prove.) Thus $[Q(\pi) : Q]$ is not a finite extension.

The following proposition is just what we were expecting.

PROPOSITION 2 Let $p(x) \in F[x]$ be an irreducible polynomial of degree $n > 0$ and let α be a root of $p(x)$ in some extension K of F. Then $[F(\alpha) : F] = n$.

Proof. The n elements of the set $S = \{1, \alpha, \ldots, \alpha^{n-1}\}$ clearly span $F(\alpha)$ since every element in $F(\alpha)$ can be expressed as a linear combination of members of S. We must show that S is a linearly independent set. Suppose that $a_0 + a_1\alpha + \ldots + a_{n-1}\alpha^{n-1} = 0$. We will show that each $a_i = 0$. Let $f(x) = a_0 + a_1x + \ldots + a_{n-1}x^{n-1}$. Then $f(\alpha) = 0$. Let $g(x) \in F[x]$ be the greatest common divisor of the polynomials $f(x)$ and $p(x)$. Since we can find $s(x)$ and $t(x)$ in $F(x)$ such that $g(x) = s(x)f(x) + t(x)p(x)$, we see that α is a root of $g(x)$ as well. Thus $\deg(g) > 0$. Since $p(x)$ is irreducible, and $g(x)$ divides $p(x)$, it must be the case that $p(x) = cg(x)$, for some $c \in F$, which in turn implies that $p(x)$ divides $f(x)$. Since $\deg(f) < \deg(p)$, $f(x)$ must be identically 0 and its coefficients $a_1, a_2, \ldots, a_{n-1}$ must all be zero. Thus the set $\{1, \alpha, \ldots, \alpha^{n-1}\}$ is linearly independent. ▲

Adjoining a root of an irreducible polynomial in $F[x]$ to F results in a finite extension of F. Conversely, every element in a finite extension K of F is the root of an irreducible polynomial in $F[x]$. To see this, suppose that $[K : F] = n$ and $\alpha \in K$. The terms $1, \alpha, \alpha^2, \ldots, \alpha^n$ are $n + 1$ elements of the n dimensional vector space K and hence linearly *dependent*. Thus we can find elements $a_0, \ldots, a_n$ in F, not all zero, such that $a_0 + a_1\alpha + \ldots + a_n\alpha^n = 0$. So α is a root of the polynomial $f(x) = a_0 + a_1x + \ldots + a_nx^n$. If $f(x)$ itself is not irreducible, then α must be the root of some irreducible factor of $f(x)$. An irreducible polynomial of minimal degree satisfied by α is called a **minimal polynomial of α**.

DEFINITION 2. Let $F \subseteq K$ be an extension and suppose that $\alpha \in K$ is the root of a polynomial $f(x) \in F[x]$. An irreducible polynomial of *minimal* degree in $F[x]$ satisfied by α is called a **minimal polynomial of α**.

PROPOSITION 3 Any two minimal polynomials differ by a constant factor.

The proof is straightforward and is left as an exercise. The expression "**the minimal polynomial**" will be used for the unique **monic** minimal polynomial of α. Note that if $\alpha \in K$ is not 0, its minimal polynomial has a nonzero constant term.

EXAMPLE 4

Let $\alpha = 1 + 2^{1/3} + 2^{2/3} \in \mathbf{Q}(2^{1/3})$. We now find the minimal polynomial in $\mathbf{Q}[x]$ satisfied by α. Since $[\mathbf{Q}(2^{1/3}) : \mathbf{Q}] = 3$, the four elements 1, α, α^2, and α^3 must be linearly dependent. Thus we can find rational numbers, $a_0, a_1, a_2,$ and a_3, not all equal to 0, such that $a_0 + a_1\alpha^1 + a_2\alpha^2 + a_3\alpha^3 = 0$. Expanding each of the powers of α, we have

$$a_0 + a_1(1 + 2^{1/3} + 2^{2/3}) + a_2(5 + 4 \cdot 2^{1/3} + 3 \cdot 2^{2/3})$$
$$+ a_3(19 + 15 \cdot 2^{1/3} + 12 \cdot 2^{2/3}) = 0.$$

Collecting the coefficients of 1, $2^{1/3}$ and $2^{2/3}$, we obtain

$$(a_0 + a_1 + 5a_2 + 19a_3) + (a_1 + 4a_2 + 15a_3)2^{1/3} + (a_1 + 3a_2 + 12a_3)2^{2/3} = 0.$$

Since minimal polynomials can differ by a multiplicative constant, we can set $a_0 = 1$. The resulting system of equations is

$$a_1 + 5a_2 + 19a_3 = -1$$
$$a_1 + 4a_2 + 15a_3 = 0$$
$$a_1 + 3a_2 + 12a_3 = 0.$$

The unique solution to this system is $a_1 = 3, a_2 = 3$ and $a_3 = -1$. Thus a minimal polynomial for α is $p(x) = 1 + 3x + 3x^2 - x^3$. A quick check will verify that $p(\alpha) = 0$. "The" minimal polynomial for α is the monic polynomial $x^3 - 3x^2 - 3x - 1$. ∎

The next theorem holds the key to solving the classical problems mentioned at the beginning of this section.

THEOREM 4 Let K be a finite extension of E and let E be a finite extension of F. Then K is a finite extension of F and $[K : F] = [K : E][E : F]$.

Proof. Let $\{x_1, \ldots, x_m\}$ be a basis for K over E and let $\{y_1, \ldots, y_n\}$ be a basis for E over F. Each element $z \in K$ can be expressed uniquely as $z = a_1x_1 + \ldots + a_mx_m$ with $a_i \in E$ for $i = 1, \ldots, m$. In turn, each $a_i \in E$ can be expressed uniquely as $a_i = b_{i,1}y_1 + \ldots + b_{i,n}y_n$, with each $b_{i,j} \in F$. Substituting for each a_i, we find that

$z = \sum_{i=1}^{m}\left(\sum_{j=1}^{n}b_{i,j}y_jx_i\right)$. Thus we see that K, considered as a vector space over F, is spanned by the nm elements in the set $S = \{y_jx_i : i = 1,\ldots, m \text{ and } j = 1,\ldots, n\}$. Now we must show that the elements in S are linearly independent. Suppose that $\sum_{i,j} c_{i,j}y_jx_i = \sum_{i=1}^{m}\left(\sum_{j=1}^{n}c_{i,j}y_j\right)x_i = 0$. Since the x_i are linearly independent, the sum $\sum_{j=1}^{m}c_{i,j}y_j$ must equal 0 for each i from 1 to m. Since the y_j are linearly independent, each $c_{i,j}$ must equal zero. Thus the set S is linearly independent and spans K. The degree of K over F is therefore mn. ▲

EXAMPLE 5

We show that $[Q(\sqrt{3}, \sqrt{2}) : Q] = 4$. First note that $\sqrt{3} \notin Q(\sqrt{2})$. If it were, there would be rational numbers a and b such that $\sqrt{3} = a + b\sqrt{2}$. Squaring both sides, we would have $3 = a^2 + 2ab\sqrt{2} + 2b^2$. We could then solve for $\sqrt{2}$ as a rational number, which is impossible. Thus the polynomial $x^2 - 3$ is irreducible over $Q(\sqrt{2})$ and $[Q(\sqrt{3}, \sqrt{2}) : Q(\sqrt{2})] = 2$. Since $[Q(\sqrt{2}) : Q] = 2$, Theorem 4 tells us that $[Q(\sqrt{3}, \sqrt{2}) : Q] = 4$. The set $\{1, \sqrt{3}, \sqrt{2}, \sqrt{6}\}$ is a basis for $Q(\sqrt{3}, \sqrt{2})$ over Q. ■

EXAMPLE 6

Let $Q_1 = Q(2^{1/3}) \cap Q(\sqrt{3}, \sqrt{2})$. The degree $[Q_1 : Q]$ must divide both $[Q(\sqrt{3}, \sqrt{2}) : Q]$, which equals 4, and $[Q(2^{1/3}) : Q]$, which equals 3. Thus $[Q_1 : Q] = 1$ and $Q_1 = Q$. ■

Let $F \subseteq K$ be a field extension. An element a in K is **algebraic** over F if it is the root of a polynomial $f(x)$ in $F[x]$. The next proposition says that if K is characteristic zero and K is an extension field obtained by adjoining two algebraic elements to F, then K can be formed by adjoining just one algebraic element.

PROPOSITION 5 Let F be a field with characteristic zero and suppose that a and b are algebraic over F. There exists $c \in F(a, b)$ such that $F(c) = F(a, b)$.

Proof. Suppose that the minimal polynomials of a and b over F are $f(x)$ and $g(x)$ respectively, with degrees m and n, respectively. Let K be an extension field of F in which both $f(x)$ and $g(x)$ have all their roots. Suppose that the roots of $f(x)$ in K are $\{a_1, \ldots, a_m\}$ and that $a = a_1$. Similarly, suppose that the roots of $g(x)$ are $\{b_1, \ldots, b_n\}$ and that $b = b_1$. The set S of values assumed by the expression $\frac{a_i - a}{b - b_j}$ for $i \geq 1$ and $j > 1$ is a finite set. Since F has characteristic 0, F is an infinite set. So there must be some element of F not equal to any of the elements in S. Let d be such an element. For $i \geq 1$ and for $j > 1$, we have $d \neq \frac{a_i - a}{b - b_j}$. Equivalently, $a_i \neq a + d(b - b_j)$ for $i \geq 1$ and for $j > 1$. Let $c = a + db$. We shall show that $b \in F(c)$, which in turn implies that $a \in F(c)$. Let $h(x) = f(c - dx)$. Then $h(b) = f(c - db) = f(a) = 0$. Note that $h(x) \in F(c)[x]$ and that b is a root of both $h(x)$ and $g(x)$. However, $g(x)$ and $h(x)$ share no other root because, if $h(b_j) = 0$ for some $j > 1$, then $c - db_j = a_i$ or $a + d(b - b_j) = a_i$, which contradicts our choice of d. Since $g(x)$ is the minimal polynomial of b and hence irreducible, $x - b$ divides $g(x)$ but $(x - b)^2$ does not. Thus the greatest common divisor of $g(x)$ and $h(x)$ is $x - b$, and so $x - b \in F(c)[x]$. Thus $b \in F(c)$ and, since $a = c - db$, $a \in F(c)$. So we have shown that $F(a, b) \subseteq F(c)$. Clearly, $F(c) \subseteq F(a, b)$. Therefore $F(c) = F(a, b)$. ▲

EXAMPLE 7

Let $K = Q(\sqrt{3}, \sqrt{2})$. The minimum polynomials for $\sqrt{2}$ and $\sqrt{3}$ are $x^2 - 2$ and $x^2 - 3$, respectively. The set S described in the Proposition 5 is the set $\{0, \frac{\sqrt{2}}{\sqrt{3}}\}$. Since d can be any element of Q that is not in S, we set $d = 1$ and $c = \sqrt{3} + \sqrt{2}$. Now we show that $Q(\sqrt{3}, \sqrt{2}) = Q(\sqrt{3} + \sqrt{2})$. Clearly, since $\sqrt{3} + \sqrt{2} \in Q(\sqrt{3}, \sqrt{2})$, we have $Q(\sqrt{3} + \sqrt{2}) \subseteq Q(\sqrt{3}, \sqrt{2})$. So what we need to show is that $\sqrt{3}$ and $\sqrt{2}$ are each elements of $Q(\sqrt{3} + \sqrt{2})$. Note that $c^2 = 5 + 2\sqrt{3}\sqrt{2}$ and $(c^2 - 5)/2 = \sqrt{2}\sqrt{3}$. Thus $\sqrt{2}\sqrt{3} \in Q(\sqrt{3} + \sqrt{2})$ as is the product $\sqrt{2}\sqrt{3}(\sqrt{3} + \sqrt{2}) = 3\sqrt{2} + 2\sqrt{3}$. The difference $3\sqrt{2} + 2\sqrt{3} - 2c = \sqrt{2}$ is a member of $Q(\sqrt{3} + \sqrt{2})$ and so both $\sqrt{3}$ and $\sqrt{2}$ are members of $Q(\sqrt{3} + \sqrt{2})$. Thus $Q(\sqrt{3}, \sqrt{2}) = Q(\sqrt{3} + \sqrt{2})$. ■

THEOREM 6 (*Primitive Element Theorem*) Suppose F has characteristic zero. If $F \subseteq K$ is a finite extension, then $K = F(c)$ for some c in K that is algebraic over F.

Proof. We prove the theorem by induction on $[K : F]$. If $[K : F] = 1$, then $K = F$ and the theorem is true with $c = 1$. Suppose $[K : F] > 1$ and that the theorem holds for all extensions of degree less than $[K : F]$. Let $a \in K$ be any element of K not in F. If $[K : F(a)] = [K : F]$, the theorem holds with $a = c$. If $[K : F(a)] < [K : F]$, by the induction hypothesis, there exists $b \in K$ such that $K = F(a)(b) = F(a, b)$. By Proposition 5, we can find c such that $K = F(c)$. ▲

SUMMARY

If E is a field extension of F, we can regard E as a vector space over the field F. The **degree of the extension**, denoted $[E : F]$, is the dimension of the vector space E over F. When $E = F(\alpha)$ and α is the root of an irreducible polynomial $p(x)$ in $F[x]$, the degree of the extension is equal to the degree of the polynomial $p(x)$, as expected. This simple number gives us information about other elements in $F(\alpha)$. For instance, every element of $F(\alpha)$ is itself the root of a **minimal polynomial** in $F[x]$ of degree less than or equal to the degree of $p(x)$. Theorem 4 tells us that $[K : F] = [K : E][E : F]$. This arithmetic result gives us an easy way of showing that, for instance, $Q(\sqrt{2}) \cap Q(2^{1/3}) = Q$. We saw that if K is a field extension of F obtained by adjoining two **algebraic** elements a and b of K to F, then K could be expressed more simply as $F(c)$, a field obtained by adjoining just one algebraic element. Reasoning inductively, we obtained the **Primitive Element Theorem**. In our next section, we shall see a direct connection between the results obtained here, especially Theorem 4, and three classical problems inherited from antiquity.

6.2 Exercises

1. Find the degree of each of the following extensions.
   i. $Q(\sqrt{3})$ over Q
   ii. $Q(\sqrt{3}, i)$ over $Q(i)$
   iii. $Q(i\sqrt{3}, i)$ over $Q(i\sqrt{3})$
   iv. $Q(\sqrt{\omega})$ over $Q(\omega)$, where ω is a complex cube root of unity.
   v. $Q(\sqrt{2}, \sqrt{3}, \sqrt{6})$ over $Q(\sqrt{2})$
2. Show that both $\sqrt{2}$ and i are in $Q(i + \sqrt{2})$ and conclude that $[Q(i + \sqrt{2}) : Q] = 4$. Find the minimal polynomial of $i + \sqrt{2}$ over Q.
3. What is the degree of the extension $Q \subseteq E$, where E is the subfield of C with all the roots of $x^6 - 1$ adjoined?
4. Let p be a prime number. What is the degree of the extension $Q \subseteq E$, where E is the subfield of C with all the roots of $x^p - 2$ adjoined?
5. An extension $F \subseteq K$ is called an **algebraic extension** of F if every element in K is algebraic over F. Prove that if $F \subseteq K$ is a finite extension, then K is an algebraic extension of F.
6. Suppose $a \neq 0$ and that a is algebraic over F with minimal polynomial $p(x)$. Find the minimal polynomial of $1/a$.
7. Suppose that a and b are algebraic over F and that $a \neq 0$. Prove that $a \pm b$, ab, and $1/a$ are also algebraic.
8. Let K be an extension of F. Let k be the set of elements in K that are algebraic over F. Prove that k is a field.
9. Let K be an algebraic extension of E and let E be an algebraic extension of F. Prove that K is an algebraic extension of F.
10. Prove that any two minimal polynomials differ by a constant factor.

To the Teacher

In this section, we have a confirmation of a hunch developed in high school: You can't express the cube root of a prime in terms of square roots. Why? Intuitively, when you square $\sqrt[3]{a}$ and square it again and again, you get $\sqrt[3]{a}$ raised to a power 1 or 2 because $2^n \bmod 3$ is always 1 or 2, and never 0. In this section, we resolved the issue. The cube root would lie in an extension of degree 3 while anything expressible in terms of square roots would lie in extension of degree 2^n. How this same consideration answered historically interesting and persistent questions is explored in the next section. We urge the perspective teacher to explore it. It brings together algebra, geometry, and trigonometry across a long spectrum of time.

6.3 RULER AND COMPASS CONSTRUCTIONS

Euclid's *Elements*, written c. 300 B.C.E., has been one of the most influential treatises in the history of mathematics. More than a compendium of geometrical theorems and their proofs, it paved the way for the axiomatic thinking that underlies modern mathematics. It remained required reading for American secondary school students into the twentieth century. In Book X, Euclid classified the line segments that could be constructed with these simple tools: a fixed unit length, a compass and a straightedge. Euclid's goal was to give precise relations between the sides and various diagonals of the "Platonic" solids—the tetrahedron, cube, octahedron, dodecahedron, and icosahedron. He accomplished this goal. But three related problems, formulated by ancient geometers as construction problems, remained unresolved until modern times. The problems are **to double the cube, to square the circle, and to trisect the angle**. The goal of this section is to interpret and to resolve these ancient problems with the algebraic techniques developed in this chapter.

Our tool kit for classical geometry will be a locking[1] compass, a straightedge, a pencil, and unit length for reference. With these tools we can construct and, more importantly, reconstruct line segments with an assortment of different and interesting lengths. For instance, we can draw a line L and place a point O on that line. Because our compass is locking, we can lock its radius onto the unit length, put its foot on O, and mark off a unit length on L at P_1 to the right of O. Then we can put its foot at P_1 and mark off a point P_2 to the right of P_1, etc. This way, for each integer $n > 0$, we can mark off a point P_n on L that is a distance n units away from O. We can go to the left of O and generate a similar series of points $N_1, N_2, \ldots$. (See Diagram 1.) The distance between any two marks that are produced by our tools via finitely many repeatable procedures is a **constructible distance**. So far, we know that all integer distances are constructible distances.

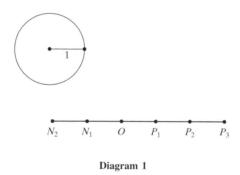

Diagram 1

More interesting, Diagram 2b shows that if a is a constructible distance, then $\sqrt{a}$ is a constructible distance. The construction uses the following facts.

a. If a and b are constructible distances, then so is $a + b$.

b. Given any two points A and B on a line L, we can construct their midpoint P.

c. Through a point P on a line L, we can construct a line through P perpendicular to L. (See Diagram 2a.)

To reproduce Diagram 2b, start with a line segment OP of length a and construct a line segment OQ of length $a + 1$. Mark the midpoint of OQ at R and construct a circle C of

---

[1] With a locking compass, we can lift a constructible length to another position. Actually, the ancient tool box only had floppy compasses. But it can be proved that any figure or length constructible with a locking compass is constructible with a floppy one. The constructions with a locking compass often have fewer steps.

radius $(a+1)/2$ centered at R. Construct a line perpendicular to OQ through P. Label its intersection with the circle C by T. The triangles OPT and TPQ are similar because the measures of $\angle OTP$ and $\angle TQP$ are equal. If we let b denote the length of PT, then using similarity, we have $a/b = b/1$ or $b^2 = a$. Thus the length of PT is $\sqrt{a}$.

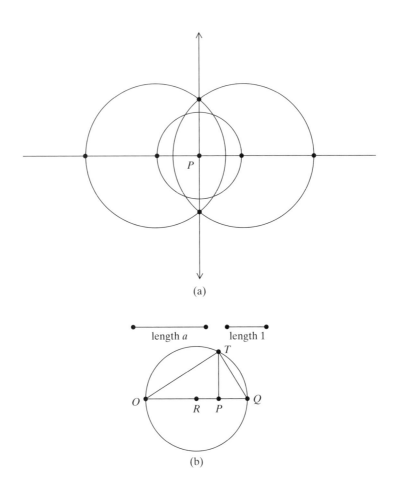

Diagram 2 Perpendicular lines and square roots

Since we know that distances of integer lengths are constructible, it follows that $\sqrt{n}$ is constructible for $n = 1, 2, 3$, etc. The construction in Diagram 3a shows that, given a line L_1 through the points A and B and another point P off L_1, we can construct a line L_2 through P that is perpendicular L_1. (Start with any circle centered at P that intersects L_1 in two places.) From the construction in Diagram 2a, it follows that we can construct a line L_3 through P perpendicular to L_2 and hence parallel to L_1. We need the construction in Diagram 3a to complete Diagrams 3b and 3c. In Diagram 3b, OP has length 1, OB has length b, OA has length a and BC is parallel to PA. The segment OC has length ab. In Diagram 3c, again OP has length 1, OB

has length b, OA has length a. Here PC is parallel to BA and the triangles OPC and OBA are similar. Segment OC has length a/b.

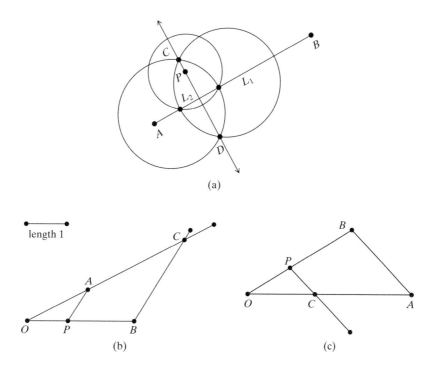

Diagram 3

So, starting with constructible lengths a and b, we can construct $a + b$, $a - b$, $a \cdot b$, a/b, and $\sqrt{a}$. Our tool kit serves as a 5-function calculator for constructible distances! (It struggles a bit with subtraction $a - b$ if $a < b$, but we can interpret appropriately.) Our goal is to obtain an algebraic characterization of all numbers that can occur as constructible distances. To that end, we'll perform our constructions in the Cartesian plane so that our marks will have coordinates. Let's summarize what we can construct and how we can construct it.

0. We start by drawing a line L_x in the plane to serve as the x-axis and marking a point O on L_x to serve as the origin.
1. We can construct a line L_y perpendicular to L_x at O. We can (potentially) mark off the infinite rectangular lattice $Q \times Q$ of all points with rational coordinates. The distance between any two marked points is a constructible distance.
2. We can draw a line between any two points already marked in the plane. We can center our compass at any point marked on the plane and use any constructible distance as a radius. We can obtain **new** points in the plane by marking the intersection of any two such lines and circles. The distance between any marked points is a constructible distance.

3. Only distances obtained through step 1 and a finite number of repetitions of step 2 are constructible distances.

Notice that if $P = (x_0, y_0)$ is a point obtained in step 2, then both $|x_0|$ and $|y_0|$ are constructible since we can drop a perpendicular to either L_x or L_y.

> **DEFINITION.** A number in R is **constructible** if its absolute value is a constructible distance.

We will denote the set of constructible numbers by CN. Here's what we know about the algebraic structure of CN so far:

1. The set of constructible numbers is a field that is a subfield of R.
2. The set of constructible numbers contains the square roots of any of its positive elements.

For example, since 3 and $\sqrt{2}$ are constructible, CN must contain $3 + \sqrt{2}$ and $\sqrt{3 + \sqrt{2}}$. Thus $Q \subseteq CN \subseteq R$. In fact, for each positive integer n, $Q \subseteq Q(\sqrt{n}) \subseteq CN \subseteq R$. Just what other numbers does CN contain? All real numbers? If not, then perhaps other roots like $\sqrt[3]{2}$. To answer these questions, let's take an even closer look at exactly what we can construct. Geometrically, we generate points in the plane by intersecting pairs of lines, pairs of circles, and by intersecting circles and lines. The lines we use go through points with constructible numbers as coordinates. The circles are also centered at such points and have radii that are constructible distances. If we let F be any subfield of the field CN of constructible numbers—for instance, we can start by letting F be the field of rational numbers Q—then the following hold.

I. Let $y = mx + b$ be the equation through any two points in the plane with coordinates that are in $F \times F$. Then m and b are in F.

II. Let (x_0, y_0) be the point obtained by intersecting two lines defined by $y = m_1 x + b_1$ and $y = m_2 x + b_2$, with m_1, m_2, b_1, and b_2 in F, $m_1 \neq m_2$, Then x_0 and y_0 are in F.

III. The distance between two points with coordinates in F is the square root of a number in F.

IV. Let L be the line defined by $y = mx + c$, where m and c are in F and let C be the circle defined by $(x - a)^2 + (y - b)^2 = r^2$, where a, b and r are in F. Suppose that L and C meet at (x_0, y_0). Then x_0 and y_0 are either in F or they are the square roots of numbers in F.

V. Let C_1 and C_2 be two circles with centers in $F \times F$ and radii in F. Suppose they meet at (x_0, y_0). Then x_0 and y_0 are either in F or they are the square roots of numbers in F.

Assertions I through V are not difficult to prove. We will prove Assertion I as an example and leave the rest as exercises.

Proof of I. Suppose that a line passes through the distinct points (x_0, y_0) and (x_1, y_1) with $x_0 \neq x_1$, and x_0, x_1, y_0 and y_1 are in F. Then $m = \frac{y_1 - y_0}{x_1 - x_0} \in F$ and $b = \frac{y_0 x_1 - x_0 y_1}{x_1 - x_0} \in F$ because F is a field. ▲

Assertions IV and V are what we need. They say that once we have generated the constructible numbers in a subfield F of CN, then any constructible numbers that can be generated by lines and circles through points in $F \times F$ are either in F or they are in a degree 2 extension of F. Thus, starting with Q, every constructible number lies in a field extension K where $[K : Q]$ is a power of 2. We summarize our results in the next theorem.

THEOREM 1 Every constructible number lies in a field extension K of Q for which $[K : Q]$ is a power of 2.

Now we are ready to state and answer three famous and very old questions.

Can a cube be doubled? Interpretation: Suppose that the length of the side of a cube is 1. Let a be the side of a cube with twice the volume. Is a a constructible number, and, in particular, is $\sqrt[3]{2}$ constructible? What we are really asking here is if cube roots of all rational numbers are constructible. The answer is no because, for instance, any field extension of Q that contains $\sqrt[3]{2}$ must have a degree that is divisible by 3. All constructible numbers lie in field extensions with degrees that are powers of 2. So a cube cannot be doubled.

Can the circle be squared? Interpretation: Suppose we have a circle with radius 1. Let a be the length of the side of the square with the same area. Is a constructible? Here we are asking if the number $\sqrt{\pi}$ (and hence π itself) is a constructible number. The answer here is more difficult to come by. In 1882, Ferdinand Lindemann[2] proved that π is a transcendental number, which is to say that it is not the root of any polynomial with rational coefficients. Thus π is not contained in *any finite* field extension of Q, much less one that has the appropriate degree. So the number π is thus not constructible and a circle cannot be squared.

Can all angles be trisected with ruler and compass? An angle α is constructible if and only if $\sin(\alpha)$ and $\cos(\alpha)$ are constructible. A $60°$ angle is certainly constructible since $\sin(60°) = \sqrt{3}/2$ and $\cos(60°) = 1/2$. But $20°$ is not constructible because $\cos(20°)$ is a root of the irreducible polynomial $8x^3 - 6x - 1$ and hence in a degree 3 extension of Q. Therefore, $60°$ is an angle that cannot be trisected.

SUMMARY

The tools of ancient geometry are a straightedge and a compass. Starting with a unit length for reference, the geometer can construct other lengths. These lengths measure the line segments between points of intersection of circles and lines drawn with the straightedge and compass. In this section, we characterized these lengths algebraically. **Constructible** lengths are sums, differences, products and quotients of integer multiples of the reference unit length. Most importantly, the square root of a

---

[2] Lindemann, F. "Über die Zahl π," *Math. Ann.* **20**, 213–225, 1882.

constructible number is constructible. But that's all! All constructible lengths arise this way. Interpreted algebraically, this means that any **constructible number** must be in a field extension of Q that has a degree that is a power of two, like square roots of square roots. Three ancient questions are now solved. The circle cannot be squared, the cube cannot be doubled, and in general, an angle cannot be trisected. In Section 6.5, we resolve a more recent problem, "Is the quintic solvable?"

6.3 Exercises

1. Using straightedge and compass, construct line segments of lengths $\sqrt{3}$ and $\frac{1+\sqrt{5}}{2}$.
2. Prove that $\cos(\theta)$ is constructible if and only if $\sin(\theta)$ is constructible.
3. Prove assertions II through V.
4. Prove that $8x^3 - 6x - 1$ is irreducible over Q.
5. Use the half-angle formula and the formulas for cosine of the sum and differences of angles to prove that $\cos(20°)$ is a root of $8x^3 - 6x - 1$. (Work with $20 + 40 = 60$.)
6. Use the results of Exercise 5 to show that $\cos(40°)$ is not constructible. Conclude that a regular 9-gon is not constructible.
7. Can a cube be tripled or quadrupled? How about 8-pled?
8. A regular n-gon is constructible if and only if $\cos\left(\frac{2\pi}{n}\right)$ is constructible.
    i. Prove that a regular hexagon is constructible.
    ii. Prove that a regular pentagon is constructible. (*Hint*: Consider the polynomial $4x^2 + 2x - 1$.)
9. Prove that a regular 7-gon (heptagon) is not constructible because $\cos\left(\frac{2\pi}{7}\right)$ is a root of the irreducible polynomial $8x^3 + 4x^2 - 4x - 1$.

To the Teacher

The worksheet that follows can easily be taken into the high school geometry class. For a fuller discussion of the role of such ancient techniques in the modern mathematics curriculum, see the *In the Classroom* essay, Section 6.6.

6.4 WORKSHEET 10: ON THE CONSTRUCTION OF REGULAR POLYGONS

In the *Elements*, Euclid presents constructions of the first three regular polygons, the equilateral triangle, the square and the regular pentagon. These constructions are accomplished with an unmarked straightedge and a compass. The first two are quite easy and are used frequently by Euclid in the subsequent parts of the *Elements*. The construction of the regular pentagon is much more difficult and requires a theorem from Book II of the *Elements*, namely Proposition 11. This proposition shows the relation of the pentagon to the golden ratio. Greek mathematicians did not use algebra in our sense. In many ways, constructions took the place of solving equations algebraically. Their importance went beyond the simple geometric ability to produce

Section 6.4   Worksheet 10: On the Construction of Regular Polygons

a certain figure. Problems that we would represent by an equation were considered solvable if a line segment of the appropriate length could be constructed. It took many years before the solution of polynomial equations was separated from geometric construction. Euclid closed the *Elements* in Book XIII with a study of the Platonic solids. He was interested in expressing the lengths of the various sides and diagonals of these solids in terms of the radius of a circumscribing sphere. Where we might express the length of a side of a regular dodecahedron as a number, or through an equation, Euclid constructed line segments of the proper length. The construction of the equilateral triangle, square, and regular pentagon are just the simplest examples of this quest.

A. The Equilateral Triangle

Euclid constructs an equilateral triangle with a given side in the very first proposition of the *Elements*. You will need a straightedge and a compass for the following exercises.

- a. Turn a blank sheet of paper so that its longest side is horizontal and draw a line segment about three inches long horizontally in the middle of the sheet. Label the endpoints of the segment A and B. Using the segment AB as a radius, draw a circle with center A and radius AB with the compass. Then draw a second circle with center B and radius AB. These two circles intersect at two points, one above and one below AB. Label the upper one C and the lower one D. The triangle ABC is an equilateral triangle as is the triangle ABD.
- b. What kind of figure is the quadrilateral $ACBD$?
- c. Connect C and D with a straight line. Then AB and CD are the diagonals of the quadrilateral $ACBD$. What are the relations between AB and CD? (Euclid based many of his constructions on the properties of the quadrilateral $ACBD$ and its diagonals.)

B. The Square

Euclid waits until Proposition 46 of Book I of the *Elements* to construct a square with a given side. (He needs it for Proposition 47, which is known to us as the Pythagorean Theorem.)

- a. Again turn a blank sheet of paper so that its longest side is horizontal and draw a line segment about three inches long horizontally in the middle of the sheet. Label the endpoints of the segment A and B. The next step is to construct a line at A perpendicular to AB. Extend AB in the direction from B to A. Using a radius of about AB draw a circle at A. Call E and F the points where this circle intersects the extended line AB. Now expand the compass a bit and draw circles centered on E and F. These circles will intersect above A at a point we will call G. Extend the segment AG if necessary so that one can find the point D above A such that $AD = AB$. Now repeat the construction of a segment at D perpendicular to AD. Find the point C on this line that lies above B with $DC = AD$. Let C and B be joined. Then $ABCD$ is a square.

C. The Regular Pentagon

To begin we need to analyze the regular pentagon. Draw an approximate regular pentagon and label the vertices A, B, C, D, and E with A at the top.

a. What is the sum of the measures of the interior angles of the pentagon? If you have forgotten, then draw in diagonals AC and AD and use the three triangles to compute the sum of the angles.
b. All the interior angles of the regular pentagon have the same measure. What is the measure of $\angle EAB$?
c. If you didn't before, now draw the diagonals AC and AD. This produces three isosceles triangles—$\triangle AED$, $\triangle ACD$, and $\triangle ABC$. Using the fact that these triangles are isosceles, compute the measures of the following angles:

$$m(\angle EAD) = \underline{\qquad}$$

$$m(\angle AED) = \underline{\qquad}$$

$$m(\angle CAD) = \underline{\qquad}$$

$$m(\angle ACD) = \underline{\qquad}$$

d. Draw diagonal BD, which intersects AC at a point F. What is the relation of $\triangle CFB$ to $\triangle ABC$? You may want to fill in more angle measures to figure this out. Note that $AF = AB$. Why?
e. Justify the following proportion.

$$AC : AF = AF : FC$$

f. The ratio in part e is the goal of Book II, Proposition 11 of the *Elements*.

> *Proposition* 11—To cut a given straight line so that the rectangle contained by the whole and one of the segments equals the square on the other segment.

(You may note that Euclid uses an infinitive clause for this construction. In the *Elements* the propositions are divided between constructions, which are written as infinitives, and theorems, which are written as declarative sentences.)

If we take AC as the whole segment, the area of the rectangle with side AC and FC equals the area of the square of side AF. The above ratio gives us $AC \times FC = AF^2$. Suppose we take AC to have length a and AF to have length x. Write the quadratic equation in x that the ratio gives. Solve the equation. What is the numerical ratio a/x? Can you name it?

g. We can now construct a regular pentagon with a given diagonal AB. All we need to do is to find the point given by Euclid's Proposition 11. Then the larger of the two segments determines the side of the pentagon. How does one construct the point? Draw a segment AB. Construct a segment perpendicular to AB that has A as an endpoint. Call the segment AD. Find a point C on AD such that $AC = \frac{1}{2}(AB)$. Find the point E on BC such that $CE = AC$. Then BE is the desired segment. That is the ratio $BE : AB$ found in part f. Why?

h. To complete the construction of a regular pentagon, draw a segment with length equal to BE from part g. With the compass set to a radius of AB draw circles with center B and center E. Let F be their point of intersection above BE. There is a regular pentagon with B, E, and F as three of the five vertices and BF and EF as diagonals. Complete the construction.

In the *Elements* Euclid devotes much of his effort to determining which numbers are constructible. Book X is a very long and complicated treatise on constructible segments. With the rise of algebra in Western Europe in the 1500s there was a change in the kinds of questions asked by mathematicians about lengths and numbers. While constructible lengths are developed using the four operations and square roots, algebraists expanded the kinds of numbers they were interested in to those that could be expressed with the four operations and the extraction of roots of any order. The questions then became, "What polynomials have roots that can be expressed in radical form?" In his major work *Disquisitiones*, Gauss shows that all the various roots of unity are expressible in terms of roots of integers. On the way he is able to determine which regular polygons are constructible with straightedge and compass. He showed that a regular polygon with n sides is constructible if and only if $n = 2^a(2^{2^b} + 1)$ for some integers a and b.

6.5 THE SOLVABILITY OF POLYNOMIAL EQUATIONS

In this book, we have taken the study of polynomials to a far deeper level than encountered in secondary mathematics. The questions we have asked about polynomials are familiar: How do they factor? Where are their roots? How do we find them? By considering polynomials with coefficients in a variety of different rings, we have enlarged the context and increased the complexity of the answers. Just apply any of the preceding questions to the polynomial $x^n + x^{n-1} + \ldots + 1$ and consider how your technique and conclusions compare in the context of $Z_2[x]$ versus $C[x]$. How does all of this relate to the group theory presented in Chapters 4 and 5? The study of mathematical structures like groups has often been called "Modern Algebra" to distinguish it from the algebra that springs from the quadratic formula. But group theory had its origins in the service of answering fundamental questions about solving polynomials. In 1831[3], a young mathematician, Évariste Galois, established the connection between the group structure of S_5 and the nonsolvability of the general fifth-degree polynomial. In doing so, he solved an important old problem and launched a new mathematical discipline—the study of abstract structures. In this section, we hope to give the reader a feel for how group structures relate to the study of polynomials and how information is garnered. The full story is deep and complex, beyond the scope of this introductory text. What follows will be a bit sketchy. Even so, the results are interesting and not at all trivial.

Our goal in the next few pages is to show how we can answer questions about the solvability of polynomials with the tools of group theory. To keep the treatment

---

[3]Galois' paper, *Memoire sur les conditions de resolubilite des equations par radicaux* (1831), can be found in *Ecrits et Memoires mathematiques d'Evariste Galois*, Bourgne and Azra, Eds. Paris: Gauthier-Villars, 1962.

as accessible as possible, all fields considered from here on will be subfields of the complex numbers **C**. As such, each field F will have characteristic zero and hence be an extension field of **Q** (i.e., $\mathbf{Q} \subseteq F \subseteq \mathbf{C}$).

An **automorphism** φ of a field F is an isomorphism of the form $\varphi \colon F \to F$. We denote the set of all automorphisms of a field F as **Aut(F)**. As with any isomorphism, an automorphism must be a bijection that preserves addition and multiplication, and it must take identities to identities. Since $\varphi(1 + 1 + 1 + \ldots + 1) = \varphi(1) + \varphi(1) + \ldots + \varphi(1) = 1 + 1 + \ldots + 1$, we have $\varphi(m) = m$ for each natural number m. Since inverses map to inverses, $\varphi(m/n) = m/n$ for any rational number of the subfield $\mathbf{Q} \subseteq F$. To emphasize this fact, we say that φ **fixes** the elements of **Q**.

> **DEFINITION 1.** Suppose that $F \subseteq E$ is a field extension. An automorphism $\sigma \in \mathrm{Aut}(E)$ **fixes** $x \in F$ if $\sigma(x) = x$. If σ fixes x for all $x \in F$, we say that σ **fixes** F. The set of automorphisms of E that fix F is denoted by **Aut(E, F)**.

In symbols, $\mathrm{Aut}(E, F) = \{\sigma \in \mathrm{Aut}(E) : \sigma(x) = x \text{ for each } x \in F\}$. For any field $E \subseteq \mathbf{C}$, $\mathrm{Aut}(E) = \mathrm{Aut}(E, \mathbf{Q})$. But an automorphism can fix other subfields of **C** as well.

EXAMPLE 1

Let $\varphi \colon \mathbf{C} \to \mathbf{C}$ be the map defined by $\varphi(a + bi) = a - bi$. It is easy to verify that φ is a bijection that preserves addition and multiplication. Thus $\varphi \in \mathrm{Aut}(\mathbf{C})$. Since $\mathbf{R} = \{a + bi : b = 0\}$ and since $\varphi(a + 0i) = a$, we see that φ not only fixes all the elements of **Q** but all elements of **R** as well. Thus $\varphi \in \mathrm{Aut}(\mathbf{C}, \mathbf{R})$. ∎

The following proposition relates the study of field extensions to the study of groups.

> **PROPOSITION 1** Aut(E, F) is a group under the operation of composition.

Proof. If φ and γ are automorphisms of E that fix F and if $x \in F$, then $\varphi \circ \gamma(x) = \varphi(\gamma(x)) = \varphi(x) = x$, which shows that $\mathrm{Aut}(E, F)$ is closed under composition. Composition is an associative operation. The identity map on E is certainly in $\mathrm{Aut}(E, F)$. If $\varphi \in \mathrm{Aut}(E, F)$, then its inverse φ^{-1} is an automorphism of E. Let $x \in F$. Since $\varphi(x) = x$, $\varphi^{-1}(x) = x$ and φ^{-1} is an automorphism in $\mathrm{Aut}(E, F)$. Thus $\mathrm{Aut}(E, F)$ is a group under composition. ▲

EXAMPLE 2

Let $\gamma \in \mathrm{Aut}(\mathbf{C}, \mathbf{R})$. Then $\gamma(-1) = -1$ since γ fixes members of **R**. Since γ is operation preserving, $\gamma(-1) = \gamma(i^2) = (\gamma(i))^2 = -1$ from which it follows that $\gamma(i) = i$ or $\gamma(i) = -i$. If $\gamma(i) = i$, then $\gamma(a + bi) = a + bi$ and γ is the identity map. If $\gamma(i) = -i$, then $\gamma(a + bi) = a - bi$ and $\gamma = \varphi$ as defined in Example 1. Thus there are only two automorphisms in $\mathrm{Aut}(\mathbf{C}, \mathbf{R})$, the identity map and φ. The group $\mathrm{Aut}(\mathbf{C}, \mathbf{R})$ is isomorphic to the group $\{\mathbf{Z}_2, +\}$. ∎

In what follows, the fields that we study will be associated with polynomials. They will be fields like $F[x]/\langle p(x) \rangle$ or like $F(\alpha_1, \ldots, \alpha_n)$, where the members of the set $\{\alpha_1, \ldots, \alpha_n\}$ are the roots of a polynomial in $F[x]$. The groups of automorphisms of such fields establish the basic link between the study of groups and the study of polynomials. We shall start by investigating how to deduce the structure of various automorphism groups like $\text{Aut}(F(\alpha_1, \ldots, \alpha_n), F)$. Then we shall reverse our point of view: We shall look to see what light the structure of the group can shed on the nature of the roots. It is this reversal that ushered in the modern study of abstract structures.

The next proposition provides us with a handy tool for investigating the structure of $\text{Aut}(F(\alpha_1, \ldots, \alpha_n), F)$.

PROPOSITION 2 Let F be a subfield of E. Suppose that $f(x) \in F[x]$ and that $y \in E$ is a root of $f(x)$. Let $\varphi \in \text{Aut}(E, F)$. Then $\varphi(y)$ is a root of $f(x)$.

Proof. Suppose that $f(x) = a_0 + a_1 x + \ldots + a_n x^n$ has coefficients in F and that $y \in E$. Since φ is operation preserving and fixes all the coefficients of $f(x)$, we have $0 = \varphi(0) = \varphi(a_0 + a_1 y + \ldots + a_n y^n) = a_0 + a_1 \varphi(y) + \ldots + a_n (\varphi(y))^n$. Thus $\varphi(y)$ is a root of $f(x)$. ▲

If $F \subseteq E$ is a **finite** field extension, Proposition 2 limits the number of possible automorphisms in $\text{Aut}(E, F)$. The Theorem 6 of Section 6.2 guarantees that we can find an element $a \in E$ such that $E = F(a)$ where a is algebraic over F. Let $p(x) \in F[x]$ be a minimal polynomial of a and suppose that $\deg(p(x)) = n$. Each element $y \in E$ can be expressed uniquely as $y = c_0 + c_1 a + c_2 a^2 + \ldots + c_{n-1} a^{n-1}$ where each $c_i \in F$. Let φ be in $\text{Aut}(E, F)$. Because φ fixes F, we have $\varphi(c_0 + c_1 a + c_2 a^2 + \ldots + c_{n-1} a^{n-1}) = c_0 + c_1 \varphi(a) + c_2 \varphi(a)^2 + \ldots + c_{n-1} \varphi(a)^{n-1}$. Thus any automorphism $\varphi \in \text{Aut}(E, F)$ is completely determined by the value of $\varphi(a)$. From Proposition 2, we know that $\varphi(a)$ must be a root of $p(x)$. Since $p(x)$ has at most n roots in E, there are at most n distinct automorphisms of E that fix F. We also know that $\deg(p(x)) = [E : F]$. Thus we have proved the following proposition.

PROPOSITION 3 Let $F \subseteq E$ be a finite field extension. Then $|\text{Aut}(E, F)| \leq [E : F]$.

EXAMPLE 3

Let $E = \mathbf{Q}(2^{1/3})$. So $\mathbf{Q} \subseteq E \subseteq \mathbf{R}$. Any automorphism in $\text{Aut}(E, \mathbf{Q})$ must map $2^{1/3}$ to another root of $x^3 - 2$ in $\mathbf{Q}(2^{1/3})$. But there are no other roots of $x^3 - 2$ in $\mathbf{Q}(2^{1/3})$ because the other roots are complex numbers. Thus $\text{Aut}(E, \mathbf{Q})$ has only one element, the identity map. In this example, $|\text{Aut}(E, \mathbf{Q})| < [E : \mathbf{Q}]$. ∎

EXAMPLE 4

Let $E = \mathbf{Q}(\sqrt{2})$ and $F = \mathbf{Q}$. It is easy to check that the map $\varphi(a + b\sqrt{2}) = a - b\sqrt{2}$ is an automorphism in $\text{Aut}(E, F)$. Any automorphism of E fixes elements of $\mathbf{Q}$ and is thus completely determined by where it maps $\sqrt{2}$. Since there are only two choices,

$\sqrt{2} \to \sqrt{2}$ and $\sqrt{2} \to -\sqrt{2}$, there are only two elements in Aut(E, F), the identity map id and φ. In this example, $|\text{Aut}(E, F)| = [E : F]$. ∎

EXAMPLE 5

Let $E = \mathbf{Q}(\sqrt{2}, \sqrt{3})$, which has degree 4 over $\mathbf{Q}$. The elements of E are all of the form $a + b\sqrt{2} + c\sqrt{3} + d\sqrt{2}\sqrt{3}$, which we can abbreviate as (a, b, c, d). We can think of E in two ways, either as the field $\mathbf{Q}(\sqrt{2})$ with $\sqrt{3}$ adjoined or as the field $\mathbf{Q}(\sqrt{3})$ with $\sqrt{2}$ adjoined. Let ρ be the automorphism of E that fixes $\mathbf{Q}(\sqrt{3})$ and takes $\sqrt{2}$ to $-\sqrt{2}$ so that $(a, b, c, d) \to (a, -b, c, -d)$. Similarly let σ be the automorphism of E that fixes $\mathbf{Q}(\sqrt{2})$ and takes $\sqrt{3}$ to $-\sqrt{3}$ so that $(a, b, c, d) \to (a, b, -c, -d)$. The compositions $\sigma \circ \rho$ and $\rho \circ \sigma$ are equal. In each case, $(a, b, c, d) \to (a, -b, -c, d)$. So Aut$(E, \mathbf{Q}) = \{id, \sigma, \rho, \sigma \circ \rho\}$. It is isomorphic to the Klein 4-Group. ∎

Suppose that $f(x)$ is a polynomial with coefficients in a field $F \subseteq \mathbf{C}$ and that $E = F(\alpha_1, \alpha_2, \ldots, \alpha_n)$ where the $\alpha_i \in \mathbf{C}$ are the n distinct roots of $f(x)$ that do not already lie in F. Then E is called the **splitting field of $f(x)$**. The polynomial f has all its roots in E. We say that f **splits in E** because it factors into the product of linear terms (degree one polynomials) in $E[x]$. Note that f does not split in any subfield of E. By Proposition 2, any automorphism $\sigma \in \text{Aut}(E, F)$ simply permutes the elements of the set $\{\alpha_1, \alpha_2, \ldots, \alpha_n\}$. In fact, since any element of $F(\alpha_1, \alpha_2, \ldots, \alpha_n)$ can be written as the sum of products of members of F and powers of the elements of $\{\alpha_1, \alpha_2, \ldots, \alpha_n\}$ and since σ fixes F, σ is completely determined by its value on the set $\{\alpha_1, \alpha_2, \ldots, \alpha_n\}$. We can think of S_n as the set of all permutations of the n elements in the set $\{\alpha_1, \alpha_2, \ldots, \alpha_n\}$. The next proposition links the structure of S_n to Aut(E, F).

PROPOSITION 4 Suppose that E is a splitting field of the polynomial $f(x) \in F[x]$ and consider S_n to be the group of permutations of the n distinct roots of $f(x)$. Let $\Phi: \text{Aut}(E, F) \to S_n$ be defined by letting $\Phi(\sigma)$ be the restriction of σ to the roots of $f(x)$. Then Φ is a group isomorphism from Aut(E, F) onto a *subgroup* of S_n.

Proof. The map Φ is injective because any two automorphisms of E that fix F and agree on all the roots of $f(x)$ are identical. If σ and ρ are automorphisms that fix F, and α_i is a root of $f(x)$, then $\Phi(\sigma \circ \rho)(\alpha_i) = \sigma \circ \rho(\alpha_i) = \Phi(\sigma) \circ \Phi(\rho)(\alpha_i)$. So Φ is operation preserving. ▲

Our next example shows that even though a polynomial may have n roots in E, Aut(E, F) can be a proper subgroup of S_n.

EXAMPLE 6

The field $E = \mathbf{Q}(\sqrt{2}, \sqrt{3})$, considered in Example 5, is the splitting field of the polynomial $p(x) = (x^4 - 5x^2 + 6) = (x^2 - 2)(x^2 - 3)$. But not all permutations of its four roots are possible since, for instance, $\sqrt{2}$ cannot be mapped to $\sqrt{3}$ because it must map to a root of the factor $x^2 - 2$. Thus $\text{Aut}(E, \mathbf{Q})$ is a proper subgroup of S_4. As we saw in Example 5, $\text{Aut}(E, \mathbf{Q})$ has order 4 rather than 4!. ∎

Now we consider an example where K is the splitting field of an *irreducible* polynomial.

EXAMPLE 7

The splitting field K for the polynomial $x^3 - 2$ over $\mathbf{Q}$ is the field $\mathbf{Q}(2^{1/3}, \sqrt{-3})$ and $\deg(K, \mathbf{Q}) = 6$. To show that $\text{Aut}(K, \mathbf{Q})$ is the full group S_3, we will construct an isomorphism from $\text{Aut}(K, \mathbf{Q})$ to S_3 explicitly. Let $\xi = \frac{-1+\sqrt{-3}}{2}$ be a cube root of unity and recall that $\xi^2 = \frac{-1-\sqrt{-3}}{2}$ and that ξ^2 is also a root of unity. Note that $K = \mathbf{Q}(\sqrt[3]{2}, \xi)$ and so we have $\mathbf{Q} \subseteq \mathbf{Q}(\xi) \subseteq K$. Also K is the splitting field of $x^3 - 2$ over $\mathbf{Q}(\xi)$. There are only two automorphisms of $\mathbf{Q}(\xi)$. They are the identity map and the map that sends ξ to its complex conjugate ξ^2. We can extend each of these to K in three ways, thus obtaining the six automorphisms of K.

$$\sigma_1: \xi \to \xi \quad \text{and} \quad \sqrt[3]{2} \to \sqrt[3]{2} \quad \text{(This is the identity map.)}$$
$$\sigma_2: \xi \to \xi \quad \text{and} \quad \sqrt[3]{2} \to \xi\sqrt[3]{2}$$
$$\sigma_3: \xi \to \xi \quad \text{and} \quad \sqrt[3]{2} \to (\xi^2)\sqrt[3]{2}$$
$$\sigma_4: \xi \to \xi^2 \quad \text{and} \quad \sqrt[3]{2} \to \sqrt[3]{2}$$
$$\sigma_5: \xi \to \xi^2 \quad \text{and} \quad \sqrt[3]{2} \to \xi\sqrt[3]{2}$$
$$\sigma_6: \xi \to \xi^2 \quad \text{and} \quad \sqrt[3]{2} \to (\xi^2)\sqrt[3]{2}$$

Let $a = \sqrt[3]{2}$, $b = \xi\sqrt[3]{2}$ and $c = (\xi^2)\sqrt[3]{2}$. The six automorphisms of K correspond to the following six permutations of the symbols a, b, and c. It is not difficult to check that the bijection is operation preserving.

$$\sigma_1 : \begin{pmatrix} a & b & c \\ a & b & c \end{pmatrix} \quad \sigma_2 : \begin{pmatrix} a & b & c \\ b & c & a \end{pmatrix} \quad \sigma_3 : \begin{pmatrix} a & b & c \\ c & a & b \end{pmatrix}$$
$$\xi \to \xi \qquad\qquad \xi \to \xi \qquad\qquad \xi \to \xi$$

$$\sigma_4 : \begin{pmatrix} a & b & c \\ a & c & b \end{pmatrix} \quad \sigma_5 : \begin{pmatrix} a & b & c \\ b & a & c \end{pmatrix} \quad \sigma_6 : \begin{pmatrix} a & b & c \\ c & b & a \end{pmatrix}$$
$$\xi \to \xi^2 \qquad\qquad \xi \to \xi^2 \qquad\qquad \xi \to \xi^2$$
∎

Example 6 points to a more general assertion presented in the next proposition. It is given without proof. However, we will need its conclusion later.

PROPOSITION 5 Suppose that $p(x)$ is an *irreducible* polynomial in $F[x]$ and that $K \subseteq \mathbf{C}$ is its splitting field. Then $|\text{Aut}(K, F)| = \deg[K : F]$.

Next we determine the structure of the group $\text{Aut}(K, F)$ in the case that $F \subseteq K$ is a prime radical extension so that $K = F(\alpha)$, where α is the root of $x^p - a$, and $a \in F, a \neq 1$. This is the first step to linking group theory to questions about the solvability of polynomials by radicals.

PROPOSITION 6 Let p be a prime and suppose that F is a field that contains all the pth roots of unity. Suppose also that $a \in F$ and that $f(x) = x^p - a$ is irreducible in $F[x]$. Suppose that $\alpha \in C$ is a root of $x^p - a$ and that $K = F(\alpha)$. Then $\mathrm{Aut}(K, F)$ is isomorphic to the cyclic group $\{\mathbf{Z_p}, +\}$.

Proof. Since F contains the pth roots of unity, all p roots of $x^p - a$ are in K. These roots are the of the form $\xi^i \alpha \in K$, $i = 0, \ldots, p-1$, where $\xi \neq 1$ is a pth root of unity. Every automorphism in $\mathrm{Aut}(K, F)$ must map α to $\xi^i \alpha$ for some i. For each $i = 0, \ldots, p-1$, let φ_i be the map defined by $\varphi_i(a_0 + a_1 \alpha + \ldots + a_{p-1} \alpha^{p-1}) = a_0 + a_1(\xi^i \alpha)^1 + a_2(\xi^i \alpha)^2 + \ldots + a_{p-1}(\xi^i \alpha)^{p-1}$. Clearly, φ_i preserves addition. To see that it preserves multiplication, note that $a_k b_j (\xi^i \alpha)^{k+j} = a_k b_j (\xi^i \alpha)^k (\xi^i \alpha)^j$. The inverse of φ_i is the map φ_j where $i + j = p$. Thus each φ_i is an automorphism and, since there are no others, the order of the group $\mathrm{Aut}(K, F)$ is p. Therefore, $\mathrm{Aut}(K, F)$ must be isomorphic to $\{\mathbf{Z_p}, +\}$. ▲

The above proposition shows us that the structure of the automorphism group $\mathrm{Aut}(F(\alpha), F)$ of a prime radical extension is particularly simple—it is a cyclic group. Now we turn our attention to the structure of $\mathrm{Aut}(F_n, F)$, where F_n is the last field in a prime radical tower over F. Recall that $F = F_0 \subseteq F_1 \subseteq \ldots \subseteq F_n$ is a prime radical tower if each $F_i \subseteq F_{i+1}$ is a prime radical extension. The roots of a polynomial in $F[x]$ that is solvable by radicals are found in the last field of such a tower. So knowing the structure of $\mathrm{Aut}(F_n, F)$ is the key step in relating group theory to the question of the solvability of polynomials. To that end, for each $i = 1, \ldots, n$, let $a_i \in F_{i-1}$ and let p_i be a prime such that $x^{p_i} - a_i$ is irreducible over F_{i-1}. Let $F_i = F_{i-1}(\sqrt[p_i]{a_i})$ and let $\mathbf{G_i} = \mathbf{Aut}(F_n, F_i)$ for $i = 0, \ldots, n$. The group G_n is the group of automorphisms of F_n that fix F_n. It has only one element, the identity element. We are interested in the group $G_0 = \mathrm{Aut}(F_n, F_0)$. Notice that we have $G_n \subseteq G_{n-1} \subseteq \ldots \subseteq G_0$ because if an automorphism is in G_i and fixes the elements of F_i, it must certainly fix an element of the subfield $F_{i-1} \subseteq F_i$. The next theorem uses the notation established in this paragraph.

THEOREM 7 Suppose that $F_0, F_1, \ldots, F_n$ are subfields of C and that $F_0 \subseteq F_1 \subseteq \ldots \subseteq F_n$. For each $i = 1, \ldots, n$, suppose that $F_i = F_{i-1}(\sqrt[p_i]{a_i})$ where $a_i \in F_{i-1}$, the integer p_i is a prime, and $x^{p_i} - a_i$ is irreducible over F_{i-1}. Assume that the base field F_0 contains all the pth roots of unity for $p = p_i$, $i = 1, \ldots, n$. For each $i = 0, \ldots, n-1$, G_{i+1} is a normal subgroup of G_i and the quotient group G_i / G_{i+1} is cyclic.

Proof. Let $f \in G_i$ and $g \in G_{i+1}$ be automorphisms that fix F_i and F_{i+1}, respectively. To show that G_{i+1} is a normal subgroup of G_i, we must show that $f^{-1} \circ g \circ f \in G_{i+1}$, which means that $f^{-1} \circ g \circ f$ is an automorphism of F_n that fixes elements of F_{i+1}. Let $y \in F_{i+1}$. Then $y = b_0 + b_1 \alpha + \ldots + b_{p-1} \alpha^{p-1}$, where $p = p_{i+1}$, $\alpha^p = a_{i+1}$, and $b_j \in F_i$ for $j = 0, \ldots, p-1$. Since f fixes elements of F_i, $f(y) = b_0 + b_1 f(\alpha) + \ldots + b_{p-1} f(\alpha)^{p-1}$. Since $f(\alpha)$ must be a root of $x^p - a_{i+1}$, we know that $f(\alpha) = \xi \alpha$ where $\xi \in F_{i+1}$ is a pth root of unity. Thus $f(y) \in F_{i+1}$.

Since g fixes elements of $F_{i+1}, g(f(y)) = f(y)$, and we have $f^{-1} \circ g \circ f(y) = f^{-1} \circ f(y) = y$, which proves normality. We have also shown that $f(F_{i+1}) \subseteq F_{i+1}$.

Next we show that G_i/G_{i+1} is isomorphic to a subgroup of $\text{Aut}(F_{i+1}, F_i)$. To do this we construct a homomorphism $\Phi: \text{Aut}(F_n, F_i) \to \text{Aut}(F_{i+1}, F_i)$ that has kernel $\text{Aut}(F_n, F_{i+1})$. Let $f \in \text{Aut}(F_n, F_i)$ and define $\Phi(f)$ to be the restriction of f to F_{i+1}. As noted above, $f(F_{i+1}) \subseteq F_{i+1}$ so that Φ is well-defined. The kernel of Φ is the set of all automorphisms that restrict to the identity map on F_{i+1}, that is, $\ker(\Phi) = \text{Aut}(F_n, F_{i+1}) = G_{i+1}$. By the First Isomorphism Theorem (see Section 5.2), G_i/G_{i+1} is isomorphic to a subgroup of $\text{Aut}(F_{i+1}, F_i)$. Since $\text{Aut}(F_{i+1}, F_i)$ is cyclic, all of its subgroups are cyclic. Thus G_i/G_{i+1} is cyclic. ▲

In Section 5.3, we called a group G "solvable" if we could find a finite chain of subgroups $\{e\} = G_n \subseteq G_{n-1} \subseteq \ldots \subseteq G_0 = G$ such that $G_{i+1} \triangleleft G_i$ for $i = 0, \ldots, n-1$, and such that the quotient groups G_{i+1}/G_i are cyclic. So, under the hypotheses of Theorem 7, we can say that $\text{Aut}(F_n, F)$ is a solvable group. We have proved the following statement in the special case when F contains the necessary roots of unity. (The statement actually holds true more generally.)

> If a polynomial is *solvable by radicals*, then the automorphism group of its splitting field must be a *solvable group*.

This result is just a small part of the theory of Galois that links the study of polynomials to the study of groups. But it is enough to get us a very significant result: Not all polynomials are solvable by radicals. Why? In the next example, we investigate a specific fifth degree polynomial. We will see that the automorphism group of its splitting field is S_5, which is *not* a solvable group. (Example 5 of Section 5.3 shows that S_5 is not solvable.) Notice how many concepts explored in the previous pages are utilized in Example 8.

EXAMPLE 8

Let $F \subseteq C$ be the field Q with the third and fifth roots of unity adjoined. Let $f(x) = x^5 - 6x + 3 \in F[x]$. Applying Eisenstein's irreducibility criterion with $p = 3$, we see that $f(x)$ is irreducible over Q and has no roots in Q. A fairly straightforward degree argument, guided in the exercises, shows that $f(x)$ is irreducible over F as well. An application of the Euclidean Algorithm to $f(x)$ and its derivative $f'(x) = 5x^4 - 6$ shows that their greatest common divisor is 1. (See Section 3.3.) Thus f has five distinct roots $\{\alpha_1, \alpha_2, \alpha_3, \alpha_4, \alpha_5\}$ in C. Let $K = F(\alpha_1, \alpha_2, \alpha_3, \alpha_4, \alpha_5) \subseteq C$ be the splitting field of $f(x)$. The degree of the extension $F \subseteq K$ must be divisible by 5 since $F \subseteq F(\alpha_1) \subseteq K$ and the degree of the extension $F \subseteq F(\alpha_1)$ is 5. Denote $\text{Aut}(K, F)$ by G. By Proposition 5, $|G| = [K : F]$ and so $|G|$ is divisible by 5. By Cauchy's Theorem (Theorem 7 in Section 5.2), G must have an element of order 5. Since G is a subgroup of S_5, regarded as the group of permutations of the set $\{\alpha_1, \alpha_2, \alpha_3, \alpha_4, \alpha_5\}$, G contains a 5-cycle.

Since the complex roots of a polynomial with real coefficients come in conjugate pairs, the polynomial $f(x)$ has 1, 3, or 5 real roots. If it had 5 real roots, its derivative, $f'(x) = 5x^4 - 6$, would have four real roots, which it does not. Since $f(-2) = -17$, $f(0) = 3$, $f(1) = -2$ and $f(2) = 23$, $f(x)$ has at least three real roots by the Intermediate Value Theorem. Thus $f(x)$ has two complex roots that are complex conjugates. The automorphism of C that takes $z \in C$ to its complex conjugates restricts to an automorphism of the splitting field K, leaves the real roots fixed and interchanges the complex roots. It is a transposition or 2-cycle. We saw in Example 6 of Section 5.2, that S_5 is generated by a transposition and a 5-cycle. So $G = S_5$. Since S_5 is not a solvable group, the polynomial $f(x)$ is **NOT** solvable by radicals. ∎

The implications are interesting: the roots of this polynomial are not expressible in terms of square roots of fifth roots of cube roots, etc. of rational numbers. (Ask a CAS like Maple to solve it exactly, not with numerical approximations. It gives up!) So these roots are not transcendental numbers like π or e that are not roots of any polynomial with integer coefficients, but they are also not expressible in terms of radicals as, say, the complex cube roots of unity are: $\frac{-1 \pm \sqrt{-3}}{2}$. We call numbers that are roots of polynomials with rational coefficients "algebraic numbers." The roots of a polynomial that is not solvable are indeed algebraic numbers, but they are not otherwise easy to characterize.

SUMMARY

A goal of this book is to draw the connection between old-fashioned algebra, the algebra of polynomials studied in high school, with abstract algebra, the study of algebraic structures. The goal is achieved in this chapter. The old problem of whether we can solve higher-degree polynomials in terms of radicals (as we can do with quadratics) is solved not computationally, but through group theory and ring theory.

Here is the argument in brief. If the roots of an irreducible polynomial $p(x)$ in $F[x]$ are not in F, then we can find those roots in an algebraic **field extension** K of F, where $K = F(\alpha_1, \alpha_2, \ldots, \alpha_n)$ and $\alpha_1, \alpha_2, \ldots, \alpha_n$ are the roots of $p(x)$. Any **automorphism** of K that **fixes** F must simply permute the roots of $p(x)$. The set of such automorphisms, $\text{Aut}(K, F)$, is a group under composition. This is the link. From the structure of the group $\text{Aut}(K, F)$, we can deduce whether or not the polynomial $p(x)$ is **solvable by radicals**. For $p(x)$ to be solvable by radicals, the group $\text{Aut}(K, F)$ must be a solvable group. We found that there are quintic polynomials for which $\text{Aut}(Q(\alpha_1, \alpha_2, \ldots, \alpha_5)) = S_5$, which is *not* a solvable group. Therefore, are polynomials of degree five and higher that are not solvable by radicals. With these observations, Galois ushered in a new era in the study of algebra.

The technical apparatus that we developed to study field extensions, namely our observations about the **degree** of algebraic extensions, allowed us to solve problems of "solvability" left to us by ancient Greek mathematicians. In a sense, to be "solvable" in the mathematics of Euclid was to be constructible. The tools that helped determine the nonsolvability of the quintic also determined the nonconstructability of the lengths needed to square the circle, double the cube, and trisect the angle. Thus this chapter looks back in time as well as pushes forward the context in which we can view the familiar algebra of high school.

6.5 Exercises

1. Determine the group of automorphisms that fix Q for each of the following field extensions:
   i. $Q(\sqrt{2}, \sqrt{3}, \sqrt{5})$
   ii. $Q(\sqrt[5]{2})$, where $\sqrt[5]{2}$ is a real number
   iii. $Q(\sqrt{-3})$
2. Determine $\text{Aut}(F(\sqrt[5]{2}), F)$, where $F = Q(\zeta)$ and $\zeta^5 = 1$, $\zeta \neq 1$.
3. Let p be a prime and let $\zeta \in C$ be a primitive root of unity such that $\zeta^p = 1$ and $\zeta \neq 1$. Prove that $\text{Aut}(Q(\zeta))$ is isomorphic to U_p and hence cyclic.
4. Let $K \subseteq C$ be the splitting field of $x^5 - 2$.
   i. Show that $[K : Q] = 20$. (Recall that a primitive fifth root of unity satisfies a fourth-degree irreducible polynomial.)
   ii. Deduce that $|\text{Aut}(K)| = 20$.
   iii. With Example 7 as a model, characterize each of the members of $\text{Aut}(K)$.
   iv. Again with Example 7 as a model, express each automorphism of K as a permutation of the roots of $x^5 - 2$.
   v. Show that $\text{Aut}(K)$ is not abelian.
   vi. Show that $\text{Aut}(K)$ is a solvable group.
5. Prove that the polynomial $f(x) = x^5 - 6x + 3$ of Example 8 is irreducible over F, where F is the field Q with the third and fifth roots of unity adjoined. Here are some hints:
   i. Show that $[F: Q]$ is a power of 2.
   ii. If $f(x)$ factors over F, and has a linear factor of the form $(x - \alpha)$, consider the degree implications of $Q \subseteq Q(\alpha) \subseteq F$.
   iii. Suppose $f(x)$ factored as the product of an irreducible cubic and an irreducible quadratic and α is a root of one of these. Let $K = F(\alpha)$ and consider the degree implications of $Q \subseteq Q(\alpha) \subseteq K$.

To the Teacher

We hope that what you have learned in this book provides you with a new confidence in what you can teach and a deep appreciation of the depth and richness of what you will teach. Good luck!

6.6 IN THE CLASSROOM: CONSTRUCTIONS

In this chapter we have explored which numbers are constructible and have used that knowledge to provide algebraic techniques to resolve the three famous construction problems of trisecting an angle, squaring a circle, and doubling a cube using only a straightedge and compass. For thousands of years, mathematicians were fascinated with these problems. Gauss had stated that doubling the cube and trisecting the angle with compass and straightedge were impossible but gave no proofs. In 1837 Pierre Wantzel published the proofs. Carl Louis Ferdinand von Lindemann in 1882 proved π is transcendental which was what was needed to prove that squaring the circle with a compass and straightedge was insolvable. Students today may have trouble understanding the fascination that so many people had concerning whether these constructions could actually be done using only a compass and straightedge. In fact, today's students are likely to wonder why constructions are included at all in the secondary curriculum. These students are not used to having resources restricted to them. The idea of limiting oneself to a ruler that has no markings and a compass to create geometric figures may seem silly to them. In this essay we will look at the inclusion of geometric constructions in the secondary mathematics curriculum, and the value to the curriculum beyond their role as a historical topic in mathematics.

We begin by consulting the National Council of Teachers of Mathematics' *Principles and Standards for School Mathematics*, which includes geometric constructions as a topic for the secondary curriculum in the last bulleted item of the following Geometry Standard:

> *Instructional programs from prekindergarten through grade 12 should enable all students to*
>
> o *Analyze characteristics and properties of two- and three-dimensional geometric shapes and develop mathematical arguments about geometric relationships;*
> o *Specify locations and describe spatial relationships using coordinate geometry and other representational systems;*
> o *Apply transformations and use symmetry to analyze mathematical situations;*
> o *Use visualization, spatial reasoning, and geometric modeling to solve problems. (NCTM, 2000, p. 41)*

The 9–12 grade band level's detailed expectations for this last bullet include the statement that students should *draw and construct representations of two- and three-dimensional geometric objects using a variety of tools* (NCTM, 2000, p. 308). As a major influence of state and local standards, the inclusion of constructions in this document speaks of their viewed importance as a topic in the secondary mathematics curriculum.

We now look at what is covered regarding constructions in a secondary mathematics curriculum to meet this standard and its expectations. A look at current popular textbooks reveals that work on constructions may include constructions of

- A segment congruent to a given segment
- A triangle congruent to a given triangle
- An angle bisector
- An angle congruent to a given angle
- A perpendicular bisector of a given segment
- A midpoint of a given segment
- The perpendicular to a line through a given point (on the line and not on the line)
- The parallel to a line through a given point not on the line
- An equilateral triangle

- An isosceles triangle
- A translation of a segment given a translation vector
- A rotation of a segment about a point by a given angle
- A reflection of a segment across a given line
- Centroid, incenter, circumcenter, and orthocenter of a triangle
- Circle that circumscribes a given triangle

We recall that NCTM promoted students' ability to do constructions such as the ones above with "a *variety* of tools" (NCTM, 2000, p. 308). Typically, this includes a compass and straightedge, paper folding, and increasingly some dynamic geometry software program, such as *The Geometer's Sketchpad* (Jackiw, 1995) and *Cabri Geometry II* (Texas Instruments, 1994). While easy accessibility to computers on which to use these programs may be an issue for some high schools, many schools have incorporated handheld graphing calculators into their program. Several of these graphing calculators, including the TI-89, TI-92 Plus, and Voyage 200 have applications for *The Geometer's Sketchpad* and *Cabri Geometry*, and a *Cabri Jr.* application is available for the TI-83 Plus and TI-84 Plus.

What do these programs have to offer the teacher and student? First, these programs allow students to make any construction they would have done by hand with the ease of computerized tools. There is no more slipping of the arm of the compass or puncturing the paper with the point or smearing all the pencil marks made during the construction. The use of software programs such as *The Geometer's Sketchpad* can also be beneficial to students with special needs (Shaw, Durden, and Baker, 1998). These programs have various construction tools located in the "tool bar". In Figure 1, taken from a TI-92 Plus with *The Geometer's Sketchpad*, the tools going down the right-hand side of the screen include a **Selection Arrow** tool, a **Point** tool, a **Compass** tool, a **Straightedge** tool, a **Text** tool, and a **Custom** tool. Students can use the Point, Compass, and Straightedge tools to make any constructions they could have done by hand with an actual compass and straightedge.

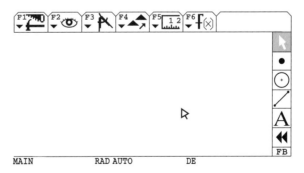

FIGURE 1

Second, once these basic compass and straightedge tools are moved to the computer, there is an opportunity to provide other built-in tools that automate a variety of constructions which are multi-step with a compass and straightedge. Scher (2000) provides an interesting account of the evolution of *The Geometer's Sketchpad* program and some thoughts of the designer and programmer, Nicholas Jackiw, on this issue. The programs provide these additional tools in menus like the one illustrated in Figure 2, showing tools for the construction of objects such as the midpoint of a segment, a parallel line, a perpendicular line, and an angle bisector.

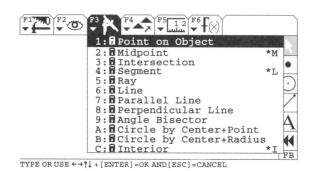

FIGURE 2

To use these tools, students must select the objects in their figure that are key to the multi-step construction when using only a compass and straightedge. For example, to use the Perpendicular Line tool on the menu in Figure 2, students must select the line or line segment and a point through which they want to construct the perpendicular. Only when appropriate objects are selected in the figure will the tools requiring those items be accessible. On the computer version of *The Geometer's Sketchpad*, accessibility of a tool in a menu is indicated by the tool name being highlighted. On the TI-92 Plus, students will get an error message, as in Figure 3, if they try to use a tool in the menu for which they do not have the appropriate objects selected. Students with knowledge of how to do these constructions by hand with a standard compass and straightedge will use the computer tools more efficiently by knowing which items must be selected for each tool. If the built-in tools in the menus are not sufficient, students can create their own or use packages of common constructions that can be added to the Custom tool.

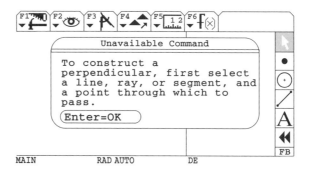

FIGURE 3

But finally, the most significant feature of programs such as *The Geometer's Sketchpad* and *Cabri Geometry* is the ability to "drag" objects and manipulate them dynamically while still preserving the defining properties of the construction. For example, if a student constructs a right triangle, no matter how the figure is moved or dragged, the right angle will always remain right. This only works if students have in fact constructed a right angle and not just made a drawing of one using the tools. If a student uses the Straightedge tool to make a triangle that appears to have a right angle by eyeballing the angle between the segments, then the angle measure will change when the vertices are dragged.

These software programs allow students to discover—through exploration and experimentation—the relationships among figures. Students can look for patterns and make conjectures about properties as they transform or distort figures interactively. Students can make one construction and drag the vertices to actually observe hundreds of examples in a short amount of time, more so than time and patience might allow by doing the constructions by hand. The range of explorations is endless. For example, younger students can construct a triangle and then use the program to measure the angles and compute the sum. As the vertices of the triangle are dragged, the measurements of the angles change but the sum remains a constant 180°. Older students can construct a triangle, construct a new triangle by joining the midpoints of its three sides and investigate the ratio of the area of the midpoint triangle to the area of the original triangle. Dragging the vertices of the original triangle will change the areas, but the ratio will remain a constant 0.25. Teachers can set up investigations that aim to develop a particular result (close-ended) or can have investigations that are more open-ended. Even when a particular result is the goal, the teacher can continue to prod the student to ask "what if" questions. For example, in the previous midpoint triangle investigation, the teacher can extend the exploration by asking, "What do you think would happen if we used polygons other than triangles? Do you think the ratio of the areas remain constant?"

When students make a conjecture and it holds as they drag their figure, they are likely to be completely satisfied with this empirical evidence of the validity of the conjecture and not see the need for more formal deductive reasoning. If these conjectures came from close-ended tasks or were results given to students as potential theorems, then students are less motivated to go beyond the confirmation given to them by the examples they generate with the computer. They are more likely to be interested in a more formal proof of a "shaky" conjecture, which is one whose validity they doubt (Zbiek et al., 2000). Keeping this in mind, as teachers, we can look to provide open-ended explorations in which students can make their own conjectures rather than giving them a conjecture to explore or leading them to one specific result.

Another interesting activity that can be done with these programs is one in which students are given a construction that has already been made with the software and asked to reproduce that construction so that their figure has all the same properties as the given one even when dragged. Laborde (2002) refers to these activities as "tasks to recreate visual phenomena." Students drag the given construction to explore its properties and discover which are invariant.

Whether students do constructions by hand or using one of these dynamic software programs, we still ask the question—why include these constructions at all in the curriculum? What benefits to the classroom and the mathematical development of their students do teachers get from inclusion of this material?

First, constructions are basically nonarithmetic mathematics. Some students that may be less proficient with the more computational mathematics may really enjoy the constructive, hands-on nature of this topic. Second, as discussed above, there is a rich opportunity for the study of constructions to engage students in problem solving, reasoning, and proof. Students are actively engaged in the topic. Sharing their conjectures with other students and attempting to convince others of their results promotes communication. Third, this topic is axiomatic in the sense that you start with a limited set of tools and rules and from them create an incredible amount of figures with interesting properties that can be explored. Fourth, this topic is an introduction to universal mathematics. When students make constructions, they use procedures that despite the neatness of their execution will produce the same results every time they are used. And finally, constructions are fun. It is a fulfilling and almost magical experience to see the wonderful figures that are living in the circles and lines that are constructed. Consider the beautiful figure in Figure 4. It is easy to construct and students will enjoy creating it. The hexagon is regular and contains many equilateral triangles, including the one drawn in dotted lines.

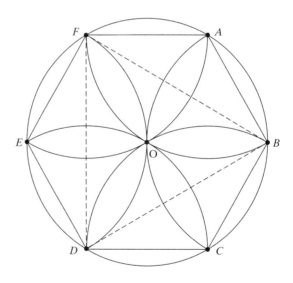

FIGURE 4

In summary, we have seen that NCTM's Geometry Standard expects that students learn constructions using a variety of tools. We have seen what kinds of constructions are being done in secondary schools to meet this standard as well as the introduction and influence of dynamic geometry software on the topic. We have also tried to answer the question of why we should do constructions at all with our students.

6.6 Exercises

1. Review all the constructions mentioned in this essay. Make sure that you can do them with a compass and straightedge and also using some dynamic software program.
2. Brainstorm a variety of ways to construct a square. For each, be sure you can prove your figure is a square. Once you have a technique, try creating a custom tool for it in one of the dynamic geometry software programs.
3. You are teaching a geometry class and discuss with students the historical problem of the impossibility of trisecting an angle. Later a student comes to you and shows how she trisected an angle using *Geometer's Sketchpad*. She constructed $\angle ACB$, measured it using the Measure menu, calculated one-third of it, and then constructed $\overrightarrow{CD}$ as a rotation of $\overrightarrow{CB}$ about point C through this angle. She is very excited and thinks she has done something no one else has been able to do. After all, you did tell her the construction was impossible. How do you respond to her?

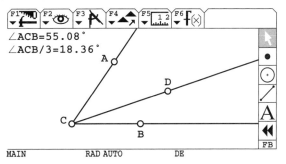

4. Tangrams are ancient oriental toys in which one uses the seven pieces, including five triangles, one square, and one parallelogram to recreate various puzzle diagrams. We include labels on all the vertices to provide the following information: $ABCD$ is a square, E is the midpoint of $\overline{BD}$, F is the midpoint of $\overline{BC}$, G is the midpoint of $\overline{CD}$, H is the midpoint of $\overline{DE}$, I is the midpoint of $\overline{GF}$ and J is the midpoint of $\overline{BE}$. Construct a tangram.

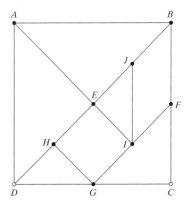

5. Start with any triangle. Join the midpoints of the triangle's three sides to construct a new triangle called the midpoint triangle and investigate the ratio of the area of the midpoint triangle to the area of the original triangle. Provide a proof for why the ratio remains a constant 0.25. Try to extend this exploration by considering other convex polygons.
6. Recreate the construction in Figure 4. Prove that *ABCDEF* is indeed a regular hexagon. Prove that $\triangle BDF$ is equilateral.
7. Construct a regular dodecagon (12 sides). Start with a regular hexagon.
8. Construct a regular octagon (8 sides). Start with a square.
9. Given a convex quadrilateral, construct a triangle of equal area and prove it. Try to extend your technique to construct a triangle of equal area to a given convex pentagon.

Bibliography

Clements, Douglas H., Michael T. Battista, Julie Sarama, and Sudha Swaminathan. "Development of Students' Spatial Thinking in a Unit on Geometric Motions and Area." *Elementary School Journal* 98, no. 2 (1997): 171–186.

Embse, Charles Vonder and Arne Engebretsen. "Using Interactive-Geometry Software for Right-Angle Trigonometry." *The Mathematics Teacher* 89, no. 7 (Oct. 1996): 602–604.

Gau, David Y. and Lindsay A. Tartre. "The Sidesplitting Story of the Midpoint Polygon." *The Mathematics Teacher* 87, no. 4 (April 1994): 249–256.

Glass, Brad. "Making Better Use of Computer Tools in Geometry." *The Mathematics Teacher* 94, no. 3 (March 2001): 224.

Jackiw, N. (1995). *The Geometer's Sketchpad*. [Computer software]. Emeryville, Califa: Key Curriculum Press.

Laborde, Collete. (2002) "The Process of Introducing New Tasks Using Dynamic Geometry into the Teaching of Mathematics." In Mathematics Education in the South Pacific. Proceedings of the Annual Conference of the Mathematics Education Research Group of Australasia Incorporated (25th, Auckland, New Zealand, July 7–10, 2002). Volume I [and] Volume II.

National Council of Teachers of Mathematics. *Curriculum and Evaluation Standards for School Mathematics*. Reston, Va.: National Council of Teachers of Mathematics, 1989.

_____. *Historical Topics for the Mathematics Classroom*. Reston, Va.: National Council of Teachers of Mathematics, 1969.

_____. *Principles and Standards for School Mathematics*. Reston, Va.: National Council of Teachers of Mathematics, 2000.

Scher, Daniel. "Lifting the Curtain: The Evolution of the Geometer's Sketchpad." *The Mathematics Educator* 10, no. 1 (Winter 2000): 42–48.

Shaw, L. S., P. Durden, and A. Baker. "Learning How Amanda, a High School Cerebral Palsy Student Understands Angles." *School Science and Mathematics* 98, no. 4: 198–204.

Texas Instruments (1994). *Cabri Geometry II*. [Computer software]. Dallas, Texas.

Zbiek, Rose Mary, and Brad Glass. "Conjecturing and Formal Reasoning about Functions in Dynamic Environment: Resolving Shaky Conjectures." In Proceedings of the Twelfth Annual International Conference on Technology in Collegiate Mathematics, edited by Gail Goodell, pp. 424–428. Reading, Mass.: Addison-Wesley Longman, 2000.

Zbiek, Rose Mary, Carmen Latterell, Brad Glass, and Teresa Finken. "Technology-Generated Examples as a Means and Not an End to More Formal Mathematical Reasoning." In Proceedings of the Eleventh Annual International Conference on Technology in Collegiate Mathematics, edited by Gail Goodell, pp. 476–480. Reading, Mass.: Addison-Wesley Longman, 2000.

6.7 FROM THE PAST: QUADRATIC EQUATIONS

Quadratic equations make their appearance early in high school algebra. Factoring, using the quadratic formula, completing the square, and guessing are all techniques that are used to solve such equations. Though in one form or another such equations have been worked on for four thousand years, it has taken most of that time just to write a quadratic equation in the algebraic form we know: $ax^2 + bx + c = 0$. We will consider some of the steps that have led to our modern formulation. This will take us to ancient Mesopotamia and their clay tablets, to ancient Greece and Euclid's *Elements*, and to modern Europe and the mathematics of Rene Descartes. What we think of as an algebraic equation began as a problem expressed in geometric terms. The history of the development of the quadratic formula is the history of a transition from geometry to algebra as new kinds of numbers (negative and complex), numbers that didn't have obvious geometrical meaning, were seen as possible solutions.

There are often disputes in the history of mathematics about when certain mathematical advances first occurred. We will be satisfied here with just bringing an early example of a quadratic equation without claiming it is the earliest. We go back to ancient Mesopotamia (modern-day Iraq) in about 1800 B.C.E. The following problem is given as a problem in geometry but to us has a formulation as a quadratic equation. This problem is found on a baked clay tablet along with other similar problems designed as exercises for students. The problem reads:

> I have the area (of a rectangle), 60. I have subtracted the width from the length, 7. Find the length and the width.

Restated in our terms the problem asks for the length and width of a rectangle whose area is 60 and whose length minus width equals 7. The solution proceeds with no symbols, only numbers and operations. The algorithm for solving this problem is presented through the particular example. Surviving texts contain many variations on this kind of problem. The solution reads:

> Take half of 7 by which the length exceeds the width and multiply it by itself. $\left(\frac{7}{2}\right)^2 = \frac{49}{4}$. Add to this the area $\frac{49}{4} + 60 = \frac{289}{4}$ and take its (square) root, $\frac{17}{2}$. The length is this number added to $\frac{7}{2}$, $\frac{17}{2} + \frac{7}{2} = 12$. The width is this number subtracted from $\frac{17}{2}$, $\frac{17}{2} - \frac{7}{2} = 5$.

What follows is a modern algebraic derivation of the solutions.

> Let x and y be the length and width. Then $x - y = 7$ and $xy = 60$. Thus $y = x - 7$ and $x(x - 7) = 60$. This reduces to $x^2 - 7x = 60$. Add $\left(\frac{7}{2}\right)^2 = \frac{49}{4}$ to each side of the equation to complete the square, which yields $x^2 - 7x + \left(\frac{7}{2}\right)^2 = 60 + \left(\frac{7}{2}\right)^2 = \left(\frac{17}{2}\right)^2$. The left-side of the equation is a square. $\left(x - \frac{7}{2}\right)^2 = x^2 - 7x + \frac{49}{4} = \frac{289}{4}$. Taking square roots of both sides yields $x - \frac{7}{2} = \frac{17}{2}$ or $x = 12$. Subtracting 7 gives the width or $y = 12 - 7 = 5$.

Most of the problems we have on record were problems prepared for students and not practical problems written by people applying mathematics. The fact that the problems were generally expressed in geometric form limits the kinds of equations considered and their possible solutions. The solutions had to be positive numbers which expressed geometric magnitudes. For instance, typical kind of problem would give the area and perimeter of a rectangle and ask for its length and width. Suppose one asked for such a rectangle whose perimeter was to be A and its area B. The maximum area of such a rectangle is $\left(\frac{A}{4}\right)^2$. Thus if B were greater than $\left(\frac{A}{4}\right)^2$, the problem would not have a geometrically realistic solution.

The Mesopotamians used what we would view as two basic algebraic techniques to solve quadratic equations: completing the square and interchanging sum and difference.

Completing the Square:

Solve $x^2 + bx = c$. Square half the coefficient of x, namely $\left(\frac{b}{2}\right)^2$ and add it to both sides of the equation, which yields $x^2 + bx + \frac{b^2}{4} = (x + \frac{b}{2})^2 = c + \frac{b^2}{4}$. Taking square roots of both sides yields $x + \frac{b}{2} = \pm\sqrt{c + \frac{b^2}{4}}$. This reduces to $x = -\frac{b}{2} \pm \sqrt{c + \frac{b^2}{4}}$.

For the Mesopotamians only the positive square root would result in a positive value of x. Note that variations on this method would solve equations of the form $x^2 = bx + c$ and $x^2 + c = bx$, where we use only positive coefficients.

Interchanging Sum and Difference:

Given $x + y = a$, $xy = b$, find x and y.

$$(x+y)^2 = a^2 = x^2 + 2xy + y^2$$
$$a^2 - 4b = (x+y)^2 - 4xy = (x-y)^2$$
$$x - y = \sqrt{a^2 - 4b}$$
$$2x = (x+y) + (x-y) = a + \sqrt{a^2 - 4b}$$
$$x = \frac{a + \sqrt{a^2 - 4b}}{2}, y = \frac{a - \sqrt{a^2 - 4b}}{2}$$

Problems were presented as geometric problems with particular numbers for the quantities. A solution was essentially a recipe or algorithm to find the unknown quantities. There were no equations or variables. It is likely that there were geometric pictures that aided the discovery of the method of solution.

Many other ancient societies solved quadratic equations. We move to Euclid in ancient Greece for a different approach to the problem. In Book II of the *Elements* Euclid presents several propositions that have connection to quadratic equations.

Euclid's Proposition 5.

If a straight line be cut into equal and unequal segments, the rectangle contained by the unequal segments of the whole together with the square on the straight line between the points of section is equal to the square on the half.

The connection of this proposition to quadratic equations is not obvious. This proposition guarantees that a certain relation exists between areas of rectangles and squares formed by segments of a given line segment. Let the original straight line have length a. Then the line is bisected, giving two segments of length $a/2$. Finally, the line is cut into two unequal segments, of length x and $a - x$. The rectangle "contained by the unequal segments" is a rectangle whose sides are x and $a - x$. Its area is $x(a - x)$. The straight line between the points of section is the line segment between the midpoint and the point that cuts the line into the segments of length x and $a - x$. This segment has length $\frac{a}{2} - x$. Its square has area $\left(\frac{a}{2} - x\right)^2$. Finally, the square on the half has area $\left(\frac{a}{2}\right)^2$. The proposition claims that for all x the following relation holds between these areas:

(1) $$x(a - x) + \left(\frac{a}{2} - x\right)^2 = \left(\frac{a}{2}\right)^2.$$

Expanding the left side, we have

$$x(a - x) + \left(\frac{a}{2} - x\right)^2 = ax - x^2 + \left(\frac{a}{2}\right)^2 - 2\left(\frac{a}{2}\right)x + x^2$$

$$= ax - ax - x^2 + x^2 + \left(\frac{a}{2}\right)^2 = \left(\frac{a}{2}\right)^2.$$

So far this is just an identity. How can it solve a quadratic equation? Suppose the problem was to cut a segment of length a into segments of length x and $a - x$ such that the area of the rectangle with these two sides was equal in area to a given square, say a square of side b. Then the geometric relation could be translated to the equation $x(a - x) = b^2$ or $x^2 - ax + b^2 = 0$. The problem is to find the length x. If we replace $x(a - x)$ with b^2 in expression (1), we have

(2) $$b^2 + \left(\frac{a}{2} - x\right)^2 = \left(\frac{a}{2}\right)^2.$$

Equation (2) is of the form of the Pythagorean Theorem and we can construct a right triangle given the hypotenuse and one of the sides. If we take one of the sides of a right triangle to be b and the hypotenuse to be $\frac{a}{2}$, then we can construct the other side, which equals $\frac{a}{2} - x$. Subtracting this length from $\frac{a}{2}$, we have constructed length x and thus solved the quadratic equation $x^2 - ax + b^2 = 0$.

Historians of mathematics are currently engaged in a debate as to the algebraic content of Euclid's Proposition 5. Some see it as simply algebra presented geometrically while others see this proposition as part of Euclid's plan to present a complete picture of the Platonic solids. Either way, later mathematicians used this proposition in the solution quadratic equations viewed algebraically. In fact, most of the algebra from the time of Euclid up to 1600 is expressed in geometric terms. The quantities considered are regarded as geometric quantities—lengths, areas, and volumes. An equation had a related geometric picture. It is with Rene Descartes that algebra begins to separate itself from geometry.

Rene Descartes (1596–1650) was a French scientist and philosopher. His chief work, *Discourse on Method* (1637), is a presentation of what Descartes viewed as the correct method for obtaining the truth about the world and humankind. It is primarily a treatise on philosophy and logic. But to illustrate his "method", Descartes included several appendices that showed his method in action. One of these, *La Geometrie*, has been enormously influential on the development of mathematics. In it, Descartes begins the separation of algebra from geometry. He still considers quantities as geometric magnitudes. For example, a might be the length of a line segment. Traditionally a^2 or aa would represent the area of the square of side a. But using a simple straight-edge and compass construction, Descartes produced a line whose length, a^2, satisfies the proportion $1 : a = a : a^2$. Thus all quantities could be treated as lengths of lines. Before Descartes, mathematicians had difficulty with quantities like a^5 since it represented a geometrical quantity of a dimension that was not understood, beyond area and volume. Descartes showed that it could be viewed simply as the length of a line. With another construction he begins with line segments of length a and b and constructs a line segment of length ab. With yet another construction, he could extract square roots of these quantities. He had taken the essential steps in turning a geometric relation into an algebraic equation. Since quantities like a, a^2, ba, and b^5 were all lengths, they could be added and subtracted, and relations among them could be expressed as polynomial equations. For awhile, mathematicians would keep the geometric content in mind but it became apparent that one could simply deal with a polynomial equations. With these concepts in mind, Descartes could solve an equation of the form: $z^2 = -az + b^2$. He expresses the solution as

$$z = -\frac{1}{2}a + \sqrt{\frac{1}{4}a^2 + b^2}, \qquad (3)$$

which is only one of the roots obtained by applying the quadratic formula to the quadratic equation $z^2 + az - b^2 = 0$. He ruled out the negative root that we would express as $z = -\frac{1}{2}a - \sqrt{\frac{1}{4}a^2 + b^2}$. He does not accept negative numbers since all his numbers are lengths of lines. Thus he must express his quadratic equations in different forms. He follows his first example with the equation $z^2 = az - b^2$ for which he gives two solutions: $z = \frac{1}{2}a + \sqrt{\frac{1}{4}a^2 - b^2}$ and $z = \frac{1}{2}a - \sqrt{\frac{1}{4}a^2 - b^2}$, which are both positive. The important point is that he gave us a quadratic formula for finding the solutions to an equation in terms of the coefficients of the polynomial in question.

After Descartes, mathematicians expanded the kinds of numbers allowed as coefficients and as solutions to polynomial equations. Negative numbers and complex numbers would soon be folded in so that a single version of the quadratic formula held in all cases. Within a century of Descartes, Euler was perfectly comfortable with $x = \frac{-b \pm \sqrt{b^2 - 4ac}}{2a}$ as the solutions to $ax^2 + bx + c = 0$, regardless of the nature (real, negative, or complex) of the coefficients a, b, and c and whether or not the equation referred to a geometric situation.

Bibliography

Descartes, Rene. *The Geometry of Rene Descartes*, translated by David Eugene Smith and Marcia L. Latham. New York: Dover, 1954.

6.7 Exercises

1. Translate the following Mesopotamian problem into modern algebraic language and show how the solution was developed:

    I have added up the area and the side of my square: $3/4$. You write down 1, the coefficient. You break off half of 1. $1/2$ $1/2$ and $1/2$ you multiply: $1/4$. You add $1/4$ to $3/4$: 1. This is the square of 1. From 1 you subtract $1/2$, which you multiplied. $1/2$ is the side of the square.

2. Use the method of interchanging sum and difference to find two numbers whose sum is 22 and whose product is 120.
3. Use the method of interchanging sum and difference to find two numbers whose product is 35 and whose difference is 2.
4. Draw a line segment approximately 2″ long and another approximately 5″ long. Using Euclid's Proposition 5 from above construct the point A on the longer segment such that the rectangle formed by the two segments A cuts the longer segment into has the same area as the square whose side is the approximately 2″ segment.
5. We saw that Euclid's Proposition 5 from Book II could be used to solve the quadratic equation $x^2 - ax + b^2 = 0$. Proposition 6 of Book II can be used to solve a different form of quadratic equation. Read the following proposition and determine the form of the quadratic equation that it can be applied to.

    Proposition

    If a straight line be bisected and a straight line be added to it in a straight line, the rectangle contained by the whole with the added straight line and the added straight line together with the square on the half is equal to the square on the straight line made up of the half and the added straight line.

6. Suppose that we start with the quadratic equation $Ax^2 + Bx + C = 0$. Write the quadratic formula using the coefficients A, B, and C. Then rewrite the equation in the form given by Descartes, namely, $z^2 = -az + b^2$. Show that Descartes version of the quadratic formula is the same as the one we use now.

Chapter 6 Highlights

Section 6.5, *The Solvability of Polynomial Equations*, summarizes and uses almost all of what came before it in our book. We hope that you have enjoyed the journey. It has taken you many places, from very ancient number theory and geometry, to the more recent venue of abstract algebra, and possibly even into the high school classroom. The exploration of the evolving concepts of number and structure has been our common theme. Our history sections show that this exploration has intrigued the human mind for a very long time. We hope that the classroom sections help you to share that exploration across many levels of endeavor.

Chapter Questions

1. Give the definition of the following terms or expressions.
    i. extension field
    ii. radical extension
    iii. solvable by radicals
    iv. degree of the extension
    v. minimal polynomial
    vi. automorphism

2. If a polynomial in $Q[x]$ has a root in $Q(\sqrt[3]{2})$, does it factor completely when it is considered to be a polynomial in $Q(\sqrt[3]{2})[x]$?

3. Show that the fields $Q(\sqrt[3]{2})$ and $Q(\sqrt[3]{5})$ are not isomorphic.

4. Characterize the elements of an extension field of Q that contains the roots of both the polynomials $x^3 - 5$ and $x^2 - 5$.

5. Find a radical tower over Q in which the polynomial $x^3 - 3x + 6$ has all its roots.

6. Find a radical tower over Q in which the polynomial $x^4 - x^2 - 6$ has all its roots.

7. Find the minimal polynomial of $\sqrt{3} + i$ over Q.

8. Find the minimal polynomial of $\sqrt{5} + \sqrt{2}$ over Q.

9. Find the degree of each of the following extensions.
    i. $Q(\sqrt[3]{3})$ over Q
    ii. $Q(\sqrt{2} + i)$ over Q
    iii. $Q(\sqrt{5} + \sqrt{2}, i)$ over Q

10. Suppose that p is a prime and that ω is a complex root of $x^p - 1$. What is the degree of $Q(\omega)$ over Q?

11. Suppose that p is a prime and let $a \in Q$ be such that $x^p - a$ is irreducible in $Q[x]$. Find the smallest value of n such that all the p roots of $x^p - a$ are in an extension of Q of degree n.

12. Can a tetrahedron "be doubled"? Interpret and explain.

13. Prove that $\cos(\theta)$ is constructible if and only if $\cos(\theta/2)$ is constructible.

14. Determine the group of automorphisms that fix Q for each of the following field extensions:

   i. $Q(\sqrt{2}, i)$
   ii. $Q(\sqrt[5]{11})$, where $\sqrt[5]{11}$ is a real number
   iii. $Q(\xi)$, where ξ is a 5th root of unity

15. Let $F \subseteq C$ be the field Q with the third and fifth roots of unity adjoined. Let K be the field F with all the roots of $p(x) = x^5 - 10x + 5$ adjoined. Find Aut(K, F).

16. Let $p(x) = x^5 - 2$ and let K be the splitting field of $p(x)$. Find Aut(K, F).

Appendix

This appendix contains a summary of basic notation and facts about logic and proof, sets, functions, and binary operations.

Propositions and Predicates

A **proposition** or **statement** is a sentence to which we can assign a truth value of either *true* or *false*, but not both. For example, "$2 + 3 = 6$" is a statement that is false but "$a^2 + b^2 = c^2$" is not a statement since its truth value depends on the values of a, b, and c. We shall use uppercase letters to represent propositions. These letters are called **logical variables**. We can combine simple propositions to form compound propositions. If P and Q are propositions, the **conjunction** of P and Q is the proposition "P and Q" and is denoted $P \wedge Q$. Similarly, the proposition "P or Q" is the **disjunction** of P and Q and is denoted $P \vee Q$. The connector "or" should always be taken in the *inclusive* sense, meaning "P or Q or both." The truth values of compound statements depend on the truth values of their logical variables. The statement "P and Q" is true only when both P and Q are true; it is false otherwise. The statement "P or Q" is true when either P or Q (or both) are true; it is false otherwise. For instance, if P is the statement, "7 is prime," and Q is the statement, "5 is even," then $P \vee Q$ is true since at least one of P and Q is true, but $P \wedge Q$ is false since at least one of P and Q is false.

The **negation** of a proposition P, denoted by $\sim P$ and read "not P," results in a proposition with the opposite truth value of that of P. For instance, the negation of the false statement, "5 is even," is the true statement, "5 is not even." If we let P be the proposition, "It is snowing," and Q be the proposition, "It is dark outside," the proposition $\sim(P \wedge Q)$ is, "It is not the case that it is both snowing *and* dark outside." If $\sim(P \wedge Q)$ is true, then either it is not snowing or it is not dark outside; that is, $\sim P \vee \sim Q$ is true. If $\sim(P \wedge Q)$ is false, then $P \wedge Q$ is true and both P and Q are true. So $\sim P \vee \sim Q$ is false. So no matter what the truth values of P and Q are, the proposition $\sim(P \wedge Q)$ has the *same* truth value as $\sim P \vee \sim Q$. The two statements are **logically equivalent**. This means that if we assign the same truth values to their logical variables, the two statements will have the same truth value. The logical equivalence $\sim(P \wedge Q) = \sim P \vee \sim Q$ is one of **De Morgan's laws** that tells how to negate conjunctions and disjunctions.

PROPOSITION 1 *(De Morgan's Laws)*

(a) $\sim(P \wedge Q) = \sim P \vee \sim Q$
(b) $\sim(P \vee Q) = \sim P \wedge \sim Q$

Note that $\sim(\sim P)$ is logically equivalent to P for all propositions P.

From two statements P and Q, we can form other compound statements. One that we encounter often is the statement "If P then Q," which is called a **conditional statement** or **implication**. It is denoted by $P \to Q$. The statement P is called the **hypothesis** and the statement Q is called the **conclusion**. There are many equivalent ways to word a conditional statement. The following is a list of some of the statements equivalent to "If P then Q":

P implies Q.

P only if Q.

Q if P.

Q whenever P.

P is sufficient for Q.

Q is necessary for P.

The truth value of the conditional statement $P \to Q$ depends on the truth values of the logical variables P and Q. The statement says that if P holds, then Q will hold as well. It does not promise anything about Q when P is false. Therefore, the only time when the conditional $P \to Q$ is false is when P is true and Q is false. The negation of $P \to Q$ is $P \wedge \sim Q$. For example, let P be the statement "Mary is wearing green socks" and let Q be the statement "Mary passed algebra." If Mary wears green socks all semester but does not pass algebra, then our conditional statement, "If Mary wears green socks, then Mary passed algebra," is false. If however Mary does not wear green socks, then she may or may not have passed algebra. Neither situation contradicts the implication $P \to Q$. The negation of $P \to Q$ is the statement, "Mary wears green socks and Mary has not passed algebra."

The conditional statement $P \to Q$ is also equivalent to the form $\sim Q \to \sim P$, which is called the **contrapositive** of the conditional statement. The contrapositive statement, "If Mary did not pass algebra, then Mary did not wear green socks" has the same meaning as the conditional statement "If Mary wears greens socks, then Mary passed algebra." If we simply swap the hypothesis and conclusion of a conditional statement $P \to Q$, we get its **converse**, $Q \to P$. The converse of a conditional statement is **not** logically equivalent to the conditional statement. For example, consider the conditional statement, "If Mr. Smith is running for office, then he is a Democrat." The converse of this statement is the statement, "If Mr. Smith is a Democrat, then he is running for office." The converse is true in the case that Mr. Smith is not a Democrat, yet is running for office. However, the original conditional is not true in this situation.

The biconditional statement $P \leftrightarrow Q$ is defined as $(P \to Q) \wedge (Q \to P)$. For $P \leftrightarrow Q$ to be true, both $P \to Q$ and $Q \to P$ must be true. The biconditional $P \leftrightarrow Q$ is translated, "P if and only if Q," and sometimes abbreviated as "P iff Q." For example, the statement "7 is prime if and only if 5 is odd," is a true statement because the implication, "If 7 is prime, then 5 is odd," is true and its converse, "If 5 is odd, then 7 is prime," is also true. The statement, "5 is even if and only if 9 is prime," is also true. However, the statement, "5 is even if and only if 7 is prime," is false. For $P \leftrightarrow Q$ to be true, P and Q must both be true, or P and Q

must both be false. The negation of $P \leftrightarrow Q$ is $(P \wedge \sim Q) \vee (Q \wedge \sim P)$. The words "$P$ is necessary and sufficient for Q" are also used for $P \leftrightarrow Q$. The implication $P \rightarrow Q$ is phrased, "P is sufficient for Q," and the implication $Q \rightarrow P$ is phrased, "P is necessary for Q."

An assertion that contains one or more variables is called a **predicate**; its truth value is predicated on the values assigned to its variables. A predicate P containing n variables, $x_1, x_2, \ldots, x_n$ is called an n-predicate and is denoted by $P(x_1, x_2, \ldots, x_n)$. For example, $P(a, b, c) : a^2 + b^2 = c^2$ is a 3-predicate. Before we can assign values to the variables appearing in a predicate, we must specify a **universe of discourse**—that is, a nonempty set of values in which the variables may assume their values. One way for a predicate to become a proposition is to assign a value to each of its variables. For example, if we take the universe of discourse to be the set of integers for the predicate $P(a, b, c) : a^2 + b^2 = c^2$, then $P(1, 2, 3) : 1^2 + 2^2 = 3^2$ is a false statement, while $P(-3, 4, 5) : (-3)^2 + 4^2 = 5^2$ is a true statement.

Another way to obtain a proposition from a predicate is to use **quantifiers** for its variables. There are two types of quantifiers: a **universal quantifier** and an **existential quantifier**. Universal quantification asserts that the predicate is a true proposition for any values assigned to its variables. A universal quantifier is typically characterized by phrases such as "for every," "for all," "every" and "for each" and is denoted by the symbol $\forall$. Existential quantification asserts that there exists at least one value in the universe of discourse for which the predicate is true. An existential quantifier is identified by phrases that include "there exists," "for some," "some," and "there is/are" and is denoted by the symbol $\exists$.

For example, suppose that the universe of discourse is the set of real numbers. The quantified predicate, $\forall x[x^2 > 0]$, says, "For all real values of x, $x^2 > 0$." It is a false statement because it fails to hold when $x = 0$. The quantified statement, $\exists x[x^2 > 0]$, says, "There exists at least one value of x such that $x^2 > 0$." It is a true statement because we can let $x = 2$ and obtain a true statement. The statement, $\exists x \forall y[x > y]$ asserts that there is some value of x such that $x > y$ for all values of y. This is a false statement since there is no number that is greater than every other number. However, the statement $\forall y \exists x[x > y]$ is a true statement. It asserts that for each value of y there is another number x that is greater than y. (The x in question can be different for each y.) So be careful. The order in which quantifiers appear makes a big difference.

The negation of $\forall x[x^2 > 0]$ is the statement $\exists x[x^2 \not> 0]$, or $\exists x[x^2 \leq 0]$. The negation of the proposition $\exists x[x^2 > 0]$ is $\forall x[x^2 \not> 0]$, or $\forall x[x^2 \leq 0]$. In general, the negation of $\forall x P(x)$ is $\exists x[\sim P(x)]$, and the negation of $\exists x P(x)$ is $\forall x[\sim P(x)]$. Thus the negation of $\exists x \forall y[x > y]$ is $\forall x \exists y[x \leq y]$.

Exercises on Propositions and Predicates

1. Determine which of the following are statements.

    (a) 527 is an even integer.
    (b) $x < x^2$.

2. Negate the following statements.

   (a) 2 is even and 2 is prime.
   (b) It is raining and I do not have my umbrella.
   (c) For all real numbers x, $x^2 > x$.
   (d) There exists a real number x such that $x \cdot 2 = 1$.
   (e) $\forall x \exists y \, [x + y = 0]$.

3. Write out the converse, contrapositive, and negation of each of the following.

   (a) If you don't do your homework, then you can't go outside.
   (b) If Sally wins, then Mary loses.
   (c) If John wins, then both Mary loses and the school closes.

The Language of Proof

Let's begin with a typical theorem and look at what needs to be done before proving it.

THEOREM. Let x, y and z be real numbers and $x \neq 0$. Assume that x and z are rational numbers and that $xy = z$. Then y is rational.

First, note the universe of discourse. It is the set of real numbers. Next we transcribe it into a quantified predicate so that we know exactly what implication we want to prove and can identify our hypotheses and conclusion.

> For all real values of x, y and z, if $x \neq 0$, x and z are rational, and $xy = z$, then y is rational.

So, as an implication $P \rightarrow Q$, P is the premise "$x \neq 0$, both x and z are rational, and $xy = z$," and Q is the conclusion, "y is rational." The next step is critical. **Know your definitions!** A *rational* number is a real number than can be expressed as the quotient of integers, $\frac{a}{b}$, with $b \neq 0$.

Now for the proof. When we want to prove a conditional statement, we assume the hypothesis is true and show that the conclusion must also be true. Such a proof is called a **direct proof** of the conditional statement $P \rightarrow Q$. We use the hypothesis, and definitions, and previous propositions to create a series of statements that are equivalent to P and lead to the conclusion Q.

Proof. Suppose that x, y and z are any rational numbers such that x and z are rational, $x \neq 0$, and $xy = z$. (Here we restated our hypotheses.) Since x is rational and $x \neq 0$, $x = \frac{m}{n}$ for some nonzero integers m and n. Since z is rational, $z = \frac{s}{t}$ for some integers s and t, with $t \neq 0$. (Here we have used the definition of a rational number.) Since $\frac{m}{n} y = \frac{s}{t}$, we have that $y = \frac{ns}{mt}$. (Here we did the crucial step: we substituted and solved for y.) Since ns and mt are integers, and $mt \neq 0$, y can be expressed as the quotient of integers and is therefore rational. ▲

Although the foregoing theorem was not difficult to prove, its proof had typical features of all proofs.

- Know the structure of the implication you must prove.
- Know your universe of discourse.
- Know your definitions.
- Start with a simple statement that says what objects you will be talking about. In this case, we started with a set of any three numbers that satisfy the hypothesis. If the proof works for ANY x, y, and z, it works for ALL x, y and z.

Often when given a conditional statement $P \to Q$ to prove, we may find it easier to prove the equivalent contrapositive statement, $\sim Q \to \sim P$. Sometimes, this is called an **indirect proof**. Here's an example.

THEOREM. Suppose that n is a integer and that n^2 is odd. Then n is odd.

So we must prove the statement that if n^2 is odd, then n is odd. The contrapositive of the statement is, "If n is not odd (i.e., even), then n^2 is not odd (i.e., even)."

Proof. Let n be an even integer. Let p be the integer such that $n = 2p$. Then $n^2 = 4p^2$. Since $4p^2$ is a multiple of 2, n^2 is even. ▲

Another commonly used technique for indirectly proving a conditional statement $P \to Q$ is a **proof by contradiction**. The idea of a proof by contradiction is to assume that the statement P that you want to prove is false and reach a contradiction based on this assumption. The contradiction is the negation of a fact known to be true. When that happens, it means that the assumption that P is false is incorrect, and so P must therefore be true. Here's a classic example.

THEOREM. If x is any rational number, then $x^2 \neq 2$.

The statement that we want to prove is that for all rational numbers x, $x^2 \neq 2$. We will start the proof by asserting its negation, namely that there exists a rational number x such that $x^2 = 2$. We will end up contradicting the fact that every rational number can be expressed in lowest terms. Note that if a rational number $\frac{m}{n}$ is expressed in lowest terms, m and n cannot both be even.

Proof. Suppose that x is a rational number and that, $x = \frac{m}{n}$, where m and n are two integers with no common factors. Suppose that $x^2 = 2$. Then $\frac{m^2}{n^2} = 2$ and $m^2 = 2n^2$. Since m^2 is even, it must be the case that m is even (Exercise 5). So $m = 2p$ for some integer p. Thus $4p^2 = 2n^2$ or $2p^2 = n^2$, and n^2 is even. But this implies that n is even. This contradicts the fact that x is in lowest terms and m and n cannot both be even. Thus x^2 cannot equal 2. ▲

Often, mathematical theorems are stated as biconditional statements. For example, you might be asked to prove the following theorem.

THEOREM. An integer n is even if and only if n^2 is even.

We must prove two implications "If an integer n is even, then n^2 is even" and its converse, "If n is an integer and n^2 is even, then n is even." Remember that we have the option of proving the contrapositive of either statement. (It is easier to prove the contrapositive of "If n is an integer and n^2 is even, then n is even.") Thus a typical strategy for proving $P \leftrightarrow Q$ is to prove $P \rightarrow Q$ and $\sim P \rightarrow \sim Q$.

Sometimes, mathematical statements are not true! To prove that a statement is not true, we often provide a **counterexample**.

Statement: Every rational number has a rational square root.

Counterexample: 2 is a rational number and there is no rational number x such that $x^2 = 2$.

Statement: The product of any two irrational numbers is irrational.

Counterexample: $\sqrt{2}$ and $\sqrt{8}$ are both irrational numbers but $\sqrt{2}\sqrt{8} = \sqrt{16} = 4$, which is rational.

Exercises on the Language of Proof

1. Prove directly: The product of two odd integers is odd.
2. Prove using the contrapositive: If the product of two integers x and y is odd, then both x and y are odd.
3. Prove using a proof by contradiction: If x is rational and y is irrational, then $x + y$ is irrational.
4. Prove that each of the following statements is false by giving a counterexample.

    (a) The square root of any integer is irrational.
    (b) All odd numbers are divisible by 3.
    (c) If p is an odd prime number, then $p + 2$ is also prime.

5. Prove the following statement: "An integer n is even if and only if n^2 is even."

Sets

A **set** is a collection of objects. The objects in a set are also called its **elements, members,** or **points**. It is customary to use capital letters to designate sets. The symbol $\in$ denotes membership in a set. Thus we write $a \in A$ to mean that object a is a member of the set A, and we write $a \notin B$ to mean that object a is not a member of set B. For example, if we let $\mathbf{Z}$ be the set of all integers, then $2 \in \mathbf{Z}$, but $\frac{1}{2} \notin \mathbf{Z}$.

There are various ways to indicate the elements belonging to a set. We can describe the elements of a set in words. For example, we can let A be "the set of all integers greater than 2 and less than 8." The set A described has only five elements: 3, 4, 5, 6, and 7. We can list the elements of a set in braces, making use of ellipses to indicate a pattern when needed. For example, $A = \{3, 4, 5, 6, 7\}$ or $B = \{0, 2, -2, 4, -4, 6, -6, \ldots\}$.

We note that the set {3, 1, 5, 6, 7} is the same as {7, 1, 3, 6, 5} since the listed order of the elements does not change the contents of the set. We can also use **set builder notation**, which uses a predicate $P(x)$. The set $S = \{x : P(x)\}$ is the set of all elements x in the universe of $P(x)$ such that $P(x)$ is a true statement. The set A of all integers greater than 2 and less than 8 can be denoted in set builder notation as $\{x : x$ is an integer and $2 < x < 8\}$. The set of even integers can be denoted as $\{x : x = 2m$ for some integer $m\}$.

We will refer to the following sets frequently:

Z, the set of all integers, $\{0, 1, -1, 2, -2, 3, -3, \ldots\}$;

N, the set of all natural numbers, $\{1, 2, 3, \ldots\}$;

W, the set of all whole numbers, $\{0, 1, 2, 3, \ldots\}$;

Q, the set of all rational numbers, $\{\frac{a}{b} : a, b \in \mathbf{Z}, b \neq 0\}$;

R, the set of all real numbers;

R$^+$, the set of strictly positive real numbers, $\{x \in \mathbf{R} : x > 0\}$;

The **empty(null) set** contains no elements and is denoted by $\emptyset$ or $\{\}$.

If A and B are sets such that every element of A is an element of B, then A is a **subset** of B, which we denote by $A \subseteq B$. To prove that $A \subseteq B$, we need to show that each element x of A is also in B. We note that $A \subseteq B$ leaves open the possibility that the two sets are **equal**. In fact, to prove that $A = B$, we must show $A \subseteq B$ and $B \subseteq A$. If A is a subset of B, but B is not a subset of A, then there is some element in B that is not in A. In this case, we say that A is a **proper subset** of B. Note that every set is a subset of itself and that the empty set $\emptyset$ is a subset of every set.

EXAMPLE

Let $A = \{x \in \mathbf{R} : 0 \leq x \leq 1\}$, $B = \{x \in \mathbf{R} : -1 \leq x \leq 1\}$, and $C = \{x \in \mathbf{R} : x^2 - x \leq 0\}$. The set A is a proper subset of B. The predicates describing sets A and C are quite different, but the sets A and C are equal. Here's a proof.

Proof. First we show that $A \subseteq C$. Let $x \in A$. Since $0 \leq x \leq 1$, $x^2 \leq x$ and therefore $x^2 - x \leq 0$. Since $x^2 - x \leq 0$, $x \in C$. Thus $A \subseteq C$. Now we show that $C \subseteq A$. Let $x \in C$. Then $x^2 - x = x(x - 1) \leq 0$. So either $x \leq 0$ and $(x - 1) \geq 0$, or $x \geq 0$ and $(x - 1) \leq 0$. Since the first case in impossible, it must be that $x \geq 0$ and $(x - 1) \leq 0$. So $0 \leq x \leq 1$ and $x \in A$. Thus $C \subseteq A$. Since $A \subseteq C$ and $C \subseteq A$, we conclude that $A = C$. ∎

Set Operations

We now describe how to build new sets from given ones. The set of elements that the two sets A and B have in common is called the **intersection** of A and B, and denoted $A \cap B$. More precisely, $A \cap B = \{x : x \in A$ and $x \in B\}$. The **union** of A and B, denoted $A \cup B$ is the set of all elements that are in either set A or in set B. That is, $A \cup B = \{x : x \in A$ or $x \in B\}$. The word "or" is being used inclusively, meaning that if x is in $A \cup B$, x can be an element of both A and B. Hence we have $A \cap B \subseteq A \cup B$. If two sets do not share any elements, they are said to be **disjoint**. In this case, $A \cap B = \emptyset$. For example, consider the following subsets of $\mathbf{R}$: $A = \{x : 0 < x < 2\}$, $B = \{x : 1 < x < 3\}$, and $C = \{x : 2 < x < 3\}$. Then $A \cup B = \{x : 0 < x < 3\}$, $A \cap B = \{x : 1 < x < 2\}$, and since $A \cap C = \emptyset$, we say that sets A and C are disjoint.

Using these basic definitions, the following proposition contains some important set identities referred to as the **idempotent**, the **commutative**, the **associative**, and the **distributive properties**, respectively, of the intersection and union of sets. The proofs are left to the reader as exercises.

PROPOSITION 2 Let A, B and C be any sets, then

1. $A \cap A = A$ and $A \cup A = A$;
2. $A \cap B = B \cap A$ and $A \cup B = B \cup A$;
3. $(A \cap B) \cap C = A \cap (B \cap C)$ and $(A \cup B) \cup C = A \cup (B \cup C)$;
4. $A \cup (B \cap C) = (A \cup B) \cap (A \cup C)$ and $A \cap (B \cup C) = (A \cap B) \cup (A \cap C)$.

We can extend our notions of intersection and union to any finite collection of sets, $\{A_1, A_2, \ldots, A_n\}$. The union of the collection is the set of all elements that are in *at least* one of the sets $A_1, \ldots, A_n$. The intersection of the collection is the set of elements that are in *all* of the sets $A_1, \ldots, A_n$. That is,

$$A_1 \cup A_2 \cup \ldots \cup A_n = \bigcup_{i=1}^{n} A_i = \{x : x \in A_i \text{ for some } i\}, \text{ and}$$

$$A_1 \cap A_2 \cap \ldots \cap A_n = \bigcap_{i=1}^{n} A_i = \{x : x \in A_i \text{ for all } i\}.$$

We can make another new set from given sets A and B by forming the **complement of B relative to A** or the **set difference of A and B**, denoted $A \backslash B$ and read "A minus B." It is the set of all elements of A that are not in B. That is, $A \backslash B = \{x : x \in A \text{ and } x \notin B\}$. For example, if $A = \{a, b, c, d, e\}$ and $B = \{a, d, f, r\}$, then $A \backslash B = \{b, c, e\}$ and $B \backslash A = \{f, r\}$. In general, given two sets S and T, $S \backslash T \neq T \backslash S$. In fact, these two sets are disjoint. (We leave the proof as an exercise.)

Sometimes all the sets in question are understood to be subsets of some larger set. Such a set is called the **universal set** U. If we find the complement of B relative to U, we refer to it simply as the **complement of B** and denote it by B' rather than $U \backslash B$. We have $B' = \{x : x \in U \text{ and } x \notin B\}$. It is easy to see that $(B')' = B$. The following proposition illustrates how complements interact with intersections and unions.

PROPOSITION 3 (*De Morgan's Laws for Sets*). For all sets A and B,

(a) $(A \cup B)' = A' \cap B'$;   (b) $(A \cap B)' = A' \cup B'$.

Proof. (b) Let x be any element of $(A \cap B)'$. Since x fails to be in the intersection of A and B, x fails to be in at least one of A or B. Thus x is an element of $A' \cup B'$

and we have that $(A \cap B)' \subseteq A' \cup B'$. Now let x be any element of $A' \cup B'$. Then either x is not an element of A or it is not an element of B. In either case, x is not an element of $A \cap B$ and hence is in $(A \cap B)'$. Thus $A' \cup B' \subseteq (A \cap B)'$ and we conclude that $(A \cap B)' = A' \cup B'$. ▲

Our final example of a new set derived from two given sets is the Cartesian product. If A and B are two nonempty sets, then the **Cartesian product** of A and B is denoted by $A \times B$ and is defined by

$$A \times B = \{(a, b) : a \in A \text{ and } b \in B\}.$$

For example, if $A = \{a, b, c\}$ and $B = \{1, 2\}$, then $A \times B = \{(a, 1), (a, 2), (b, 1), (b, 2), (c, 1), (c, 2)\}$. The order that the sets are listed in the Cartesian product is important and determines the set from which the entries in the ordered pairs come. The Cartesian product $B \times A$ contains the same number of ordered pairs as $A \times B$, but with entries swapped: $B \times A = \{(1, a), (2, a), (1, b), (2, b), (1, c), (2, c)\}$. We note that $A \times B \neq B \times A$. We may extend the notion of the Cartesian product to collections of more than two sets. Suppose $A_1, A_2, \ldots, A_n$ is a collection of nonempty sets. The Cartesian product $A_1 \times A_2 \times \cdots \times A_n$ consists of all possible ordered n-tuples and is defined by

$$A_1 \times A_2 \times \cdots \times A_n = \{(a_1, a_2, \ldots, a_n) : a_i \in A_i \text{ for all } 1 \leq i \leq n\}.$$

The Cartesian product $\mathbf{R} \times \mathbf{R} \times \mathbf{R}$ is the set of all ordered triples of real numbers, the familiar 3-space of calculus.

Exercises on Sets and Set Operations

1. List all the elements in each of the following sets, using braces and ellipses where necessary.

    (a) $\{x : x \text{ is a natural number divisible by } 5\}$
    (b) $\{x : x \text{ is a negative odd integer}\}$
    (c) $\{x : x \text{ is an even prime number}\}$

2. Redefine each of the following sets using set builder notation.

    (a) $\{-2, -4, -6, \ldots\}$
    (b) $\{0, 3, -3, 6, -6, 9, -9, \ldots\}$
    (c) $\{1, \frac{1}{2}, \frac{1}{3}, \frac{1}{4}, \ldots\}$

3. Determine which of the following pairs of sets are equal.

    (a) $A = \{x : x \text{ is an integer divisible by both 3 and 2}\}$ and $B = \{6, 12, 18, 24, \ldots\}$
    (b) $X = \{x : x \text{ is real and } x^2 < x\}$ and $Y = \{x : x \text{ is real and } 0 < x < 1\}$

4. Prove that the following sets are equal.
   $A = \{x : x \text{ is real and } 0 \leq x < 4\}$ and $B = \{x : x = y^2 \text{ for } y \text{ real and } -2 < y < 2\}$.
5. Determine whether each of the following inclusions is proper.

   (a) $A \subseteq B$, where $A = \{x : x \text{ is an odd prime}\}$ and $B = \{x : x \text{ is an integer not divisible by } 2\}$.
   (b) $X \subseteq Y$, where $X = \{x : x^2 \text{ is an integer divisible by } 9\}$ and $Y = \{x : x \text{ is an integer divisible by } 3\}$.

6. Prove if $A \subseteq B$ and $B \subseteq C$, then $A \subseteq C$.
7. Let U be the set of letters of the alphabet. Let $A = \{a, b, c, \ldots, l\}$, $B = \{h, i, j, \ldots, q\}$ and $C = \{o, p, q, \ldots, z\}$. Find the elements in each of the following sets.

   (a) $A \cap B$
   (b) $A \cup C$
   (c) $A \cap (B \cup C)$
   (d) $(A \cap B) \cup C$
   (e) $A' \cap B'$
   (f) $(A \cap B)'$
   (g) $A \backslash B$
   (h) $B \backslash A$
   (i) $A \backslash (B \backslash C)$
   (j) $A \backslash (C \backslash B)$

8. Prove that $A \cup A' = U$ and $A \cap A' = \emptyset$.
9. Prove $A \cup (B \cap C) = (A \cup B) \cap (A \cup C)$ from part (4) of Proposition 2.
10. Prove part (a) of Proposition 3.
11. Let $A = \{1, 2\}$, $B = \{a, b, c\}$ and $C = \{x, y\}$. List all the elements in each of the following sets.

    (a) $A \times C$
    (b) $B \times B$
    (c) $A \times B \times C$

12. List some of the elements in each of the following sets. Express each as a Cartesian product or indicate why this cannot be done.

    (a) $\{(x, y) : x \text{ is a digit and } y \text{ is a letter}\}$
    (b) $\{(x, y) : x \in N \text{ and } y = x + 1\}$

13. When we teach young children how to add, we often write story problems such as "John has 2 apples and Ami has 3 apples. How many apples do they have together?" Think about how a young student would figure this out. Write a definition involving sets for addition of two positive numbers a and b.

Functions

Let A and B be sets. A **function** f from A to B is a subset of the Cartesian product $A \times B$ in which each a in A appears exactly once as the first entry in a pair $(x, y) \in f$. We write $f : A \to B$ and we call A the **domain** of f, and B the **codomain** of f. If $(x, y) \in f$, we write $y = f(x)$ and y is called the **function value** of f at x. Such a function is also called a **mapping** of A into B. For example, let $A = \{a, b, c\}$ and $B = \{s, t, u\}$. Let $f = \{(a, s), (b, s), (c, t)\}$ so that $f: A \to B$. Then f is a function from A to B because each element of A appears exactly once as a first entry. We have $f(a) = f(b) = s$ and $f(c) = t$. However, the set $g = \{(s, a), (s, b), (t, c)\}$ does not define a function from B to A because s appears twice as a first coordinate and u does not appear at all. Both of these assignments can be illustrated by arrow diagrams as in Figure 1.

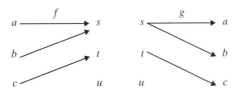

FIGURE 1

Consider another example. Let $f: \mathbf{R} \to \mathbf{R}$ be the set $f = \{(x, y): x \in \mathbf{R}$ and $y = 2x^3 - 1\}$. Then $f(3) = 53$ and $f(0) = -1$, and in general, $f(x) = 2x^3 - 1$. With such a function, we usually omit the reference to the Cartesian product and define the function simply by the formula or rule by which we compute $f(x)$. That is, we simply write $f(x) = 2x^3 - 1$ rather than $f = \{(x, y): x \in \mathbf{R}$ and $y = 2x^3 - 1\}$.

A function $f: A \to B$ is **one-to-one** or **injective** if, whenever a and b are elements of A and $a \neq b$, then $f(a) \neq f(b)$, or, equivalently, if $f(a) = f(b)$, then $a = b$. A function $f: A \to B$ is **onto** or **surjective** if, for any value y in B, there is at least one element x in A for which $f(x) = y$. The function f in Figure 1 is not injective because $a \neq b$ but $f(a) = f(b)$. It is not surjective because for there is no value of $x \in A$ such that $f(x) = u$. The function $f: \mathbf{R} \to \mathbf{R}$ defined by $f(x) = 2x^3 - 1$ is injective because if $2a^3 - 1 = 2b^3 - 1$, then some algebra tells us that $a = b$. It is surjective because given any $y \in \mathbf{R}$, we can let $x = \left(\dfrac{y+1}{2}\right)^{1/3}$ and then $f(x) = y$. A function that is both injective and surjective is called a **bijection** or a **one-to-one correspondence**.

Suppose that $g: A \to B$ and $f: B \to C$ are functions. The **composition** of f with g is a function from A to C denoted by $f \circ g$, where $(f \circ g)(x)$ is defined to be $f(g(x))$ for each x in A. In order for $f \circ g$ to be defined, the codomain of g must be contained in the domain of f. For example, suppose f and g are functions from $\mathbf{R}$ to $\mathbf{R}$ defined by $f(x) = x^2 + 3$ and $g(x) = 4x - 5$. Then $(f \circ g)(x) = f(g(x)) = f(4x - 5) = (4x - 5)^2 + 3 = 16x^2 - 40x + 28$. But $(g \circ f)(x) = g(f(x)) = g(x^2 + 3) = 4x^2 + 7$. So although both $f \circ g$ and $g \circ f$ are defined, they are not equal.

We now relate the concepts of injectivity and surjectivity to composition.

PROPOSITION 4 Suppose that $g: A \to B$ and $f: B \to C$ are functions.

(a) If both f and g are injective, then $f \circ g$ is injective.
(b) If both f and g are surjective, then $f \circ g$ is also surjective.

Proof.

(a) Suppose that s and t are in A and are not equal. Since g is injective, $g(s)$ is not equal to $g(t)$. Since f is injective, $f(g(s))$ is not equal to $f(g(t))$. Thus $(f \circ g)(s)$ is not equal to $(g \circ f)(t)$, and $f \circ g$ is injective.

(b) Let z be any element of C. We must find x in A such that $(f \circ g)(x) = z$. Since f is surjective, there is an element y in B such that $f(y) = z$. Since g is surjective, there is an element x in A such that $g(x) = y$. Thus $(f \circ g)(x) = f(g(x)) = f(y) = z$, and we have established the surjectivity of $f \circ g$. ▲

If $f : A \to B$ is a bijection, then every a in A has exactly one b in B to which it is assigned and every b in B has exactly one a in A that is assigned to it. Therefore, we can interchange the domain and range of f and produce a bijection from B to A. This function is called the **inverse of f** and is denoted f^{-1}. In other words, given a bijection $f : A \to B$, we define the function $f^{-1} : B \to A$, by the following rule: If z is an element of B, then $f^{-1}(z) = w$, where $f(w) = z$. For example, let $f : R \to R$ be the bijection defined by $f(x) = 2x^3 - 1$. Then $f^{-1}(x) = \left(\dfrac{x+1}{2}\right)^{1/3}$.

PROPOSITION 5 Suppose that $f : A \to B$ is a bijection and f^{-1} is its inverse. For each x in B we have $(f \circ f^{-1})(x) = x$ and for each x in A we have $(f^{-1} \circ f)(x) = x$.

Proof. Let x be an element of B and let $f^{-1}(x) = z$. By the definition of an inverse function, we have $f(z) = x$. Thus $(f \circ f^{-1})(x) = f(f^{-1}(x)) = f(z) = x$. Now if x is in A and $f(x) = z$, then $f^{-1}(z) = x$ and so $(f^{-1} \circ f)(x) = f^{-1}(f(x)) = f^{-1}(z) = x$. ▲

Suppose that $f : A \to B$ and let S be a subset of A. The set $f(S) = \{y : y = f(x)$ for some $x \in S\}$ is called the **direct image** of S under f. If $A = S$, then $f(A)$ is called the **image of f** or the **range of f**. For example, let $f : R \to R$ be defined by $f(x) = x^2$ and let $S = [-1, 2] = \{x \in R : -1 \leq x \leq 2\}$. Then $f(S) = [0, 4] = \{x \in R : 0 \leq x \leq 4\}$. Now let T be a subset of B. The set $f^{-1}(T) = \{x \in A : f(x) \in T\}$ is called the **preimage** of T under f. For example, if $T = [1, 4] = \{x \in R : 1 \leq x \leq 4\}$, then $f^{-1}(T) = [-2, -1] \cup [1, 2]$. **Note:** For a function like $f(x) = x^2$ that is *not* a bijection, the inverse function $f^{-1} : R \to R$ is *not* defined and we cannot compute $f^{-1}(1)$. However, the preimage of the *set* $\{1\}$ is defined: $f^{-1}(\{1\}) = \{-1, 1\}$. Be careful to distinguish between the concepts of inverse function and preimage.

Exercises on Functions

1. Determine which of the following rules define functions from the set $\{a, b, c, d\}$ into the set $\{s, u, v\}$.

(a)

(b)

(c)

(d) $f : R \to Z$, where $f(x)$ is the integer nearest x

2. Let $A = \{a, b, c\}$ and let $B = \{s, t\}$. List all the functions from A to B.
3. Determine the injectivity and the surjectivity of the following functions. Indicate which (if any) are bijections.

   (a) $f : R \to R$ defined by $f(x) = (2x + 1)/3$
   (b) $f : Z \to Z$ defined by $f(n) = 3n + 1$
   (c) $f : R \to R^+$ defined by $f(x) = x^2 + 2$

4. Define a function from N to N that is

   (a) injective but not surjective
   (b) surjective but not injective
   (c) bijective but not the identity function

5. Let $f : R \to R$ be defined by $f(x) = (3x - 1)/5$ and let $g : R \to R$ be defined by $g(x) = (x^2 + 1)/2$. Find $f \circ g$ and $g \circ f$.
6. Provide a counterexample to the converse of part (b) of Proposition 4.
7. Find the inverse of each of the following bijections.

   (a) $f : R \to R$ defined by $f(x) = (1 - 2x)/3$
   (b) $g : R^+ \to \{x : x > 1/2\}$ defined by $g(x) = (x^2 + 1)/2$

8. Prove that if f is a bijection, then $(f^{-1})^{-1} = f$.
9. Let $f : R \to R$ be defined by $f(x) = 2x - 1$. Let $A = \{x : -3 \leq x \leq 5\}$ and let $B = \{x : 0 \leq x \leq 2\}$. Find $f(A)$, $f^{-1}(B)$, $f(R^+)$ and $f^{-1}(R^+)$.
10. Give an example to show that, in general, $f(f^{-1}(A)) \neq A$. Under what conditions would equality hold?

Binary Operations

Our work with groups in this text requires us to have an understanding of a type of function called a binary operation. A **binary operation** $*$ on a set S is a function f_* from $S \times S$ into S. It takes an element (a, b) in $S \times S$ and assigns it to a single element of S, $f_*(a, b)$, that we often denote $a * b$. Addition and multiplication are familiar binary operations on $\mathbf{R}$. However, division is not a binary operation on $\mathbf{R}$. We cannot define it for any pair $(a, 0) \in \mathbf{R} \times \mathbf{R}$ because $a \div 0$ is not defined. A less familiar binary operation might be defined on $\mathbf{Z}$ as follows: $f_*(m, n) = 2mn$. In this case, $f_*(3, 5) = 3 * 5 = 2 \cdot 3 \cdot 5 = 30$.

Suppose that $*$ is a binary operation on a set S and that K is a subset S. We can certainly compute $a * b$ for any pair of elements a and b in K. But $*$ is not a binary operation on K unless the subset K is **closed** under $*$, which means that given any $a, b \in K$, we must have $a * b \in K$. If this happens then the binary operation $*$ on S is a binary operation when it is restricted to K. We call it the **induced binary operation** of $*$ on K. For example, the set of odd integers is not closed under the binary operation of addition since $3 + 5 = 8$ and 8 is not odd. Therefore, addition restricted to the subset of odd integers is not a binary operation. Addition on the set of even integers is a closed and therefore addition restricted to the subset of even integers is an induced binary operation.

Binary operations can arise in quite abstract settings as well. The table in Figure 2 defines a binary operation $*$ on the set $\{a, b, c\}$. To obtain $x * y$, we find the entry in the row labeled x and the column labeled y. Thus $b * c = a$. Such a table is called a **Cayley table** for the operation $*$.

$$
\begin{array}{c|ccc}
* & a & b & c \\
\hline
a & a & b & c \\
b & b & c & a \\
c & c & a & b
\end{array}
$$

FIGURE 2

A binary operation $*$ on a set A is said to be **commutative** if for every pair of elements a and b in A, we have $a * b = b * a$. A binary operation $*$ on a set A is said to be **associative** if for every three elements $a, b,$ and c in A, we have $a * (b * c) = (a * b) * c$. The Cayley table given in Figure 2 is symmetric; that is, the entry in the (x, y) position is the same as the entry in the (y, x) position. Thus for any pair (x, y), we have $x * y = y * x$. This means that the operation $*$ is commutative. We can also verify that for any of the 27 possible triples (x, y, z) that $x * (y * z) = (x * y) * z$. This means that the operation $*$ is associative.

Consider the following examples.

- Let $\mathbf{R}$ be the set of real numbers, and define the binary operation # on $\mathbf{R}$ by $x \# y = |x - y|$. The operation # is commutative because $|x - y| = |y - x|$, but it is not associative. For example, $2 \# (2 \# 3) = |2 - |2 - 3|| = 1$, whereas $(2 \# 2) \# 3 = ||2 - 2| - 3| = 3$.

- Let X be a nonempty set, and let $C(X)$ be the set of functions of the form $f: X \to X$. The operation of composition is associative. To see this, let f, g, and h be any functions in $C(X)$, and let x be an element of X. Then we have $((f \circ g) \circ h)(x) = (f \circ g)(h(x)) = f(g(h(x))) = f((g \circ h)(x)) = (f \circ (g \circ h))(x)$. Thus $(f \circ g) \circ h = f \circ (g \circ h)$. This proves associativity. However, composition is not commutative.

If we look again at the binary operation $*$ defined by the Cayley table in Figure 2, we find that for each x in A, we have $a * x = x$ and $x * a = x$. The same role is played by the number 0 in real number addition and the number 1 for real number multiplication. For any binary operation $*$ on a set A, we say that an element e of A is a **right identity** element if for each x in A, we have $x * e = x$. An element e of A is a **left identity** element if for each x in A, we have $e * x = x$. We say that e is an **identity** element if e is both a right and left identity element. For the example in Figure 2, the element a is the identity element of the operation $*$ as defined on set A. If X is a nonempty set, and $C(X)$ is the set of functions of the form $f: X \to X$, then the function $f_{id}: X \to X$ defined by $f_{id}(x) = x$ for all x in X is the identity element of $C(X)$ under the binary operation of function composition. Sometimes, however, we may have a right identity element that is not a left identity element or vice versa. Consider the binary operation of subtraction on $\mathbf{Z}$. The number 0 is a right identity element since $x - 0 = x$ for all $x \in \mathbf{Z}$, but 0 is not a left identity element since $0 - x = -x$, not x.

Exercises on Binary Operations

1. Let $A = \{a, b, c\}$, and let $*$ be the binary operation defined in the accompanying table. Determine the associativity and commutativity of $*$. Determine also whether $*$ has a right or a left identity element.

$*$	a	b	c
a	b	c	a
b	c	a	b
c	a	b	c

2. Repeat Exercise 1 for the operation # defined in the accompanying table.

#	a	b	c
a	a	b	c
b	b	b	a
c	a	c	b

3. Determine the commutativity and the associativity of the following binary operations defined on $\mathbf{Z}$.

   (a) $f_*(m, n) = 2mn$
   (b) $f_\#(m, n) = 2m + n$

4. Determine whether the operations given in Exercise 3 have identity elements.
5. Determine whether each of the following definitions of an operation $*$ defines a binary operation on the given set. If it does not, explain why.

   (a) On $\mathbf{Z}^* = \{x \in \mathbf{Z} : x \geq 0\}$, define $m * n = m - n$.
   (b) On $\mathbf{R} \backslash \{0\}$, define $x * y = x - y$.

6. Is the set of prime numbers closed under multiplication? Under addition?
7. Let $A = \{a, b, c, d\}$, and let S be the collection of subsets of A that contain b. Show that S is closed under $\cap$.

Bibliography

Books

1. Andrews, George. *Number Theory*. New York: Dover, 1994.
2. Armstrong, M. A. *Groups and Symmetry*, UTM, New York: Springer-Verlag, 1988.
3. Artin, Michael. *Algebra*, Englewood Cliffs, N.J.: Prentice Hall, 1991.
4. Bressoud, D. M. *Factorization and Primality Testing*. New York: Springer-Verlag, 1989.
5. Burton, David. *Elementary Number Theory*, 5th edition. New York: McGraw-Hill, 2001.
6. Dummit, David and Foote, Richard. *Abstract Algebra*, 2nd edition. Upper Saddle River, N.J.: Prentice Hall, 1999.
7. Fraleigh, John B. *A First Course in Abstract Algebra*; historical notes by Victor Katz, 7th edition. Reading, Mass.: Addison-Wesley, 2002.
8. Gallian, Joseph A. *Contemporary Abstract Algebra*, 5th edition. Lexington, Mass.: D. C. Heath, 2001.
9. Katz, Victor. *A History of Mathematics*, 2nd edition. Reading, Mass. Addison-Wesley.
10. Koblitz, N. *A Course in Number Theory and Cryptography*. New York: Springer-Verlag, 1987.
11. I. Niven, H. Zuckerman, and H. Montgomery. *An Introduction to the Theory of Numbers*, 5th edition. New York: John Wiley & Sons, 1991.
12. Ore, O. *Number Theory and Its History*. New York: McGraw-Hill, 1949.
13. Sundstrom, Ted. *Mathematical Reasoning: Writing and Proof*. Upper Saddle River, N.J.: Prentice Hall, 2002.
14. Tucker, Alan. *Applied Combinatorics*, 4th edition. New York: John Wiley & Sons, 2001.
15. Vorobiev, N. *Fibonacci Numbers*. Basel, Switzerland: Birkhauser, 2002.

Articles

The following articles are drawn primarily from the journals of the Mathematical Association of America (MAA).

I. Number Theory

1. Cassinet, J. "The first arithmetic book of Francisco Maurolico," in C. Hay (ed.), *Mathematics from Manuscript to Print 1300–1600* (Oxford, 1988), 162–179.
2. Davis, D. and Shisha, O. "Simple Proofs of the Fundamental Theorem of Arithmetic," *Mathematics Magazine*, Vol. 54, No. 1, pp. 18, 1981.
3. Hoggatt, V. and Bicknell, M. "Some Congruences of the Fibonacci Numbers Modulo a Prime p," *Mathematics Magazine*, 47 (1974) 210–214.
4. Levine, L. "Fermat's Little Theorem: A Proof by Function Iteration," *Mathematics Magazine*, Vol. 72, No. 4, pp. 308–309, 1999.
5. Osborn, Roger. "A 'Good' Generalization of the Euler-Fermat Theorem," *Mathematics Magazine*, 47 (1974) 28–31.

6. Osler, T. "Fermat's Little Theorem From the Multinomial Theorem," *College Mathematics Journal*, Vol. 33, No. 3, p. 239, 2002.
7. Pomerance, Carl. "Lecture Notes on Primality Testing and Factoring," *MAA Notes* 4, Math. Assoc. of America (Washington), 1984.
8. Stevens, G. "Forward and Backward with Euclid," *College Mathematics Journal*, Vol. 12, No. 5, pp. 302–306, 1981.

II. Modular Arithmetic and Number Systems

1. Gallian, Joseph and Van Buskirk, James. "The Number of Homomorphisms from Z_m into Z_n," *American Mathematical Monthly*, Vol. 91, No. 3 (March 1984), pp. 196–197.
2. Gilmer, Robert. "Zero Divisors in Commutative Rings," *American Mathematical Monthly*, Vol. 93, No. 5 (May 1986), pp. 382–387.
3. Kalman, Dan. "Fractions with Cycling Digit Patterns," *College Mathematics Journal*, Volume 27, Number 2, pages 109–115 (1996).
4. Kleiner, Israel. "Field Theory: From Equations to Axiomatization, Part II," *American Mathematical Monthly*, Vol. 106, No. 9 (November 1999), pp. 859–863.
5. Meijer, A.R. "Groups, Factoring, and Cryptography," *Mathematics Magazine*, Volume 69, Number 2, pp. 103–109 (1996).
6. Mollin, Richard Anthony. "A Brief History of Factoring and Primality Testing B.C. (Before Computers)," *Mathematics Magazine*, Volume 75, Number 1, pp. 18–29 (2002).
7. Stillwell, John. "What Are Algebraic Integers and What are They Good For?" *American Mathematical Monthly*, Vol. 101, No. 3 (March 1994), pp. 266–270.

III. Polynomials

1. Chapman, Scott T. "A Simple Example of Non-Unique Factorization in Integral Domains," *American Mathematical Monthly*, Vol. 99, No. 10 (December 1992), pp. 943–946.
2. Eggleton, R. B., Lacampagne, C. B., and Selfridge, J. L. "Euclidean Quadratic Fields," *American Mathematical Monthly*, Vol. 99, No. 9 (November 1992), pp. 829–837.
3. Isaacs, I. M. "Solution of Polynomials by Real Radicals," *American Mathematical Monthly*, Vol. 92, No. 8 (October 1985), pp. 571–575.
4. Rosen, Michael. "Niels Hendrik Abel and Equations of the Fifth Degree," *American Mathematical Monthly*, Vol. 102, No. 6 (June–July 1995), pp. 495–505.
5. Spearman, Blair K. and Williams, Kenneth. "Characterization of Solvable Quintics $x^5 + ax + b$," *American Mathematical Monthly*, Vol. 101, No. 10 (December 1994), pp. 986–992.
6. Wavrik, John J. "Computers and the Multiplicity of Polynomial Roots," *American Mathematical Monthly*, Vol. 89, No. 1 (January 1982), pp. 34–56.

IV. A First Look at Group Theory

1. Bogart, Kenneth B. "An Obvious Proof of Burnside's Lemma," *American Mathematical Monthly*, Vol. 98, No. 10 (December 1991), pp. 927–928.
2. Cannonito, F. B. "On Inequivalent Group Extensions," *American Mathematical Monthly*, Vol. 97, No. 4 (April 1990), pp. 317–319.
3. Fournelle, Thomas A. "Symmetries of the Cube and Outer Automorphisms of S_6," *American Mathematical Monthly*, Vol. 100, No. 4 (April 1993), pp. 377–380.

4. Gallian, Joseph. "Another Proof that A_5 Is Simple," *American Mathematical Monthly*, Vol. 91, No. 2 (February 1984), pp. 134–135.

5. Gupta, Narain. "On Groups in Which Every Element has Finite Order," *American Mathematical Monthly*, Vol. 96, No. 4 (April 1989), pp. 297–308.

6. Isaacs, I. M. and Robinson, G. R. "On a Theorem of Frobenius: Solutions of $x^n = 1$ in Finite Groups," *American Mathematical Monthly*, Vol. 99, No. 4 (April 1992), pp. 352–355.

7. Isaacs, I. M. and Zieschang, Thilo. "Generating Symmetric Groups," *American Mathematical Monthly*, Vol. 102, No. 8 (October 1995), pp. 734–738.

8. Jungnickel, Dieter. "On the Uniqueness of the Cyclic Group of Order n," *American Mathematical Monthly*, Vol. 99, No. 6 (June–July 1992), pp. 545–548.

9. Larsen, Mogen Esrom. "Rubik's Revenge: The Group Theoretical Solution," *American Mathematical Monthly*, Vol. 92, No. 6 (June–July 1985), pp. 381–390.

10. MacCluer, C. R. "A Simple Proof of the Fundamental Theorem of Finite Abelian Groups," *American Mathematical Monthly*, Vol. 91, No. 1 (January 1984), p. 52.

11. Poonen, Bjorn, "Congruences Relating the Order of a Group to the Number of Conjugacy Classes," *American Mathematical Monthly*, Vol. 102, No. 5 (May 1995), pp. 440–442.

12. Reid, J. D. "On Finite Groups and Finite Fields," *American Mathematical Monthly*, Vol. 98, No. 6 (June–July 1991), pp. 549–551.

13. Richards, I. M. "A Remark on the Number of Cyclic Subgroups of a Finite Group," *American Mathematical Monthly*, Vol. 91, No. 9 (November 1984), pp. 571–572.

14. Schattschneider, Doris. "The Taxicab Group," *American Mathematical Monthly*, Vol. 91, No. 7 (August–September 1984), pp. 423–428.

15. Surowski, David. "The Uniqueness Aspect of the Fundamental Theorem of Finite Abelian Groups," *American Mathematical Monthly*, Vol. 102, No. 2 (February 1995), pp. 162–163.

16. Turner, Edward C. and Gold, Karen. "Rubik's Groups," *American Mathematical Monthly*, Vol. 92, No. 9 (November 1985), pp. 617–629.

V. New Structures from Old

1. Cross, James T. "The Euler φ-Function in the Gaussian Integers," *American Mathematical Monthly*, Vol. 90, No. 8 (October 1983), pp. 518–528.

2. Johnson, Steve. "Factor Rings of Integers," *American Mathematical Monthly*, Vol. 96, No. 6 (June–July 1989), pp. 521–522.

3. Lawrence, J. and Simons, G. "Equations in Division Rings—A Survey," *American Mathematical Monthly*, Vol. 96, No. 3 (March 1989), pp. 220–232.

4. Leary, F. C. "Rings with Invertible Regular Elements," *American Mathematical Monthly*, Vol. 96, No. 10 (December 1989), pp. 924–926.

5. Owings, James C. "A Combinatorial Proof of a Criterion for Normality," *American Mathematical Monthly*, Vol. 92, No. 9 (November 1985), pp. 662–663.

VI. Looking Forward and Back

1. Bashmakova, I. G. and Rudakov, A. N. "The Evolution of Algebra 1800–1870," *American Mathematical Monthly*, Vol. 102, No. 3 (March 1995), pp. 266–270.

2. Duncan, Donald and Barnier, William. "On Trisection, Quintisection, ... etc.," *American Mathematical Monthly*, Vol. 89, No. 9 (November 1982), p. 693.

3. Hungerford, Thomas W. "A Counterexample in Galois Theory," *American Mathematical Monthly*, Vol. 97, No. 1 (January 1990), pp. 54–57.
4. Kappe, Luise-Charlotte and Warren, Bette. "An Elementary Test for the Galois Group of a Quartic Polynomial," *American Mathematical Monthly*, Vol. 96, No. 2 (February 1989), pp. 133–137.
5. Janusz, Gerald and Rotman, Joseph. "Outer Automorphisms of S_6," *American Mathematical Monthly*, Vol. 89, No. 6 (June–July 1982), pp. 407–410.

Selected Answers

ANSWERS 1.1

1. **i.** The answer is yes because any nonempty set of positive integers has a smallest member by the Well-Ordering Principle. The smallest member is 1 because we can write 1 as $1 = 139 \cdot 397 - 102 \cdot 541$.
    ii. No. If m/n is in the set, then $m/2n$ is also in the set. So there is no smallest member. The Well-Ordering Principle does not apply because the set in question is not a subset of the integers.

3. Let $n = 1$. Then $\frac{1-r^{n+1}}{1-r} = \frac{1-r^2}{1-r} = \frac{(1-r)(1+r)}{1-r} = (1+r)$ since $r \neq 1$. Assume $1 + r + r^2 + \ldots + r^n = \frac{1-r^{n+1}}{1-r}$. Then $1 + r + r^2 + \ldots + r^n + r^{n+1} = \frac{1-r^{n+1}}{1-r} + r^{n+1} = \frac{1-r^{n+1}}{1-r} + \frac{(1-r)r^{n+1}}{1-r} = \frac{1-r^{n+1}}{1-r} + \frac{r^{n+1}-r^{n+2}}{1-r} = \frac{1-r^{n+2}}{1-r}$.

5. If there is one person in the room, there are 0 handshakes. Assume that if n people are in the room, there are $\frac{n(n-1)}{2}$ handshakes. If person number $n+1$ enters the room, then n more handshakes will occur, making the total $\frac{n(n-1)}{2} + n$. Now $\frac{n(n-1)}{2} + n = \frac{n(n-1)}{2} + \frac{2n}{2} = \frac{n^2-n}{2} + \frac{2n}{2} = \frac{n^2+n}{2} = \frac{n(n+1)}{2}$.

7. Reflexivity: For all pairs (x, y), $0 = 2(x - x) = (y - y)$. So $(x, y)R(x, y)$.
    Symmetry: Assume $(x, y)R(s, t)$. Then $2(x - s) = (y - t)$. So $2(s - x) = (t - y)$ and $(s, t)R(x, y)$.
    Transitivity: Let $(x, y)R(s, t)$ and $(s, t)R(u, v)$. Then $2(x - s) = (y - t)$ and $2(s - u) = (t - v)$. Adding left and right sides, we get $2(x - u) = (y - v)$. Thus $(x, y)R(u, v)$. The equivalence class $[(1, 1)]$ consists of all points on the line $(y - 1) = 2(x - 1)$.

9. Reflexivity: xRx because x is in the same member of $\mathcal{C}$ as itself.
    Symmetry: If xRy, then yRx since y and x are in the same member of $\mathcal{C}$.
    Transitivity: If xRy and yRz, then x and y are in the same set in C and also y and z are in the same set in C. Since y is in exactly one subset of C, x and z must be in the same subset. Therefore, xRz.

11. Let $P(n)$ be the statement, "If $S \subseteq N$ contains any integer that is less than or equal to n, then S has a smallest member." By proving that $P(n)$ is true for all n, we prove that every nonempty set of natural numbers has a least element, which is the Well-Ordering Principle. Here's the proof by induction: $P(1)$ is true because if a set contains the natural number 1, its smallest member is 1. Assume that $P(n)$ is true for the integer n. Let S be a set that contains the integer $n+1$. If S contains no integer less than $n+1$, then $n+1$ is its smallest member. If S does contain an integer less than $n+1$, then it

certainly contains an integer that is less than or equal to n. By the induction hypothesis, S has a least member.

ANSWERS 1.2

1. Adding $-a$ to both sides, we obtain $-a + (a+b) = -a + (a+c)$. By the associative law, this is equivalent to $(-a+a)+b = (-a+a)+c$. Since $-a$ and a are additive inverses, we have $0+b = 0+c$. Since 0 is the additive identity, we obtain $b = c$.

3. First note that $-(-a) + (-a) = 0$. Adding a to both sides, we have $(-(-a) + (-a)) + a = 0 + a = a$. By associativity, we have $-(-a) + ((-a) + a) = a$ and so $-(-a) + 0 = a$. Thus $-(-a) = a$.

5. Assume that $ab = ac$ and that $a \neq 0$. By subtracting ac from both sides, we obtain $ab - ac = 0$. By distribution, we have $a(b-c) = 0$. Since $a \neq 0$, $b - c = 0$. Adding c to both sides, we have $b = c$. Here we need the fact that for integers, if $ab = 0$, either a or b (or both) must be 0.

7. **i.** $335 = 19 \cdot 17 + 12$
   ii. $-335 = -20 \cdot 17 + 5$
   iii. $21 = 1 \cdot 13 + 8$
   iv. $13 = 1 \cdot 8 + 5$

9. If $a = qb + r$, then $-a = (-q-1)b + (b-r)$. (Note that $0 \leq (b-r) < b$.)

11. Proof by Induction:
    Base Case: Let $n = 0$. Then we have $2^{n+1} + 3^{3n+1} = 2 + 3 = 5$ and 5 certainly divides 5. Assume that 5 divides $2^{n+1} + 3^{3n+1}$ so that there is an integer q such that $5q = 2^{n+1} + 3^{3n+1}$. Now consider $2^{n+2} + 3^{3n+4}$. Note that $2^{n+2} + 3^{3n+4} = 2 \cdot 2^{n+1} + 27 \cdot 3^{3n+1} = 2 \cdot 2^{n+1} + 2 \cdot 3^{3n+1} + 25 \cdot 3^{3n+1}$. This equals $2(2^{n+1} + 3^{3n+1}) + 25 \cdot 3^{3n+1} = 2(5q) + 5 \cdot 5 \cdot 3^{3n+1} = 5(2q + 5 \cdot 3^{3n+1})$. Thus 5 divides $2^{n+1} + 3^{3n+1}$, which proves that 5 divides $2^{n+1} + 3^{3n+1}$ for all $n \geq 0$.

ANSWERS 1.3

1. **i.** 2
   ii. 17
   iii. 1
   iv. 1

3. By Theorem 1 we know that there are integers m and n such that $am + bn = \gcd(a,b)$. If x divides both a and b, then x divides both summands on the left side and thus it divides their sum.

5. By Theorem 1, we can find integers s, t, p and q such that $1 = sx + tm$ and $1 = py + qm$. Then $1 = spxy + (pyt + tqm + sxq)m$. Again by Theorem 1, $\gcd(xy, m) = 1$.

7. i. $23 = 1 \cdot 13 + 10$
         $13 = 1 \cdot 10 + 3$
         $10 = 3 \cdot 3 + \mathbf{1}$
    ii. $1234 = 10 \cdot 123 + 4$
         $123 = 30 \cdot 4 + 3$
         $4 = 1 \cdot 3 + \mathbf{1}$
    iii. $442 = 1 \cdot 289 + 153$
         $289 = 1 \cdot 153 + 136$
         $153 = 1 \cdot 136 + \mathbf{17}$
         $136 = 8 \cdot 17 + 0$
9. First note that if n is odd, both $3n$ and $3n+2$ are odd numbers. The first step of Euclid's Algorithm, applied to $3n+2$ and $3n$, is as follows: $3n+2 = 1 \cdot 3n + 2$. Thus $\gcd(3n+2, 3n)$ is either 2 or 1. But it cannot be 2 since both $3n+2$ and $3n$ are odd.
11. $x = 5t$ and $y = 18t - 3$ for any positive value of t.
13. i. $q_2 = 1$, $q_3 = 1$ and $q_4 = 3$. Thus $s_4 = 4$ and $t_4 = -7$. The sum $4 \cdot 23 - 7 \cdot 13 = 1$.
    ii. $q_2 = 10$, $q_3 = 30$ and $q_4 = 1$. Thus $s_4 = 31$ and $t_4 = -311$. The sum $31 \cdot 1234 - 311 \cdot 123 = 1$.
    iii. $q_2 = 1$, $q_3 = 1$ and $q_4 = 1$. Thus $s_4 = 2$ and $t_4 = -3$. The sum is $2 \cdot 442 - 3 \cdot 289 = 17$.

ANSWERS 1.4

1. $12347983 = 281 \cdot 43943$ and both factors are prime numbers.
3. i. Let n_i be the maximum of m_i and k_i. Then $\text{lcm}(a,b) = p_1^{n_1} p_2^{n_2} \cdots p_n^{n_n}$.
    ii. $2^5 3^5 5^1 7^2 11^2 13^3$
5. i. Let n be a composite number, p be a prime that divides n, and q be another prime that divides n. Suppose that $p > \sqrt{n}$ and $q > \sqrt{n}$. Then $p \cdot q > \sqrt{n} \cdot \sqrt{n} = n$, which is a contradiction.
    ii. 541 is prime.
7. $2 \cdot 3 \cdot 5 \cdot 7 \cdot 11 \cdot 13 \cdot 17 \cdot 19 \cdot 23 \cdot 29 \cdot 31 + 1 = 200560490131$
9. $2 = 1 + 1$ $14 = 13 + 1$
    $4 = 3 + 1$ $16 = 13 + 3$
    $6 = 3 + 3$ $18 = 17 + 1$
    $8 = 5 + 3$ $20 = 19 + 1$
    $10 = 7 + 3$ $22 = 19 + 3$
    $12 = 11 + 1$ $24 = 19 + 5$
11. Let p be any prime number. Since p is prime, the only positive factors of p are 1 and p. Suppose that $\sqrt{p}$ is rational. Then $\sqrt{p} = \frac{a}{b}$, where a and b are relatively prime nonzero integers, and $\left(\frac{a^2}{b^2}\right) = p$ so $a^2 = p \cdot b^2$. Since p divides the right side of the equation, it must also divide the left side of the

equation, and since p is a prime, it must divide a (Euclid's Lemma). So we can rewrite a as $p \cdot n$, which gives us the equation $p^2 \cdot n^2 = p \cdot b^2$. Dividing both sides by p, we get $p \cdot n^2 = b^2$ and so, by the same argument, p must divide b. Thus a and b are not relatively prime, which is a contradiction.

ANSWERS 1.6

1. **i.** First express the numbers with a common denominator: $\frac{1}{2} = \frac{3}{6}$ and $\frac{1}{3} = \frac{2}{6}$. Then $\frac{3}{6} = 1 \cdot \frac{2}{6} + \frac{1}{6}$ and $\frac{2}{6} = 2 \cdot \frac{1}{6} + 0$. So the common measure of $\frac{1}{2}$ and $\frac{1}{3}$ is $\frac{1}{6}$. This means that both $\frac{1}{2}$ and $\frac{1}{3}$ are **integer** multiples of $\frac{1}{6}$ and that $\frac{1}{6}$ is the largest such rational number.
   ii. $\frac{3}{8} = \frac{9}{24}$ and $\frac{5}{6} = \frac{20}{24}$. Now $\frac{20}{24} = 2 \cdot \frac{9}{24} + \frac{2}{24}$ and $\frac{9}{24} = 4 \cdot \frac{2}{24} + \frac{1}{24}$. Thus $\frac{1}{24}$ is the largest common measure.
3. **i.** $\{0; 1, 2, 3\} = \frac{7}{10}$ and $\{3; 1, 2, 1, 2, 1, 2\} = \frac{153}{41}$
5. **i.** $\{0; 2, 1, 5, 2\}$, **ii.** $\{2; 11\}$, **iii.** $\{1; 4, 1, 1, 1, 2\}$
7. $\{1; 1, 1, 1, \ldots\}$
9. For both i and ii, notice that $a_{n-1} + \frac{1}{a_n} = a_{n-1} + \frac{1}{(a_{n-1})+1} = a_{n-1} + \frac{1}{(a_n-1)+\frac{1}{1}}$.

ANSWERS 1.7

1. No, since 15 is divisible by 5.
3. $a^{12} = (a^2)^6$ and by Fermat's Theorem, $(a^2)^6 - 1$ is divisible by 7. Similarly, $(a^3)^4 - 1$ is divisible by 5. Since 5 and 7 are relatively prime, $a^{12} - 1$ is divisible by 35.
5. Since $91 = 13 \cdot 7$ and 91 divides $3^{90} - 1$, we cannot use Fermat's Theorem to test for primes because there are nonprime values of p for which the conclusion holds for some values of a. But if there is any a for which the result does not hold, we are guaranteed that p is not prime.
7. 63504
9. **i.** $\tau(2) = 2$ and $\sigma(2) = 3$; $\tau(10) = 4$ and $\sigma(10) = 18$; $\tau(28) = 6$ and $\sigma(28) = 56$
   ii. The positive divisors of n are all of the form $p_1^{x_1} p_2^{x_2} \cdots p_q^{x_q}$, where $0 \leq x_i \leq n_i$. Thus there are $n_i + 1$ possibilities for the exponent of p_i.
   iii. Proof by induction on the number q of distinct prime factors of n. If $q = 1$, then $n = p^{n_1}$ for some prime p and positive n_1. Its divisors are $1, p, \ldots, p^{n_1}$. Their sum is $\frac{1-p^{n_1+1}}{1-p}$. Now suppose the assertion holds for numbers that factor into powers of $q - 1$ distinct primes and assume that n factors as $p_1^{n_1} p_2^{n_2} \cdots p_q^{n_q}$. By the induction hypothesis, the sum of the

factors of the form $p_1^0 p_2^{i_2} \cdots p_q^{i_q} = \prod_{i=2}^{q} \frac{1-p_i^{n_i+1}}{1-p_i}$. Let S be the set of all factors of the form $p_1^0 p_2^{i_2} \cdots p_q^{i_q}$. Then $\sigma(n) = \sum_{i=0}^{n_1} \sum_{a \in S} p_1^i a = \sum_{i=0}^{n_1} p_1^i \left(\sum_{a \in S} a \right) = \left(\frac{1-p_1^{n_1+1}}{1-p_1} \right) \left(\prod_{i=2}^{q} \frac{1-p_i^{n_i+1}}{1-p_i} \right) = \prod_{i=1}^{q} \frac{1-p_i^{n_i+1}}{1-p_i}$.

    iv. $\tau(n) = 72$; $\sigma(n) = 191{,}319{,}912{,}000$

11. Note that $(2^n - 1)$ and (2^{n-1}) are relatively prime. Since $(2^n - 1)$ is prime, $\sigma(2^n - 1) = 2^n$. Also, $\sigma(2^{n-1}) = 2^n - 1$. So $\sigma((2^n - 1)(2^{n-1})) = (2^n - 1)(2^n)$. The sum of the divisors strictly less than $(2^n - 1)(2^{n-1})$ is $(2^n - 1)(2^n) - (2^n - 1)(2^{n-1}) = (2^n - 1)(2^{n-1})$. Thus if $(2^n - 1)$ is prime, $(2^n - 1)(2^{n-1})$ is a perfect number. The first four perfect numbers are 6, 28, 496, and 8128. Looking further, 19156194260823610729479337808430363813099732 1548169216 is a perfect number! (Let $n = 89$.)

ANSWERS 2.1

1. $[13]_9 = \{\ldots, -14, -5, 4, 13, 22, 31, 40, \ldots\}$, $[3]_{10} = \{\ldots, -17, -7, 3, 13, 23, 33, 43, \ldots\}$, $[4]_{11} = \{\ldots, -18, -7, 4, 15, 26, 37, 48, \ldots\}$
3. Since $m|(a-b)$ and $m|(c-d)$, it follows that $m|((a+c)-(b+d))$.
5. Since $103 \equiv 3 \bmod 5$, we have $103^{45} \equiv 3^{45} \bmod 5$. Since $3^4 \bmod 5 = 1$, we have $3^{45} \bmod 5 = (3^4)^{11} 3^1 \bmod 5 = 3$.
7. The terminal digit of a must be 1, 2, 3 or 4. Thus the terminal digit of a^2 must be 1, 4, 9, or 6. So $a^2 \bmod 5$ must be 1 or 4. With the same reasoning, $a^4 \bmod 5 = 1$.
9. The solution sets are
    i. empty
    ii. $[1]_8 \cup [3]_8 \cup [5]_8 \cup [7]_8$
    iii. empty
    iv. $[3]_5$
    v. $[6]_{11}$
11. The solution sets are
    i. $[4]_{15} \cup [9]_{15} \cup [14]_{15}$
    ii. $[6]_{35} \cup [13]_{35} \cup [20]_{35} \cup [27]_{35} \cup [34]_{35}$
    iii. $[15]_{19}$
13. i. $7x \equiv 5 \bmod 11$; an inverse of 7 is 8 since $56 \equiv 1 \bmod 11$. Thus $8 \cdot 5 \equiv 40 \bmod 11$ and $40 \equiv 7 \bmod 11$. So $x = [7]_{11}$.
    ii. $8x \equiv 2 \bmod 6$; no inverse; $x = [1]_6$ and $x = [4]_6$.
    iii. $5x \equiv 3 \bmod 12$; the inverse of 5 is 5 since $25 \equiv 1 \bmod 12$. So $25x \equiv 15 \bmod 12$ and $15 \equiv 3 \bmod 12$. Thus $x = [3]_{12}$.
15. i. 4; ii. 1; iii. 2
17. $[206]_{210}$
19. The number of coins is 3930.

ANSWERS 2.4

1.

$Z_3, +$	0	1	2
0	0	1	2
1	1	2	0
2	2	0	1

$Z_3, *$	0	1	2
0	0	0	0
1	0	1	2
2	0	2	1

$Z_4, +$	0	1	2	3
0	0	1	2	3
1	1	2	3	0
2	2	3	0	1
3	3	0	1	2

$Z_4, *$	0	1	2	3
0	0	0	0	0
1	0	1	2	3
2	0	2	0	2
3	0	3	2	1

3. i. 0; ii. 3; iii. 2; iv. 2

5. i. 4; ii. 3; iii. 4; iv. 10; v. 8; vi. $\{2, 5, 8, 11\}$

7. Let p be prime. Then for all x such that $0 < x < p$, $\gcd(x, p) = 1$ and $ax \equiv 1$ mod p has a solution.

9. Suppose that a is a nonzero element in Z_m. Let $d = \gcd(a, m)$ and $y = m/d$. If $d > 1$, then a is a zero divisor because $ay = 0$ in Z_m but $y \neq 0$. Conversely, if $d = 1$, and $ay \equiv 0$ mod m, then $m | y$. So $y = 0$ in Z_m and a is not a zero divisor.

11. No. For instance, in Z_6, $2 \cdot 4 = 2 \cdot 1$ but, canceling the 2, $4 \neq 1$.

13. Let $m = 9$. Then $4 + 4 + 4 = 1 + 1 + 1$ but $4 \neq 1$.

ANSWERS 2.5

1. iii. $(-a)(-b) + (-a)(b) = (-a)(-b + b) = (-a)0 = 0$. Since, by part ii, $(-a)b = -(ab)$, we also have that $-(ab)$ is the additive inverse of both ab and $(-a)(-b)$. Thus these two are equal since their additive inverses are unique.

iv. $(-1)a = -(1 \cdot a) = -a$, again applying part ii

3. Suppose that a and b are multiplicative inverses of x in a ring R. Then $(ax)b = a(xb)$ by associativity. So $(ax)b = 1b = b = a(xb) = a1 = a$. Thus the two inverses are equal.

5. The symbol 2 has a multiplicative inverse in Z_5 and in Z_{15} but not in Z_4 or Z_{20}. Thus 2 can be cancelled from $2x = 2y$ in Z_5 and in Z_{15} but not in the others.

7. The units in Z_m are the elements that are relatively prime to m. If $\gcd(x, m) = 1$, then there are integers u and v such that $ux + vm = 1$. Thus $ux = 1 - vm$ and $ux = 1$ in Z_m. So x is a unit. If $\gcd(x, m) \neq 1$, then x is a zero divisor and hence not a unit. (See Exercise 4.)

9. Let $p(x) = a_n x^n + \ldots + a_0$ and $q(x) = b_k x^k + \ldots + b_0$ be two nonzero polynomials with coefficients in a ring R. Suppose that neither a_n nor $b_k = 0$. Their product is $a_n b_k x^{n+k} + \ldots + a_0 b_0$. If R is an integral domain, $a_n b_k \neq 0$ and so the $p(x)q(x) \neq 0$. Thus $R[x]$ is an integral domain. Conversely, suppose R is not an integral domain, and let a and b be zero divisors in R. Let $p(x) = a$ and $q(x) = b$. Then $p(x)q(x) = 0$ and $R[x]$ is not an integral domain.

11. In Z_5, $x = 2$. In Z_7, there is no solution.

13. The multiplicative inverse of a nonzero element $a + b\sqrt{5}$ is $\frac{a - b\sqrt{5}}{a^2 - 5b^2}$.

15. No, because $(2 + i)(2 - i) = 4 + 1 = 0 \bmod 5$.

17. The sum of any polynomial with itself p times results in the zero polynomial.

ANSWERS 2.6

1. $z + w = 3 - 2i$; $zw = 5 - 5i$; $z/w = (-1/5 - 7i/5)$; $w^2 = 3 + 4i$; $z^3 = -26 + 18i$

3. Both are sides are equal to $(ac + ae - bd - bf) + (ad + af + bc + be)i$.

5. $i^0 = 1, i^1 = i, i^2 = -1, i^3 = -i$. Thus $i^n = i^{n \bmod 4}$.

7. If $z = a + bi$, then both $|z|^2$ and $\bar{z}z$ are equal to $a^2 + b^2$.

9. Since $\sqrt{3} + i = 2\left(\cos\left(\frac{\pi}{6}\right) + i \sin\left(\frac{\pi}{6}\right)\right)$, $(\sqrt{3} + i)^5 = 2^5\left(\cos\left(\frac{5\pi}{6}\right) + i \sin\left(\frac{5\pi}{6}\right)\right) = -16\sqrt{3} + 16i$.

Since $(1 + i) = \sqrt{2}\left(\cos\left(\frac{\pi}{4}\right) + i \sin\left(\frac{\pi}{4}\right)\right)$, $(1 + i)^n = (\sqrt{2})^n \left(\cos\left(\frac{\pi n}{4}\right) + i \sin\left(\frac{\pi n}{4}\right)\right)$, for $n = 1, 2, \ldots$.

11. The sixth roots of unity are as follows:

$\cos(0) + \sin(0)i = 1$

$\cos\left(\frac{2\pi}{6}\right) + \sin\left(\frac{2\pi}{6}\right)i = \frac{1}{2} + \frac{i\sqrt{3}}{2}$

$\cos\left(\frac{4\pi}{6}\right) + \sin\left(\frac{4\pi}{6}\right)i = -\frac{1}{2} + \frac{i\sqrt{3}}{2}$

$\cos\left(\frac{6\pi}{6}\right) + \sin\left(\frac{6\pi}{6}\right)i = -1$

$\cos\left(\frac{8\pi}{6}\right) + \sin\left(\frac{8\pi}{6}\right)i = -\frac{1}{2} - \frac{i\sqrt{3}}{2}$

$\cos\left(\frac{10\pi}{6}\right) + \sin\left(\frac{10\pi}{6}\right)i = \frac{1}{2} - \frac{i\sqrt{3}}{2}$

13. $1 + i = \sqrt{2}\left(\cos\left(\frac{\pi}{4}\right) + i\sin\left(\frac{\pi}{4}\right)\right)$. Its cube roots are $2^{1/6}\left(\cos\left(\frac{\pi}{12}\right) + i\sin\left(\frac{\pi}{12}\right)\right)$, $2^{1/6}\left(\cos\left(\frac{9\pi}{12}\right) + i\sin\left(\frac{9\pi}{12}\right)\right)$, and $2^{1/6}\left(\cos\left(\frac{17\pi}{12}\right) + i\sin\left(\frac{17\pi}{12}\right)\right)$, where $2^{1/6}$ denotes the real sixth root of 2.

15. We can check that for any $k > 0$, $x^k - 1 = (x-1)(x^{k-1} + x^{k-2} + \ldots + 1)$ by multiplying. If $n = pq$, then $x^{pq} - 1 = (x^p)^q - 1 = (x^p - 1)((x^p)^{q-1} + (x^p)^{q-2} + \ldots + 1)$.

17. $z = \frac{-3i}{2} \pm \frac{\sqrt{-9-8i}}{2}$

ANSWERS 3.1

1. $x^2 + 3x + 1$ and $x^2 + 1$
3. a. sum: $2x^2 + 6x + 3$;   product: $6x^3 + 13x^2 + 9x + 2$
   b. sum: $2x^2 + x + 3$;   product: $x^3 + 3x^2 + 4x + 2$
   c. sum: $2x^2 + x$;   product: $x^3 + 2x^2 + 2$
   d. sum: $(2 + 3i)x^2 + 6x + 3i$ product: $(6 + 9i)x^3 + (3 + 4i)x^2 + 9ix - 2$
5. i. quotient: $x^2 + x + 1$;   remainder: 0
   ii. quotient: $x^{n-1} + x^{n-2} + \ldots + x + 1$;   remainder: 0
   iii. quotient: $2x^2 + 4x + 4$;   remainder: 2
   iv. quotient: $x^3 + x^2 + 1$;   remainder: 0
   v. quotient: $x^2 + x + 2 + 2i$ remainder: $3 + 2i$
7. i. yes; ii. no; iii. no
9. 0 (Use the Remainder Theorem and evaluate the polynomial at $x = 1$.)
11. $3x^2 + 3x$. Its roots are 0, 1, 2, 3, 4, and 5.
13. Monic: $x^2 + 1$, $x^2 + x + 2$, and $x^2 + 2x + 2$; Not monic: $2x^2 + 2$, $2x^2 + 2x + 1$, and $2x^2 + x + 1$.
15. Let $n = 1$. A polynomial of the form $x - a$ has exactly one root, namely a. Assume that the theorem is true for polynomials of degree less than n and suppose that $p(x)$ is a polynomial of degree $n > 1$. If $p(x)$ has no roots, then we are done since $0 < n$. If $p(x)$ has a root at x_0, then $p(x) = g(x)(x - x_0)$ and $\deg(g(x)) = n - 1$. By the induction hypothesis, $g(x)$ has at most $n - 1$ roots. Any root of $p(x)$ is either equal to x_0 or it is a root of $g(x)$. Thus $p(x)$ has at most n roots.

ANSWERS 3.2

1. $(x^5 + x^4 + x^3 - 2x^2 - 2x - 2) = x(x^4 + x^3 - x^2 - 2x - 2) + (2x^3 - 2)$,
   $(x^4 + x^3 - x^2 - 2x - 2) = (x/2 + 1/2)(2x^3 - 2) + (-x^2 - x - 1)$,
   $(2x^3 - 2) = (-2x + 2)(-x^2 - x - 1) + 0$.

Thus $(-x^2 - x - 1) = \gcd(f, g)$ and therefore $x^2 + x + 1$ is "the" $\gcd(f, g)$.
3. **i.** $x - 2$
   ii. $x + 4$
   iii. 1
5. **i.** $x^2(x + 1)^2$
   ii. $x(x^3 + x^2 + 1)$
   iii. $x(x + 1)^3$
   iv. $x^4 + x^3 + x^2 + x + 1$
7. **i.** $b^2 - 4ac = 0$, $(x + 4)^2$
   ii. irreducible because $b^2 - 4ac = 2$ and 2 does not have a square root in Z_5
   iii. irreducible because $b^2 - 4ac = 3$ and 3 does not have a square root in Z_5
9. By induction. Briefly, it's true if $p(x)$ has degree 0 or 1. Let $n > 1$ and assume true for polynomials of degree less than n. If $p(x)$ is not irreducible, apply induction hypothesis to its factors.

ANSWERS 3.4

1. i is primitive and ii has content 3.
3. Suppose that $f(x)$ was not primitive. Then $f(x) = tf_1(x)$, where $f_1(x) \in Z[x]$ is primitive and $t \neq \pm 1$. So $f(x)g(x) = tf_1(x)g(x)$ is not primitive.
5. **i.** $x^3 + x + 1$ is irreducible in $Z_2[x]$ since neither 0 nor 1 is root.
   ii. Irreducible. First check that $x^4 + x^2 + x + 1$ has no root in $Z_3[x]$. Then equate coefficients to see that it has no factorization of the form $(x^2 + ax + b)(x^2 + cx + d)$ in $Z_3[x]$.
7. **i.** Irreducible by using Eisenstein with $p = 5$.
   ii. $(x^2 + 2)(x^2 + 1)$
   iii. Irreducible by using Eisenstein with $p = 2$
   iv. Irreducible; no roots and no factorization of the form $(x^2 + ax + b)(x^2 + cx + d)$

ANSWERS 3.5

1. **i.** real part: $x^2 + 1 + 2x - y^2$; imaginary part: $-2 + 2y + 2xy$
   ii. real part: $y^4 - 5y^2 - 6x^2y^2 + x^4 + 5x^2 + 6$; imaginary part: $-4xy^3 + 10xy + 4x^3y$
   iii. real part: $x^3 - 3xy^2$; imaginary part: $-y^3 + 3x^2y + 1$
3. $|(s + t) - t| \leq |s + t| + |t|$, which implies that $|s| \leq |s + t| + |t|$, which in turn implies that $|s + t| \geq |s| - |t|$.
5. $p(1 + i) = 1 - (1 + i)^4 = 5$ but $p(0) = 1$.
7. Induction on the degree of the polynomial and Corollary 1.

9. Solve $0 = T = -r^2\sin^2(t) + 2r\sin(t) + r^2\cos^2(t) - r\cos(t) - 1$ and $0 = U = 2r^2\cos(t)\sin(t) - r\sin(t) - 2r\cos(t) + 1 = (2r\cos(t) - 1)(r\sin(t) - 1)$
If, from $U = 0$, $(r\sin(t) - 1) = 0$, then $T = 0$ reduces to $r\cos(t) = 0$ or $r\cos(t) = 1$. Since $r \neq 0$, $t = \frac{\pi}{2}$ or $\frac{\pi}{4}$. If $t = \frac{\pi}{2}$, then $r = 1$. If $t = \frac{\pi}{4}$, then $r = \sqrt{2}$. Thus the roots are $z = \sqrt{2}(\cos(\frac{\pi}{4}) + i\sin(\frac{\pi}{4}))$ and $z = i\sin(\frac{\pi}{2})$.

ANSWERS 3.6

1. $\omega = \frac{-1+\sqrt{-3}}{2}$, $\omega^2 = \frac{-1-\sqrt{-3}}{2}$
   i. $A = 4$, $B = 2$; $u = 4^{1/3}$, $v = 2^{1/3}$, both real;
   $x_1 = 4^{1/3} + 2^{1/3}$, $x_2 = \omega 4^{1/3} + \omega^2 2^{1/3}$, $x_2 = \omega^2 4^{1/3} + \omega 2^{1/3}$
   or approximately
   $x_1 = 2.847$, $x_2 = -1.423 + 0.284i$, $x_3 = -1.423 - 0.284i$

2. i. Write $x^4 = -8x^2 - 8x - 2 = -2(4x^2 + 4x + 1) = (\sqrt{-2})^2(2x+1)^2$.
   Thus we can reduce the quartic to two quadratics:
   $$x^2 = \sqrt{-2}(2x+1)$$
   $$x^2 = -\sqrt{-2}(2x+1).$$
   Thus we must solve $x^2 - 2\sqrt{2}ix - \sqrt{2}i = 0$ and $x^2 + 2\sqrt{2}ix + \sqrt{2}i = 0$.
   The solutions are $x = \sqrt{2}i \pm \sqrt{\sqrt{2}i - 2}$ and $x = -\sqrt{2}i \pm \sqrt{-\sqrt{2}i - 2}$.

ANSWERS 3.7

1. no
3. Since c is a root of $f(x) - f(c)$, $(x - c)$ is a factor of $f(x) - f(c)$.
5. i. Since $x \equiv -2 \mod (x+2)$, replace x by -2 to obtain -5.
   ii. Replace x^3 by $(-1-x)$ twice to obtain $2x^2 + x$.
   iii. Replace x^2 by -1 to obtain $1 + 2x$.
7. i. $3x^2 + 2x + 2$
   ii. $4x^2 + x + 1$
   iii. $1 + x + x^2$
9.

$\cdot$	0	1	x	$1+x$
0	0	0	0	0
1	0	1	x	$1+x$
x	0	x	$1+x$	1
$1+x$	0	$1+x$	1	x

11.

·	0	1	2	x	$1+x$	$2+x$	$2x$	$1+2x$	$2+2x$
0	0	0	0	0	0	0	0	0	0
1	0	1	2	x	$1+x$	$2+x$	$2x$	$1+2x$	$2+2x$
2	0	2	1	$2x$	$2+2x$	$1+2x$	x	$2+x$	$1+x$
x	0	x	$2x$	2	$x+2$	$2x+2$	1	$x+1$	$2x+1$
$1+x$	0	$1+x$	$2+2x$	$x+2$	$2x$	1	$1+2x$	2	x
$2+x$	0	$2+x$	$1+2x$	$2x+2$	1	x	$1+x$	$2x$	2
$2x$	0	$2x$	x	1	$1+2x$	$1+x$	2	$2+2x$	$x+2$
$1+2x$	0	$1+2x$	$2+x$	$x+1$	2	$2x$	$2+2x$	x	1
$2+2x$	0	$2+2x$	$1+x$	$2x+1$	x	2	$x+2$	1	$2x$

13. Let $\alpha = [x]$; i.e., the equivalence class of x in $F = \mathbb{Z}_2[x]/\langle p(x)\rangle$. Then α is a root of $p(x)$. Use long division to divide $p(x)$ by $x - \alpha$ to obtain $p(x) = (x - \alpha)(x^2 + \alpha x + 1 + \alpha^2)$. Check to see if any element of $\mathbb{Z}_2[x]/\langle p(x)\rangle$ is a root of $(x^2 + \alpha x + 1 + \alpha^2)$. Both α^2 and $\alpha^2 + \alpha$ are roots. Thus $p(x) = (x - \alpha)(x - \alpha^2)(x - (\alpha^2 + \alpha))$.

15. (Use the Euclidean Algorithm.) The inverse of $1 + 2^{1/3} - 3(2^{2/3})$ is
$-\frac{(7 + 17 \cdot 2^{1/3} + 4 \cdot 2^{2/3})}{87}$

17. i. First note that $\alpha^2 + \beta\alpha = -\beta^2$. Then $\xi^2 = \frac{4\alpha^2 + 4\alpha\beta + \beta^2}{\beta^2} = \frac{4(-\beta^2) + \beta^2}{\beta^2} = -3$.

  ii. Use the quadratic formula to solve for a root α of $x^2 + \beta x + \beta^2 = 0$ and simplify.

ANSWERS 4.1

1. i. No, subtraction is not associative since $(1 - 2) - 3 \neq 1 - (2 - 3)$.

  ii. Yes; multiplication is associative; 1 is the identity element; $2 \cdot 4 = 1$.

*	1	2	4
1	1	2	4
2	2	4	1
4	4	1	2

  iii. No; $4 \cdot 2 = 3$ and 3 is not a member of the set. The operation is not closed.

3. It is given (and you can check) that the operation is associative. The identity element is c. (Note the columns and rows labeled c.) Each column and row has c as an entry. So every element has an inverse.

5. i. Proof by induction. For $n = 1$, we have $(a^1)^{-1} = a^{-1} = (a^{-1})^1$. Now assume the result is true for $n - 1$. We know that $(a^n)^{-1} a^n = e$. Consider the product $(a^{-1})^n a^n = (a^{-1})^{n-1} a^{-1} a a^{n-1} = (a^{-1})^{n-1} e a^{n-1} = (a^{-1})^{n-1} a^{n-1}$. By the induction hypothesis $(a^{n-1})^{-1} = (a^{-1})^{n-1}$ and thus $(a^{-1})^n a^n = (a^{-1})^{n-1} a^{n-1} = (a^{n-1})^{-1} a^{n-1} = e$. The equality $(a^n)^{-1} a^n = e = (a^{-1})^n a^n$ gives the desired result.

ii. Fix a positive integer j. We will prove first that $a^i a^j = a^{i+j}$ for i a nonnegative integer. If $i = 0$, we have $a^i = e$. So $a^i a^j = ea^j = a^{0+j}$. Now assume that the result is true for $i \leq k-1$ and write $a^k = aa^{k-1}$. Then $a^k a^j = (aa^{k-1})a^j = a(a^{k-1}a^j) = aa^{k+j-1} = a^{k+j}$. (The rest of the proof follows from part i).

7. $(a^{-3}b^2) \to -3a + 2b$
   $(5a - 2b) \to a^5 b^{-2}$

9. Part v. Note that $(xy)(y^{-1}x^{-1}) = x(yy^{-1})x^{-1} = xex^{-1} = xx^{-1} = e$. Also, $(xy)(xy)^{-1} = e$. Thus by part ii, $(xy)^{-1} = y^{-1}x^{-1}$.

11. The multiplicative inverse of 3 in U_8 is 3. Multiplying both sides of $3x = 7$ by 3 yields $3(3x) = x = 3(7) = 21 = 5 \mod 8$. Answer: $x = 5$.

13. $U_{10} = \{1, 3, 7, 9\}$

·	1	3	7	9
1	1	3	7	9
3	3	9	1	7
7	7	1	9	3
9	9	7	3	1

$U_{12} = \{1, 5, 7, 11\}$

·	1	5	7	11
1	1	5	7	11
5	5	1	11	7
7	7	11	1	5
11	11	7	5	1

Every element in U_{12} is its own inverse. In U_{10}, only 1 and 9 are their own inverses.

15. There are four possible tables. In only one of these, $x^2 = e$ for $x = a, b,$ and c. In all three others, exactly one of a, b, or c had the property that $a^2 = e$.

17. By induction. Note that $abab = a(ba)b = a(ab)b = a^2b^2$. Choose an example from $\mathbf{M}_{2,2}$.

19. By hypothesis, $aabb = abab$. We can multiply each side on the left by a^{-1} and on the right by b^{-1} to obtain $ab = ba$.

ANSWERS 4.2

1. The easiest way to solve $R_1 * x = F_b$ is to look across the row that corresponds to R_1 until F_b appears and look for the element at the top of the column. It is F_c. Thus $x = F_c$. Another way to solve for x is to find R_1^{-1}, which is R_2. Then $R_2 * R_1 * x = R_0 * x = x = R_2 * F_b = F_c$.

Selected Answers   407

3. The operation table without any flips is

*	R_0	R_1	R_2
R_0	R_0	R_1	R_2
R_1	R_1	R_2	R_0
R_2	R_2	R_0	R_1

If you try to write out a table with only flips and R_0 the operation table will not be closed since, for example, $F_a * F_b = R_1$.

A table with R_0 and a single flip will be closed. For example,

*	R_0	F_a
R_0	R_0	F_a
F_a	F_a	R_0

5.

*	R_0	R_1	R_2	R_3	H	V	D_1	D_2
R_0	R_0	R_1	R_2	R_3	H	V	D_1	D_2
R_1	R_1	R_2	R_3	R_0	D_1	D_2	V	H
R_2	R_2	R_3	R_0	R_1	V	H	D_2	D_1
R_3	R_3	R_0	R_1	R_2	D_2	D_1	H	V
H	H	D_1	V	D_2	R_0	R_2	R_3	R_1
V	V	D_2	H	D_1	R_2	R_0	R_1	R_3
D_1	D_1	V	D_2	H	R_1	R_3	R_0	R_2
D_2	D_2	H	D_1	V	R_3	R_1	R_2	R_0

7. Starting with H and R_2, the only other symmetries needed to fill out a group table are V and R_0. Starting with H and R_1, all the other symmetries are required.

9. Let the vertices of the regular hexagon be *ABCDEF* and let O be the center of the hexagon. The 12 symmetries consist of six rotations about O of 0°, 60°, 120°, 180°, 240°, and 300° and their six compositions with the flip about the line through A and O. These compositions are themselves flips about axes through O and pairs of opposite vertices or through O and the midpoints of pairs of opposite sides. We call this group of symmetries D_6.

11. A regular icosahedron has 12 vertices and five faces that share a vertex. There are 20 faces. Let A be a fixed vertex. Then a symmetry can send A to one of 12 vertices and follow that with one of 5 rotations. Thus there are 60 symmetries of the icosahedron. The regular dodecahedron has 20 vertices and 3 regular pentagons share each vertex. Thus any symmetry moves a fixed vertex into one of 20 positions followed by three possible rotations for a total of 60 symmetries. An icosahedron can be obtained from a dodecahedron by placing a dot in the middle of each of its twelve faces and then connecting

the dots in the adjoining faces to form triangles. These triangles form the faces of the icosahedron. Similarly, a dodecahedron can be obtained from an icosahedron by placing and connecting dots in the centers of its triangular faces.

ANSWERS 4.4

1. For example, let p be a prime. Let Q_p be the set of all nonzero rational numbers that can be expressed as $1/n$, where n is any integer power of p. Then Q_p is a proper subgroup of $\{Q \setminus \{0\}, \cdot\}$.
3. Let F be the flip of the triangle about the vertical axis and let R be the rotation of $60°$ about the center of the triangle. Then the symmetries of the equilateral triangle can be expressed as $\{I, R, R^2, F, RF, R^2F\}$. Proper, nontrivial subgroups are $\{I, F\}$, $\{I, R, R^2\}$, $\{I, RF\}$, and $\{I, R^2F\}$.
5. Let H_1 and H_2 be subgroups of G. $H_1 \cap H_2$ is not empty because $e \in H_1 \cap H_2$. If g and h are any elements of $H_1 \cap H_2$, then they are members of H_1 and H_2. Thus gh and g^{-1} are also in $H_1 \cap H_2$ and so $H_1 \cap H_2$ is a subgroup of G. More, generally, if $\{H_c : c \in C\}$ is a collection of subgroups of G, then $e \in \bigcap_{c \in C} H_c$. If g and h are members of $\bigcap_{c \in C} H_c$, then gh and g^{-1} are also in $\bigcap_{c \in C} H_c$, which is thus a group.
7. Note that $\det(I) = 1$. If $\det(A) = 1$ and $\det(B) = 1$, then $\det(A^{-1}) = (\det(A))^{-1} = 1$ and $\det(A)\det(B) = \det(AB) = 1$.
9. Since H is nonempty, it has an element a. Let $x = a$ and let $y = a$ as well. By the given property, the product $a^{-1}a = e$ is in H. Let $b \in H$. By the defining property, $b^{-1}e = b^{-1}$ is in H. Finally, if a and b are in H, then a^{-1} is in H. Thus $(a^{-1})^{-1}b = ab \in H$. Hence H is closed and is a subgroup.
11. $\{0, 3, 6, 9\}$
13. $\langle 5 \rangle = \{1, 5\}$
15. The elements of $\langle A \rangle$ are matrices of the form $\left\{ \begin{bmatrix} 2^n & 0 \\ 0 & 3^n \end{bmatrix} : n \in Z \right\}$.
17. $\langle H, V \rangle = \{R_0, R_2, H, V\}$
19. Let $c = \gcd(m, n)$. Then $\langle m, n \rangle = \langle c \rangle$.
21. The order of 3 is 7 in Z_7. The order of 7 is 8 in Z_8. The order of 6 is 4 in Z_8. The order of m in Z_n is given by $n/\gcd(m, n)$.
23. $m/\gcd(m, i)$
25. U_{20} has 8 elements and none of them is a generator. Thus U_{20} is not cyclic.
27. All elements have order 2. Thus $U_8 \times Z_2$ is not cyclic.
29. The subgroups of U_8 are $G_1 = \{1\}$; $G_2 = \{1, 3\}$, $G_3 = \{1, 5\}$, $G_4 = \{1, 7\}$, and $G_5 = U_8$. The subgroups of Z_5 are $H_1 = \{0\}$ and $H_2 = Z_5$. The subgroups of $U_8 \times Z_5$ are the groups $G_i \times H_j$.
31. Let $n = |a|$. Since $(a^{-1})^n = (a^n)^{-1} = e^{-1} = e$, the order of the inverse of a is n.
33. See the solution to Problem 23.

ANSWERS 4.5

1. The operation table for the set of rotations under the operation $*$ is

$*$	R_0	R_1	R_2
R_0	R_0	R_1	R_2
R_1	R_1	R_2	R_0
R_2	R_2	R_0	R_1

   From the table R_0 is the identity element and R_2 is the inverse of R_1 and vice versa. The bijection defined by $R_0 \leftrightarrow 0$, $R_1 \leftrightarrow 1$, $R_2 \leftrightarrow 2$ takes the above table to the table for $\{Z_3, +\}$.

3. The following one-to-one correspondence is an isomorphism: $(0,0) \leftrightarrow R_0$, $(1,0) \leftrightarrow H$, $(0,1) \leftrightarrow V$, $(1,1) \leftrightarrow R_2$.

5. There is only one isomorphism class for groups of order 3 and one for groups of order 5. There are two nonisomorphic groups of order 4, the Klein 4-Group and $\{Z_4, +\}$.

7. Consider $f(e) = e' = f(x \cdot x^{-1}) = f(x) \cdot f(x^{-1})$. Since $f(x) \cdot f(x^{-1}) = e'$ and inverses are unique we have $f(x^{-1}) = (f(x))^{-1}$.

9. Let G and H be of the same order n. Let $G = \langle x \rangle$ and $H = \langle y \rangle$. Define $f : G \to H$ by $f(x^k) = y^k$. Then f is an isomorphism.

11. For $n \in Z_9$ let $f(n) = n \bmod 3$.

13. Suppose $f \colon Z_m \to Z_n$ is a group homomorphism. Let $q = f(1)$ and let $x \in Z_m$. Then $x = 1 + 1 + \ldots + 1$ and $f(x) = q + q + \ldots + q = qx \bmod n$. However, the map $x \to 2x$ from Z_3 to Z_5 is not a homomorphism because $f(0) = f(1+1+1) = (2+2+2) \bmod 5 = 1$ but $f(0) = 0$.

15. Suppose the order of $f(x)$ is m. Suppose that m does not divide n. Then we can use the Division Algorithm to express n as $qm + r$, where $0 < r < m$. Since $f(x^n) = (f(x))^n = f(e) = e'$, we have

    $$f(x^{mq+r}) = ((f(x))^m)^q (f(x))^r = (e')^q (f(x))^r = (f(x))^r = e'.$$

    But m is the smallest positive exponent that satisfies $(f(x))^m = e'$ while r is smaller than m, which is a contradiction.

17. $f(x) = f(y)$ if and only if $f(x)f(y)^{-1} = e_H$. Since f is operation preserving, $f(x)f(y)^{-1} = e_H$ if and only if $f(xy^{-1}) = e_H$. Now $f(xy^{-1}) = e_H$ if and only if $xy^{-1} \in \ker(f)$. So if $\ker(f) = \{e_G\}$ and $f(x) = f(y)$, then $xy^{-1} = e_G$ and $x = y$. Thus f is injective. Conversely, if f is injective and $f(x) = e_H$, then $x = e_G$. So $\ker(f) = \{e_G\}$.

ANSWERS 4.6

1. Let $|G| = n$. Then $|b| = n$ if and only if the subgroup generated by b, namely $\langle b \rangle$, has n elements. Furthermore, $\langle b \rangle$ has n elements if and only if $\langle b \rangle = G$.

3. The set of generators of U_7 are $\{3,5\}$ and of U_{13} are $\{2,6,7,11\}$, U_{20} does not have a generator.
5. **a.** $\{2,-2\}$
   b. $\{1,3,7,9,11,13,17,19\}$
   c. $\{a, a^2, a^4, a^7, a^8, a^{11}, a^{13}, a^{14}\}$
   d. $\{a^3, a^6, a^9, a^{12}\}$
7. $\{e\}, \langle a \rangle$
9. The order of a^{20} is 7. The subgroup is $\{a^{20}, a^{12}, a^4, a^{24}, a^{16}, a^8, a^{28} = e\}$.
11. Suppose that $G = \langle a \rangle$ is a cyclic group of order m. Let d be a divisor of m and let $i = m/d$. Then $|\langle a^i \rangle| = d$. The generators of $\langle a^i \rangle$ are elements of the form $a^{k \cdot i}$, where $\gcd(k, d) = 1$. There are exactly $\phi(d)$ such elements in $\langle a^i \rangle$. Furthermore, if a^j is any element of G of order d, then $d = m/\gcd(j, m)$ or $\gcd(j, m) = m/d = i$. Since i divides j, $a^j \in \langle a^i \rangle$.
13. A generator for $\langle a^m \rangle \cap \langle a^n \rangle$ is a^k where $k = \text{lcm}(m, n)$. Note that $\langle a^m \rangle \cup \langle a^n \rangle$ is a subgroup of $\langle a \rangle$ only if m divides n or vice versa, in which case, its generator is a^j where $j = \min(m, n)$.
15. **Proof.** Let p be a prime. Fermat's Little Theorem states that if a is not a multiple of p, then $a^{p-1} \equiv 1 \bmod p$. Thus the polynomial $x^{p-1} - 1 \in Z_p[x]$ has $p - 1$ distinct roots in the field Z_p. Its roots are all the members of the set $U_p = \{1, 2, \ldots, p-1\}$.

Step 1: Show that $x^d - 1$ is a factor of $x^{p-1} - 1$ if and only if d is a factor of $p - 1$.

More generally, $x^d - 1$ is a factor of $x^n - 1$ if and only if d is a factor of n.

Proof by induction on n: The statement is true for $n = 1, 2, \ldots, d$. Suppose $n > d$ and that the statement is true for all values less than n. Then $x^n - 1 = (x^d - 1)x^{n-d} + (x^{n-d} - 1)$ as in the first step of the long division process. So $x^n - 1$ is divisible by $x^d - 1$ if and only if the remainder $(x^{n-d} - 1)$ is divisible by $(x^d - 1)$. Now apply the induction hypothesis and note that $n - d$ is divisible by d if and only if n is divisible by d.

Step 2: Show that if d is a factor of $p - 1$, then $x^d - 1$ has d distinct roots in U_p.

By step 1, $x^{p-1} - 1 = (x^d - 1)(x^{p-d-1} + \ldots + 1)$. Since $x^{p-1} - 1$ has $p - 1$ roots, and $(x^{p-d-1} + \ldots + 1)$ has at most $p - d - 1$ roots, $x^d - 1$ has d roots.

Now suppose that q is prime and that q^m divides $p - 1$. There are $q^m - q^{m-1}$ numbers $x \in U_p$ that are relatively prime to q^m since $\varphi(q^m) = q^m - q^{m-1}$.

Step 3: Suppose that $x \in U_p$ and $x^{q^m} = 1$. Suppose that d is the smallest positive integer such that $x^d = 1$. Show that $d | q^m$. (Use the Division Algorithm.)

$$\text{If } q^m = td + r, \quad \text{then } x^{td+r} = 1 = x^r.$$

Step 4: Show that there are $q^m - q^{m-1}$ members of U_p that are roots of $x^{q^m} - 1$ but that are not roots of $x^d - 1$ for any $d < q^m$.

We know that $x^{q^m} - 1$ has q^m roots by step 1. The proper divisors of q^m that are less than q^m are $q, q^2, \ldots, q^{m-1}$. Thus the factors $x^{q^i} - 1$ each have q^i roots. So the number roots of $x^{q^m} - 1$ that are also roots of $x^{q^i} - 1$ for $i < 1$ is $q + (q^2 - q) + (q^3 - q^2) + \ldots + (q^{m-1} - q^{m-2}) = q^{m-1}$. (*Note:* In each parenthesis, we subtracted the number of roots that were roots of lower degree.)

In the language of groups, Step 4 means that if x is one of the $q^m - q^{m-1}$ member of U_p described in Step 4, then the order of x is q^m. Now suppose that $p - 1 = q_1^{m_1} q_2^{m_2} \cdots q_n^{m_n}$, where each q_i is a distinct prime. For each i, let x_i be an element of order $q_i^{m_i}$. Let $x = x_1 x_2 \cdots x_n$.

Step 5: Show that the order of $x = x_1 x_2 \cdots x_n$ is $p - 1$.

17. U_{10} is cyclic of order 4 and 3 is a generator, but 10 is not prime.

ANSWERS 4.7

1. Using the following notation for the elements of S_3, the group table is below:

$$\sigma_1 = \begin{pmatrix} 1 & 2 & 3 \\ 1 & 2 & 3 \end{pmatrix} \quad \sigma_2 = \begin{pmatrix} 1 & 2 & 3 \\ 2 & 3 & 1 \end{pmatrix} \quad \sigma_3 = \begin{pmatrix} 1 & 2 & 3 \\ 3 & 1 & 2 \end{pmatrix}$$

$$\sigma_4 = \begin{pmatrix} 1 & 2 & 3 \\ 1 & 3 & 2 \end{pmatrix} \quad \sigma_5 = \begin{pmatrix} 1 & 2 & 3 \\ 3 & 2 & 1 \end{pmatrix} \quad \sigma_6 = \begin{pmatrix} 1 & 2 & 3 \\ 2 & 1 & 3 \end{pmatrix}$$

$\circ$	σ_1	σ_2	σ_3	σ_4	σ_5	σ_6
σ_1	σ_1	σ_2	σ_3	σ_4	σ_5	σ_6
σ_2	σ_2	σ_3	σ_1	σ_6	σ_4	σ_5
σ_3	σ_3	σ_1	σ_2	σ_5	σ_6	σ_4
σ_4	σ_4	σ_5	σ_6	σ_1	σ_2	σ_3
σ_5	σ_5	σ_6	σ_4	σ_3	σ_1	σ_2
σ_6	σ_6	σ_4	σ_5	σ_2	σ_3	σ_1

3. $\pi(6) = 4$. The chart of the permutation is $\begin{pmatrix} 1 & 2 & 3 & 4 & 5 & 6 \\ 5 & 3 & 6 & 1 & 2 & 4 \end{pmatrix}$. Starting with 6, the cycle for π looks like (6, 4, 1, 5, 2, 3).

5. $\begin{pmatrix} 1 & 2 & 3 & 4 & 5 & 6 & 7 & 8 \\ 5 & 3 & 1 & 4 & 6 & 2 & 8 & 7 \end{pmatrix}$

7. Let $\beta = (1,2,3)(4,5)$ and $\alpha = (1,2)(3,4)(5)$. Then $\beta\alpha = (1,3,5,4)(2)$, and $\alpha\beta = (1)(2,4,5,3)$.

9. (1,2,3,4,5)   24 different permutations (even)
   (1,2,3,4)(5)   30 (odd)
   (1,2,3)(4,5)   20 (odd)
   (1,2,3)(4)(5) 20 (even)
   (1,2)(3,4)(5) 15 (even)
   (1,2)(3)(4)(5) 10 (odd)
   (1)(2)(3)(4)(5) 1 (even)
   There are 60 even permutations.

11. See Exercise 2 for a subgroup of order 8. Since the identity is even, no subgroup can have only odd permutations.

13. The orders of the elements of S_5 are given for a permutation of each cycle form.

Cycle Structure	Order
(1,2,3,4,5)	5
(1,2,3,4)(5)	4
(1,2,3)(4,5)	6
(1,2,3)(4)(5)	3
(1,2)(3,4)(5)	2
(1,2)(3)(4)(5)	2
(1)(2)(3)(4)(5)	1

15. Given two 2-cycles, (a,b) and (c,d) with all the elements different, $(a,b)(c,d) = (a,c,b)(a,c,d)$. Given two 2-cycles of the form (a,b) and (a,c), their product is the 3-cycle (a,c,b). Any element of A_n that is the product of an even number of 2-cycles can be thus be represented as a product of 3-cycles.

ANSWERS 5.1

1. If G is abelian, H is a subgroup, and $a \in G$, then $aH = \{ah : h \in H\} = \{ha : h \in H\} = Ha$.

3. H is both a left and right coset by itself. There is one other left and one other right coset of H. (They are identical.)

$$(1,3)H = \{(1,3),(1,2),(2,3)\} = (1,2)H = (2,3)H$$
$$H(1,3) = \{(1,3),(1,2),(2,3)\} = H(1,2) = H(2,3)$$

5. The other left and right cosets of S (those that differ from S itself) are equal:

$$VS = \{V, H, D_1, D_2\} = SV.$$

7. **Proposition 1.** (For right cosets) Let G be a group and let H be subgroup of G. Let a and b be elements of G.

   1. There is a one-to-one correspondence from H to Ha.
   2. Either $Ha = Hb$ or $Ha \cap Hb = \emptyset$.

   Proposition 2. (For right cosets) Let H be a subgroup of G and let a and b be elements of G. Then $Ha = Hb$ if and only if $ab^{-1} \in H$.

   Corollary 3. (For right cosets) $Ha = H$ if and only if $a \in H$.

9. Euler's Theorem states that if a and m are relatively prime then $a^{\phi(m)} \equiv 1 \pmod{m}$. Suppose that a and m are relatively prime. Suppose that $0 < a < m$. (If not, replace a by $a \bmod m$.) Then $a \in U_m$. Since $|U_m| = \varphi(m)$, we have that $a^{\phi(m)} = 1$ in U_m by Corollary 5.

11. The result holds because R_2 (rotation through 180 degrees) commutes with every element in D_4.

13. $\{0, 2, 4\}, \{1, 3, 5\}$

15. Suppose that $aH = Ha$. Then for any $h \in H$, $ah = h_1 a$ for some $h_1 \in H$. So we have that $aha^{-1} = h_1 aa^{-1} = h_1$. Thus $aHa^{-1} \subseteq H$. To see that $H \subseteq aHa^{-1}$, note that $h = a(a^{-1}ha)a^{-1}$. Now suppose that $aHa^{-1} = H$. Then for any $h \in H$, $aha^{-1} = h_1$ for some $h_1 \in H$. Thus $ah = h_1 a$ and $aH \subseteq Ha$. Similarly, $Ha \subseteq aH$.

17. Suppose that the order of G is m, and that m is not a prime. Let x be an element that is not the identity. If the order of x is less than m, then we are done since $\langle x \rangle$ is a nontrivial, proper subgroup. So suppose that x has order m. Let q be a proper factor of m greater than 1. (This exists because m is not a prime.) Then the element x^q has order m/q and hence $\langle x^q \rangle$ is a nontrivial, proper subgroup of G.

19. Assume that A_4 has a subgroup H of order 6.
    a. The permutations of order 3 are of the form $(a, b, c)(d)$. To see that are 8 such permutations, note that there are 4 ways to pick the 3 elements in the 3-cycle and 2 ways to arrange them. The elements of order 2 are of the form $(a, b)(c, d)$. There are six ways to pick the two symbols for the first 2-cycle and 1 way to then fill in the second 2-cycle. But we must divide by 2 since the cycles commute, making a total of 3.
    b. Let a be any element of A_4 of order 3. Since H has six elements, aH and a^2H each have six elements. Either H and aH are the same set or together they have 12 distinct elements and together contain all of A_4. Similarly with a^2H. Thus either a^2H and aH coincide or one of them is the same as H.
    c. If $H = aH$ then ae is in H. If $aH = a^2H$, then there is an element b of H such that $ab = a^2$ which implies that $a = b$ and hence that a is in H. Finally, assume that H and a^2H are the same. Since a is of order 3, $a^4 = a$, and since a^4 is in H, so is a.
    d. By part c, H contains all elements of order 3. There are 8 of those. Thus H is not of order 6.

ANSWERS 5.2

1. The proper normal subgroups of S_3 are $\{e\}$ and $\{e, (1, 2, 3), (1, 3, 2)\}$.
3. The proper normal subgroups of D_4 are $\{R_0\}$, $\{R_0, R_2\}$, $\{R_0, R_1, R_2, R_3\}$.
5. **Proof.** Suppose that axa^{-1} and aya^{-1} are in aHa^{-1}. Then the product $axa^{-1}aya^{-1} = axya^{-1}$ is in aHa^{-1} and so aHa^{-1} is closed. Since $(axa^{-1})^{-1} = ax^{-1}a^{-1}$, the inverse of each element of aHa^{-1} is in aHa^{-1}. Thus aHa^{-1} is a subgroup of G.
7. See Proposition 1.
9. We must show that, for any rotation R and any symmetry x, $x^{-1}Rx$ is also a rotation. That is certainly true if x is a rotation. If x is a flip, then x^{-1} is a flip. Since a flip followed by a rotation followed by another flip leaves the original side forward, it is a rotation.
11. Yes. To check, we evaluate permutations of the following form.

$$(a, c, b)[(a, b)(c, d)](a, b, c) = (a, c)(b, d)$$
$$(a, d, b)[(a, b)(c, d)](a, b, d) = (a, d)(b, c)$$
$$(a, d, c)[(a, b)(c, d)](a, c, d) = (a, c)(b, d)$$
$$(b, d, c)[(a, b)(c, d)](b, c, d) = (a, d)(b, c)$$

13. The cosets in S_3/A_3 are $eA_3 = \{e, (1, 2, 3), (1, 3, 2)\}$ and $(1, 2)A_3 = \{(1, 2), (2, 3), (1, 3)\}$. Denoting these by the symbols e and $(1, 2)$ respectively, the Cayley table is as follows.

*	e	(1, 2)
e	e	(1, 2)
(1, 2)	(1, 2)	e

15. The cosets are $i + H$, for $i = 0, 1, \ldots, 5$. If we represent the cosets by the symbols 0 through 5 respectively, the Cayley table is identical to that of $\{\mathbf{Z_6}, +\}$.
17. Suppose that (a, b) is an element of $\mathbf{Z} \oplus \mathbf{Z}$. Express a as $a = 2n + i$, where $i = 0$ or 1 depending on whether a is even or odd. Let $x = b - 2n$. Then (a, b) is in the coset of (i, x). Thus every coset has an element of the form (i, x), where $i = 0$ or 1. Denote the coset $(i, x)H$ simply by (i, x) and the operation by $+$. Then $(i, x) + (j, y) = (k, r)$, where $k = (i + j) \bmod 2$ and $r = x + y - 2\left\lfloor \frac{i+j}{2} \right\rfloor$. Note that $-(0, x) = (0, -x)$ and $-(1, x) = (1, -x + 2)$.
The quotient group is **not** cyclic. Any generator would have to be of the form $(1, x)$ because $(1, 1)$ can not be obtained as the sum elements of the form $(0, y)$. However, if q is odd, the element $(0, qx)$ cannot be obtained as the sum of elements of the form $(1, x)$.
19. There is no element of U_{35} that has order 8 but 8 divides 24.

21. Yes. Let $z \in G_1 \times G_2 \times \ldots \times G_n$ be the element $(h_1, h_2, \ldots, h_i, \ldots, h_n)$ and let $x \in \tilde{G}_i$ be the element $(e_1, e_2, \ldots, e_{i-1}, g, e_{i+1}, \ldots, e_n)$. Then $z^{-1}xz = (h_1^{-1}, h_2^{-1}, \ldots, h_i^{-1}, \ldots, h_n^{-1})(e_1, e_2, \ldots, e_{i-1}, g, e_{i+1}, \ldots, e_n)(h_1, h_2, \ldots, h_i, \ldots, h_n) = (h_1^{-1}h_1, \ldots, h_i^{-1}gh_i, \ldots, h_n^{-1}h_n) = (e_1, \ldots, h_i^{-1}gh_i, \ldots, e_n) \in \tilde{G}_i$.

23. Suppose that $|G| = m$ and $|K| = n$ with m and n relatively prime. Suppose that $f : G \to H$ is a nontrivial homomorphism. Then $\ker(f) \neq G$. Thus the order of the quotient group $G/\ker(f) \neq 1$ and it divides m. By the First Isomorphism Theorem, $G/\ker(f)$ is isomorphic to a subgroup of K. But then order of $G/\ker(f)$ divides n, which is impossible if m and n are relatively prime.

ANSWERS 5.3

1. For both parts i and ii, $\gamma = (b, c, d)$.
3. For α and β equal to (a, b, c, d, e, f) and (a, c, d, b, e, f), respectively, $\gamma = (b, c, d)$ and $\gamma = (a, f, e, b)$.
5. All the conjugates of $(a, b)(c, d)$ are in K. Therefore K is a normal subgroup of S_4. Since K is normal in S_4, it is normal in A_4.
7. All the given groups are solvable.
    i. $\{e\} \triangleleft \mathbf{Z}_5$
    ii. $\{e\} \triangleleft \{e, R_2\} \triangleleft \{e, R_1, R_2, R_3\} \triangleleft D_4$
    iii. Interpret the numbers 1, 2, 3, and 4 used in Exercise 10 of Section 5.2 as the labels of the vertices of the tetrahedron. Then we have the following sequence: $\{e\} \triangleleft \{e, (1, 2)(4, 3)\} \triangleleft \{e, (1, 2)(4, 3), (1, 3)(2, 4), (1, 4)(2, 3)\} \triangleleft T$
9. Case 1. Both fix the same element.
$$(a, b)(c, d)(e) \text{ and } (a, c)(d, b)(e) \text{ with } \gamma = (b, c, d)$$
    Case 2. A different element is fixed.
    i. $(a, b)(c, d)(e)$ and $(a, b)(e, c)(d)$ with $\gamma = (c, e, d)$
    ii. $(a, c)(b, d)(e)$ and $(a, b)(c, e)(d)$ with $\gamma = (c, b)(d, e)$
    iii. $(a, d)(c, b)(e)$ and $(a, b)(c, e)(d)$ with $\gamma = (d, b, e)$
11. Suppose that α is a 5-cycle and β is a 3-cycle. It is not difficult to verify that $\alpha^i \beta^j$ has order two for some i and j such that $1 \leq i \leq 4$ and $1 \leq j \leq 2$. Thus the order of the subgroup generated by a 5-cycle and a 3-cycle must be at least 30 because the subgroup must have elements of order 5, 3, and 2. Any subgroup of order 30 must be normal in A_5 because A_5 has 60 elements. But as we saw in Example 4, A_5 has no normal subgroups. Thus the subgroups generated by a 3-cycle and a 5-cycle must be the entire group A_5.

ANSWERS 5.4

1. The proper subrings of $\mathbf{Z}_6$ are $\{0\}$, $\{0, 2, 4\}$, and $\{0, 3\}$.
3. The values of $q \in \mathbf{Z}_n$ for which $q\mathbf{Z}_n$ is a proper subring of $\mathbf{Z}_n$ are those that are not relatively prime to n. Otherwise, q is a generator of the additive group $\mathbf{Z}_n$ and hence the ring $\mathbf{Z}_n$.

5. The set $S = \{a + bi : a$ is an even integer$\}$ is subring of $\mathbf{Z}[i]$ because it is closed under addition and multiplication and contains the additive inverse of each of its elements.

7. The set S is a subring of $\mathbf{Z}[i]$. (See Exercise 5.) Let $x = a + bi$ be an element of S and let $y = c + di$ be any element of $\mathbf{Z}[i]$. The product $xy = (ac - bd) + (ad + bc)i$ is in S because both the real and imaginary parts are even.

9. **Proof.** Let x and y be any elements of $I_1 \cap I_2$. Then $-x, x + y$ and xy are in $I_1 \cap I_2$ since they are in each of I_1 and I_2. So $I_1 \cap I_2$ is a subring of R. Let $a \in R$. Then ax and xa are in both I_1 and I_2. Thus $I_1 \cap I_2$ is an ideal.

11. Additionally, q must satisfy the property that q^2 mod $n = q$ mod n. Then $f(x)f(y) = q^2 xy$ and $f(xy) = qxy$ are equal mod n.

13. Let $f\colon \mathbf{Q}[x] \to \mathbf{Q}(\sqrt{3})$ be defined by $f(p(x)) = p(\sqrt{3})$. To see that f is a ring homomorphism, let $p(x)$ and $q(x)$ be polynomials in $\mathbf{Q}[x]$. Then $p(\sqrt{3}) + q(\sqrt{3}) = (p+q)(\sqrt{3})$ and $p(\sqrt{3})q(\sqrt{3}) = pq(\sqrt{3})$. The kernel of f is the ideal $\langle x^2 - 3 \rangle$, i.e., all polynomials in $\mathbf{Q}[x]$ that are divisible by $x^2 - 3$.

15. We must show that $(3 - i)(x + iy) \neq k, k = 1, 2, \ldots, 9$ for any integers x and y. Note that $(3 - i)(x + iy) = (3x + y) + (3y - x)i$. The imaginary part of the latter must be 0, and thus $x = 3y$. Substituting for x in the real part, we find that $3x + y = 9y + y = 10y$ and $10y$ cannot equal any of the integers $1, 2, \ldots, 9$ for any value of y.

17. Note: to add and multiply members of $\mathbf{Z} \times \mathbf{Z}$, add and multiply component wise so that, for instance, $(a, b) + (x, y) = (a + x, b + y)$. The cosets of $\mathbf{Z} \times \mathbf{Z}/I$ are $(0, 0) + I$, $(0, 1) + I$, $(1, 0) + I$ and $(1, 1) + I$. In the tables, we will simply write $(0, 0), (0, 1), (1, 0)$ and $(1, 1)$.

+	(0, 0)	(0, 1)	(1, 0)	(1, 1)
(0, 0)	(0, 0)	(0, 1)	(1, 0)	(1, 1)
(0, 1)	(0, 1)	(0, 0)	(1, 1)	(1, 0)
(1, 0)	(1, 0)	(1, 1)	(0, 0)	(0, 1)
(1, 1)	(1, 1)	(1, 0)	(0, 1)	(0, 0)

×	(0, 0)	(0, 1)	(1, 0)	(1, 1)
(0, 0)	(0, 0)	(0, 0)	(0, 0)	(0, 0)
(0, 1)	(0, 0)	(0, 1)	(0, 0)	(0, 1)
(1, 0)	(0, 0)	(0, 0)	(1, 0)	(1, 0)
(1, 1)	(0, 0)	(0, 1)	(1, 0)	(1, 1)

19. If $I = \langle 3x, 2 \rangle = \langle p(x) \rangle$, then $p(x)$ must be the constant polynomial $p(x) = 2$. But all polynomials in $\langle 2 \rangle$ have even coefficients and $3x$ is not in $\langle 2 \rangle$. So $\langle 3x, 2 \rangle$ is not a principal ideal.

21. **Proof.** If $p(x)$ is not irreducible, then $p(x) = q(x)t(x)$, where neither $q(x)$ nor $t(x)$ is a constant polynomial. Then the product $q(x)t(x) \in \langle p(x) \rangle$ but neither $q(x)$ nor $t(x)$ is in $\langle p(x) \rangle$. Conversely, suppose that $p(x)$ is irreducible and

$f(x)g(x) \in \langle p(x) \rangle$. Then either $f(x)$ or $g(x)$ is divisible by $p(x)$. Thus either $f(x) \in \langle p(x) \rangle$ or $g(x) \in \langle p(x) \rangle$. Thus $\langle p(x) \rangle$ is a prime ideal.

ANSWERS 5.6

1. i. $5 + 5i = (1 + i)(1 + 2i)(1 - 2i)$
   ii. $13 = (3 + 2i)(3 - 2i)$
   iii. $2 - i$ is irreducible since $N(2 - i) = 5$, which is prime.
   iv. $2 - 2i = (1 - i)^2(1 + i)$
   v. $2 + 3i$ is irreducible since $N(2 + 3i) = 13$
   vi. $3 + 4i = (1 - 2i)(-1 + 2i)$

3. i. $-14 + 7\sqrt{-5} = (4 + \sqrt{-5})(-1 + 2\sqrt{-5})$
   ii. $26 = 2 \cdot 3$, 2 and 13 are irreducible in $\mathbf{Z}(\sqrt{-5})$
   iii. $20 = 2^2 \cdot 5$
   iv. $7 + \sqrt{-5} = (1 + \sqrt{-5})(2 - \sqrt{-5})$
   v. $-4 + 2\sqrt{-5} = (1 + \sqrt{-5})^2$

5. The norm in $\mathbf{Z}(\sqrt{-5})$ is $N(a + b\sqrt{-5}) = a^2 + 5b^2$. So $N(1 + 2\sqrt{-5}) = 21 = (3)(7)$. Neither 3 nor 7 is a possible value of N. Thus $1 + 2\sqrt{-5}$ is irreducible. However, $1 + 2\sqrt{-5}$ divides 21 since $(1 + 2\sqrt{-5})(1 - 2\sqrt{-5}) = 21$. For $1 + 2\sqrt{-5}$ to be prime, it would have to divide either 3 or 7, which it does not.

7. i. $\frac{5+7i}{2+i} = \frac{17}{5} + \frac{9}{5}i$. The integers closest to the two rational numbers are 3 and 2. Thus we let the quotient be $3 + 2i$. Then $r = 1$ and $5 + 7i = (3 + 2i)(2 + i) + 1$.
   ii. $\frac{9+i}{3-2i} = \frac{25}{13} + \frac{21}{13}i$. The quotient is $2 + 2i$ and $9 + i = (2 + 2i)(3 - 2i) + (-1 - i)$.

9. i. $5 + i = (1 - i)(1 + 3i) + (1 - i)$ and $(1 + 3i) = (1 - i)(-1 + 2i)$, thus the gcd is $1 - i$.
   ii. $4 + 7i = 2(3 + 4i) + (-2 - i)$; $3 + 4i = (-2 - i)(-2 - i) + 0$. Thus $(-2 - i)$ is the gcd.

11. The $\gcd(5 + i, 1 + 3i) = 1 - i$, which is a generator of the ideal.

13. Let $I = \langle 2 + i \rangle$. The cosets are $\{0 + I, 1 + I, 2 + I, i + I, 1 + i + I\}$.

·	$0+I$	$1+I$	$2+I$	$i+I$	$1+i+I$
$0+I$	$0+I$	$0+I$	$0+I$	$0+I$	$0+I$
$1+I$	$0+I$	$1+I$	$2+I$	$i+I$	$1+i+I$
$2+I$	$0+I$	$2+I$	$1+i+I$	$1+I$	$i+I$
$i+I$	$0+I$	$i+I$	$1+I$	$1+i+I$	$2+I$
$1+i+I$	$0+I$	$1+i+I$	$i+I$	$2+I$	$1+I$

It is a field since $(2 + i)$ is irreducible.

15. Let I be an ideal in $\mathbf{Z}[i]$. We must show that $I = \langle x \rangle$ for some x in $\mathbf{Z}[i]$. Let A be the set of all integers that occur as values of $N(y)$ as y goes through the nonzero elements of $\mathbf{Z}[i]$. If we assume that I is not $\langle 0 \rangle$, then A is a nonempty set of positive integers and has a least element, say p. Let x be an element of I that satisfies $N(x) = p$. Let z be any other element of I. Divide z by x. We have $z = qx + r$ with $N(r) < N(x)$. But $r = z - qx$, which is an element of I. Thus r must be 0 and z is in the ideal $\langle x \rangle$, which must equal I.

17. Let $x \in I_i$ and let $y \in I_j$, where $i \geq j$ and let $z \in R$. Then x and y are both in I_i and therefore $x + y$ and zx and zy are all in I_i and hence in the union of all the ideals. The union is thus an ideal. It is proper because $1 \notin I_j$ for any j and therefore not in the union.

ANSWERS 6.1

1. Certainly, $\mathbf{Q} \subseteq F \subseteq \mathbf{R}$. Since $x^2 - p$ is irreducible in $\mathbf{Q}[x]$, $\mathbf{Q}[x]/\langle p(x) \rangle$ is a field. Since F is isomorphic to $\mathbf{Q}[x]/\langle p(x) \rangle$, F is a field. Note that the inverse of $a + b\sqrt{p}$ is $\frac{a - b\sqrt{p}}{a^2 - pb^2}$.

3. $\mathbf{Q}(\alpha_i), i = 1, 2, \ldots, 5$, where the α_i are the distinct roots of $x^5 - 2$.

5. In each case, elements are linear combinations of terms of the form $\alpha^i \beta^j$.

7. The field consists of all rational functions of e.

9. Refer back Ferrari's method in Section 3.6.

ANSWERS 6.2

1. i. 2; **ii.** 2; **iii.** 2; **iv.** 2; **v.** 2

3. The degree is 2 because $E = \mathbf{Q}(\sqrt{-3})$.

5. Suppose that $[F : K] = n$ and that $y \in K$. The elements $y^0, y^1, \ldots, y^n$ are not linearly independent. So we can find constants $a_0, a_1, \ldots, a_n$ in F such that $a_0 + a_1 y + \ldots + a_n y^n = 0$. Thus y is a root of the polynomial $a_0 + a_1 x + \ldots + a_n x^n \in F[x]$.

7. Follows immediately from Proposition 5

9. Follows from Theorem 4 and Exercise 5

ANSWERS 6.3

1. Suggestions:
   i. See Figure 1 that follows. Both are unit circles, each centered on the circumference of the other. The length of AB is $\sqrt{3}$.

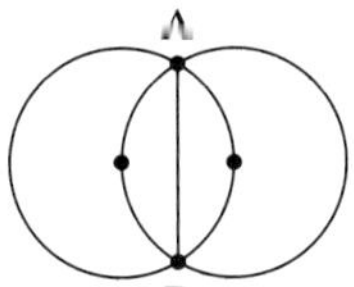

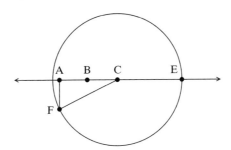

FIGURE 1

ii. AC is a unit length. Bisect it at B and construct $\triangle FAC$, a right triangle with hypotenuse $\frac{\sqrt{5}}{2}$. (Details not shown.) Construct a circle centered at C of radius $\frac{\sqrt{5}}{2}$. It intersects the line through AC at E. The segment BE has length $\frac{1+\sqrt{5}}{2}$. See Figure 2.

FIGURE 2

3. **II.** The x-coordinate of the intersection is $x = \frac{b_2-b_1}{m_1-m_2}$, clearly in the field containing m_1, m_2, b_1 and b_2. Similarly for the y-coordinate.
   III. Use the distance formula.
   IV. For x_0, solve the equation $(x-a)^2 + (mx+c-b)^2 = r^2$, which is quadratic in x. Similarly for y_0.
   V. To solve eq1: $(x-a_1)^2 + (y-b_1)^2 = r_1^2$ and eq2: $(x-a_2)^2 + (y-b_2)^2 = r_2^2$ simultaneously, first subtract the equations to obtain eq3 = eq1 − eq2, which is a linear expression in x and y. Then solve the system {eq1, eq3}.
5. $\cos(60) = \frac{1}{2} = \cos(40+20) = \cos(40)\cos(20) - \sin(40)\sin(20) = (\cos^2(20) - \sin^2(20))\cos(20) - 2\sin(20)\cos(20)\sin(20) = \ldots = 4\cos^3(20) - 3\cos(20)$.
7. A cube can be 8-pled by doubling its side. But to quadruple its volume, the side must be multiplied by $\sqrt[3]{4}$, which is not constructible. Similarly, since $\sqrt[3]{3}$ is not constructible, the cube cannot be tripled.
8. **i.** Since $\cos(60) = \frac{1}{2}$, a regular hexagon can be constructed.
   ii. We offer two different solutions.
   First Solution. $\cos\left(\frac{2\pi}{5}\right)$ is the real part of a primitive fifth root of unity. Let x be such a root, namely $x = \cos\left(\frac{2\pi}{5}\right) + i\sin\left(\frac{2\pi}{5}\right)$. Then x satisfies the

equation $x^4 + x^3 + x^2 + x + 1 = 0$. The expression on the left is a factor of the polynomial $x^5 - 1$. Thus we have $x^5 = 1$ or $x^4 = \frac{1}{x}$ and $x^3 = \frac{1}{x^2}$. Thus the equation $x^4 + x^3 + x^2 + x + 1 = 0$ can be rewritten as $\left(x + \frac{1}{x}\right) + \left(x^2 + \frac{1}{x^2}\right) + 1 = 0$. Squaring, $\left(x + \frac{1}{x}\right)^2 = x^2 + 2 + \frac{1}{x^2}$. Finally, our original equation $x^4 + x^3 + x^2 + x + 1 = 0$ becomes $\left(x + \frac{1}{x}\right)^2 + \left(x + \frac{1}{x}\right) - 1 = 0$. Thus the quantity $x + \frac{1}{x}$ is constructible. Let $y = x + \frac{1}{x}$. Then x satisfies the quadratic equation $x^2 - yx + 1 = 0$ and hence x constructible.

Second Solution. Let $ABCDE$ be a regular pentagon. Draw in the diagonals AC, AD, and BD. Let F be the point of intersection of AC and BD. Using the fact that the interior angles of the regular pentagon are $108°$, we find that $\triangle ACD$ and $\triangle ABF$ are similar. We have the following equality of lines: $AB = BC = CD = AF$, $BF = FC$ and $AC = AD$. Using the similarity of the triangles, we have $AC : CD = AB : BF$. Also, $AC = AF + FC$. Let $AB = 1$ and $FC = x$. Then we have $(1 + x) : 1 = 1 : x$. Thus x satisfies the quadratic equation $x^2 + x - 1 = 0$ and is thus a constructible number. The measure of angle ACD is $\frac{2\pi}{5}$. Using the Law of Cosines on $\triangle ACD$ yields $AD^2 = AC^2 + CD^2 - 2(AC)(CD)\cos\left(\frac{2\pi}{5}\right)$. This translates to $(1 + x)^2 = (1 + x)^2 + 1^2 - 2(1 + x)(1)\cos\left(\frac{2\pi}{5}\right)$. Thus $\cos\left(\frac{2\pi}{5}\right) = \frac{1}{2(1+x)}$. Since x is constructible, so is $\cos\left(\frac{2\pi}{5}\right)$.

9. Using Maple, we can easily compute the expansion $\sin(7x) = 64 \sin(x) \cos^6(x) - 80 \sin(x) \cos^4(x) + 24 \sin(x) \cos^2(x) - \sin(x)$. If we let x have the value $\frac{2\pi}{7}$, then this expansion reduces to $0 = 64 \sin\left(\frac{2\pi}{7}\right) \cos^6\left(\frac{2\pi}{7}\right) - 80 \sin\left(\frac{2\pi}{7}\right) \cos^4\left(\frac{2\pi}{7}\right) + 24 \sin\left(\frac{2\pi}{7}\right) \cos^2\left(\frac{2\pi}{7}\right) - \sin\left(\frac{2\pi}{7}\right)$. Factoring out $\sin\left(\frac{2\pi}{7}\right)$ leaves a cubic equation in $\cos^2\left(\frac{2\pi}{7}\right)$, namely $64x^3 - 80x^2 + 24x - 1 = 0$ with $x = \cos^2\left(\frac{2\pi}{7}\right)$. So x is a root of an irreducible cubic and hence not constructible. Thus $\cos\left(\frac{2\pi}{7}\right)$ can not be constructible because if it was constructible, its square, namely x, would also be constructible.

ANSWERS 6.5

1. i. (See Example 5) $Z_2 \times Z_2 \times Z_2$
   ii. $\{id\}$ (See Example 3)
   iii. $\{Z_2, +\}$
2. $\{Z_5, +\}$, $i \to \varphi_i$, $\varphi_i(2^{1/5}) = 2^{1/5} \xi^i$

3. The complex pth roots of unity are $\{\xi_k = e^{(\frac{2\pi i}{p})k} : k = 1, \ldots, p-1\}$. $\text{Aut}(Q(\zeta)) = \{\varphi_k : k = 1, \ldots, p-1\}$ where $\varphi_k(\xi_1) = \xi_k$. The bijection $\varphi_k \leftrightarrow k$ is operation preserving since $\varphi_k \circ \varphi_j(\xi_1) = \xi^{kj \bmod(p)}$.

4. i. K must contain the real number $\sqrt[5]{2}$ and ξ, a primitive fifth root of unity. Thus $K = Q(\sqrt[5]{2}, \xi)$. The element $\sqrt[5]{2}$ satisfies an irreducible polynomial of degree 5 and ξ satisfies an irreducible polynomial of degree 4. Since both 4 and 5 divide the degree of the extension, the degree must be 20.

   ii. Apply Proposition 5.

   iii. Let $\sqrt[5]{2}$ denote the unique real fifth root of 2 and ξ a complex fifth root of unity. The five roots of $x^5 - 2$ are $a = \sqrt[5]{2}, b = \xi\sqrt[5]{2}, c = \xi^2\sqrt[5]{2}, d = \xi^3\sqrt[5]{2}$, and $e = \xi^4\sqrt[5]{2}$. Any automorphism f is determined by $f(\xi)$ and $f(\sqrt[5]{2})$, where $f(\sqrt[5]{2})$ can equal any of the roots of $x^5 - 2$ and $f(\xi)$ can equal any of the roots of $x^4 + x^3 + x^2 + x + 1$, i.e., the complex fifth roots of unity.

   iv. The following table gives the permutation of the roots that each possible automorphism induces. It is **not** a Cayley table. The row and column labels give the assignment of ξ and $\sqrt[5]{2}$, respectively.

	$\sqrt[5]{2} \to \sqrt[5]{2}$	$\sqrt[5]{2} \to \xi\sqrt[5]{2}$	$\sqrt[5]{2} \to \xi^2\sqrt[5]{2}$	$\sqrt[5]{2} \to \xi^3\sqrt[5]{2}$	$\sqrt[5]{2} \to \xi^4\sqrt[5]{2}$
$\xi \to \xi$	$\begin{pmatrix} a\ b\ c\ d\ e \\ a\ b\ c\ d\ e \end{pmatrix}$	$\begin{pmatrix} a\ b\ c\ d\ e \\ b\ c\ d\ e\ a \end{pmatrix}$	$\begin{pmatrix} a\ b\ c\ d\ e \\ c\ d\ e\ a\ b \end{pmatrix}$	$\begin{pmatrix} a\ b\ c\ d\ e \\ d\ e\ a\ b\ c \end{pmatrix}$	$\begin{pmatrix} a\ b\ c\ d\ e \\ e\ a\ b\ c\ d \end{pmatrix}$
$\xi \to \xi^2$	$\begin{pmatrix} a\ b\ c\ d\ e \\ a\ c\ e\ b\ d \end{pmatrix}$	$\begin{pmatrix} a\ b\ c\ d\ e \\ b\ d\ a\ c\ e \end{pmatrix}$	$\begin{pmatrix} a\ b\ c\ d\ e \\ c\ e\ b\ d\ a \end{pmatrix}$	$\begin{pmatrix} a\ b\ c\ d\ e \\ d\ a\ c\ e\ b \end{pmatrix}$	$\begin{pmatrix} a\ b\ c\ d\ e \\ e\ b\ d\ a\ c \end{pmatrix}$
$\xi \to \xi^3$	$\begin{pmatrix} a\ b\ c\ d\ e \\ a\ d\ b\ e\ c \end{pmatrix}$	$\begin{pmatrix} a\ b\ c\ d\ e \\ b\ e\ c\ a\ d \end{pmatrix}$	$\begin{pmatrix} a\ b\ c\ d\ e \\ c\ a\ d\ b\ e \end{pmatrix}$	$\begin{pmatrix} a\ b\ c\ d\ e \\ d\ b\ e\ c\ a \end{pmatrix}$	$\begin{pmatrix} a\ b\ c\ d\ e \\ e\ c\ a\ d\ b \end{pmatrix}$
$\xi \to \xi^4$	$\begin{pmatrix} a\ b\ c\ d\ e \\ a\ e\ d\ c\ b \end{pmatrix}$	$\begin{pmatrix} a\ b\ c\ d\ e \\ b\ a\ e\ d\ c \end{pmatrix}$	$\begin{pmatrix} a\ b\ c\ d\ e \\ c\ b\ a\ e\ d \end{pmatrix}$	$\begin{pmatrix} a\ b\ c\ d\ e \\ d\ c\ b\ a\ e \end{pmatrix}$	$\begin{pmatrix} a\ b\ c\ d\ e \\ e\ d\ c\ b\ a \end{pmatrix}$

   v. Let $\alpha = \begin{pmatrix} a & b & c & d & e \\ a & e & d & c & b \end{pmatrix}$ and $\gamma = \begin{pmatrix} a & b & c & d & e \\ b & c & d & e & a \end{pmatrix}$.
   Then $\alpha \circ \gamma = \begin{pmatrix} a & b & c & d & e \\ e & d & c & b & a \end{pmatrix}$ but $\gamma \circ \alpha = \begin{pmatrix} a & b & c & d & e \\ b & a & e & d & c \end{pmatrix}$.

   vi. The permutations listed in rows 1 and 4 in the above table give the group D_5 which is the group of symmetries of the pentagon in terms of the permutation of its vertices labeled a, b, c, d, e clockwise. Since it has order 10 in a group of order 20, it is a normal subgroup of $\text{Aut}(K)$ and the quotient groups $\text{Aut}(K)/D_5$ is isomorphic to Z_2. The first row is a cyclic group of order 5, C_5, normal in D_5 with quotient group D_5/C_5 again isomorphic to Z_2. Thus $\text{Aut}(K)$ is solvable:

$$\{id\} \subseteq C_5 \subseteq D_5 \subseteq \text{Aut}(K).$$

5. **i.** The degree 2 extension $Q(\sqrt{-3})$ contains all the cube roots of unity but no complex fifth roots of unity, i.e., no roots of $p(x) = x^4 + x^3 + x^2 + x + 1$. Adjoining a root of $p(x)$ to $Q(\sqrt{-3})$ (an hence all roots of $p(x)$) must result in an extension of degree 2 or 4.

 ii. $x^5 - 6x + 3$ is irreducible in $Q[x]$ by Eisenstein. If α is a root of $x^5 - 6x + 3$ in C, then $[Q: Q(\alpha)] = 5$. If $\alpha \in F$, then $Q \subseteq Q(\alpha) \subseteq F$, which implies that 5 divides $[Q: F]$. But $[Q: F]$ is a power of 2.

 iii. Suppose that α is a root of an irreducible polynomial in $F[x]$ of degree 2 or 3. Then the only possible divisors of $[Q: F(\alpha)]$ are 2 and 3. But $Q \subseteq Q(\alpha)] \subseteq F(\alpha)$, implies that 5 divides $[Q: F(\alpha)]$.

Appendix Exercise Solutions

Exercises on Propositions and Predicates.

1. (a) yes (b) no
2. (a) 2 is odd or 2 is not prime.
   (b) It is not raining or I have my umbrella.
   (c) There exists a real number x such that $x^2 \leq x$.
   (d) For all real numbers x, $x \cdot 2 \neq 1$.
   (e) $\exists x \forall y [x + y \neq 0]$.
3. (a) Converse: If you can't go outside, then you don't do your homework.
   Contrapositive: If you can go outside, then you do your homework.
   Negation: You do not do your homework and you can go outside.
   (b) Converse: If Mary loses, then Sally wins.
   Contrapositive: If Mary does not lose, then Sally does not win.
   Negation: Sally wins and Mary does not lose.
   (c) Converse: If Mary loses and the school closes, then John wins.
   Contrapositive: If Mary does not lose or the school does not close, then John does not win.
   Negation: John wins and either Mary does not lose or the school does not close.

Exercises on the Language of Proof.

1. Let m and n be two odd integers. Then there exist integers k and l such that $m = 2k + 1$ and $n = 2l + 1$. Then $mn = (2k+1)(2l+1) = 4kl + 2k + 2l + 1 = 2(2kl + k + l) + 1$. Since $2kl + k + l$ is an integer, mn is odd.
2. Suppose m or n is even. Then either $m = 2k$ or $n = 2l$ for some integers k and l. Then either $mn = 2kn$ or $mn = 2ml$. In either case, mn is even.
3. Assume that x is rational, that y is irrational, and that $x + y$ is rational. Then there exists integers $m, n, a,$ and b with $n, b \neq 0$ such that $x = \frac{m}{n}$ and $x + y = \frac{a}{b}$. Substituting and solving for y, we get $y = \frac{a}{b} - \frac{m}{n} = \frac{an - bm}{bn}$. Since $an - bm$

and bn are integers and $bn \neq 0$, y is rational. This contradicts the assumption that y is irrational and therefore $x + y$ cannot be rational and must be irrational.

4. **(a)** 4 is an integer but $\sqrt{4} = 2$ is rational.
   (b) 7 is an odd number, but 7 is not divisible by 3.
   (c) 7 is an odd prime number, but 9 is not prime.
5. Let n be an even integer. Then there exists an integer k such that $n = 2k$. This implies that $n^2 = 2(2k^2)$. Since $2k^2$ is an integer, n^2 is even. Now let n be an odd integer. There exists an integer k such that $n = 2k + 1$. Then $n^2 = (2k + 1)^2 = 4k^2 + 4k + 1 = 2(2k^2 + 2k) + 1$. Since $2k^2 + 2k$ is an integer, n^2 is odd.

Exercises on Sets and Set Operations.

1. **(a)** $\{5, 10, 15, 20, \ldots\}$
   (b) $\{\ldots, -7, -5, -3, -1\}$
   (c) $\{2\}$
2. **(a)** $\{x : x \text{ is a negative even integer}\}$
   (b) $\{x : x \text{ is an integer divisible by 3}\}$
   (c) $\{x : x = \frac{1}{n}, \text{where } n \in N\}$
3. **(a)** Not equal. $0 \in A$, but $0 \notin B$.
   (b) Equal. First we show that $X \subseteq Y$. Let $x \in X$. Then $x^2 - x = x(x - 1) < 0$. So either $x < 0$ and $(x - 1) > 0$, or $x > 0$ and $(x - 1) < 0$. Since the first case is impossible, it must be that $x > 0$ and $(x - 1) < 0$. So $0 < x < 1$ and $x \in Y$. Thus, $X \subseteq Y$. Now we show that $Y \subseteq X$. Let $x \in Y$. Since $0 < x < 1$, $x^2 < x$ and $x \in X$. Thus $Y \subseteq X$. Since $X \subseteq Y$ and $Y \subseteq X$, we conclude that $X = Y$.
4. Let $x \in A$ and let $y = x^{1/2}$. Then $x = y^2$ and $0 \leq y < 2$. Thus $x \in B$ and $A \subseteq B$. Now let $x \in B$ and choose y such that $x = y^2$. (The existence of y is guaranteed by the predicate defining B.) Since $-2 < y < 2$, $0 \leq y^2 < 4$ and thus x is an element of A and $B \subseteq A$. Since both $A \subseteq B$ and $B \subseteq A$, we have that $A = B$.
5. **(a)** Proper. $9 \in B$ but $9 \notin A$.
   (b) Not Proper. $X = Y$.
6. Let $a \in A$. Then $a \in B$ since $A \subseteq B$. Since $B \subseteq C$ and $a \in B$, we have $a \in C$. Thus, we have shown that $A \subseteq C$.
7. **(a)** $\{h, i, j, k, l\}$ **(b)** $\{a, b, c, \ldots, l, o, p, q, \ldots, z\}$
   (c) $\{h, i, j, k, l\}$ **(d)** $\{h, i, j, k, l, o, p, q, \ldots, z\}$
   (e) $\{r, s, t, u, v, w, x, y, z\}$ **(f)** $\{a, b, c, d, e, f, g, m, n, o, p, q, \ldots, z\}$
   (g) $\{a, b, c, d, e, f, g\}$ **(h)** $\{m, n, o, p, q\}$
   (i) $\{a, b, c, d, e, f, g\}$ **(j)** A
8. $A \cup A' = U$: First let $x \in A \cup A'$. Then $x \in A$ or $x \in A'$. In either case, $x \in U$ and $A \cup A' \subseteq U$. Now let $x \in U$. Then x is either in A or not. So $x \in A$ or

$x \in A'$ and thus, $x \in A \cup A'$. Therefore, $U \subseteq A \cup A'$. Since we have shown that $A \cup A' \subseteq U$ and $U \subseteq A \cup A'$, we have $A \cup A' = U$.

$A \cap A' = \emptyset$: Suppose on the contrary that $A \cap A' \neq \emptyset$. Then there exists an $x \in A \cap A'$. This implies that $x \in A$ and $x \in A'$. However, A' consists of elements that are in U and not in A and we therefore have a contradiction to the statement that $x \in A$ and $x \notin A$. So it must be the case that $A \cap A' = \emptyset$.

9. First, let $x \in A \cup (B \cap C)$. Then $x \in A$ or $x \in B \cap C$. In the case that $x \in A$, we have that $x \in A \cup B$ and $x \in A \cup C$, and thus, $x \in (A \cup B) \cap (A \cup C)$. In the case that $x \in B \cap C$, we have that $x \in B$ and $x \in C$. Thus $x \in A \cup B$ and $x \in A \cup C$ and so $x \in (A \cup B) \cap (A \cup C)$. Therefore, we have shown that $A \cup (B \cap C) \subseteq (A \cup B) \cap (A \cup C)$.

Now, we let $x \in (A \cup B) \cap (A \cup C)$. Then $x \in A \cup B$ and $x \in A \cup C$. We must show that $x \in A \cup (B \cap C)$. We consider two cases. Suppose $x \in A$. Then $x \in A \cup (B \cap C)$ as needed. Suppose $x \notin A$. Since $x \in A \cup B$ and $x \in A \cup C$ and $x \notin A$, we must have $x \in B$ and $x \in C$. Thus, $x \in B \cap C$ and we have $x \in A \cup (B \cap C)$. In both cases we have shown that $x \in A \cup (B \cap C)$ and this proves that $(A \cup B) \cap (A \cup C) \subseteq A \cup (B \cap C)$.

Since we have proven that $A \cup (B \cap C) \subseteq (A \cup B) \cap (A \cup C)$ and $(A \cup B) \cap (A \cup C) \subseteq A \cup (B \cap C)$, we have that $A \cup (B \cap C) = (A \cup B) \cap (A \cup C)$.

10. First we let $x \in (A \cup B)'$. Since x fails to be in the union of A and B, x fails to be in both A and in B. Thus x is an element of $A' \cap B'$ and we have that $(A \cup B)' \subseteq A' \cap B'$. Now let $x \in A' \cap B'$. Then x is not an element of A and it is not an element of B. This means that x is not an element of $A \cup B$ and hence is in $(A \cup B)'$. Thus $A' \cap B' \subseteq (A \cup B)'$ and we conclude that $(A \cup B)' = A' \cap B'$.

11. (a) $\{(1,x),(1,y),(2,x),(2,y)\}$
    (b) $\{(a,a),(a,b),(a,c),(b,a),(b,b),(b,c),(c,a),(c,b),(c,c)\}$
    (c) $\{(1,a,x),(1,a,y),(1,b,x),(1,b,y),(1,c,x),(1,c,y),(2,a,x),(2,a,y),$
         $(2,b,x),(2,b,y),(2,c,x),(2,c,y)\}$

12. (a) $\{(0,a),(0,b),\ldots,(9,y),(9,z)\} = A \times B$, where $A = \{0,1,2,\ldots,9\}$ and B is the set of letters of the alphabet.
    (b) $\{(1,2),(2,3),(3,4),\ldots\}$. This set cannot be expressed as a Cartesian product because 2 is both a first coordinate and a second coordinate, but (2, 2) is not in the set.

13. A young elementary student would lay out John's two apples and Ami's three apples, put them together, and count how many there were in all. If we let the set $J = \{J_1, J_2\}$ represent the set of John's two apples and $A = \{A_1, A_2, A_3\}$ represent the set of Ami's two apples, then the student has found the union of the two sets $J \cup A = \{J_1, J_2, J_3, A_1, A_2\}$ and found how many elements are in $J \cup A$.

Definition: Given two positive integers a and b, if set A and set B are two disjoint sets for which the number of elements in A is a and the number of elements in B is b, then $a + b$ is equal to the number of elements in $A \cup B$.

Selected Answers

Exercises on Functions.

1. **(a)** Function
   (b) Not a function
   (c) Not a function
   (d) Not a function
2. There are eight functions from A to B.

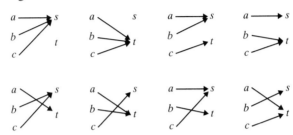

3. **(a)** bijective **(b)** injective only **(c)** neither
4. **(a)** $f(n) = 2n$ **(b)** $f(n) = \lfloor n/2 \rfloor + 1$ **(c)** $f(n) = n+1$ for n odd and $f(n) = n-1$ for n even
5. $(f \circ g)(x) = (3x^2 + 1)/10$, $(g \circ f)(x) = (9x^2 - 6x + 26)/50$
6.

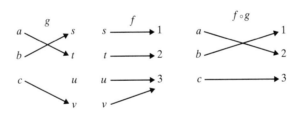

7. **(a)** $f^{-1}(x) = (1-3x)/2$ **(b)** $g^{-1}(x) = (2x-1)^{1/2}$
8. Suppose that $(f^{-1})^{-1}(x) = y$. Then $f^{-1}(y) = x$. Since $f^{-1}(y) = x$, $f(x) = y$. So $(f^{-1})^{-1}(x) = f(x)$.
9. $f(A) = \{y : -7 \leq y \leq 9\}$; $f^{-1}(B) = \{x : \frac{1}{2} \leq x \leq \frac{3}{2}\}$; $f(\mathbf{R}^+) = \{y : y > -1\}$; $f^{-1}(\mathbf{R}^+) = \{x : x > \frac{1}{2}\}$
10. Let $f : \mathbf{R} \to \mathbf{R}$ be defined by $f(x) = x^2$ and let $A = \{x : -1 < x < 4\}$. Then $f^{-1}(A) = \{x : 0 \leq x < 2\}$ and $f(f^{-1}(A)) = \{x : 0 \leq x < 4\} \neq A$. Equality holds for surjective functions.

Exercises on Binary Operations.

1. The operation $*$ is associative and commutative. The element c is both a left identity element and a right identity element.
2. The operation $\#$ is not associative because $(b\#c)\#a = a$, but $b\#(c\#a) = b$. It is not commutative since $c\#b = c$ but $b\#c = a$. The element a is a left identity element. There is no right identity element.

3. (a) Both commutative and associative
   (b) Neither commutative nor associative
4. (a) None; (b) $m = 0$ is a left identity element. There is no right identity element.
5. (a) Does not define a binary operation on $\mathbf{Z}^*$.
   (b) Does not define a binary operation on $\mathbf{R}\setminus\{0\}$.
6. The primes are not closed under either multiplication or addition.
7. S is closed under $\cap$ since the intersection of any two sets containing b must also contain b.

Index

A

Abel, Niels, 111, 146, 334
Abelian group, 189, 194, 266
Absolute value, 93
Abstract algebra, 187
AC method, 175
Additive inverse, 10–11
Additive property, 10
Adelman, L., 73
Algebra, *See also* High school algebra curriculum
    fundamental theorem of, 106, 136–146, 177
        defined, 183
        Gauss's proof of 1799, 136, 141, 144
    modern, use of term, 351
Algebra (Bombelli), 105
Algebraic elements, 340–341
Algebraic fractions, 328
Aliza problem, 105
Analytic Art, The (Veta), 178
Apotome, 104–105
Argument, 94
"Around the World Puzzle," 294
Artis Magna Sive de Reglis Algebraicis (Ars Magna or The Great Art) (Cardano), 104, 146, 180
Associative binary operations, 388–389
Associative property, 10, 382
Automorphisms, 262–263, 352–353, 357–358

B

Bar code, 70–71
Base field, prime radical tower, 333
Biconditional statements, 376–377, 379
Bijection, 385–387, *See also* Operation-preserving bijection
Binary operations, 388–390
    associative, 388–389
    Cayley tables, 388–389
    commutative, 388–389
    defined, 388
    exercises on, 389–390
    identity element, 389
    induced, 388
    left identity element, 389
    right identity element, 389
Binomium, 104–105
Bombelli, Rafael, 105

C

Cabri Geometry II (Texas Instruments), 361
Cardano, Girolamo, 103–107, 146
Cardano's method, 149, 334
Cartesian product, 383
Cauchy's Theorem, 282
Cayley, Arthur, 264–265
Cayley tables, 189, 207, 264–265, 388
    for Group of Parade Commands, 190
    symmetries, 249–252
        associativity, 250
        closure, 250
        existence of an identity, 250
        existence of inverses, 250
        noncommutativity, 250–251
        order of elements, 251
        solving equations, 251–252
        subgroup structures, 253–254
Cayley's Theorem, 237
Characteristic, 89, 108
Characteristic 0 (zero), 89, 108
Check digit, 71
Chinese Remainder Theorem, 67–68
    for polynomials, 163–164
Clock arithmetic, 59, 70
Codomains, 385
Commutative binary operations, 388–389
Commutative property, 10, 382
Commutative rings, 83, 88, 107
    polynomials with coefficients in, 111
Compass tool, Geometer's Sketchpad, 361
Complements, 382
Complex conjugate, 93
Complex Conjugate Root Theorem, 169
Complex number system, constructing from real numbers, 92–93

428 Index

Complex numbers:
  birth of, 103–106
  field of, 92–100
  plotting, 93–94
  and polar coordinates, 94–97
  and the quadratic formula, 97
Complex variable, 138
Composite numbers, 27
Conclusion, 376
Conditional statements, 376
Congruence classes, 63, 107
Congruences, 68, 107
  polynomials, 150–163, 183
Congruent mod p(x), 159
Congruent modulo *m*, 60, 68
Conjugacy class, 285–286, 292, 324
Conjugate, use of term, 285–286, 292
Conjunction, 375
Constructible distance, 343–345
Constructible lengths, 347
Constructible numbers, 346, 347–348
Constructions, 359–366
Content, of polynomials, 131, 136
Content standards, 48
Continued fractions, 38–41
Contradiction, proof by, 379
Contrapositive, 376
Converse, 376
Cosets, 271–278, 323–324
  right and left, 271, 276
Counterexample, 380
Counting, and groups, 240–246
Cube, 343
Cubic and quartic equations, solving, 146–150
Curriculum and Evaluation Standards for School Mathematics (NCTM), 48
Custom tool, *Geometer's Sketchpad*, 361–362
Cycle decomposition, 232–233, 236–238, 237, 244, 266
Cycle notation, for permutations, 231
Cycles:
  defined, 231
  disjoint, 232–234
  nondisjoint, 232
  product of two, 231–232
  q-cycle, 231
Cyclic groups, 209, 213, 224–230, 266

D
de Moivre's formula, 95, 97, 142, 147, 176
De Morgan's laws, 375–376
  for sets, 382–383
Dedekind, Richard, 30
Degree:
  of algebraic extensions, 357
  of an extension, 336–342
  of a field extension, 336, 338
del Ferro, Scipione, 103–104
Derivatives, 128–130
Descartes' Criterion, 135, 169
Descartes, Rene, 136, 367, 370–371
  Discourse on Methods (Descartes), 370
Descartes' Rule of Signs, 173, 177
Dihedral group, 198, 200
Diophantine equations, 22–24, 56, 66, 179
Diophantus of Alexandria, 179–180
Direct image, 386
Direct product, 212–213
Direct proof, 378
Direct sum, 212
Discourse on Methods (Descartes), 370
Disjoint cycles, 232–234
Disjoint sets, 382
Disjunction, 375
Disquisitiones Arithmeticae (Gauss), 59
Disquistiones (Gauss), 351
Distributive law, 10
Distributive property, 382
Divisibility, 11
Divisibility Tests, 50, 71–73
Division Algorithm, 12, 17–20, 25, 42, 56, 70, 107, 308, 311
  for polynomials, 114–115, 117, 152, 183
  for rational numbers, 26–27
Dodecahedron, 343
Domains, 385
  integral, 86, 89, 108
  principal ideal domain (P.I.D.s), 309–311, 324
Doubling a cube (problem), 343, 347

E
Eisenstein's Criterion, 133–134, 173, 183
Elements:
  algebraic, 340–341
  irreducible, 307–311, 324

prime, 324
[illegible]
Elements (Euclid), 36–37, 52, 178, 342, 348–349, 351, 367–368
Empty(null) set, 381
Encryption, 48
Equal sets, 381
Equilateral triangle:
  construction of, 349
  symmetry of, 196–200
Equivalence classes, 5–6
Equivalence relation, 4
Euclid, 358, 368–369
  Proposition 5, 369–370
Euclidean algorithm for polynomials, 120–128
Euclid's Algorithm, 15–16, 24, 25, 37, 41, 42, 56, 66
  defined, 38
  for polynomials, 114, 117, 120–128, 183
  versitility of, 36–41
Euclid's Lemma, 18, 25, 28, 56, 307
Euclid's Theorem, 52, 56
Euler, Leonhard, 106, 371
Euler phi-function, 43, 47, 56
Euler's Theorem, 43, 45, 46–47
  proof of, 45–46
Even permutations, 235, 237
Existential quantifier, 377
Extension, 327
  degree of, 336–342
  field, 327
    degree of, 336–342
  finite, 338
Extension fields, 327–336, 334

F
Factor Theorem, 169, 176
Factoring, 130–136
Factorization, 1, 134–136
False roots, 104, 136
Fermat, Pierre de, 51–53
Fermat primes, 56
Fermat's Little Theorem, 43, 46–47, 53, 55, 56, 275
Ferrari, 148
Fibonacci numbers, 33–36
Fibonacci sequence, 34
Field extension, 327, 358
  degree of, 336–342

finite, 353
Fields, 87, 88, 108
  defined, 87
Finite continued fraction approximation, 40
Finite extension, 338
Finite field extension, 353
Finite groups, 189, 209
Fiore, Antonio Maria, 104
Five Books of Zetetics (Vieta), 182
FOIL, 167, 317
Formulas:
  de Moivre's, 95, 97, 142, 147, 176
  quadratic, 97, 111–112, 173, 289, 293, 316, 335
Frieze:
  basic unit of design in, 202
  defined, 201
  patterns, 205–206
  symmetry of, 202
  tasks, 202–207
Frieze groups, 201–207
Functions, 384–387
  bijection, 385–387
  codomain, 385
  composition, 385
  defined, 384–385
  domain, 385
  exercises on, 386–387
  function value, 385
  injective, 385
  inverse, 386
  mapping, 385
  one-to-one, 385
  one-to-one correspondence, 385
  onto, 385
  surjective, 385
Fundamental Theorem of Algebra, 106, 136–146, 177
  defined, 183
  Gauss's proof of 1799, 136, 141, 144
Fundamental Theorem of Algebra, The (Fine/Rosenberg), 136
Fundamental Theorem of Arithmetic, 28–29, 31, 56, 92
  defined, 28

G
Galois, Évariste, 111, 146, 258–264, 319–323, 334, 351, 357
Galois Theory, 259

Gardner, Martin, 50
Gauss, Carl Friedrich, 59, 106, 141–145, 351, 359
  proof of 1799, 136, 141, 144
Gaussian integers, 295, 306–315
Gauss's Lemma, 131, 134, 183
General linear group of degree 2, 189
Generator, 209
Generators, 209, 224
  of cyclic groups of rotations, 255
Generators of dihedral groups, finding, 256
Geometer's Sketchpad, The (Jackiw), 246, 361, 365
  Compass tool, 361
  construction tools, 361–362
  Custom tool, 361–362
  dragging objects/manipulating dynamically, 363
  Perpendicular Line tool, 362
  Point tool, 361
  relationships among figures, 363
  Selection Arrow tool, 361
  Straightedge tool, 361, 363
  Text tool, 361
Geometry, and groups, 196–200
Geometry Standard, NCTM, 246–247, 360, 364
Glide reflection, 205
Glide symmetry, 205
Goldbach Conjecture, 32
Golden Ratio, 37–38, 41
Golden Rectangle, 37
Great Art, The, See *Artis Magnae Sive de Regulis Algebraicis*
Greatest common divisor (gcd), 15–16, 25
  of polynomials, 120, 125, 183
Group theory, 187–269
Groups, 319–323
  abelian group, 189, 194, 266
  additive group, 191
  binary operation, 193
  conjugacy class, 285–286, 292, 324
  cosets, 271–278, 323–324
  and counting, 240–246
  cyclic, 209, 213, 224–230, 266
  defined, 187, 196, 264, 266
  dihedral, 198
  finite, 189, 209
  frieze, 201–207
  general linear group of degree 2, 189
  and geometry, 196–200
  Group of Parade Commands, 190
  homomorphism, 220–222
    defined, 220, 222
    trivial, 220
  identity element, 187, 194
  infinite, 189
  isomorphism, 216–222, 222
    defined, 217, 222
    isomorphism classes, 222
    operation-preserving bijection, 217
  multiplicative notation, 191
  nonabelian group, 189, 194, 266
  normal subgroups, 271, 278–285, 323
  nth alternating, 236–237
  order, 194, 266
  permutation, 230–239
  properties, 193–194
  quotient, 222
  quotient groups, 278–285
  simple, 288, 292
  solvable, 287, 292, 324
  subgroups, 207–213, 266
  symmetry, 266
  theory of, 258–265

H

High school algebra curriculum, 126–128, 342
  addition and multiplication of vectors and matrices, 90–92
  addition and multiplication tables of mod n arithmetic, 81–82
  clock arithmetic, 70–71
  complex numbers, 97–100
  conjugacy, 293
  cubic and quartic equations, 149–150
  curriculum contents, 118, 165
  cyclic groups, 229–230
  Division Algorithm, 14–15
  for rational numbers, 26–28
  elementary mathematics, 8–9
  Euclid's Algorithm, 42
  Euler's Theorem, 48
  exercises to reinforce Fundamental Theorem, 145–146
  factorization, 135–136
  groups, 195–196, 201
  high school curriculum, 118

Index   431

ideals, 304
Irreducible elements, 312
isomorphism, 223–224
learning objectives, 145–146
mathematical pattern recognition, 9
modular arithmetic, 70–71
permutations, 238–239
polynomial equations, solving, 293
polynomials, 126–128
prime elements, 313
primes, 32–33
quotient groups, 285
rational numbers, 9
regular polygons, construction of, 348–351
remainders, 160–163
roots and factoring:
   elementary algebra, 165–166
   exercises for exploring ideas more fully, 175
   factoring using the AC method, 175
   high school curriculum, 165
   Intermediate Algebra, 168–169
   teachers' vantage point, 170–174
solvable by radicals, 335–336
subgroups, 215–216
symmetry, 246–258
Homomorphism, 220–222, 266, 295
   defined, 220, 222
   trivial, 220
Hypothesis, 376

I
Icosahedron, 343
Ideal rings, 271, 296
Ideals, 324
   maximal, 301–303, 324
   prime, 301, 303, 324
   principal, 296, 324
Idempotent property, 382
Identity element of G, 187, 194
Identity permutations, 235
Image, 386
Imaginary part, 92–93
Imaginary quantities, 106
Implication, 376
Index, subgroups, 275
Indirect proof, 379
Induced binary operations, 388
Induction proofs, 2

Infinite groups, 189
Injective functions, 385
Integer coefficients, 130, 136
Integer multiples, 26
Integer values, 22
Integral domains, 86, 89, 108
Intermediate Value Theorem, 358
intersect command, 166
Intersection, 381–382
Introduction to Algebra (Euler), 106
Inverse, 188
   additive, 10–11
   identity elements, 194
   multiplicative, 66, 85–86
Inverse function, 386
Irrational number, 40
Irreducible, defined, 307
Irreducible elements, 307–311, 324
Irreducible polynomials, 125, 135, 328, 355
   of minimal degree satisfied by a, 338–339
Isomers, 245
Isomorphism, 216–222, 266
   defined, 217, 222
   isomorphism classes, 222
   operation-preserving bijection, 217, 222

J
Jackiw, Nicholas, 362

K
Kernel, 222, 295, 299
Klein 4-Group, 190–191, 219

L
Lagrange Interpolation, 163–165
Lagrange's Theorem, 274–276
Last field, prime radical tower, 333
Leading coefficient, 113
Least common multiple (lcm), 25
Left coset, 271, 276
Left identity element, 389
Liber Abacci (Fibonacci), 33
Lindemann, Carl Louis Ferdinand von, 347, 359
Linear congruence, 61, 68
Locking compass, 343
Logical variable, 375
Logically equivalent statements, 375
Long division algorithm, 14

432 Index

M
Mapping, 385
"Mathematical Games" section (Gardner) of Scientific American magazine, 50
Matrices, 92
Matrix addition, 90, 191
Matrix-vector multiplication, 221
Maximal ideals, 301–303, 324
Members, of a set, 380
Memoire sur les conditions de resolubilite des equations par radicaux (Galois), 351*fn*
Mesopotamians, 367–368
  completing the square, 368
  interchanging sum and difference, 368
Minimal polynomial, 338–339, 341
MIRAs, 246
Modern algebra, use of term, 351
Modular arithmetic, 59–109, 107
  congruent modulo, 60
  linear congruence, 61
  modulus, 60
  notation, 60–61
Modulus, 60, 93–94
Monic minimal polynomial of a, 339
Monic, use of term, 113
Multinomial coefficients, 54
Multiple factor, 129
Multiple root, 129
Multiples, 11
Multiplicative function, 47
Multiplicative inverse, 66, 85–86
Multiplicative property, 10
Multiplicity of a root, 129

N
National Council of Teachers of Mathematics (NCTM):
  Principles and Standards for School Mathematic, 6–7, 26, 48, 50, 246, 315–316
  problem solving standard, 49
  worthwhile mathematical tasks (standard), 49
Natural numbers, 1–3, 8, 14, 25, 29, 36, 84, 306, 309, 316, 335
  multiplication of, 1
Negation of a proposition, 375
Negative numbers, product of, 100–102

New York State Archives URL, 150
Nonabelian group, 189, 194, 266
Nondisjoint cycles, 232
Nonempty set of natural numbers, 1–2
Nonnegative number, 12
Normal subgroups, 271, 278–285, 323
Notation:
  Cardano, 180–181
  cycle, for permutations, 231
  Diophantus, 103
  exponential, 98
  modern, 103, 181
  modular arithmetic, 60–61
  rings, 84–85
  set builder, 381
  Vieta, 181
n-predicate, 377
nth alternating group, 236–237
nth roots of unity, 96
nth symmetric group, 231
Number and Operations Standard, NCTM, 50
Number theory, 31, 107
  defined, 1
  and primes, 11
  topics in, 1–6

O
Octahedron, 343
Odd permutations, 235, 237
One-to-one correspondence, 385
One-to-one functions, 385
Operation table, 189
Operation-preserving bijection, 217, 222, 266
Order of G, 191
Order of subgroups, 211, 266

P
Pacioli, Luca, 103
Partition, 6
Peano, Giuseppe, 30
Perfect numbers, 48, 51–52
  defined, 52
Permutation groups, 230–239
Permutations, 259–265, 320–323
  arrays of, 259–264
  cycle notation for, 231
  defined, 230
  disjoint cycles, 232
  even, 235, 237

identity, 235
nondisjoint cycles, 232
odd, 235, 237
Perpendicular Line tool, *Geometer's Sketchpad*, 362
Pigeonhole principle, 242
Pisa, Leonardo da, 33
Platonic solids, 343, 349, 370
Plotting complex numbers, 93–94
Point tool, *Geometer's Sketchpad*, 361
Points, of a set, 380
Polar coordinates, and complex numbers, 94–97
Polynomial arithmetic, 111–120
Polynomial equations, solvability of, 351–358
Polynomials, 111–186
  Chinese Remainder Theorem for, 163–164
  with coefficients in commutative rings, 111, 117
  congruences, 150–163, 183
  content of, 131, 136
  cubic and quartic equations, solving, 146–150
  degree of, 113, 117
  derivatives, 128–130
  Division Algorithm for, 114–115, 117, 152, 183
  division of, 152
  Euclid's Algorithm for, 114, 117, 120–128
  in $F[x]$, 121–125
  factoring, 130–136
  factoring of, 118, 136
  false roots, 136
  Fundamental Theorem of Algebra, 136–146
  defined, 183
  greatest common divisor (gcd) of, 120, 125, 183
  ideals, 313–315
  irreducible, 125, 135
  Lagrange Interpolation, 163–165
  polynomial arithmetic, 111–120
  polynomial division, 118
  primitive, 131, 134
  pth cyclotomic polynomial, 134
  rational roots of, 135
  reducible, 123

  roots and factoring, 165–178, 262
  Vieta and symbolic algebra, 178–182
Positive integers, 12
Predicates, 377
  exercises on, 377–378
Preimage, 386
Presentation, 262–263
Primality testing, 47
Prime elements, 324
Prime ideal, 301, 303, 324
Prime Number Theorem, 31
Prime numbers, 11, 27, 31
  acquiring/reinforcing skills through, 32–33
  and number sense, 32
  and problem solving, 32
  Sieve of Eratosthenes, 31
  Twin Prime Conjecture, 32
  and unique factorization, 27–31
Prime radical extension, 332, 334
Prime radical tower, 332
Primes, 306, 311
  defined, 307
Primitive Element Theorem, 341
Primitive polynomials, 131, 134
Principal ideal domain (P.I.D.s), 309–311, 324
Principal ideals, 297, 324
Principal representative, 152
Principle of Induction, 2
Principles and Standards for School Mathematic (National Council of Teachers of Mathematics [NCTM]), 6–7, 26, 48, 50, 246, 315–316, 360
Problem solving, 48–51
  getting students started on the problem, 51
  prior knowledge needed to do the problem, 50–51
  problem, defined, 49–50
  Social Security Number Problem, 50
  standard, 49
  worthwhile mathematical tasks (standard 1), 49
Process standards, 48
Product groups, 211–213
Product of two cycles, 231–232
Professional Standards for Teaching Mathematics (NCTM), 48

Proof:
  by contradiction, 379
  direct, 378
  indirect, 379
  language of, 377–378
Proper subgroup, 208
Proper subring, 295
Proper subset, 381
Propositions, 375–378
  exercises on, 377–378
  negation of, 375
pth cyclotomic polynomial, 134
pth roots of unity, 133–134

Q
q-cycle, 231
Quadratic equations, 367–371
Quadratic formula, 97, 111–112, 173, 289, 293, 316, 335
  and complex numbers, 97
Quantifiers, 377
Quotient, 114
Quotient groups, 222, 271, 278–285
Quotient rings, 271, 295, 299–300, 303
  of R modulo I, 300

R
Radical extension, 332, 334
Radical tower, 332, 334–335
Rational coefficients, 130
Rational functions, 328
Rational numbers, 378
  Division Algorithm for, 26–27
Rational Root Theorem, 169–170, 176–177
Rational roots of a polynomial, 135
Rationalizing the denominator, use of term, 316–317
Ratios of natural numbers, 335
Real numbers, 92–93
Real part, 92–93
Real-valued functions, 138
Rectangle, symmetry of, 202–207
Reducible polynomials, 123
Reflexive property, equivalence relation, 4–5
Regular pentagon, construction of, 349–350
Regular polygons:
  construction of, 348–351
  groups of symmetries of, 246–248
Relation, defined, 4
Relatively prime integers, 17

Remainder, 114
Remainder Theorem, 116, 117, 169, 183
Right coset, 271, 276
Right identity element, 389
Ring homomorphism, 297–299
Ring of 2 X 2 matrices over integers (worksheet), 305–306
Ring Z_m, 75–82
Rings, 82–88
  commutative, 83, 88–89, 107
  defined, 82, 88, 107
  and fields, 87–88
  homomorphisms, 297–299, 303
  ideal, 271, 296
  integral domains, 86, 89, 108
  notation, 84–85
  quotient, 271, 295, 299–300, 303
  subrings, 295, 303
  units, 86, 88, 107, 113
  with unity, 82–83, 88, 107
  without unity, 83, 88
  zero divisors, 85, 88, 107, 113
Rivest, R. L., 73
Root, 111
root command, 166
Root Theorem, 116, 117, 169, 183
Roots and factoring, 165–178
Roots of unity, 96
  nth roots of unity, 96
  pth roots of unity, 133–134
RSA encryption, 48
RSA Public Key Cryptography, 73–75
Ruler and compass constructions, 342–348

S
Selection Arrow tool, *Geometer's Sketchpad*, 361
Set builder notation, 381
Set of rational numbers, 6
Sets, 380–384
  De Morgan's laws for, 382–383
  disjoint, 382
  exercises on, 383–384
  operations, 381–382
    exercises on, 383–384
Shamir, A., 73
Sieve of Eratosthenes, 30, 31
Simple finite continued fraction, 38–39
Simple groups, 288, 292
Social Security Number Problem, 50

Solvable by radicals, use of term, 333, 334–335, 357–358
Solvable groups, 287, 292, 324, 357
Splitting field, 354
Square, construction of, 349
Squaring a circle (problem), 343, 347
Statements, 375
  biconditional, 376–377, 379
  conditional, 376
    proving, 378
Straightedge tool, *Geometer's Sketchpad*, 361, 363
Strong Principle of Induction, 3
Subfield, 158, 295
Subgroups, 207–213, 266
  defined, 208, 213
  index, 275
  normal, 271, 278–285, 323
  order of, 211, 213
  product groups, 211–213
  proper, 208
  trivial, 208–209, 213
Subrings, 295–297, 303
  ideal, 296
  proper, 295
  trivial, 295
Subsets, 381
  closed under ∗, 388
  proper, 381
Substitutions, 259–260, 320–323
Subtraction operation, 82
Sum, 112
Summa de Arithmetica, Geometrica, Proportioni et Proportionalita (Pacioli), 103
Surjective functions, 385
Symmetric property, equivalence relation, 4–5
Symmetries, 192
  Cayley tables, 249–252
    associativity, 250
    closure, 250
    existence of an identity, 250
    existence of inverses, 250
    noncommutativity, 250–251
    order of elements, 251
    solving equations, 251–252
    subgroup structures, 253–254
  counting, 247–249

generators of cyclic groups of rotations, 255
generators of dihedral groups, finding, 256
of regular polygons, groups of, 246–248
representations of, as permutations of vertices, 252–253
Symmetry groups, 266
Symmetry of a regular figure, defined, 196, 200
Synthetic division, 160
Systems of numbers, 59–109

T
Tartaglia, Niccolo, 104
Teachers, *See* High school algebra curriculum
Tetrahedron, 343
Text tool, *Geometer's Sketchpad*, 361
Theory of Equations (Uspenski), 142
TI-83 Plus, 361
TI-84 Plus, 361
TI-89, 318, 361
  EXACT mode in REAL Complex Format, 173–174, 176
  EXACT mode in RECTANGULAR Complex Format, 176
TI-92 Plus, 318, 361–362
Transitive property, equivalence relation, 4–5
Transpositions, 234, 266
Trisecting an angle (problem), 343, 347
Trivial homomorphism, 220
Trivial subgroups, 208–209, 213
Twin Prime Conjecture, 32
2-cycles, 234

U
Understanding and Amendment of Equations (Vieta), 181–182
Union, 381–382
Unique factorization, 27–31
  theorem for polynomials, 124–125
Units, rings, 86, 88, 113
Unity:
  rings with, 82–83, 88, 107
  rings without, 83, 88
Universal quantifier, 377
Universal set, 382
Universe of discourse, 377

V

Vector space, 337
Vectors, 337
Vieta, Francois, 178–182
Voyage 200, 361

W

Wantzel, Pierre, 359
Well-defined operations, 6
Well-Ordering Principle, 1–2, 12–13, 17, 125
  defined, 2
Worksheets:
    on the construction of regular polygons, 348–351
    derivatives, 128–130
    Divisibility Tests, 50, 71–73
    Fibonacci numbers, 33–36
    frieze groups, 201–207
    groups and counting, 240
    Lagrange Interpolation, 163–165
    polynomial ideals, 313–315
    on the ring of 2×2 matrices over the integers, 305–306
    RSA Public Key Cryptography, 73–75

Z

Zero, 111
zero command, 166
Zero divisors, 85, 88, 107, 113